Essential Topics in Biotechnology

Volume I

Essential Topics in Biotechnology
Volume I

Edited by **Joy Adam**

R CALLISTO REFERENCE

New York

Published by Callisto Reference,
106 Park Avenue, Suite 200,
New York, NY 10016, USA
www.callistoreference.com

Essential Topics in Biotechnology: Volume I
Edited by Joy Adam

International Standard Book Number: 978-1-63239-320-3 (Hardback)

Printed in the United States of America.

Contents

Preface

Human mind does not rest, is never satisfied, and constantly craves for improvement. And on a different note, it may take years to coin a term for a particular activity. Biotechnology is that term which sums up man's constant need for improvising everything. According to Oxford Science dictionary, biotechnology is the exploitation of biological processes for industrial and other purposes, especially the genetic manipulation of microorganisms for the production of antibiotics, hormones, etc.

The definition though indeed true, doesn't show the complete picture. Biotechnology has been used by man for thousands of years in agriculture, medicine, and food production amongst other activities. In tracing the history of the subject, it has been found that farmers bred their crops with other plants after modifying their genetics and introduced them in new environments. Though it may not have been done with scientific intent, it was definitely an observation and experimentation process done for improving crops and soil fertility and hence, food production. Fermentation of beer, selective breeding of plants and crops resulting in peace rose and hybrid corn amongst others, are initial examples of using biotechnology in different fields. The subject played an important role in the development of antibiotics and discovery of penicillin, which were landmarks in the medicinal field.

Modern biotechnology which started after the success of gene splicing, soon became not only a feasible business option because of rising revenue percentage, its utilities resulted in major industrial applications like agriculture, medicine, environmental uses, and non-industrial crops, including biofuels, biodegradable plastics, vegetable oils, etc. Moreover, several branches of biotechnology have also developed like bioinformatics, green biotechnology, etc. Consequently, it has emerged as a viable career course in the recent times, further solidifying its future.

I would like to thank my publisher for giving me this incredible opportunity and my family for supporting me at every step.

Editor

Effective Combination of Photodynamic Therapy and Imiquimod 5% Cream in the Treatment of Actinic Keratoses: Three Cases

Laura Held,[1] **Thomas Kurt Eigentler,**[1] **Ulrike Leiter,**[1]
Claus Garbe,[1] **and Mark-Jürgen Berneburg**[1,2]

[1] *Department of Dermatology, Center for Dermatooncology, University Hospital Tübingen, 72076 Tübingen, Germany*
[2] *Department of Dermatology, Eberhard Karls University, Liebermeisterstraße 25, 72076 Tübingen, Germany*

Correspondence should be addressed to Mark-Jürgen Berneburg; mark.berneburg@med.uni-tuebingen.de

Academic Editor: Tim Maisch

Background. The therapy for actinic keratoses includes photodynamic therapy (PDT) and imiquimod 5% cream. The sequential use of both could result in better clinical outcomes. *Objectives.* To enhance efficacy of therapies while improving tolerability, convenience, and patient adherence with a scheme combining two concomitant or sequential AK treatments. *Methods.* All patients underwent one session of conventional PDT. Two weeks after, the PDT imiquimod 5% cream was applied to the treatment area once daily for three days per week. One course continued for four weeks followed by a clinical evaluation and decision about further treatment. Patients who had not cleared all of their AK lesions in the treatment area in course 1 participated in a second 4-week course of treatment. *Limitations.* Small size of population. *Results.* Three participants were enrolled. Two patients showed complete clinical clearance of AKs. The effect was also noted after long-term followup, at months seven and eleven. No subject discontinued for an adverse event. There were severe local skin reactions in two participants which were severe erythema, scaling, and crusting. One patient showed no response to the therapy. *Conclusions.* Photodynamic therapy followed by imiquimod was well tolerated and improved reduction of actinic keratoses. This initial proof-of-concept should be studied in larger clinical trials.

1. Introduction

Actinic keratoses (AKs) are common lesions which are induced by chronic sunlight exposure [1]. The lesions are frequently found on the hands, forearms, face, and scalp. A presentation of multiple AKs in a large area results in a field cancerization. AKs represent an in situ cancer and they belong to the group of nonmelanoma skin cancer (NMSC). The majority of the lesions may clear spontaneously; some lesions undergo transformation into invasive squamous cell carcinoma (SCC). The incidence of AK is increasing; the prevalence varies depending on the population. The risk factors for the development of AKs include high age, fair skin type, immunosuppression, and cumulative ultraviolet (UV) exposure. The link between the risk factors and AKs is well established and supported by a large body of epidemiological data.

A variety of treatment strategies are available for AKs; they have specific risks and benefits and include topical agents and surgical procedures [2]. Field-directed therapies use topical agents to treat multiple AKs over a large treatment area and require from weeks to months of use. To enhance efficacy of therapies while improving tolerability, convenience, and patient adherence scheme combining two concomitant or sequential AK treatments (photodynamic therapy, imiquimod, diclofenac, cryotherapy, 5-fluorouracil, salicylic acid, and tretinoin) [3–10] has already been assessed in clinical trials.

To further improve the duration and long-term results of the treatment, we also evaluated an alternative dosing regimen including sequential photodynamic therapy (PDT) and imiquimod. PDT is noninvasive targeted treatment which uses visible light to activate a photosensitizing agent, resulting in the formation of cytotoxic reactive oxygen species. This procedure leads to diseased tissue destruction. The therapy is highly effective in the treatment of AKs [11]. Imiquimod 5% cream regulates the level of cytokines and modifies immune response. The approved regimen in the United States for the treatment of AKs is the twice weekly use

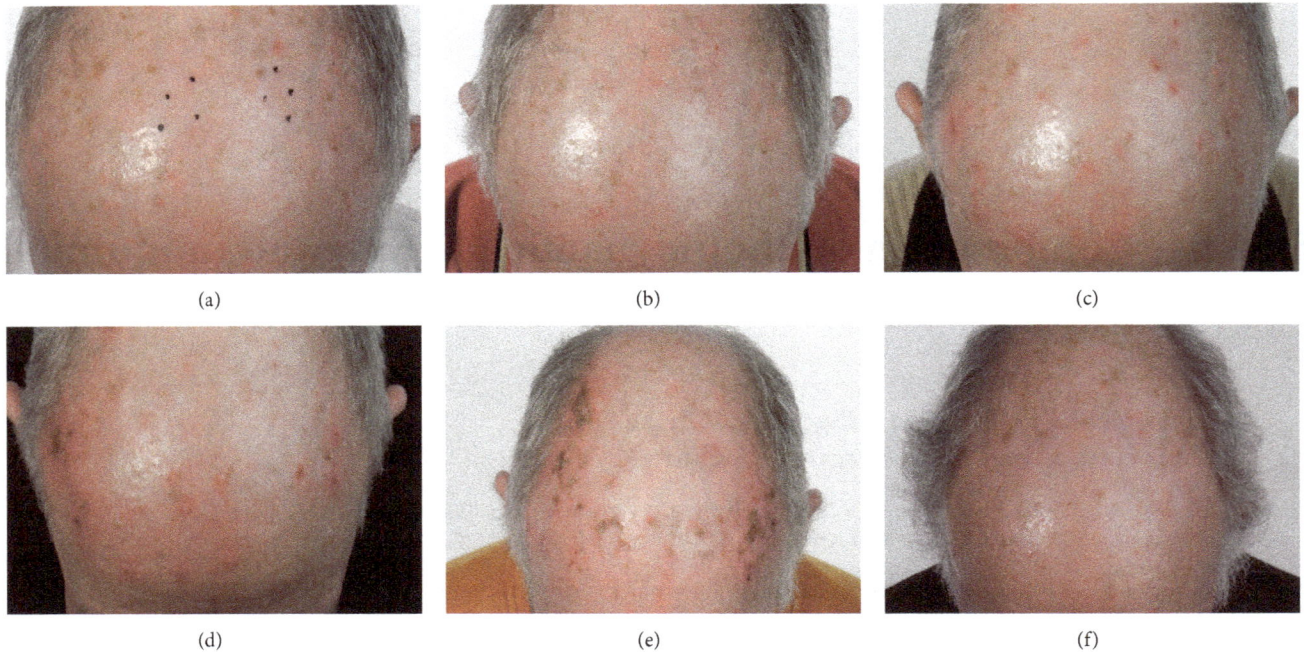

FIGURE 1: Clinical followup, Case 1. (a) Status before PDT; several hyperkeratotic lesions, hardly visible, and palpable (b) two weeks after PDT, before imiquimod; (c) two weeks after imiquimod; (d) four weeks after imiquimod, end of first course; (e) four weeks after imiquimod, end of second course; (f) seven months after the end of treatment: no visible or palpable AKs and less hyperpigmentation.

of imiquimod 5% cream for 16 weeks [12]. Imiquimod 5% cream is used as a treatment for AKs, superficial BCCs, and external genital warts. More than 95% of patients develop local skin responses such as erythema and crusting [13]. Different regimens like interval, pulse, and cycle therapies with shorter duration of treatment than conventional treatment periods have also shown to provide comparable efficacy while minimizing the local skin toxicity [14].

Due to different mechanisms of action, better clinical long-term results with less toxicity are expected from applying the PDT and imiquimod 5% cream sequentially.

The aim of this investigation was to determine the efficacy, tolerability, safety, and cosmetic outcome of PDT with 5% imiquimod cream sequential treatment in patients with AKs and to give the background for the initiation of future studies in order to confirm the results and evaluate long-term therapeutic benefits.

2. Patient's Selection and Methods

2.1. Eligibility. Patients were eligible if they were at least 18 years old and had clinically typical visible AK lesions located anywhere on the head or hands. Hyperkeratotic or hypertrophic AK lesions were not excluded. Patients were excluded if they were organ transplant recipients or if they had any dermatological disease or condition in the treatment or surrounding area.

2.2. Treatment Procedure. The lesions were gently abraded using a curette (Stiefel) so that the surface crust was removed.

5-aminolevulinic acid (ALA) was applied for 4 h, respectively, for AK. The cream was all the time occluded. In addition, regional anesthesia was applied for 30 min prior to irradiation. The Aktilite CL128 laser was used for irradiation in all patients.

2.3. Investigation Design. Baseline screening was conducted before PDT. The lesions were measured and the exact location of AKs was documented with digital photography. Two weeks after the single session of PDT, patients started to apply imiquimod 5% cream to the treatment area once daily, three days per week. Imiquimod was left on the skin for at least eight hours before being washed off. Treatment continued for four weeks (course 1) followed by a clinical evaluation and decision about the further procedure. Patients without clinically visible AK lesions in the treatment area were considered to be completely clear and their treatment ended. Patients who had not cleared all of their AK lesions in the treatment area at the end of course 1 participated in a second four-week course of treatment. Patients were clinically evaluated at least six months after the end of the treatment—one patient withdrew from further treatment because of a lack of response to the treatment. All patients gave written informed consent to participate in this investigation.

3. Case Reports

3.1. Case 1. Mr. JR is male patient, 64 years old with Fitzpatrick skin phototype II. He presented with numerous hyperkeratotic lesions on his scalp (Figure 1(a)). Several

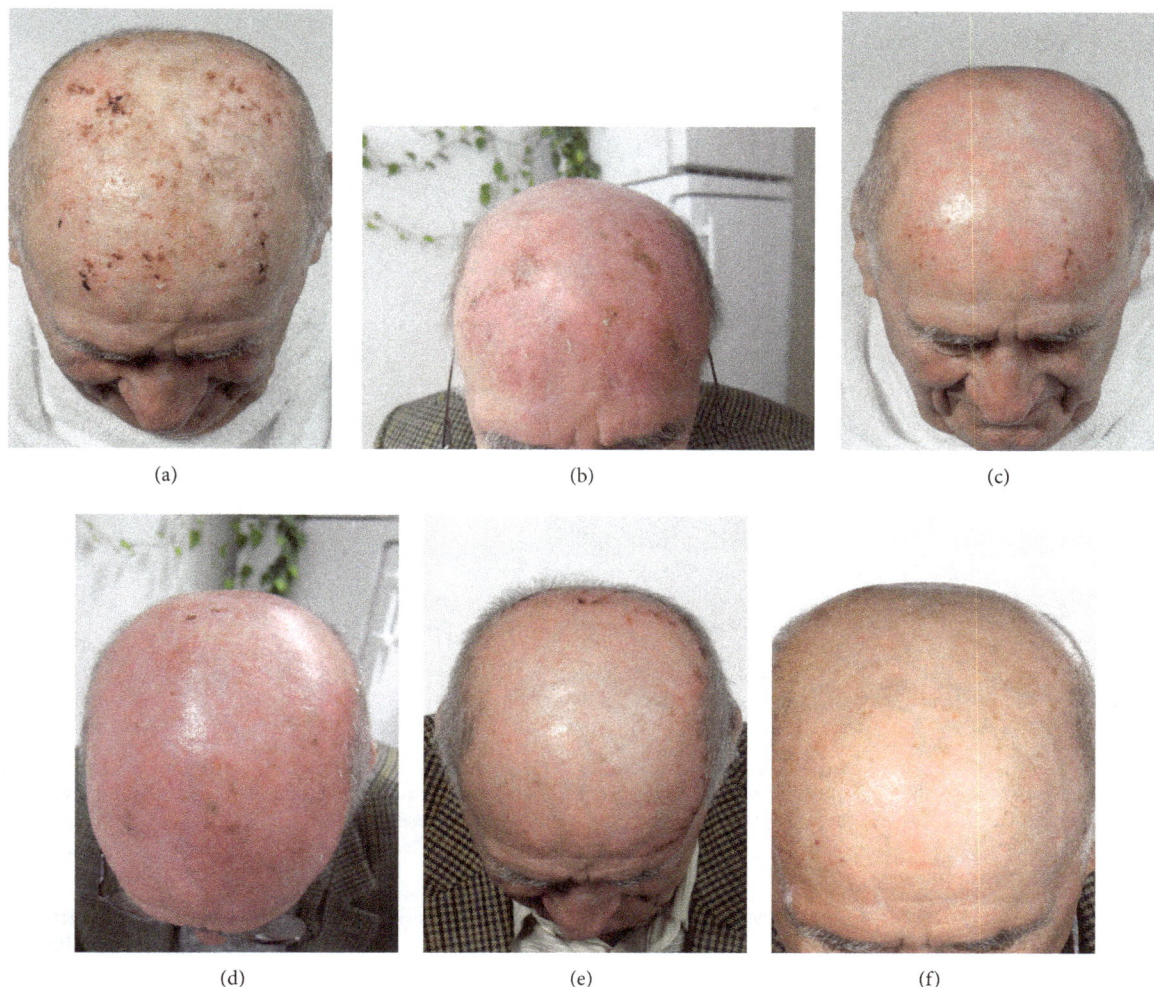

FIGURE 2: Clinical followup, Case 2. (a) Two weeks after PDT, before imiquimod; (b) two weeks after imiquimod; (c) end of first course; (d) two weeks after imiquimod, second course; (e) four weeks after imiquimod, end of second course; (f) eleven months after the end of treatment: even surface and no visible or palpable AKs.

actinic keratoses were clinically identified. Two well-defined target lesions were selected on the forehead for examination. After a single session of PDT (Figure 1(b)), the patient treated the target lesions and each hyperkeratotic region with an application of imiquimod three times per week for one course in accordance with the treatment protocol. The follow-up control two weeks after the imiquimod treatment showed mild erythema, erosion, and some desquamation (Figure 1(c)). These side effects were never disturbing to the patient. Four weeks after the usage of imiquimod, the local side effects (LSE) were more intense-severe erythema, scaling, and crusting. Mr. JR reported an additional development of numerous small vesicles in the treatment area. The patient felt moderately affected by the treatment (Figure 1(d)). The second course of treatment was repeated. The local side effects were intense and the patient was still moderately affected by the treatment (Figure 1(e)). The actinic keratoses healed completely after both courses of treatment and the cosmetic results were excellent. After seven months of followup, Mr. JR

still had no apparent actinic lesions on his scalp or forehead (Figure 1(f)).

3.2. *Case 2.* Mr. HF is a male patient, 82 years old with Fitzpatrick skin phototype III. Numerous hyperkeratotic lesions were localized on his scalp. Two target lesions on his forehead were chosen for examination. The reaction to the PDT-treatment was pronounced (Figure 2(a)). After the healing of the wounds, imiquimod was applied to the target lesions and surrounding tissue in accordance with the treatment protocol. The patient reported mild erythema, erosion, and scaling during the follow-up control two weeks after the imiquimod treatment (Figure 2(b)). After completing the first session four weeks later, the local side effects had decreased with only moderate erythema, scaling, and crusting. The patient felt moderately affected by the treatment (Figure 2(c)). After reevaluation, the second course of treatment was repeated and this resulted in milder LSE, and the patient was hardly affected by the treatment (Figures

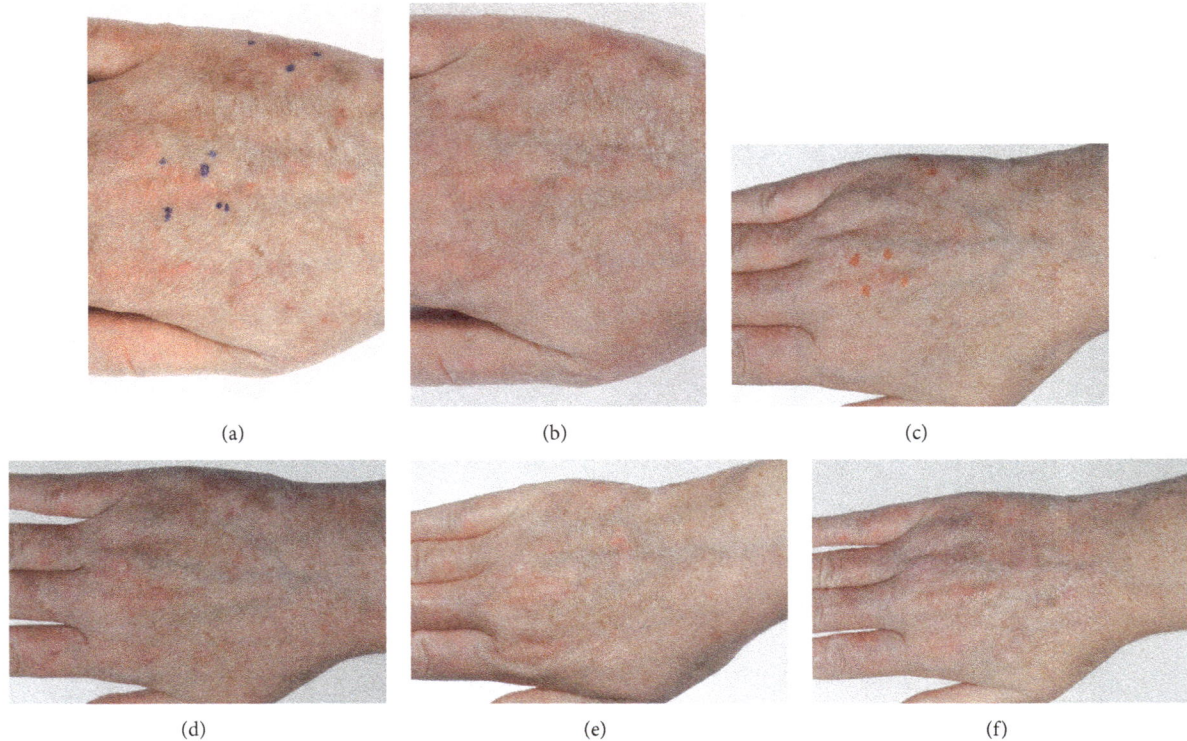

FIGURE 3: Clinical followup, Case 3. (a) Before PDT; (b) two weeks after PDT; (c) two weeks after imiquimod; (d) four weeks after imiquimod, end of first course; (e) two weeks after imiquimod, second course; (f) four weeks after imiquimod, end of second session: less scaling of the skin, more homogenous surface, the skin texture appeared even better, and AKs still present.

2(d)-2(e)). Eleven months later, Mr. HF still had no apparent actinic lesions on his scalp or forehead and he was highly satisfied with the clinical and cosmetic outcome (Figure 2(f)).

3.3. Case 3. Mrs. AD is a female patient, 42 years old with Fitzpatrick skin phototypes II-III. Numerous hyperkeratotic lesions were localized on her hands (Figure 3(a)). Two target lesions on her right hand were chosen for examination. Following by the PDT, imiquimod was applied to the target lesions and surrounding tissue in accordance with the treatment protocol. The patient reported some erythema, edema, and scaling directly after the PDT (Figure 3(b)). But there were no local side effects noted in the follow-up control two weeks after the imiquimod treatment (Figure 3(c)). After completing the first session four weeks later, there were still no local side effects observed (Figure 3(d)). The patient was not affected at all by the treatment. Mrs. AD reported, however, subjective improvement of the skin—she noted less scaling of the skin. After reevaluation, the second course of treatment was repeated. No local side effects were noted (Figures 3(e)-3(f)). When compared to baseline photographs, the skin showed a more homogeneous surface and the skin texture appeared even better. However, the actinic keratoses were still present (Figures 3(e)-3(f)).

4. Discussion

The findings of this investigation demonstrate that the sequential application of PDT and imiquimod 5% cream is a well-tolerated method in the treatment of actinic keratoses with high efficacy and safety and with an excellent cosmetic outcome. Sequential application of PDT and imiquimod has not been widely evaluated in the literature. There are only two known published studies that have compared the sequential application of PDT and imiquimod 5% cream [4, 15]. Nevertheless, the regimen differs from our evaluation. Serra-Guillen et al. compared imiquimod 5% cream, single-session PDT, and the sequential application of both [15]. They showed that the sequential application of both therapeutic modalities provides a better clinical and histological response than monotherapy either with PDT or with imiquimod. It also produces less intense local reactions and better tolerance and satisfaction than imiquimod monotherapy. With 105 patients completing the study, this was a representative evaluation. Nevertheless, the study protocol differs from the protocol in the present investigation; the imiquimod cream was administered four weeks after the PDT and the patients completed only one session of imiquimod. The time until the final evaluation was very short being only one month.

Shaffelburg designed a randomized, vehicle-controlled, split-face study to explore the safety and efficacy of photodynamic therapy followed by imiquimod [4]. Facial actinic keratoses were treated with PDT at baseline and at month one. At month two, imiquimod 5% cream was applied to one-half of the face and the vehicle to the other half, twice-times per week. The duration of imiquimod therapy was long at 16 weeks. The time until the final evaluation was long at

12 months. The method was well tolerated and improved reduction of actinic keratoses resulted.

Both studies showed good results in AK treatment. In the present investigation, two out of three patients showed in long-term followup a high benefit from this therapy modality with only one patient being a nonresponder. This could be explained by high UV-damage of the skin of the patient. The patient was very young being 42 years old, and she had reported skin problems for approximately five years. Various topical treatments (imiquimod and diclofenac in hyaluronic acid as monotherapy) were tried without any side effects nor any response in the past. This could be an explanation for the lack of the additional benefit from the combined therapy. In addition, the AKs were localized on the hands and the treatment of them seems to be more complicated. The present investigation showed that a single-session PDT and a period of eight weeks of imiquimod administration would be enough to obtain good clinical results. Nevertheless, this investigation analyzed only three cases and this was a big limitation. For confirmation of the results and evaluation of long-term therapeutic benefits, further studies are needed.

In conclusion, two patients with long-term followup showed no relapse of AKs. Sequential therapy consisting of PDT and imiquimod 5% cream seems to provide good results in the treatment of actinic keratoses.

Abbreviations

AKs: Actinic keratosis
BCC: Basal cell carcinoma
LSE: Local side effects
NMSC: Nonmelanoma skin cancer
PDT: Photodynamic therapy
SCC: Squamous cell carcinoma
UV: Ultraviolet.

Acknowledgments

This paper is not under consideration for publication elsewhere. All the authors have agreed to its submission and have declared that there is no conflict of interests. There are no financial disclosures from any author to make. Informed consent was obtained from all the study subjects. This study was supported with a research Grant from MEDA Pharmaceuticals. One author has been a speaker for MEDA Pharmaceuticals.

References

[1] W. H. W.H. Organization, *INTERSUN. The Global UV Project: A Guide and Compendium*, WHO, Geneva, Switzerland, 2003.

[2] G. Martin, "The impact of the current united states guidelines on the management of actinic keratosis: is it time for an update?" *Journal of Clinical and Aesthetic Dermatology*, vol. 3, no. 11, 2010.

[3] D. J. Gilbert, "Treatment of actinic keratoses with sequential combination of 5-fluorouracil and photodynamic therapy.," *Journal of Drugs in Dermatology*, vol. 4, no. 2, pp. 161–163, 2005.

[4] M. Shaffelburg, "Treatment of actinic keratoses with sequential use of photodynamic therapy; and imiquimod 5% cream.," *Journal of Drugs in Dermatology*, vol. 8, no. 1, pp. 35–39, 2009.

[5] M. Mastrolonardo, "Topical diclofenac 3% gel plus cryotherapy for treatment of multiple and recurrent actinic keratoses," *Clinical and Experimental Dermatology*, vol. 34, no. 1, pp. 33–35, 2009.

[6] N. M. Price, "The treatment of actinic keratoses with a combination of 5-fluorouracil and imiquimod creams.," *Journal of Drugs in Dermatology*, vol. 6, no. 8, pp. 778–781, 2007.

[7] L. Bercovitch, "Topical chemotherapy of actinic keratoses of the upper extremity with tretinoin and 5-fluorouracil: a double-blind controlled study," *British Journal of Dermatology*, vol. 116, no. 4, pp. 549–552, 1987.

[8] J. Jorizzo, J. Weiss, and G. Vamvakias, "One-week treatment with 0.5% fluorouracil cream prior to cryosurgery in patients with actinic keratoses: a double-blind, vehicle-controlled, long-term study.," *Journal of Drugs in Dermatology*, vol. 5, no. 2, pp. 133–139, 2006.

[9] E. Stockfleth, T. Zwingers, and C. Willers, "Recurrence rates and patient assessed outcomes of 0.5% 5-fluorouracil in combination with salicylic acid treating actinic keratoses," *European Journal of Dermatology*, vol. 22, no. 3, pp. 370–374, 2012.

[10] E. Stockfleth, H. Kerl, T. Zwingers, and C. Willers, "Low-dose 5-fluorouracil in combination with salicylic acid as a new lesion-directed option to treat topically actinic keratoses: histological and clinical study results," *British Journal of Dermatology*, vol. 165, no. 5, pp. 1101–1108.

[11] M. Freeman, C. Vinciullo, D. Francis et al., "A comparison of photodynamic therapy using topical methyl aminolevulinate (Metvix®) with single cycle cryotherapy in patients with actinic keratosis: a prospective, randomized study," *Journal of Dermatological Treatment*, vol. 14, no. 2, pp. 99–106, 2003.

[12] Graceway Pharmaceuticals, *Aldara Package Insert*, Graceway Pharmaceuticals, Bristol, Tenn, USA, 2007.

[13] R. Strohal, H. Kerl, and L. Schuster, "Treatment of actinic keratoses with 5% topical imiquimod: a multicenter prospective observational study from 93 Austrian office-based dermatologists," *Journal of Drugs in Dermatology*, vol. 11, no. 5, pp. 574–578, 2012.

[14] S. J. Salasche, N. Levine, and L. Morrison, "Cycle therapy of actinic keratoses of the face and scalp with 5% topical imiquimod cream: an open-label trial," *Journal of the American Academy of Dermatology*, vol. 47, no. 4, pp. 571–577, 2002.

[15] C. Serra-Guillen, E. Nagore, L. Hueso et al., "A randomized pilot comparative study of topical methyl aminolevulinate photodynamic therapy versus imiquimod 5% versus sequential application of both therapies in immunocompetent patients with actinic keratosis: clinical and histologic outcomes," *Journal of the American Academy of Dermatology*, vol. 66, no. 4, pp. 131–137, 2012.

2

Immunomodulatory Effect of Continuous Venovenous Hemofiltration during Sepsis: Preliminary Data

Giuseppe Servillo,[1] Maria Vargas,[1] Antonio Pastore,[1] Alfredo Procino,[2] Michele Iannuzzi,[1] Alfredo Capuano,[2] Andrea Memoli,[2] Eleonora Riccio,[2] and Bruno Memoli[2]

[1] *Medical Intensive Care Unit, Department of Surgical and Anesthesiological Sciences, University Federico II of Naples, 80129 Naples, Italy*
[2] *Department of Nephrology, University Federico II of Naples, 80129 Naples, Italy*

Correspondence should be addressed to Eleonora Riccio; elyriccio@libero.it

Academic Editor: Tiziano Verri

Introduction. Severe sepsis and septic shock are the primary causes of multiple organ dysfunction syndrome (MODS), which is the most frequent cause of death in intensive care unit patients. Many pro- and anti-inflammatory mediators, such as interleukin-6 (IL-6), play a strategic role in septic syndrome. Continuous renal replacement therapy (CRRT) removes in a nonselective way pro- and anti-inflammatory mediators. *Objective.* To investigate the effects of continuous venovenous hemofiltration (CVVH) as an immunomodulatory treatment of sepsis in a prospective clinical study. *Methods.* High flux hemofiltration (Qf = 60 ml/Kg/hr) was performed for 72 hr in thirteen critically ill patients suffering from severe sepsis or septic shock with acute renal failure (ARF). IL-6 gene expression was measured by real-time PCR analysis on RNA extracted from peripheral blood mononuclear cell before beginning of treatment (T0) and after 12, 24, 48, and 72 hours (T1–4). *Results.* Real-time PCR analysis demonstrated in twelve patients IL-6 mRNA reduction after 12 hours of treatment and a progressive increase after 24, 48, and 72 hours. *Conclusions.* We suggest that an immunomodulatory effect might exist during CVVH performed in critically ill patients with severe sepsis and septic shock. Our data show that the transcriptional activity of IL-6 increases during CVVH.

1. Introduction

Sepsis is a great public health problem and represents the leading cause of death in intensive care units of developed countries. Different studies have reported a high population-based incidence of sepsis, and high percentages of intensive care unit (ICU) beds have been occupied by those patients [1–4].

An increase of cytokines has been evidenced in plasma of patients presenting septic shock and has been associated with poor prognosis [5]. Microbial products are responsible for the induction of inflammation. This leads to local microvascular injury with potential dissemination and malignant sequelae at different organ levels. These effects are described as multiple organ failure (MOF). While there is a high concentration of proinflammatory mediators (IL-1, IL-6, IL-8, and TNF-α) at the site of infection, body's stress response contributes to an opposing systemic reaction by producing anti-inflammatory mediators (IL-4, IL-10, and IL-13). Often the production of proinflammatory cytokines is followed by secretion of inhibitors, such as soluble receptors or receptor antagonists. Bone et al. called this response "compensated antiinflammatory response syndrome" (CARS) [6]. Several lines of evidence suggest that compensated response and "systemic inflammatory response syndrome" (SIRS) may often coexist in the same patient but in different compartments [7]. In the "peak concentration hypothesis," peaks of proinflammatory cytokines and peaks of anti-inflammatory cytokines coexist and the concept of cutting peaks, for example, through hemofiltration, may contribute to bring the patient to a nearly normal immunohomeostasis. The therapeutic approaches focused on decreasing cytokine production or neutralizing the effects of cytokines; continuous renal replacement therapy (CRRT) may play a role in the downregulation of the inflammatory response [8] by nonselective extracorporeal removal, mainly by absorption, of cytokines and other mediators, restoring the hemodynamic and the immunologic homeostasis. However, the hypothesis that hemofiltration can remove inflammatory mediators and its role in the sepsis therapy is still conflictual.

A phase II clinical study showed no reduction of plasma levels of circulating IL-6, IL-8, IL-10, and TNF-α or occurrence of MOF in twelve patients with early septic shock or septic organ dysfunction who received 48 hours of continuous venovenous hemofiltration (CVVH) [9]. Ronco et al. estimated that CVVH with replacement fluid of 35 mL/Kg improves survival in critically ill patients with acute renal failure (ARF) admitted in two intensive care units (ICUs) [10]. However, it is still uncertain whether beneficial effect of high volume hemofiltration is the result of intensified blood purification or an effect on nonspecific cell-mediated immunity. Therefore, Dellinger et al. in the international guidelines for management of severe sepsis and septic shock recommended CVVH as the elective therapy for ARF in septic shock, allowing the management of fluid balance in hemodynamically unstable septic patients [11].

Concentration of cytokines has been measured in many studies, but few assays sized neither the bioactivity of the cytokines nor their net effects on immune functions [12]. Circulating cytokines may just be the "tip of the iceberg," implying that neither their presence nor their absence can reflect the complex interplay at the tissue level [13]. Despite the fact that their peak concentration may reflect an exacerbated production, these levels do not necessarily stand for enhanced bioactivity. To determine the balance of inflammatory response measurement of cytokines bioactivity may be superior to their absolute concentration. Moreover, variation of a single cytokine in the complex immune response syndrome could also reflect the change of other mediators. Therefore, we decided to measure interleukin-6 gene expression in mononuclear cells (PBMCs) harvested from septic patients, as a marker for immune function and proinflammatory bioactivity. Persisting high levels of IL-6, in fact, correlate with poor outcome in sepsis, and its concentration correlates with the concentration of other inflammation markers.

Aim of the present observational prospective study is to investigate the effect of high volume continuous hemofiltration on the transcriptional activity of PBMC, as a marker of immunomodulatory effect of this treatment in septic process.

2. Materials and Methods

2.1. Patients. The study was performed in the intensive care unit of the University of Naples "Federico II" after local ethics committee approval. Informed consent was obtained from patients' next of kin. From January 2007 to January 2008, we enrolled medical and surgical patients from our intensive care unit suffering from severe sepsis or septic shock with coexisting ARF, according to the American College of Chest Physicians/Society of Critical Care Medicine Consensus Conference criteria. Exclusion criteria were

age >80 years, acute bleeding, and immunodepression. Table 1 shows clinical characteristics of our patients. All patients received conventional intensive care therapies in accordance with the international sepsis guidelines [14], as intubation and mechanical ventilation with a tidal volume of 6 mL/kg and an upper limit plateau pressure <30 cmH$_2$O, sedation by continuous infusion according to clinical requirements, and intravenous antibiotics as indicated by microbiological resistance testing. All patients received inotropic support in addition to other vasoactive drugs, as clinically indicated, and fluid therapy by using crystalloids and colloids. Once the decision was made to proceed with renal replacement therapy, a double lumen catheter was inserted in femoral vein and continuous hemofiltration was started.

2.2. Treatments. Ultrafiltration rate (Qf) was 4 l/hr. Blood flow rate was more than 200 mL/min. For ultrafiltration, we used a polyethersulfone filter (Acquamax HF12) with a surface area of 1.20 m^2 and a maximal transmembrane pressure (TMP) of 600 mmHg. Anticoagulation was obtained with heparin 6.0 U/Kg/hr. In continuous venovenous hemofiltration, solute transport is achieved by pure convection. Solute flux across the membrane is proportional to the ultrafiltration rate (Qf) and the ratio between the concentration of solute in the ultrafiltrate and in the plasma water (sieving coefficient, S), since clearance is calculated from product Qf × S. When S is proximal to 1 (for solutes freely crossing the membrane), clearance is assumed to be Qf. Therefore, since ultrafiltration rate corresponds to clearance in continuous hemofiltration, it may be used as a surrogate of treatment dose. Two different machines were used for the study, both equipped with calibrated peristaltic blood pumps and fluid balance systems with calibrated scales. Replacement solution was added in the postdilutional mode. We used Accusol with potassium (Baxter Healthcare), which had a pH of 7.4, 35 mmol/L of bicarbonate, and 2 mmol/L of potassium, associated with other ionic compounds. Theoretical osmolarity of this solution was of 296 mOsm/L. Blood samples were obtained before the beginning of treatment (T0) and after 12, 24, 48, and 72 hours (T1–4). SAPS 3 [15, 16] and daily SOFA [17] were measured for each patient.

2.3. IL-6 Gene Expression Analysis. Blood samples (20 mL of heparinized blood) were collected from all patients to obtain plasma samples and isolate peripheral blood mononuclear cells (PBMCs). These cells were isolated by Ficoll-Hypaque (Flow Laboratories, Irwine, UK) gradient density centrifugation (400 ×g for 30 min). Then PBMCs were incubated in culture tubes (Falcon) in quantities of 3 × 106/mL and cultured for 24 h at 37°C in 5% CO$_2$ saturated humidity incubator. After this step, cell-free supernatants were collected by centrifugation. PBMCs were pulverized with a blender and lysed using TRIzol reagent. Total RNA was extracted by the single-step method, using phenol and chloroform/isoamylalcohol. Four micrograms of total RNA were subjected to cDNA synthesis for 1 h at 37°C using the "Ready-To-Go You-Prime First-Strand Beads" Kit (Amersham Pharmacia Biotech Little Chalfont Buckinghamshire, UK) in a reaction containing 0.5 μg oligo-dT (Amersham

TABLE 1: Characteristics of patients at baseline: gender, sex, underlying disease, origin of sepsis, and microbiology.

Patient	Gender	Age	Underlying disease	Origin of sepsis	Microbiology
N1	F	74	Gastric cancer, gastrectomy, and pulmonary thromboembolism	Abdominal	*Candida albicans*
N2	M	80	Left inferior pulmonary lobectomy, mesenteric infarction, and resection of small bowel	Abdominal	*Bacteroides fragilis*
N3	F	79	Pulmonary thromboembolism	Pulmonary	*Pseudomonas aeruginosa*
N4	F	60	Posttraumatic subdural haemorrhage	Pulmonary	*Acinetobacter baumannii*
N5	M	34	Multiple thoracic and abdominal trauma	Pulmonary	*Acinetobacter baumannii*
N6	F	51	Substitution of mitral valve after rheumatic fever cardiopathy	Mediastinal abscess	*Candida albicans* *Escherichia coli*
N7	M	56	Acute respiratory failure, lung cancer	Pulmonary	*Pseudomonas aeruginosa*
N8	M	55	Evacuation of cerebral haemorrhage	Pulmonary	*Klebsiella pneumoniae*
N9	M	51	Ligature of left external carotid, draining of facial, neck, and mediastinic abscess	Mediastinal abscess, infection of tracheostomy site	*Klebsiella pneumoniae* *Candida albicans*
N10	M	48	Multiple abdominal and pelvic trauma	Urinary	*Escherichia coli*
N11	F	59	Perforation of the colon, colonic resection, and peritonitis	Abdominal	*Escherichia coli, Enterobacter spp.*
N12	F	62	Acute respiratory failure, COPD	Pulmonary	*Pseudomonas aeruginosa Acinetobacter baumannii*
N13	M	46	Multiple thoracic abdominal and pelvic trauma	CVC	Staphylococcus coagulase negative

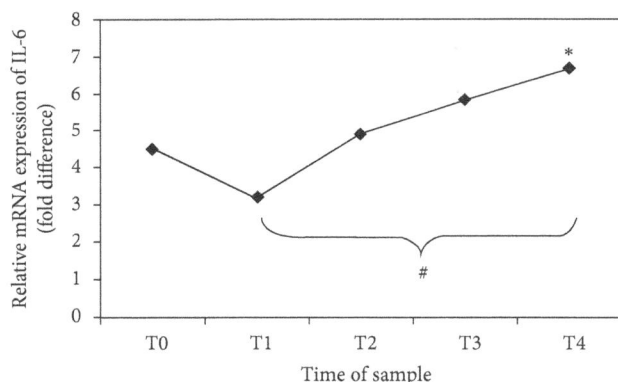

FIGURE 1: Gene expression of IL-6 in PBMCs obtained by real-time PCR analysis. The points show the relative expression of the mRNA abundance of IL-6. The values on the y-axis represent arbitrary units derived from the mean expression value for the gene of septic patients compared with the healthy subjects set at 1 (not reported in the figure). $^*P = 0.031$ trend of relative mRNA expression of IL-6 (ANOVA). $^\#P = 0.030$ T1 versus T4 (Bonferroni post hoc correction).

Pharmacia Biotech Little Chalfont Buckinghamshire, UK) [18].

Real-time quantitative PCR analysis for IL-6 gene was performed using ABI Prism 7500 (Applied Biosystem, Foster City, CA, USA) and the $5'$ exonuclease assay (TaqMan technology). This assay uses a specific oligonucleotide probe, annealing between the two primer sites, which is labeled with a reporter fluorophore and a quencher. Cleavage of the probe by exonuclease activity of Taq polymerase during strand elongation releases the reporter from the probe resulting in an

increase in reporter emission intensity owing to its separation from the quencher. This increment in net fluorescence is monitored in real time during PCR amplification. We have described the cDNA synthesis. Now, the cDNA was used for real-time PCR realized on 96-well optical reaction plates with cDNA equivalent to 20 ng RNA in a volume of $5\,\mu$L reaction containing 1x TaqMan Universal MasterMix, optimized concentrations of FAM-labelled probe, and specific forward and reverse primer for the gene selected from Assay on Demand (Applied Biosystems Foster City, CA, USA). Controls included RNA subjected to RT-PCR without reverse transcriptase and PCR with water replacing cDNA. The results were analyzed using a comparative method, and the values were normalized and converted into fold change based on a doubling of PCR product in each PCR cycle, according to the manufacturer's guidelines.

2.4. Statistics. From Table 2, it can be deduced that results obtained from genetic expression of IL-6 are continuous variables function of times T0, T1, T2, T3, and T4. Such continuous variables are expressed as mean ± SD calculated with software "Statistica7" (Statsoft). Various results for the same genetic expression of IL-6 are confronted using ANOVA followed by Bonferroni, as post hoc test. Statistical significance is at the $P < 0.05$ level.

3. Results

Results are expressed as number, percentage, or mean ± SD, as appropriate. Fifteen patients were considered eligible. Two patients developed severe hemodynamic instability during

TABLE 2: IL-6 gene expression from real-time PCR analysis. Mean values. P values result of ANOVA followed by Bonferroni as post hoc test using the following matricesT1 versus T0, T2 versus T0, T3 versus T0, and T4 versus T0.

Patient	T0	T1	T2	T3	T4
N1	3.5	1	3.6	5.15	6.7
N2	3	1.9	3.1	4.8	7
N3	4.6	2.8	3.9	5	6.2
N4	12	10.4	10	—	—
N5	9.8	8.5	10.3	11.6	—
N6	3.2	2.1	3.5	3.9	4.2
N7	1	1.8	—	—	—
N8	2.5	1	2.8	—	—
N9	3.1	2.4	2.7	5.8	9.3
N10	4.5	3	5.2	5.8	6.7
N11	4.2	2.9	5.1	6.2	6.8
N12	3.7	1.9	4.2	5.1	6.4
N13	3.9	2.2	4.5	5.4	6.9
Mean ± SD	4.53 ± 3.1	3.2 ± 2.8	4.91 ± 2.5	5.87 ± 2.1	6.68 ± 1.3
P		<0.05	>0.05	<0.05	<0.05

replacement therapy, and CVVH was discontinued. Therefore, thirteen patients were enrolled. The origin of sepsis was mostly medical but in some patients postsurgical and trauma related. Table 2 shows the results of real-time PCR analysis on blood samples and their mean values ± SD. In twelve out of thirteen patients, transcriptional activity behaviour is similar; that is, IL-6 mRNA is present in high amounts before filtration (T0), reduces after 12 hours of treatment, and progressively increases after 24, 48, and 72 hours. On the other hand, samples from patient number 4 show a different behaviour: transcriptional activity of IL-6 in this patient seems only to progressively reduce. Figure 1 shows the trend of IL-6 gene expression variation in our thirteen patients ($^*P = 0.031$) from T0 to T4 and a statistically significant increase between T1 versus T4 ($P = 0.030$). Table 3 shows clinical data, SAPS 3, and daily SOFA for each patient. Figure 2 shows trend of SOFA score modification in our thirteen patients.

4. Discussion

Our prospective and clinical study, conducted on critically ill septic patients with ARF treated using high flux hemofiltration (Qf = 60 mL/Kg/hr) with a filter of polyethersulfone, has shown a progressive increase in transcriptional activity of IL-6 produced by PBMCs. In the early 90s, Bellomo et al. conducted a study on cytokines removal from circulation with dialysis [19]. The authors showed that CVVHD attenuated progression of ARF and removed inflammatory cytokines from blood of septic patients. After these results, some authors [9, 20, 21] hypothesized a role of CVVH not only as CRRT but also as immunomodulatory therapy for septic patients. In their "peak concentration hypothesis" [22], Ronco et al. suggest that the effect of nonspecific removal of unbound mediators and cytokines from blood compartment is a resetting to a lower level of immunedysregulation,

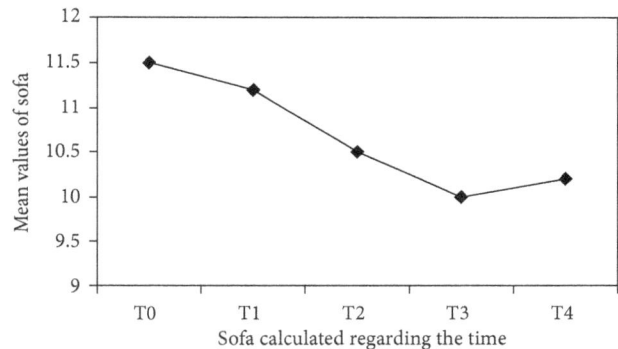

FIGURE 2: Mean values of daily SOFA measured for each patient. The points on y-axis represent that the mean values reach from this score during the time of treatment. The x-axis reported the different time of measurement.

both in prevalently proinflammatory and counterinflammatory phases of sepsis [10]. In their opinion, Ronco et al. affirmed that by eliminating the peaks, the system might restore its ability to achieve immune homeostasis [22]. More recently, Honoré et al. [20] shift their interest to the effect of high volume hemofiltration (HVHF) on the interstitium and tissue level. The authors conclude that both mediators and promediators are removed at interstitial and at tissue levels, secondary to removal from blood compartment, to a point where some pathways are completely shut down and cascades are blocked. At this point, called "Threshold point," tissue damage and organ injury are stopped in their progression. Sander et al. [21] suggest that the entire amount of cytokines potentially eliminated through hemofiltration is probably lower when compared to endogenous production. Furthermore, hemofiltration membranes per se may increase cytokine production by activating mononuclear cells. In 2007, Horner et al. [23] concluded that plasma from septic

TABLE 3: SAPS 3 admission score, SOFA score calculated at 0, 12, 24, 48, and 72 hours, and outcome of each patient.

Patient	SAPS 3	SOFA 0	SOFA 1	SOFA 2	SOFA 3	SOFA 4	Outcome
N1	55 (26%)	6	6	7	7	8	Died
N2	27 (1%)	9	10	10	8	9	Died
N3	75 (66%)	11	9	9	11	10	Died
N4	56 (28%)	12	13	13	Stop CVVH	—	Died
N5	47 (13%)	15	12	12	13	Stop CVVH	Survived
N6	80 (74%)	16	18	12	16	18	Died
N7	68 (52%)	12	15	—	—	—	Died
N8	90 (85%)	13	15	15	—	—	Died
N9	57 (30%)	10	8	7	7	7	Survived
N10	52 (32%)	8	8	12	10	10	Survived
N11	64 (44%)	12	11	10	10	10	Survived
N12	67 (50%)	12	9	9	9	10	Survived
N13	56 (28%)	14	12	10	9	10	Survived
Mean ± SD	61.1 ± 15.6	11.5 ± 2.8	11.2 ± 3.4	10.5 ± 2.4	10 ± 2.8	10.2 ± 3.1	

patients contains substances mediating a substantially different immune response pattern compared with plasma from critically ill nonseptic patients. In an experimental study, Toft et al. [24] showed that infusion of endotoxin in pigs induced a significant increase in IL-6 and IL-10 levels in peripheral blood, but there was no difference in cytokine levels between CVVH-treated and nontreated septic animals. Thus, the author concluded that CVVH did not decrease plasma levels of cytokines, although it had been described that proinflammatory as well as anti-inflammatory cytokines are excreted in the ultrafiltrate. Moreover, De Vriese et al. [25] suggest that a decrease in concentration of cytokines in plasma was possible for a few hours after a change of filter, and in fact the filter rapidly became saturated with cytokines.

Our prospective and clinical studies conducted on septic patients aim to evaluate the biological activity of leukocytes, as IL-6 production by PBMCs, during CVVH. We have applied to thirteen septic patients a CVVH with high flux (60 mL/Kg/hr), Qf of 4 L/hr, and polyethersulphone filter which allowed us to obtain a blood flow ≥ 200 mL/min. Our data show that transcriptional activity of an inflammatory cytokine such as IL-6 significantly increases during CVVH. In sepsis, however, impaired regulation may cause an excessive anti-inflammatory response, which generates monocyte "immunoparalysis" and exposes the host to further infections. Both processes (inflammation and anti-inflammation) are designed to act in response to specific stimuli in a well-balanced fashion defined as immunohomeostasis. Continuous therapies have been shown to provide clinical benefits beyond those expected by the simple substitution of renal function. In this view, it is possible to hypothesize that avoidance of peaks of mediator in blood may contribute to a partial restoration of the immunohomeostasis. When sepsis is viewed as a syndrome of immune suppression perspective, in which the immune effectors cells become dysfunctional and are no longer capable of guaranteeing a normal immune surveillance, we can hypothesize that increase of IL-6 transcriptional activity could be due to enhanced capability of immune cells to provide protection against external harmful stimuli. We therefore think that during CVVH, there might be a reactivation of immune capability of the leukocytes to produce inflammatory cytokines, restoring an immunologic balance.

In conclusion, we suggest that there might be an immunomodulatory effect of CVVH during sepsis. However, since response to sepsis should be viewed in a network perspective, within an array of interdependent mediators, it is important to further confirm this hypothesis.

Conflict of Interests

The authors declare no conflict of interests.

References

[1] D. C. Angus, W. T. Linde-Zwirble, J. Lidicker, G. Clermont, J. Carcillo, and M. R. Pinsky, "Epidemiology of severe sepsis in the United States: analysis of incidence, outcome, and associated costs of care," *Critical Care Medicine*, vol. 29, no. 7, pp. 1303–1310, 2001.

[2] R. Beale, K. Reinhart, F. M. Brunkhorst et al., "Promoting global research excellence in severe sepsis (PROGRESS): lessons from an international sepsis study," *Infectio*, vol. 37, no. 3, pp. 222–232, 2009.

[3] C. Alberti, C. Brun-Buisson, H. Burchardi et al., "Epidemiology of sepsis and infection in ICU patients from an international multicentre cohort study," *Intensive Care Medicine*, vol. 28, no. 2, pp. 108–121, 2002.

[4] E. Silva, A. Pedro Mde, A. C. Sogayar et al., "Brasilian sepsis epidemiological study (BASES study)," *Critical Care Medicine*, vol. 8, no. 4, pp. R251–R260, 2004.

[5] M. R. Pinsky, J.-L. Vincent, J. Deviere, M. Alegre, R. J. Kahn, and E. Dupont, "Serum cytokine levels in human septic shock; Relation to multiple-system organ failure and mortality," *Chest*, vol. 103, no. 2, pp. 565–575, 1993.

[6] R. C. Bone, C. J. Grodzin, and R. A. Balk, "Sepsis: a new hypothesis for pathogenesis of the disease process," *Chest*, vol. 112, no. 1, pp. 235–243, 1997.

[7] C. Ronco, M. Bonello, V. Bordoni et al., "Extracorporeal therapies in non-renal disease: treatment of sepsis and the peak concentration hypothesis," *Blood Purification*, vol. 22, no. 1, pp. 164–174, 2004.

[8] A. F. Grootendorst, "The potential role of hemofiltration in the treatment of patients with septic shock and multiple organ dysfunction syndrome," *Advances in Renal Replacement Therapy*, vol. 1, no. 2, pp. 176–184, 1994.

[9] L. Cole, R. Bellomo, G. Hart et al., "A phase II randomized, controlled trial of continuous hemofiltration in sepsis," *Critical Care Medicine*, vol. 30, no. 1, pp. 100–106, 2002.

[10] C. Ronco, R. Bellomo, P. Homel et al., "Effects of different doses in continuous veno-venous haemofiltration on outcomes of acute renal failure: a prospective randomised trial," *The Lancet*, vol. 355, no. 9223, pp. 26–30, 2000.

[11] R. P. Dellinger, J. M. Carlet, H. Masur et al., "Surviving Sepsis Campaign guidelines for management of severe sepsis and septic shock," *Intensive Care Medicine*, vol. 34, pp. 17–60, 2008.

[12] W. Ertel, J.-P. Kremer, J. Kenney et al., "Downregulation of proinflammatory cytokine release in whole blood from septic patients," *Blood*, vol. 85, no. 5, pp. 1341–1347, 1995.

[13] J.-M. Cavaillon, C. Munoz, C. Fitting, B. Misset, and J. Carlet, "Circulating cytokines: the tip of the iceberg?" *Circulatory Shock*, vol. 38, no. 2, pp. 145–152, 1992.

[14] R. P. Dellinger, J. M. Carlet, and H. Masur, "Surviving sepsis campaign guidelines for management of severe sepsis and septic shock," *Critical Care Medicine*, vol. 32, no. 10, pp. 858–873, 2004.

[15] P. G. H. Metnitz, R. P. Moreno, E. Almeida et al., "SAPS 3-From evaluation of the patient to evaluation of the intensive care unit. Part 1: objectives, methods and cohort description," *Intensive Care Medicine*, vol. 31, no. 10, pp. 1336–1344, 2005.

[16] R. P. Moreno, P. G. H. Metnitz, E. Almeida et al., "SAPS 3 - From evaluation of the patient to evaluation of the intensive care unit. Part 2: development of a prognostic model for hospital mortality at ICU admission," *Intensive Care Medicine*, vol. 31, no. 10, pp. 1345–1355, 2005.

[17] J.-L. Vincent, R. Moreno, J. Takala et al., "The SOFA (Sepsis-related Organ Failure Assessment) score to describe organ dysfunction/failure," *Intensive Care Medicine*, vol. 22, no. 7, pp. 707–710, 1996.

[18] B. Memoli, C. Libetta, T. Rampino et al., "Hemodialysis related induction of interleukin-6 production by peripheral blood mononuclear cells," *Kidney International*, vol. 42, no. 2, pp. 320–326, 1992.

[19] R. Bellomo, P. Tipping, and N. Boyce, "Continuous veno-venous hemofiltration with dialysis removes cytokines from the circulation of septic patients," *Critical Care Medicine*, vol. 21, no. 4, pp. 522–526, 1993.

[20] P. M. Honoré, O. Joannes-Boyau, and B. Gressens, "Blood and plasma treatments: the rationale of high-volume hemofiltration," *Contributions to Nephrology*, vol. 156, pp. 387–395, 2007.

[21] A. Sander, W. Armbruster, B. Sander, A. E. Daul, R. Lange, and J. Peters, "Hemofiltration increases IL-6 clearance in early systemic inflammatory response syndrome but does not alter IL-6 and TNFα plasma concentrations," *Intensive Care Medicine*, vol. 23, no. 8, pp. 878–884, 1997.

[22] C. Ronco, C. Tetta, F. Mariano et al., "Interpreting the mechanisms of continuous renal replacement therapy in sepsis: the peak concentration hypothesis," *Artificial Organs*, vol. 27, no. 9, pp. 792–801, 2003.

[23] C. Horner, S. Schuster, J. Plachy et al., "Hemofitration and immune response in severe sepsis," *Journal of Surgical Research*, vol. 142, pp. 59–65, 2007.

[24] P. Toft, V. Brix-Christensen, J. Baech et al., "Effect of hemofiltration and sepsis on chemotaxis of granulacyties and release of IL-8 and IL-10," *Acta Anaesthesiologica Scandinavica*, vol. 46, no. 2, pp. 138–144, 2002.

[25] A. S. De Vriese, F. A. Colardyn, J. J. Philippé, R. C. Vanholder, J. H. De Sutter, and N. H. Lameire, "Cytokine removal during continuous hemofiltration in septic patients," *Journal of the American Society of Nephrology*, vol. 10, no. 4, pp. 846–853, 1999.

3

Two-Photon Photodynamic Therapy by Water-Soluble Self-Assembled Conjugated Porphyrins

Kazuya Ogawa[1,2] **and Yoshiaki Kobuke**[1,3]

[1] *Graduate School of Materials Science, Nara Institute of Science and Technology, 8916-5 Takayama, Ikoma, Nara 630-0101, Japan*
[2] *Interdisciplinary Graduate School of Medicine and Engineering, University of Yamanashi, 4-3-11 Takeda, Kofu, Yamanashi 400-8511, Japan*
[3] *Institute of Advanced Energy, Kyoto University, Gokasho, Uji, Kyoto 611-0011, Japan*

Correspondence should be addressed to Kazuya Ogawa; kogawa@yamanashi.ac.jp and Yoshiaki Kobuke; kobuke@iae.kyoto-u.ac.jp

Academic Editor: Kristjan Plaetzer

Studies on two-photon absorption (2PA) photodynamic therapy (PDT) by using three water-soluble porphyrin self-assemblies consisting of ethynylene-linked conjugated *bis* (imidazolylporphyrin) are reviewed. 2PA cross-section values in water were obtained by an open aperture Z-scan measurement, and values were extremely large compared with those of monomeric porphyrins such as hematoporphyrin. These compounds were found to generate singlet oxygen efficiently upon one- as well as two-photon absorption as demonstrated by the time-resolved luminescence measurement at the characteristic band of singlet oxygen at 1270 nm and by using its scavenger. Photocytotoxicities for HeLa cancer cells were examined and found to be as high as those of hematoporphyrin, demonstrating that these compounds are potential candidates for 2PA-photodynamic therapy agents.

1. Introduction

Photodynamic therapy (PDT) is a gentle treatment modality for cancers based on the localization of a photosensitizer such as Photofrin (a mixture of hematoporphyrin oligomers) in the cancer cell followed by photoactivation [1]. In the photoreaction, the photosensitizer is promoted by photoirradiation to the excited triplet state from the excited singlet state through intersystem crossing (ISC) and transfers the excited energy to ground state oxygen ($^3O_2$) generating the singlet oxygen ($^1O_2$) which destroys the cancer. One of the problems in current PDT is the limitation of the penetration depth of light at 630 nm to restrict the treatment of deep cancer. However, the absorption by tissues is much lower in the near-infrared (NIR) region between 700–1300 nm, which is called as an optical window of biological tissue [2]. Thus, the use of light in the NIR enables the deep part cancer treatment.

Two photon absorption (2PA) is a nonlinear optical process, in which two-photons are absorbed simultaneously at wavelength practically in the NIR region even where no one-photon absorption exists to promote a molecule to the excited state corresponding to the combined energy of the two-photons. Moreover, the quadratic dependence of 2PA on the laser intensity allows a high spatial selectivity by using a focused laser beam. Therefore, PDT using 2PA is better for treating the deeper cancer with a three-dimensional selectivity. In 1990s, the two photon absorption photodynamic therapy (2PA-PDT) has been proposed and studied by some research groups [3–7]. However, these studies could not attract a lot of attention because 2PA efficiencies of photosensitizers used in those studies were low with 2PA cross-section values below 50 GM (1 GM equals to 10^{-50} cm^4 s molecule^{-1} photon^{-1}). For example, the 2PA cross-section value of protoporphyrin IX is known to be only ~2 GM [8] and other organic molecules also exhibited small values less than 1000 GM measured by femtosecond pulses. Photofrin was also investigated as a candidate for 2PA-PDT [9]. The $\sigma^{(2)}$ value of Photofrin was determined as 7.4 GM at 850 nm, and they conducted cell experiments. The total energy required for the 50% cell death was

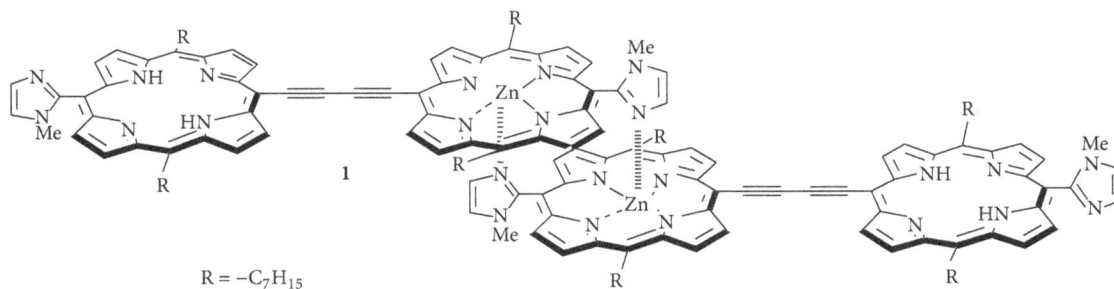

SCHEME 1: Supramolecular porphyrin array **1**.

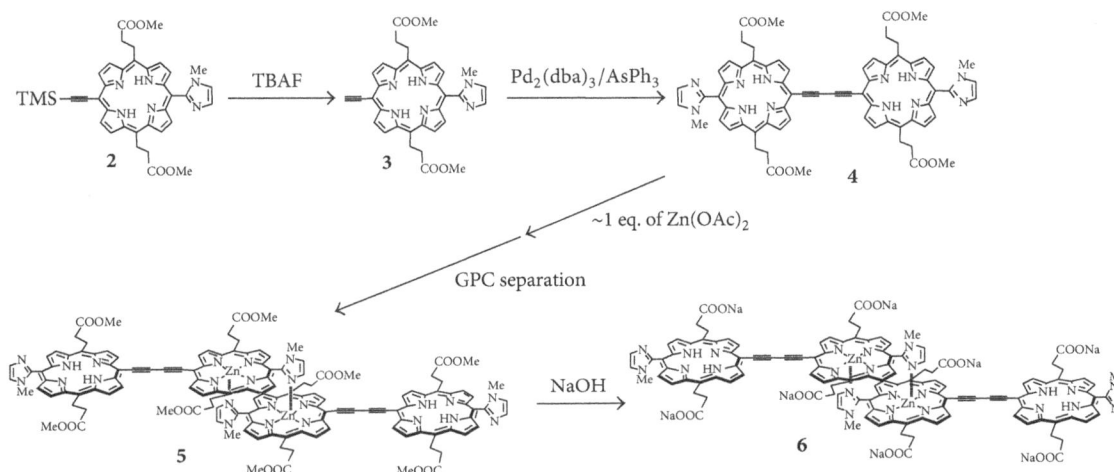

SCHEME 2: Synthetic routes of water-soluble butadiyne-linked self-assembly **6**.

$6,300 \, \text{J cm}^{-2}$, which required 4 hours irradiation, demonstrating that Photofrin was unsuitable for 2PA-PDT, and new sensitizers having much higher $\sigma^{(2)}$ values would be requested. After that, we [10–12] and some research groups [13–16] have reported 2PA-PDT studies using photosensitizers with much higher $\sigma^{(2)}$ values. 2PA-PDT employing energy transfer from a two photon absorbing dye having the $\sigma^{(2)}$ value of 217 GM to pheophorbide as a PDT photosensitizer was reported [13]. In this case, the 2PA-PDT effect was observed when two photon irradiation of Hela cells was treated overnight. Anderson reported *in vitro* 2PA-PDT as well as closure of blood-vessel by two photon excitation of butadiynylene-connected zinc-porphyrin dimer [14].

In 2003, we reported that the self-assembled conjugated porphyrin **1** (Scheme 1) through zinc-imidazolyl coordinations exhibiting a large two photon absorption cross-section value $(\sigma^{(2)})$ of 7,600 GM, which was the largest among the reported values measured using femtosecond pulses [16, 17]. This value is three or four orders of magnitude larger than that of protoporphyrin IX or Photofrin. Further, **1** was found to generate singlet oxygen with high efficiency in toluene, indicating an appropriate candidate for 2PA-PDT. Thus, we started the 2PA-PDT study with the water-soluble conjugated porphyrins. In this paper, we will report on our recent studies

on the 2PA-PDT, including the syntheses of water-soluble porphyrin self-assemblies, their two photon absorption properties, singlet-oxygen generation, and photocytotoxicity.

2. The First 2PA-PDT System Based on Self-Assembled Porphyrin Array 1 [10]

In order to solubilize porphyrin **1** in water, methoxycarbonylethyl groups, which would be hydrolyzed to give carboxyl groups, were introduced instead of heptyls at two *meso*-positions in each porphyrin. As shown in Scheme 2, *bis*(imidazolylporphyrin) **4** bridged by a butadiyne linkage was synthesized from TMS-deprotected compound **3** by a Pd(0)-mediated coupling reaction with 47% yield. The reaction of **4** with one equivalent of zinc acetate gave a complementary dimer of monozinc-*bis*(imidazolylporphyrin) **5**, which was isolated using gel permeation chromatography (GPC). Finally, the methyl ester groups were hydrolyzed by sodium hydroxide to obtain water-soluble self-assembly **6**.

The low yield of **6** (~12%) was attributed to the monometalation process, giving a mixture of starting *bis*(free base) porphyrin **4**, the desired monozinc complex **5**, and dizinc complex. The yield was further considerably decreased during GPC separation. The low yield is obviously disadvantageous for the practical use. In order to improve

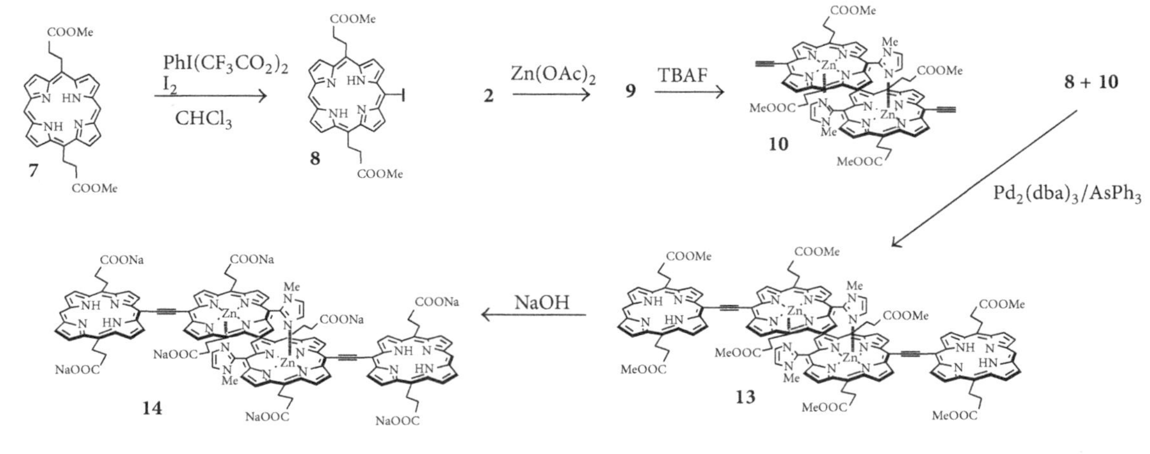

SCHEME 3: Synthetic routes of water-soluble monoacetylene-linked self-assembly **14**.

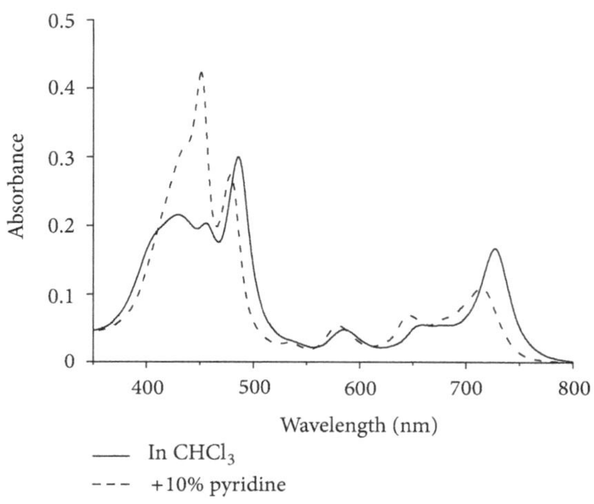

(a) **5**

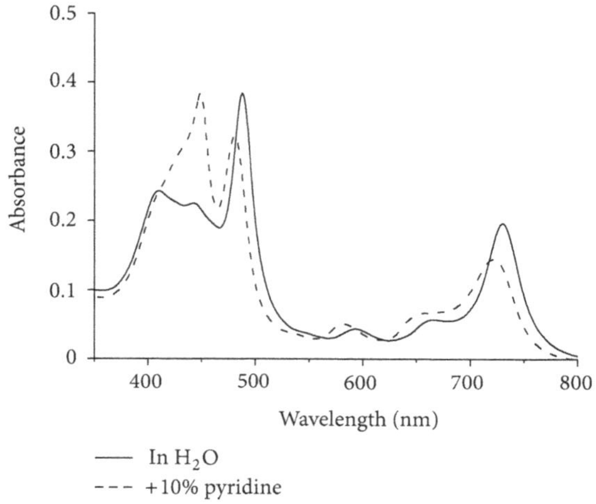

(b) **6**

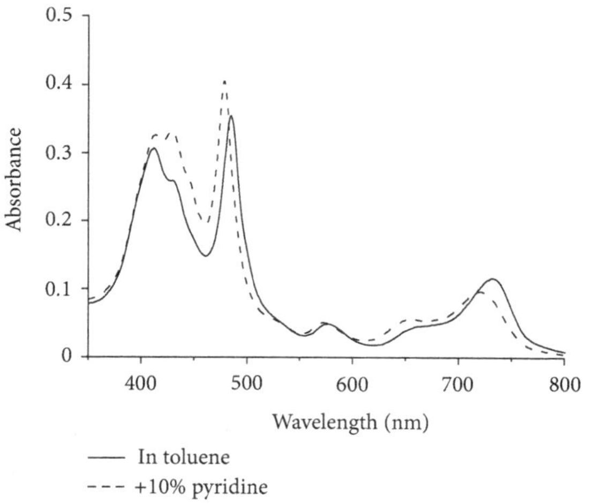

(c) **13**

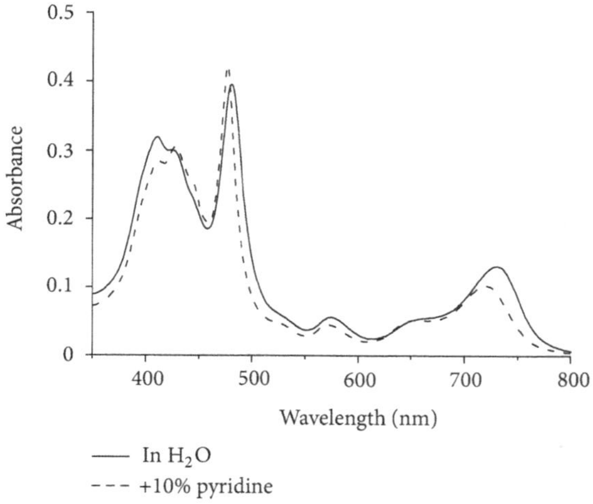

(d) **14**

FIGURE 1: UV/Visible absorption spectra; (a) **5** in $CHCl_3$ (solid line) and in $CHCl_3$/pyridine (dashed line), (b) **6** in H_2O (solid line) and in H_2O/pyridine (dashed line), (c) **13** in toluene (solid line) and in toluene/pyridine (dashed line), and (d) **14** in H_2O (solid line) and in H_2O/pyridine (dashed line). All concentrations were adjusted to ca. 0.5 μM.

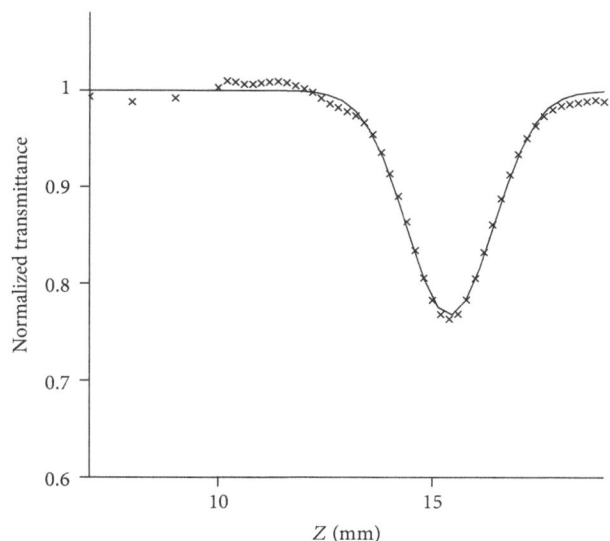

FIGURE 2: Typical open-aperture Z-scan trace (×) of **6** in water.

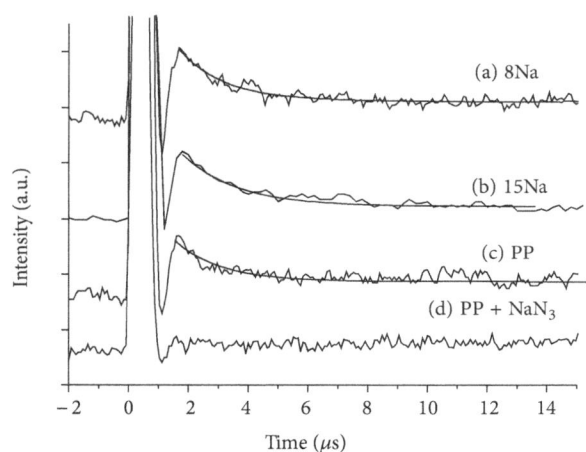

FIGURE 3: Time-resolved emission profiles at 1270 nm in H_2O; (a) **6**, (b) **14**, (c) protoporphyrin (**PP**), and (d) **PP** with NaN_3.

this problem, we designed a monoacetylene-linked self-assembly. In this case, the one-step heterocoupling reaction of monomeric zincporphyrin with free base porphyrin can be employed to produce directly the desired monoacetylene-linked, monozinc-freebase bis-porphyrin. As shown in Scheme 3, for the heterocoupling reaction, acetylenic porphyrin **10** was prepared from porphyrin **2** by zinc insertion, followed by deprotection of the TMS group. The starting porphyrin **7** was iodinated with $PhI(CF_3CO_2)_2$ and iodine to give the coupling counterpart **8**. The heterocoupling reaction of **8** and **10** was conducted using $Pd_2(dba)_3/AsPh_3$ as a catalyst system similar to the synthesis of **4**. The ester hydrolysis of **13** was performed to allow **14** in a manner similar to the case of **6**. The total yield of **14** was ~40% based on monomer **7**, significantly improved compared to that of **6**.

The absorption spectrum of ester form **5** in chloroform is shown in Figure 1(a) (solid line). The Soret band and the

Q-band were observed at 486.5 and 726.5 nm, respectively. After adding 10% pyridine that can cleave the complementary coordination of imidazolyl to zinc, these two peaks were blue-shifted to 478.5 and 714.5 nm, respectively (bold line), suggesting dissociation to monomeric bis-porphyrin by the disappearance of the head-to-tail type exciton interaction between two bis-porphyrins. Similar spectral changes were observed for **6** in water (Figure 1(b)), demonstrating that **6** existed as the self-assembled dimer in water by the complementary coordination in contrast to the monomeric form in the presence of 10% pyridine. The similar behavior was observed for ester form **13** and its water-soluble **14** (Figures 1(c) and 1(d)), also indicating that the dimer structure is maintained for **14** in water.

The $\sigma^{(2)}$ values of **6** and **14** in water were measured by an open aperture Z-scan method at 850 nm with 150 fs pulses. Figure 2 shows a typical open-aperture Z-scan trace (×) of **6**. The detailed experimental conditions were described in [10]. The $\sigma^{(2)}$ values were determined as 7,500 for **6** and 7,900 GM for **14**, respectively. The value of 7,500 GM obtained for **6** in water is equivalent to that of **1** in chloroform, indicating no solvent effect on $\sigma^{(2)}$ values. These two values are almost identical, suggesting that the 2PA efficiency of monoacetylene bridge is equivalent to that of bisacetylene. The values are significantly large compared to those of protoporphyrin [8] and hematoporphyrin [9], indicating that **6** and **14** are possible candidates for the 2PA-PDT agent.

The major pathway in PDT is generally accepted as the Type-II reaction associated with singlet oxygen generation which involves energy transfer from the triplet state photosensitizer to ground state oxygen to give toxic singlet oxygen ($^1O_2^*$) that attacks tumor cells. Thus, the efficient generation of singlet oxygen in water is required for PDT agents. The direct evidence for singlet oxygen generation can be monitored by phosphorescence from $^1\Delta_g$ to $^3\Sigma_g$ at 1270 nm. The emission from the singlet oxygen at 1270 nm under one-photon irradiation condition was measured by time-resolved experiment. The agents in water were irradiated by non-focused 5 ns Nd:YAG-OPO pulses (10 Hz and 128 shots) with a pulse energy of around 3 mJ, and the phosphorescence was detected through an interference filter with an InP/InGaAsP detector operated at −80°C. The sample concentration was 5×10^{-5} M, and the excitation wavelengths were selected to be the same absorbance (0.8) at 556 nm for **6**, and protoporphyrin (PP), and at 550 nm for **14**. The detailed experimental conditions were also described in [10]. Figure 3 shows time-resolved emission profiles at 1270 nm for (a) **6**, (b) **14**, (c) **PP**, and (d) **PP** with NaN_3 as the quencher. The fast rise components were observed for (a), (b), and (c) after the excitation, suggesting the formation of singlet oxygen by energy transfer from the photosensitizer. The lifetime was determined as ~2 μs, which was similar to the reported value of singlet oxygen in water (1.5 ~ 4 μs) [19–22]. NaN_3, being a quencher for singlet oxygen, was added into all the solutions of **6**, **14**, and **PP** to quench the emission, demonstrating that the emission originated from singlet oxygen (typical data were shown in Figure 3(d)). Moreover, as shown in the emission spectrum (typical spectrum of **PP** was presented in

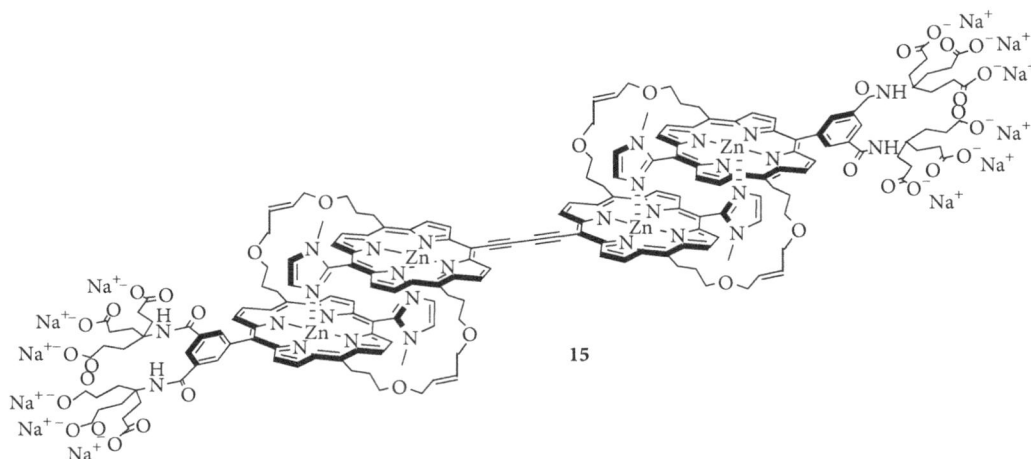

SCHEME 4: Water-soluble supramolecular porphyrin **15**.

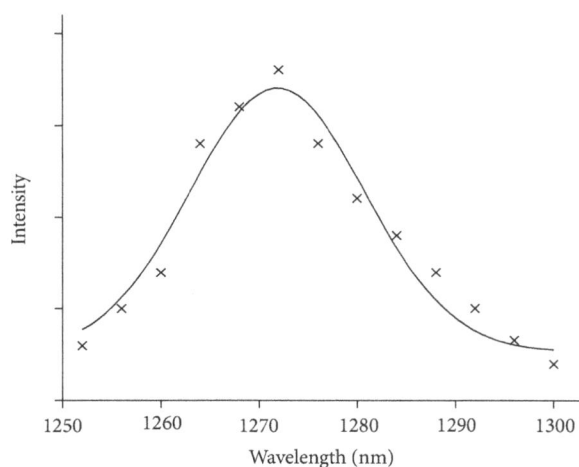

FIGURE 4: Typical emission spectrum of singlet oxygen in H_2O solution of **PP**.

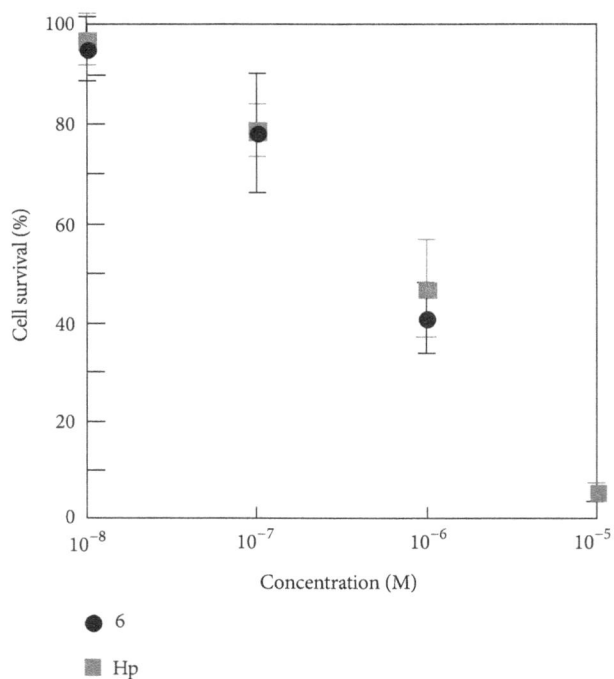

FIGURE 5: Photocytotoxicity of **6** and hematoporphyrin (Hp) for HeLa cell.

Figure 4) recorded in the range from 1250 to 1300 nm with the same equipment using a monochromator (the intensity was obtained by integrating decay profile from around 2 to 8 μs), the spectral shape with a peak maximum at around 1271 nm is similar to that for singlet oxygen as reported in the literature [20]. Samples of **6** and **14** showed almost the same emission intensity, time profile, and spectrum to **PP**.

The photocytotoxicity of the agents was examined using HeLa cells under one-photon irradiation conditions. Figure 5 shows the photocytotoxicity of **6**. The cell survival percentages after the photoirradiation was plotted against the concentration of agents. The cell was almost unchanged for concentrations lower than 10^{-8} M and cell survival decreased with increasing agent concentration. No significant difference in the photocytotoxicity was observed between **6** and hematoporphyrin (Hp), demonstrating that **6** exhibits high PDT efficiency equivalent to Hp.

The photocytotoxicity of **14** for HeLa cell was also examined by observing cell death upon photoirradiation using a microscope. A CW diode laser (671 nm) was used for excitation. The spot diameter was 30 μm with a power density of 1.8 W/cm^2, and the beam center was adjusted at the center of cell. The details were described in [10]. No cell death was observed without the agent, even after 2 hours of irradiation (total irradiation energy >12,960 J/cm^2 with a power density of 1.8 W/cm^2). On the other hand, cell death

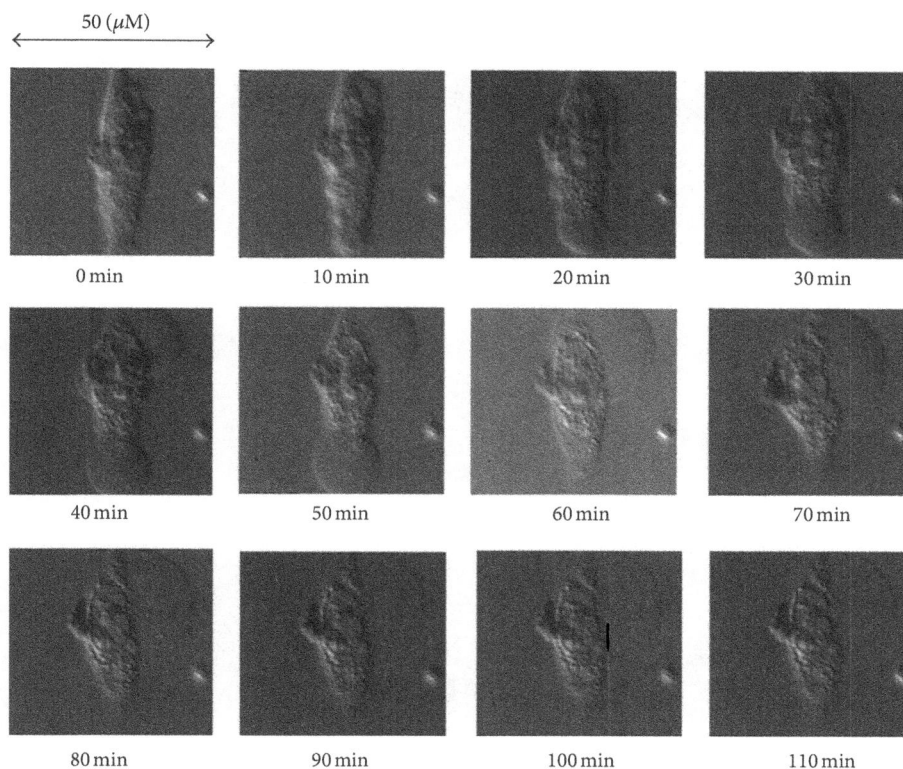

FIGURE 6: Time course of cell dying for HeLa cell upon photoirradiation with **14** (5×10^{-6} M).

TABLE 1: Photocytotoxicity of **14** for a single HeLa cell (671 nm).

Concentration	Time of cell death[*]	Irradiation energy
μM	min	J/cm^2
0	>120	>12960
1	113 ± 12	12204
5	74 ± 9	7992
10	48 ± 3	5184

[*]Time of cell death was determined by the trypan blue staining method.

was observed by administrating **14** (5×10^{-6} M). As shown in Figure 6, the leakage of the cytoplasm worsened with time course, and blebs were formed on the cell surface. Table 1 summarizes the irradiation time required for cell death at various concentrations of **14**. The exposure time until cell death was shortened with increasing concentrations of **14**. These results demonstrate that water-soluble porphyrin self-assemblies **6** and **14** are potential candidates for 2PA-PDT.

3. The Second 2PA-PDT System Using Dendritic-Type Substituent [11, 12]

Next, we reported a different approach to construct a water-soluble two photon absorbing porphyrin-based photosensitizer **15** (Scheme 4) as another potential candidate for 2PA-PDT. A butadiyne-bridged bis-porphyrin was chosen as the two photon absorbing part of this 2PA-PDT system. In contrast to the previous compounds, a dendritic-type

substituent was used as hydrophilic groups. A monomeric porphyrin having six carboxylates was attached at both ends of the butadiyne-bridged bis-porphyrin through zinc-imidazolyl coordination to allow a tetramer. The self-assembled structure was covalently fixed by olefin metathesis. [22]. In contrast to the previous compounds, the hydrophilic groups in compound **15** were larger in number and were located only at both ends of the tetramer. These factors may affect drug-delivery property into the cell.

Scheme 5 shows synthetic routes of a zinc-inserted butadiyne-bridged imidazolylporphyrin dimer **17**Zn as the 2PA component and a zinc-inserted isophthalamidoimidazolylporphyrin having 12 carboxylic acid groups **19**Zn as the water-soluble component.

The butadiyne-bridged bis-porphyrin was synthesized by Pd-catalyzed coupling of **16** using Pd$_2$(dba)$_3$ (dba = dibenzylideneacetone) and triphenylarsine to afford freebase **17** in 64% yield. Freebase **17** was treated with zinc acetate to give the 2PA component **17**Zn. In order to increase hydrophilicity, the water-soluble component **19**Zn was synthesized from **18**ZnH with a precursor of dendrimer via BOP (benzotriazol-1-yloxytris(dimethylamino)phosphoniumhexafluorophosphate) condensation in a 90% yield.

In noncoordinating solvents such as CHCl$_3$, imidazolylporphyrins **17**Zn, and **19**Zn exist as polymer (**17**Zn)$_n$ and as dimer (**19**Zn)$_2$, respectively, through the complementary coordination of imidazolyl to zinc as shown in Scheme 6. However, in coordinating solvents such as pyridine (denoted as L in Scheme 6), porphyrins **17**Zn and **19**Zn exist as their

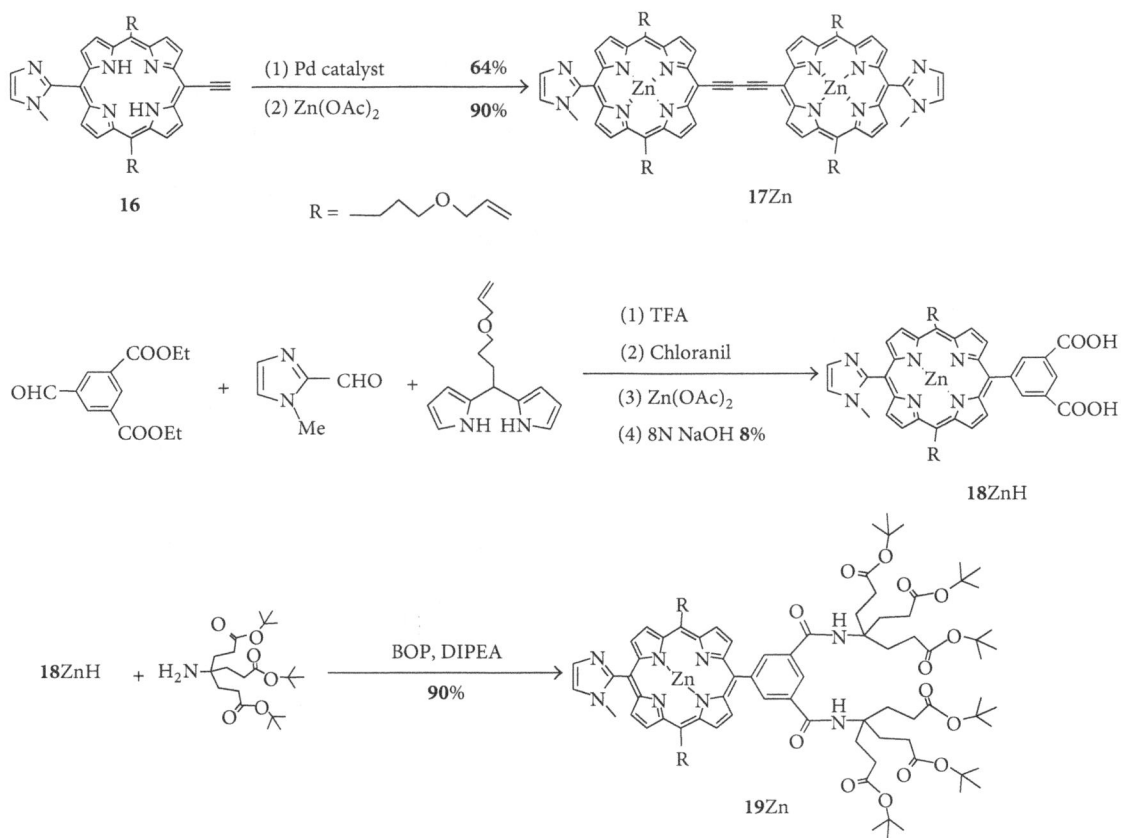

SCHEME 5: Synthetic routes of **17**Zn and **19**Zn.

monomeric form. In order to lead to the desired tetramer **20**$_1$, the initial coordination dimers of **19**Zn and **17**Zn in a 2 : 1 molar ratio were dissociated by dissolving in pyridine. Reorganization was conducted by removing pyridine to form different length arrays of **20**$_n$. The tetramer **20**$_1$ can be isolated using preparative GPC (8.7%). In order to prevent reorganization in other solution conditions, the coordination structure was fixed via metathesis of the allyl ether side chains using Grubbs catalyst to get compound **21** (80%). Compound **21** was treated with formic acid to cleave t-Bu groups giving the carboxylic acid form **21**H, and subsequent treatment with an equimolar amount of NaOH yielded the water-soluble tetramer **15** (85%). The characterizations of the compound **15** including GPC, mass, UV/vis absorption and emission spectral measurements were described in detail in [11].

The effective 2PA cross-section was measured using an open-aperture Z-scan method with nanosecond pulses. A typical Z-scan trace of compound **15** in water at 890 nm with theoretically fitted curve is shown in Figure 7. The effective 2PA spectrum of compound **15** in water is shown in Figure 8 [11, 18].

The 2PA maximum peak for compound **15** appeared at 890 nm with a value of 33,000 GM. It should be noted that it is difficult to compare this 2PA cross-section value with those of **6** and **14** obtained by femtosecond pulses. The nanosecond

values are ca. 30 times larger compared to the femtosecond values for our previously reported compounds [17]. The large discrepancy between nanosecond and femtosecond values is attributed to excited state absorption (ESA) due to the longer pulse width in nanosecond lasers as compared to those in femtosecond pulses. The effective $\sigma^{(2)}$ value of compound **15** was three orders of magnitude larger than that of H2TPP (29 GM at 780 nm) measured by employing the same nanosecond pulses [20].

Compound **15** generated singlet oxygen by one-photon irradiation as seen in Figures 3 and 4 [12]. However, this direct measurement could not be applied to the two photon conditions since the emission signal was too weak to detect. Singlet oxygen can not only be measured by the direct observation but also be determined quantitatively by using scavengers such as anthracene-9,10-dipropionic acid sodium salt (ADPA) [11, 23, 24] which reacts with oxygen to form an endoperoxide. Therefore, singlet oxygen generation by two photon irradiation was monitored as decrease in ADPA absorption. ADPA exhibits characteristic absorption peaks at 399, 378, 359, and 342 nm. A D_2O solution of ADPA and compound **15** was irradiated with focused 100 fs pulses at 890 nm with a pulse energy of 4 nJ corresponding to the peak power of 6.1 GW/cm^2 [11]. Since the emission from singlet oxygen is very week under the two photon conditions

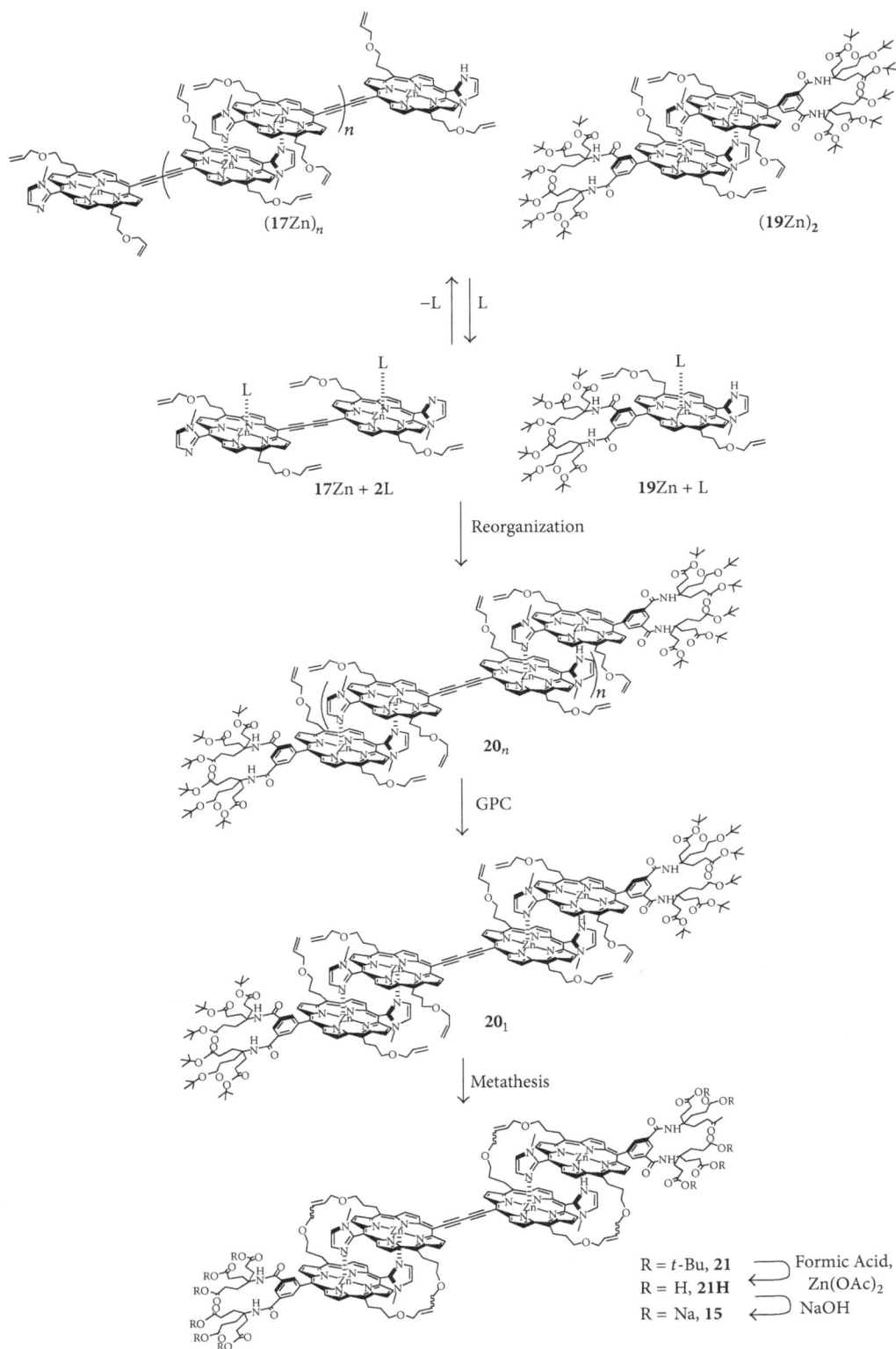

$(17Zn)_n$

$(19Zn)_2$

$-L \updownarrow L$

$17Zn + 2L$

$19Zn + L$

Reorganization

20_n

GPC

20_1

Metathesis

R = t-Bu, **21**
R = H, **21H** — Formic Acid, Zn(OAc)$_2$
R = Na, **15** — NaOH

SCHEME 6: Synthetic routes of **15**.

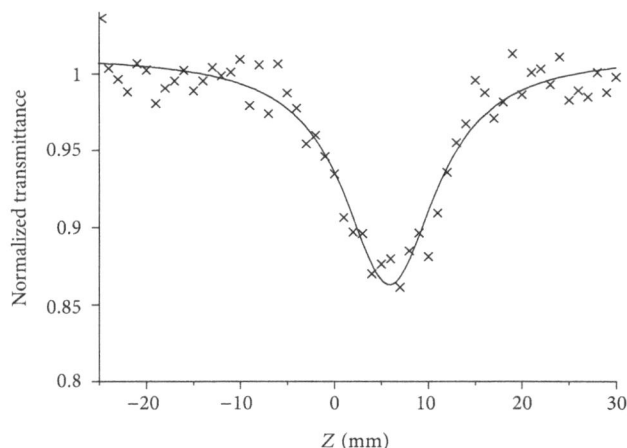

FIGURE 7: Typical open-aperture Z-scan trace ($\times$) of 0.4 mM of **15** in water.

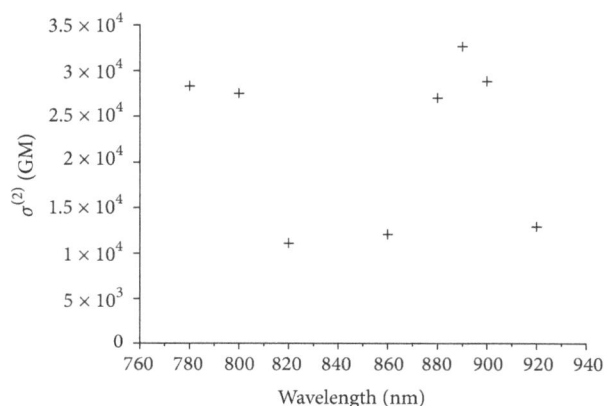

FIGURE 8: Effective 2PA spectrum of 0.4 mM of **15** in water [11, 18].

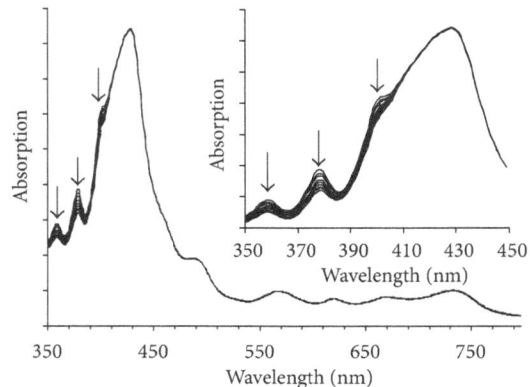

FIGURE 9: Change in the absorption spectra of ADPA with **15** upon two photon irradiation.

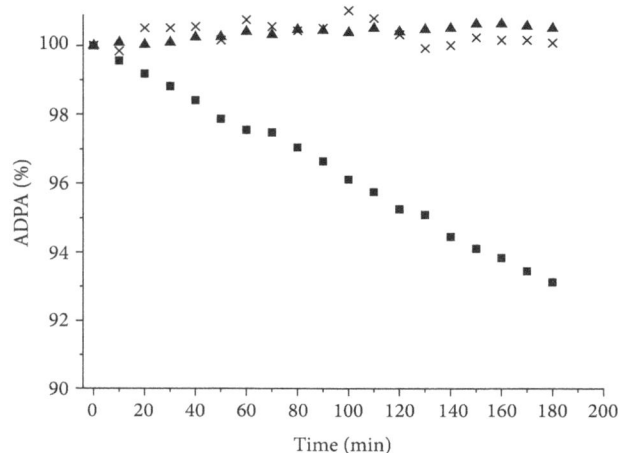

FIGURE 10: Photobleaching of the ADPA absorption peak at 379 nm (5.0×10^{-5} M) after two photon irradiation at 890 nm (no photosensitizer (triangle), TPPS ($\times$) (1.0×10^{-4} M), and **15** (square) (5.0×10^{-5} M) in D_2O.

and the lifetime is short in H_2O, D_2O was used as solvent [21]. Continuous photobleaching of anthracene absorption was observed for 3 h using 890 nm excitation. On the other hand, no change was observed in the Q-bands of compound **15** indicating that the sensitizer itself was not affected either during two photon irradiation or by singlet oxygen generation (Figure 9). Almost no decrease in the anthracene absorption was observed in the solution without **15**. The same experiment was conducted by using tetraphenylporphyrin tetrasulfonic acid (TPPS) which has very low 2PA efficiency at this wavelength region. The results were summarized in Figure 10 (no photosensitizer (triangle), TPPS ($\times$) and **15** (square)). This indicates that compound **15** is a potential agent for 2PA-PDT. The detailed experimental conditions and data were described in [11, 12].

Finally, the PDT experiment with two photon irradiation was conducted using HeLa cells. A HeLa cell incubated with **15** on a glass slide was irradiated for 5 min with 100 fs pulses at 780 nm with an average power of 2 mW which provides an average of 600 mJ/cell. Detailed experimental conditions were described in [12]. As shown in Figure 11(a), a HeLa cell at the upper site was selectively excited on the position marked by an arrow. After the irradiation, the degradation of the cell membrane was observed in the upper cell (Figure 11(b)). The lower cell which was nonirradiated was undamaged. Control experiments with Hp and without photosensitizer also resulted in no cell damage. These results suggest that compound **15** is a potential agent not only for photodynamic activity on HeLa cells but also for selective targeting of tumor cells via two photon excitation. Although femtosecond laser sources were not available in the cell experiments of compounds **6** and **14**, it would be interesting to conduct a comparative study between **15** and the previous compounds in order to determine which type of structure and hydrophilicity will give better drug delivery property.

Acknowledgments

The authors acknowledge Drs. H. Inoue and A. Ishizumi for femtosecond Z-scan measurements at Nara Institute of Science and Technology and Professors I. Okura and S. Ogura

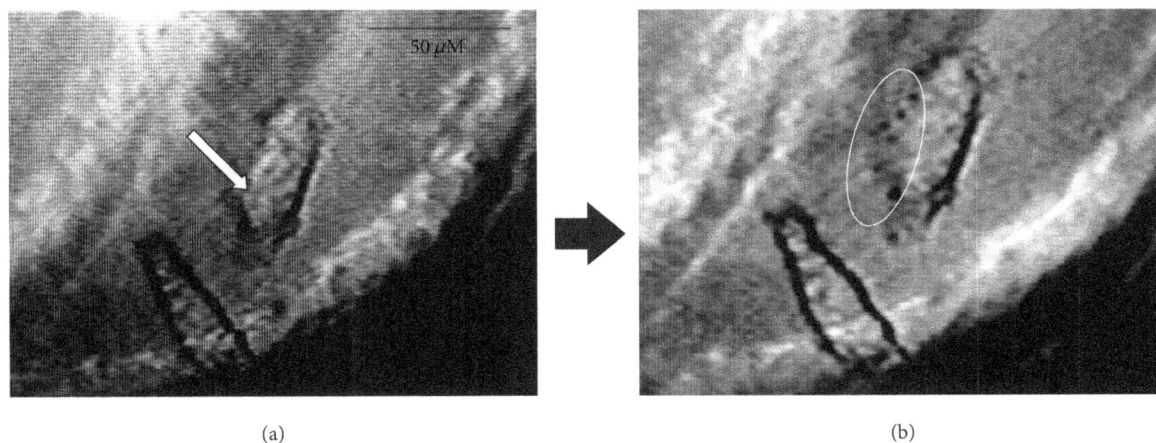

(a)   (b)

FIGURE 11: Pictures of HeLa cells incubated with **15** before (a) and after (b) two photon excitation with 100 fs pulses at 780 nm. The irradiated position is marked by a white arrow. The degradation of the cell membrane was observed as indicated by an oval.

for HeLa cell experiments at Tokyo Institute of Technology, and also Professor T. Hirano and Dr. E. Kono for PDT experiments at Hamamatsu University School of Medicine. The authors thank Drs. Y. Inaba, H. Hasegawa, and J. T. Dy at Nara Institute of Science and Technology for their efforts on these studies.

References

[1] J. Moan and Q. Peng, "An outline of the history of PDT," in *Photodynamic Therapy*, T. Patrice, Ed., The Royal Society of Chemistry, Cambridge, Mass, USA, 2004.

[2] R. W. Waynant, "Lasers in medicine," in *Electrooptics Handbook*, M. N. Ediger, Ed., Chapter 24, McGraw-Hill, New York, NY, USA, 1993.

[3] J. D. Bhawalkar, G. S. He, and P. N. Prasad, "Nonlinear multiphoton processes in organic and polymeric materials," *Reports on Progress in Physics*, vol. 59, article 1041, 1996.

[4] C. R. Shea, Y. Hefetz, R. Gillies, J. Wimberly, G. Dalickas, and T. Hasan, "Mechanistic investigation of doxycycline photosensitization by picosecond-pulsed and continuous wave laser irradiation of cells in culture," *Journal of Biological Chemistry*, vol. 265, no. 11, pp. 5977–5982, 1990.

[5] P. Lenz, "*In vivo* excitation of photosensitizers by infrared light," *Photochemistry and Photobiology*, vol. 62, no. 2, pp. 333–338, 1995.

[6] J. D. Bhawalkar, N. D. Kumar, C. F. Zhao, and P. N. Prasad, "Two-photon photodynamic therapy," *Journal of Clinical Laser Medicine and Surgery*, vol. 15, no. 5, pp. 201–204, 1997.

[7] W. G. Fisher, W. P. Partridge, C. Dees, and E. A. Wachter, "Simultaneous two-photon activation of type-I photodynamic therapy agents," *Photochemistry and Photobiology*, vol. 66, no. 2, pp. 141–155, 1997.

[8] R. L. Goyan and D. T. Cramb, "Near-infrared two-photon excitation of protoporphyrin IX: photodynamics and photoproduct generation," *Photochemistry and Photobiology*, vol. 72, no. 6, pp. 821–. 827, 2000.

[9] A. Karotki, M. Khurana, J. R. Lepock, and B. C. Wilson, "Simultaneous two-photon excitation of photofrin in relation to photodynamic therapy," *Photochemistry and Photobiology*, vol. 82, no. 2, pp. 443–452, 2006.

[10] K. Ogawa, H. Hasegawa, Y. Inaba et al., "Water-soluble bis(imidazolylporphyrin) self-assemblies with large two-photon absorption cross sections as potential agents for photodynamic therapy," *Journal of Medicinal Chemistry*, vol. 49, no. 7, pp. 2276–2283, 2006.

[11] J. T. Dy, K. Ogawa, A. Satake, A. Ishizumi, and Y. Kobuke, "Water-soluble self-assembled butadiyne-bridged bisporphyrin: a potential two-photon-absorbing photosensitizer for photodynamic therapy," *Chemistry*, vol. 13, no. 12, pp. 3491–3500, 2007.

[12] K. Ogawa, J. Dy, Y. Kobuke, S. Ogura, and I. Okura, "Singlet oxygen generation and photocytotoxicity against tumor cell by two-photon absorption," *Molecular Crystals and Liquid Crystals*, vol. 471, no. 1, pp. 61–67, 2007.

[13] S. Kim, T. Y. Ohulchanskyy, H. E. Pudavar, R. K. Pandey, and P. N. Prasad, "Organically modified silica nanoparticles co-encapsulating photosensitizing drug and aggregation-enhanced two-photon absorbing fluorescent dye aggregates for two-photon photodynamic therapy," *Journal of the American Chemical Society*, vol. 129, no. 9, pp. 2669–2675, 2007.

[14] H. A. Collins, M. Khurana, E. H. Moriyama et al., "Blood-vessel closure using photosensitizers engineered for two-photon excitation," *Nature Photonics*, vol. 2, no. 7, pp. 420–424, 2008.

[15] E. Dahlstedt, H. A. Collins, M. Balaz et al., "One- and two-photon activated phototoxicity of conjugated porphyrin dimers with high two-photon absorption cross sections," *Organic and Biomolecular Chemistry*, vol. 7, no. 5, pp. 897–904, 2009.

[16] T. Gallavardin, C. Armagnat, O. Maury et al., "An improved singlet oxygen sensitizer with two-photon absorption and emission in the biological transparency window as a result of ground state symmetry-breaking," *Chemical Communications*, vol. 48, no. 11, pp. 1689–1691, 2012.

[17] K. Ogawa, A. Ohashi, Y. Kobuke, K. Kamada, and K. Ohta, "Strong Two-Photon Absorption of Self-Assembled Butadiyne-Linked Bisporphyrin," *Journal of the American Chemical Society*, vol. 125, no. 44, pp. 13356–13357, 2003.

[18] K. Ogawa, A. Ohashi, Y. Kobuke, K. Kamada, and K. Ohta, "Two-photon absorption properties of self-assemblies of butadiyne-linked bis(imidazolylporphyrin)," *Journal of Physical Chemistry B*, vol. 109, no. 46, pp. 22003–22012, 2005.

[19] M. A. Rodgers and P. T. Snowden, "Lifetime of oxygen (O2(1.DELTA.g)) in liquid water as determined by time-resolved infrared luminescence measurements," *Journal of the American Chemical Society*, vol. 104, no. 20, pp. 5541–5543, 1982.

[20] P. R. Ogilby and C. S. Foote, "Chemistry of singlet oxygen. 42. Effect of solvent, solvent isotopic substitution, and temperature on the lifetime of singlet molecular oxygen (1Δg)," *Journal of the American Chemical Society*, vol. 105, no. 11, pp. 3423–3430, 1983.

[21] C. Schweitzer and R. Schmidt, "Physical mechanisms of generation and deactivation of singlet oxygen," *Chemical Reviews*, vol. 103, no. 5, pp. 1685–1757, 2003.

[22] P. K. Frederiksen, S. P. McIlroy, C. B. Nielsen et al., "Two-photon photosensitized production of singlet oxygen in water," *Journal of the American Chemical Society*, vol. 127, no. 1, pp. 255–269, 2005.

[23] B. A. Lindig, M. A. J. Rodgers, and A. P. Schaap, "Determination of the lifetime of singlet oxygen in D2O using 9,10-anthracenedipropionic acid, a water-soluble probe," *Journal of the American Chemical Society*, vol. 102, no. 17, pp. 5590–5593, 1980.

[24] M. A. Oar, J. M. Serin, W. R. Dichtel, J. M. J. Fréchet, T. Y. Ohulchanskyy, and P. N. Prasad, "Photosensitization of singlet oxygen via two-photon-excited fluorescence resonance energy transfer in a water-soluble dendrimer," *Chemistry of Materials*, vol. 17, no. 9, pp. 2267–2275, 2005.

4

Emerging Therapeutic Biomarkers in Endometrial Cancer

Peixin Dong,[1] **Masanori Kaneuchi,**[1] **Yosuke Konno,**[2] **Hidemichi Watari,**[2] **Satoko Sudo,**[2] **and Noriaki Sakuragi**[2]

[1] *Department of Women's Health Educational System, Hokkaido University School of Medicine, Hokkaido University, N15, W7, Sapporo 060-8638, Japan*
[2] *Department of Gynecology, Hokkaido University School of Medicine, Hokkaido University, N15, W7, Sapporo 060-8638, Japan*

Correspondence should be addressed to Peixin Dong; dongpeix@yahoo.co.jp and Noriaki Sakuragi; sakuragi@med.hokudai.ac.jp

Academic Editor: Romonia Renee Reams

Although clinical trials of molecular therapies targeting critical biomarkers (mTOR, epidermal growth factor receptor/epidermal growth factor receptor 2, and vascular endothelial growth factor) in endometrial cancer show modest effects, there are still challenges that might remain regarding primary/acquired drug resistance and unexpected side effects on normal tissues. New studies that aim to target both genetic and epigenetic alterations (noncoding microRNA) underlying malignant properties of tumor cells and to specifically attack tumor cells using cell surface markers overexpressed in tumor tissue are emerging. More importantly, strategies that disrupt the cancer stem cell/epithelial-mesenchymal transition-dependent signals and reactivate antitumor immune responses would bring new hope for complete elimination of all cell compartments in endometrial cancer. We briefly review the current status of molecular therapies tested in clinical trials and mainly discuss the potential therapeutic candidates that are possibly used to develop more effective and specific therapies against endometrial cancer progression and metastasis.

1. Introduction

Endometrial cancer (EC) is the most common gynecological malignancy among women worldwide with 287000 new cases and estimated 74000 deaths per year [1].

EC has been dichotomized into two types with distinct underlying molecular profiling, histopathology and clinical behavior: less aggressive type I and highly aggressive type II. Most ECs are type I (approximately 75%) and are estrogen-dependent adenocarcinomas with endometrioid morphology [2]. They are usually diagnosed at an early stage and have a good prognosis (a 5-year survival rate of 80–85%) after surgery [2, 3]. In contrast, type II ECs with poorly differentiated endometrioid and serous histology are associated with myometrial invasion, extrauterine spread, and a lower 5-year survival rate (35%) [3–6]. Although patients with advanced or recurrent disease typically receive adjuvant chemotherapy and radiation, they have an extremely poor prognosis. A potential strategy for the treatment of these cases is to target EC cells by blocking key signaling pathways that are necessary for tumor development.

2. Therapeutic Targets for EC

Type I EC frequently exhibits altered PI3K/PTEN/AKT/mTOR signal pathway [7–11]. Type II cancer predominantly shows mutations in p53 [12] and epidermal growth factor receptor 2 (HER-2) overexpression [13]. The upregulation of epidermal growth factor receptor (EGFR) [14, 15] and vascular endothelial growth factor (VEGF) [16], dysregulated microRNA (miRNA) [17], and activation of cancer stem cell (CSC)/epithelial-mesenchymal transition (EMT) programs are involved in oncogenesis and progression of both cancer types [18–20]. Owing to the high-frequency activation of PI3K/AKT/mTOR, EGFR/HER2 and VEGF-related pathway and their important roles in promoting EC growth and metastasis, new drug targeting these signals would be valuable to a very large number of patients with EC. Recently, clinical trials assessing the efficacy of mTOR inhibitor, EGFR/HER2 inhibitor, and antiangiogenic agent for EC have been conducted and demonstrated modest effects [21, 22] (Figure 1).

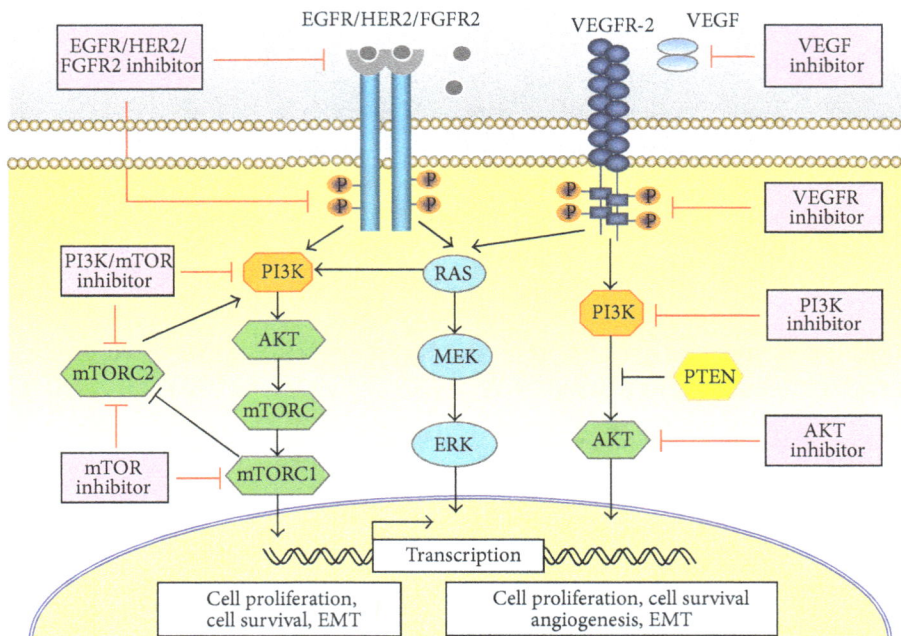

FIGURE 1: Therapeutic molecular targets for endometrial cancer. Type I endometrial cancer (EC) frequently exhibits altered PI3K/PTEN/AKT/mTOR signal pathway, whereas type II EC frequently shows mutations in p53 and HER-2 overexpression. The upregulation of EGFR and VEGF, dysregulated microRNAs, and activation of cancer stem cell (CSC)/epithelial-mesenchymal transition (EMT) programs are involved in oncogenesis and progression of both cancer types. Currently, clinical trials assessing the efficacy of mTOR inhibitor, EGFR/HER2 inhibitor, and antiangiogenic agent for EC have been conducted and demonstrated modest effects.

3. Challenges in the Molecular Therapeutics of Human Tumor

Although the therapeutic potential of targeted drugs for the treatment of human tumors appears promising, the clinical success of such drugs has been limited by key challenges, including primary/acquired drug resistance [23–25] and unexpected side effects on normal tissues due to nonspecificity [26] (Figure 2).

A portion of patients unfortunately do not respond to targeted agents (primary resistance), and the remainder might eventually acquire the resistance to targeted therapy despite an initial response. Various mechanisms of resistance have begun to be elucidated. The most frequently reported mechanism of primary resistance is genetic heterogeneity. For example, mechanisms of resistance to EGFR inhibitors are involved in point mutations, deletions, and amplifications of genomic areas of EGFR [23]. In addition to genetic alteration, epigenetic changes, such as DNA methylation at CpG islands, have been linked to the development of resistance to multiple molecular drugs [27, 28]. The generation of a population of cancer cells with stem-cell properties might provide another possible reason of resistance to EGFR inhibitor [29]. Common mechanisms of acquired resistance include secondary mutation in the target gene, activation of alternative pathway or feedback loop, and induction of EMT [23, 30]. Therefore, new therapy that concurrently attacks multiple critical pathways, inhibits the cross talk between diverse signals, and suppresses the CSC and EMT properties may be efficacious to overcome the resistance to molecular agents in EC.

Moreover, the administration of antiangiogenic agents, particularly antibodies against VEGF, leads to a more hypoxic tumor microenvironment [31], which enhances tumor cell invasion and metastasis by inducing the EMT- and CSC-like phenotype [32–34]. These works clearly suggest the need to combine antiangiogenic treatment in human tumors with new drugs targeting specific signaling pathways linked to the CSC/EMT phenotype.

Another challenge is toxicity or the side effects associated with targeted therapies, such as harmful immune responses. These include "Off-target" adverse effects caused by a drug binding to an unexpected target and "On-target" adverse effects as a result of a drug binding to its intended target that is not only present in tumor cells, but also found in normal tissue [26].

4. Potential miRNA-Based Therapies in EC

Different from gene mutations, epigenetic changes that are associated with global gene regulation such as chromatin remodeling open a new field of cancer research [35]. Epigenetic silencing of tumor suppressor genes or epigenetic

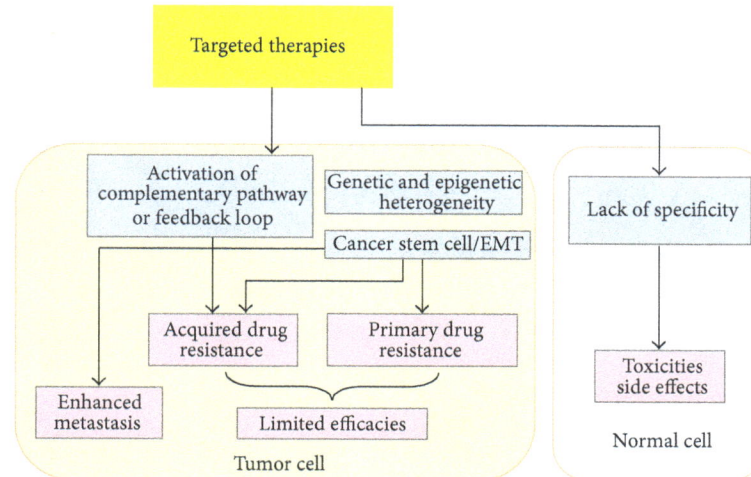

FIGURE 2: Challenges in the molecular therapeutics of human tumor. The clinical success of targeted drugs has been limited by key challenges, including primary/acquired drug resistance and unexpected side effects on normal tissues due to nonspecificity. The most frequent mechanisms of primary resistance are genetic/epigenetic heterogeneity and the existence of cancer stem cell. Acquired resistance can be caused by the secondary mutation in the target gene, activation of alternative pathway or feedback loop, and induction of EMT. Treatment of tumor cells with antiangiogenic agents can lead to a more hypoxic tumor microenvironment and enhance tumor cell invasion and metastasis by inducing the EMT- and cancer-stem-cell-like phenotype.

activation of oncogenes plays the important roles in the promotion of carcinogenesis and tumor progression [35]. Two common epigenetic changes are methylation at the promoter region and histone acetylation, which can be modulated using inhibitors of DNA methyltransferase (DNMT) and histone deacetylase (HDAC), respectively. Tumor suppressor genes including PTEN [36], DNA mismatch repair gene hMLH1 [37], adenomatous polyposis coli (APC) [38], RAS-associated domain family member protein 1 (RASSF1A) [39], and E-cadherin [40] are more frequently silenced in type I tumor than in type II tumor. DNMT and HDAC inhibitors are already in clinical use for myelodysplasia and cutaneous T-cell lymphoma [41, 42]. Preclinical study has shown that DNMT and HDAC inhibitors induce cell apoptosis and suppress the growth of EC in vivo [43]. The combination of epigenetic modifiers with chemotherapy, hormonal therapy, and targeted therapy, has been proposed [44], and this may achieve better effect than single epigenetic agent for the treatment of EC.

Another important mechanism for epigenetic regulation of gene expression is involved in noncoding RNAs, specifically small regulatory microRNA (miRNA). MiRNAs post-transcriptionally control gene expression by base pairing with the $3'$ untranslated region of target mRNAs, which triggers either mRNA translation repression or RNA degradation [45].

As miRNAs are able to bind to their mRNA targets with either perfect or imperfect complementary, one miRNA may possibly have multiple target genes and concurrently influence different cellular signaling pathways [45]. Some miRNAs can function as either promoter or suppressor participating in a wide variety of biological functions of tumor, including cell proliferation, differentiation, migration, apoptosis, and recently EMT/cancer-stem-cell-like features

[46]. Therefore, modulation of dysregulated miRNAs could be a powerful tool to correct abnormal signaling pathways related to EC.

Altered expression profiles of microRNA have been observed in EC compared with normal endometrium [47]. Several miRNAs are differentially expressed between endometrioid and serous papillary EC, indicating that they could infer mechanisms that are specific to individual tumor subtypes [48]. Among those miRNAs elevated in endometrioid EC, the expression of miR-7 can be downregulated by using anti-miRNA oligonucleotides, leading to repressed migration and invasion of EC cells [49]. On the other hand, the level of miR-194 was significantly lower in EC patients with more advanced stage, and lower expression of this miRNA was associated with worse survival [50]. We found that overexpression of miR-194 by transfection with pre-miRNA molecule inhibited EMT phenotype and EC cell invasion by targeting the oncogene BMI-1 [51]. We also identified miR-130b as one of the mutant p53-responsive 23 miRNAs, which is decreased in EC relative to adjacent normal tissue and directly targets the key EMT promoter gene ZEB1 to revert p53-mutations-induced EMT features of EC cells [52]. MiRNAs are stable in various tissues and bodily fluids [53]. This property greatly facilitates the delivery of miRNAs to recipient cells via the blood or other compartments. Collectively, targeting those miRNAs that are deeply involved in EC progression would provide a promising therapeutic option for EC.

Forced expression of tumor suppressor miRNA and suppression of oncogenic miRNA are two strategies to achieve the goal of miRNA-based cancer treatment (Figure 3). Although previous results demonstrated that restoration of tumor suppressor miR-152 effectively inhibited EC cell growth in vitro and in vivo [54], obvious challenges of

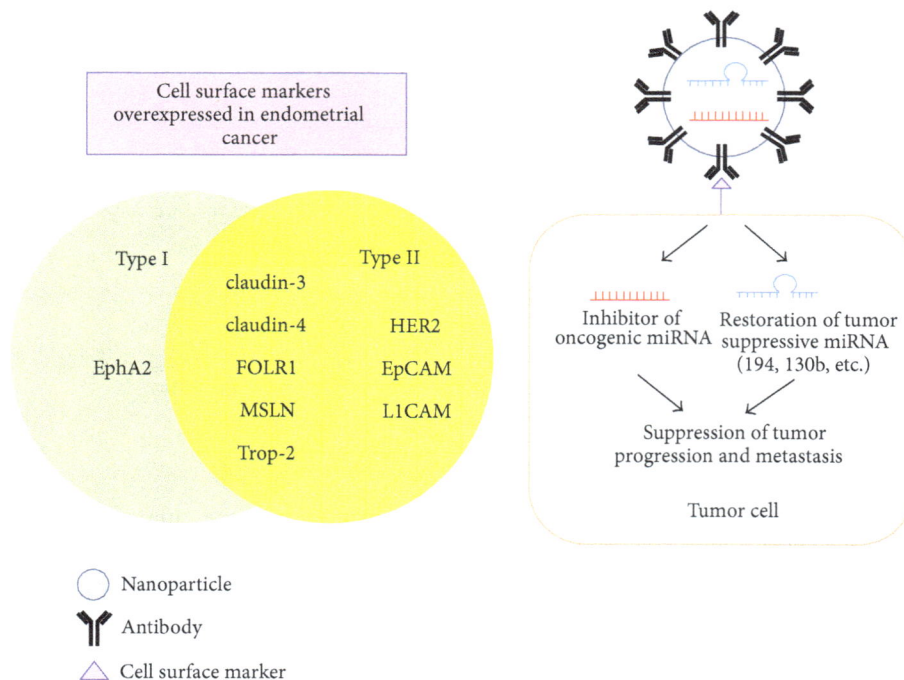

FIGURE 3: Potential miRNA-based therapies in EC. The use of antibodies against cell surface markers overexpressed in EC tissue might deliver targeted drugs to EC cells more specifically with fewer side effects on normal tissue. The nanotechnology can be used to develop a more effective delivery system for targeted agents, especially miRNA that might simultaneously modulate multiple signal pathways necessary for malignant phenotype of EC.

obtaining efficient delivery systems and tumor cell specificity must be resolved to allow clinical implementation.

The biochemical similarity between miRNA and siRNA suggests that the same delivery reagents developed for use with siRNA could be applied to the delivery of miRNA [55, 56]. Many efforts have been made to develop more effective and stable delivery systems [57]. Among them, nanoparticles confer greater miRNA stability, and the conjugation of nanoparticles to antibodies or cancer-specific ligands can notably improve their interactions with cancer cells [57]. By using the modification of GC4 single-chain fragment (a tumor-targeting human monoclonal antibody), nanoparticles injected intravenously showed greater accumulation in the tumor nodules rather than in liver and kidney. Moreover, the codelivery of three siRNAs together with miR-34a resulted in a more significant inhibition (80%) of metastatic melanoma than that obtained with siRNAs or miRNA alone [58]. These data demonstrate that the use of antibody targeting cell surface marker allows a selective delivery of miRNA into the tumor, and the combination of siRNA and miRNA could additively inhibit tumor growth and metastasis.

As mentioned, another major issue for molecular cancer therapy is toxicity. To avoid potential side effects on normal tissue, increasing attention has been directed to the identification of tumor-specific surface markers including receptors and epitopes that are highly expressed in cancer cells, but not or minimally expressed in normal cells. Some potential tumor cell surface markers overexpressed in EC compared

with normal endometrium might be used for targeted therapy (Figure 3).

Eph receptor tyrosine kinases and their ephrin ligands influence central nervous system development, stem cell niches, and cancer cells [59]. Upon the binding of EphrinA1, the EphA2 receptor becomes tyrosine phosphorylated and interacts with several proteins to elicit downstream signaling, which regulate cell adhesion, proliferation, migration, and angiogenesis [60]. Overexpression of EphA2 was found in a high proportion of endometrioid EC and correlated with advanced disease and poor prognosis, whereas its expression is present at low levels in benign endometrial tissue [61]. The microtubule inhibitor conjugated to EphA2 antibody was shown to be specifically internalized by EphA2-positive EC cells, resulting in significant growth inhibition of EC cells both *in vitro* and *in vivo* [62].

The tight junction proteins claudin-3 and claudin-4 are highly expressed in endometrioid, serous papillary, and clear-cell EC [63], but less frequently found in normal endometrium [64]. Importantly, the intratumoral injection of cytotoxic *Clostridium perfringens* enterotoxin (CPE) that interacts with claudin-3 and claudin-4 in subcutaneous serous EC xenografts led to tumor disappearance and extended survival of animals [65], indicating that targeting claudin-3 and claudin-4 by CPE or other targeted treatment may efficiently suppress the progression of EC.

Folate receptor alpha (FOLR1, a membrane-bound molecule) and mesothelin (MSLN, a glycosyl-phosphatidylinositol-linked cell surface antigen) that are

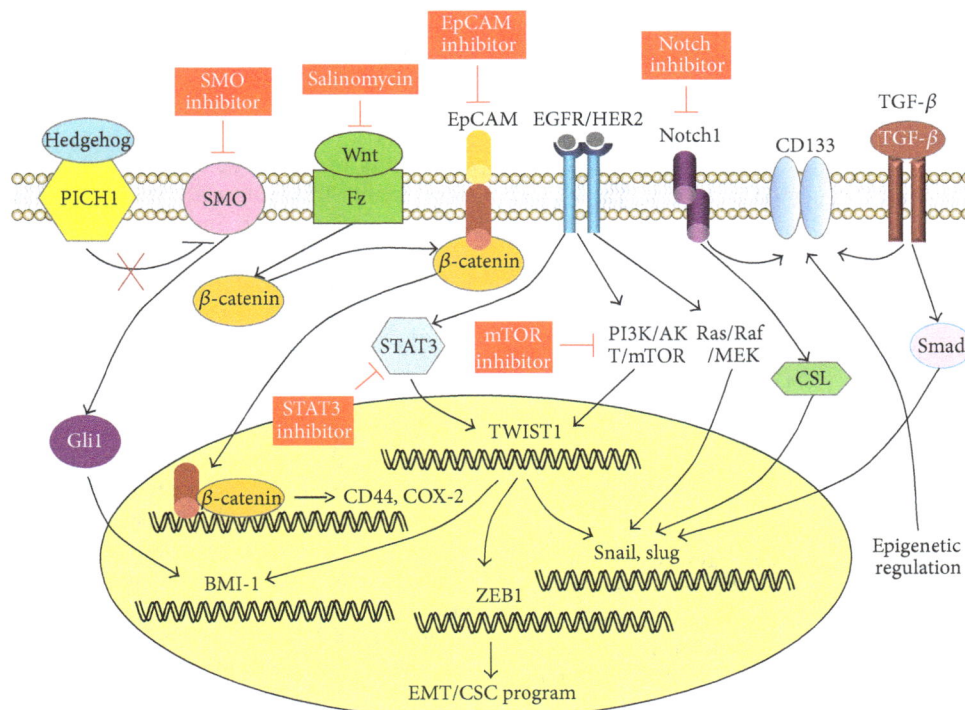

FIGURE 4: Targeting the CSC/EMT signaling pathways in EC. Tumor cells that undergo EMT not only increase their invasion ability, but also concurrently acquire cancer stem cell (CSC) properties. On the other hand, CSCs are associated with enhanced capacity to metastasize. At a molecular level, several signaling pathways involved in the self-renewal of CSCs, including Wnt/β-catenin, Hedgehog, and Notch signaling, can also induce EMT programs. Specific inhibitors targeting these CSC and EMT pathways efficiently suppress the malignant phenotype of EC cells. Other potential therapeutic candidates for EC treatment include Stattic (inhibitor of STAT3), Rapamycin (mTOR inhibitor), and CD133.

upregulated in ovarian carcinoma [66] are also upregulated in serous EC more frequently than in endometrioid EC [67]. The expression of FOLR1 cannot be observed in normal endometrium tissue [67], suggesting that FOLR1 may serve as a good tumor cell surface marker for targeted therapy, and antibodies against FOLR1 may facilitate tumor-specific cellular uptake of molecular drugs.

Trophoblast cell surface marker (Trop-2, a cell surface glycoprotein) is often overexpressed in various late stage epithelial tumor types with low or no expression in normal tissues [68]. Trop-2 is highly expressed in serous [69] and endometrioid EC [70]. Serous EC cell lines overexpressing Trop-2 show increased sensitivity to immunotherapy with hRS7, a humanized anti-Trop-2 monoclonal antibody [69]. Thus, Trop-2 would be an attractive target for EC immunotherapy.

Epithelial cell adhesion molecule (EpCAM) is overexpressed on malignant cells from a variety of different tumors and is considered as a reliable marker for tumor-initiating cells [71]. The cell surface expression of EpCAM is significantly higher among serous EC specimen compared to in normal endometrial tissue [72]. Serous EC cell lines that are positive for EpCAM exhibit high sensitivity to EpCAM antibody-mediated cytotoxicity, suggesting that EpCAM may represent a novel therapeutic target for serous EC.

In normal epithelium, the expression of L1 cell adhesion molecule (L1CAM) is undetectable. However, overexpression of L1CAM has been reported in many types of carcinomas [73]. L1CAM has been defined as a key driver for tumor cell invasion and EMT [73]. Of interest, L1CAM was absent in normal endometrium and the vast majority of endometrioid EC, but it was strongly expressed in serous and clear-cell EC [74]. The combined treatment with L1CAM antibodies and chemotherapeutic drugs in pancreatic and ovarian carcinoma model systems *in vivo* reduced tumor growth more efficiently than treatment with the cytostatic drug alone [75], indicating the value of L1CAM as a target for chemosensitizer in anticancer therapy for aggressive EC.

Taken together, antibodies against various tumor cell surface markers would provide a possibility of delivering drugs to EC cells, with fewer side effects on normal tissue. The nanotechnology or other approaches might be used to develop a more effective delivery system for targeted drugs, especially miRNAs that might simultaneously modulate a broad range of gene networks necessary for malignant phenotype of EC.

5. Targeting the CSC/EMT Signaling Pathways in EC

CSC is defined as a rare population having the ability to self-renew, initiate tumor growth, and give rise to the heterogeneous tumor cell mass [76]. Growing lines of evidence suggest that CSCs do exist and support tumor maintenance during tumor formation [77]. CSCs of EC might be located in the basal layer of endometrium and are responsible for production of EC cells [78]. Sorted CD133 (+) subpopulations from EC cell expressed higher levels of oncogene *BMI-1* [51] and showed more aggressive potential and increased tumorigenicity in nude mice than CD133 (−) cells [79]. Stem-like cell subpopulations, referred to as "side population" (SP) cells, have been isolated from EC tissue and show self-renewal capacity and enhanced tumorigenicity *in vivo* [80]. Therefore, these results suggest that selective killing of such CSCs is an appealing therapeutic prospect for EC.

Tumor cells that undergo EMT can increase their invasion ability and concurrently acquire CSC properties [81, 82]. Indeed, CSC fractions within pancreatic cancer [83] and colon cancer [84] are associated with enhanced capacity to metastasize, a process that requires considerable invasive capacity. At a molecular level, these findings are consistent with the fact that several signaling pathways involved in the self-renewal of CSCs, including Wnt/β-catenin, Hedgehog (Hh), and Notch signaling [85], can also induce EMT programs [86] (Figure 4), supporting a molecular link between EMT and CSC program in human tumor [87]. Therefore, development of specific therapies targeted at these CSC and EMT pathways raises a hope for eliminating recurrent and metastatic disease and for improvement of patient survival.

In malignant human mammary stem cells, activation of Hh signal components (SMO, PTCH1, and Gli1) increases the expression of downstream transcription factor BMI-1 and plays an important role in regulating stem cell self-renewal [88]. The overexpression of Hh-signal-related molecules is detected in EC tissue and involved in stimulated proliferation of EC cells [89]. In the same study, cyclopamine (a specific inhibitor of the SMO) has been shown to efficiently suppress the growth of EC cells [89].

Activation of Wnt/β-catenin pathway represented by the nuclear staining of β-catenin was shown to be more commonly detected in type I than type II EC [12]. More recent evidence suggests that gene sets indicating activation of Hh and Wnt/β-catenin signaling closely correlate with more aggressive EC and worse survival [90]. Wnt/β-catenin signaling was shown to induce the expression of downstream targets EpCAM and CD44 in hepatocellular carcinoma and EC, respectively [91, 92]. Salinomycin, a selective inhibitor of breast CSCs [93], was shown to induce apoptosis, inhibit Wnt/β-catenin signaling, and therefore repress the proliferation, migration, invasiveness, and tumorigenicity of SP cells obtained from invasive EC cells [94]. Thus, it is important to determine whether salinomycin alone, or in combination with other agents such as EpCAM-specific monoclonal antibody, could effectively induce apoptosis in CSC-like EC cells.

High expression of Notch1 has been detected in EC patients with poor prognosis, and treatment with a reported Notch inhibitor DAPT [95] suppresses invasiveness of EC cells [96].

Other potential therapeutic candidates for EC treatment might include Stattic, Rapamycin, and CD133. Signal transducer and activator of transcription 3 (STAT3) has been shown to transcriptionally activate the expression of EMT inducer TWIST1, resulting in promoted oncogenic properties in breast cancer [97]. Stattic (an inhibitor of STAT3) can suppress EGF-enhanced invasive behavior of EC cells [98]. Rapamycin (an mTOR inhibitor) has been used to counter the effects of *PTEN* deletion and inhibit the development of leukemia-initiating cells while preserving normal stem cell populations [99]. Targeting CD133 (+) cells by CD133 antibody-cytotoxic drug conjugates effectively inhibits the growth of hepatocellular and gastric cancer cells *in vivo* and *in vitro* [100].

The most obvious concern is whether a therapy can selectively target CSC, but not destroy normal stem cell that could share many characteristics as CSC, such as the ability to self-renew and differentiate. However, CSCs and normal stem cells display different biological behaviors, mainly due to aberrant activation of several pathways involved in proliferation, self-renewal, differentiation, and metabolism in CSCs [101, 102]. Therefore, exploiting these molecular differences could be helpful to specifically target CSCs while preserving normal stem cells. Furthermore, the combined inhibition of Hh and EGFR signaling through the use of specific inhibitors can lead to the increased rate of apoptotic death and decreased invasiveness of prostate cancer cells [103], suggesting that this treatment might be affecting the CSCs.

6. Targeting Immunosuppressive Molecular Pathways in EC

ECs are immunogenic tumors [104], and they mount potent antitumor immune responses, which might be ineffective at rejecting tumor, but might be potentially harnessed therapeutically [105]. Immune escape has been considered as the major malignant features of tumor cells. Several mechanisms are responsible for tumor immune escape, including the failure to recognize tumor cells by the immune system due to reduced major histocompatibility complex class I (MHC-I) expression, immunosuppression caused by tumor-cell-released immunosuppressive factors such as TGF-β, interleukin (IL)-10, VEGF, and cyclooxygenase-2 (COX-2), and immunoresistance resulting from the induction of EMT/CSC [104, 106, 107]. These data indicate that in addition to direct tumor cell killing, new targeted therapy might be also designed to reactivate the body's immune response against tumor cells (Figure 5).

Tumor stem cells (CD133+) have been shown to express low levels of MHC-I; however, the percentage of CD133-positive CSCs that expressed MHC-I can be significantly increased by the treatment with interferon-gamma [108], suggesting the possible use of MHC-I to generate anti-CSC immunity for human tumor including EC [106].

Some signal pathways that are activated in tumor cells are also dysregulated in immunosuppressive cells in cancer microenvironment. Immunosuppressive molecules released by tumor cells can activate STAT3 in immune cells, leading to tumour-induced immunosuppression [109]. In gastric cancer cells, oncogenic Wnt/β-catenin pathways enhance the

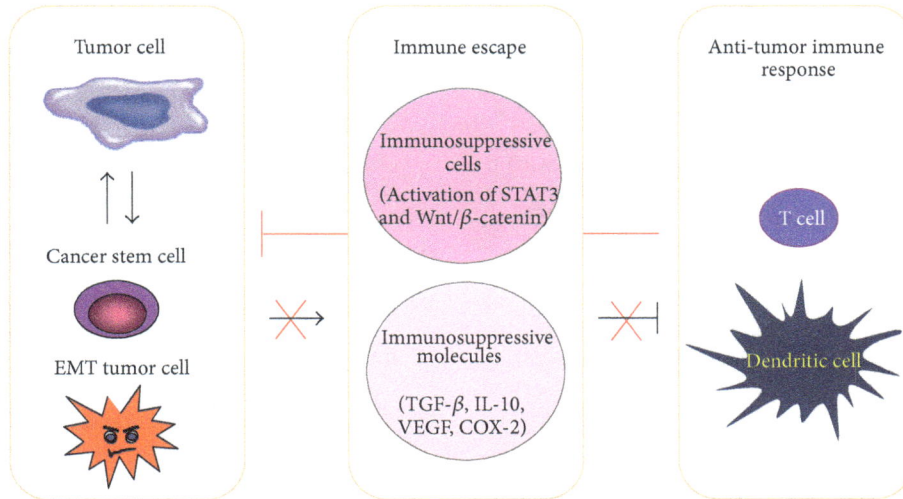

FIGURE 5: Targeting immunosuppressive molecular pathways in EC. Tumor cell induces immunosuppression by the production of immunosuppressive factors such as TGF-β, IL-10, VEGF, and COX-2. Tumor cells undergoing EMT can acquire both aggressive and immunosuppressive properties. Wnt/β-catenin pathway and STAT3-related pathway are activated in tumor cells and immunosuppressive cells and therefore they seem to be attractive targets for EC immunotherapy.

transcription of COX-2, an immunosuppressive molecule [110]. Importantly, COX-2 is upregulated and associated with VEGF expression in EC tissue [111], and selective COX-2 inhibitor etodolac exhibits antiproliferative effects on EC tissue [112], indicating that targeting COX-2 may boost immune responses towards EC and repress EC progression [113]. Although the adverse effects on normal immune cells should be avoided, targeting STAT3 or Wnt/β-catenin pathway by specific inhibitor in tumor cells and immunosuppressive cells, or along with other immunotherapy, might restore the immunocompetence of EC patients.

7. Conclusion

Currently, targeted therapies have not entered clinical practice, and clinical trials involving genetic biomarkers (mTOR, HER2, EGFR, and VEGF) administered to ECs only resulted in modest effects. Therapy targeting epigenetic regulatory mechanisms such as miRNA will need to be developed to achieve a broader impact on multiple signal pathways necessary for EC development. The use of targeted cancer therapy remains challenging because of the lack of specificity for cancer cells. Targeted agents that are specific to cell surface markers overexpressed in tumor cells would avoid potential side effects on normal tissue. More importantly, we expect that new targeted therapies that specifically attack both cancer cells and CSC-like cells can be used together with immunotherapy that stimulates a host's immune response and with other traditional treatments to achieve better clinical prognosis of EC patients in the near future.

Conflict of Interests

The authors declare no competing financial interests.

Authors' Contribution

Peixin Dong and Masanori Kaneuchi equally contributed to this paper.

Acknowledgments

This work was funded by a Grant from the Department of Women's Health Educational System, and a Grant-in-Aid from the Ministry of Health, Labour, and Welfare of Japan and Grant-in-Aid for Scientific Research (C) (23592428). The authors thank Dr. Zhujie Xu for the continuous and excellent support.

References

[1] J. Ferlay, H.-R. Shin, F. Bray, D. Forman, C. Mathers, and D. M. Parkin, "Estimates of worldwide burden of cancer in 2008: GLOBOCAN 2008," *International Journal of Cancer*, vol. 127, no. 12, pp. 2893–2917, 2010.

[2] J. V. Bokhman, "Two pathogenetic types of endometrial carcinoma," *Gynecologic Oncology*, vol. 15, no. 1, pp. 10–17, 1983.

[3] P. Singh, C. L. Smith, G. Cheetham, T. J. Dodd, and M. L. J. Davy, "Serous carcinoma of the uterus—determination of HER-2/neu status using immunohistochemistry, chromogenic in situ hybridization, and quantitative polymerase chain reaction techniques: its significance and clinical correlation," *International Journal of Gynecological Cancer*, vol. 18, no. 6, pp. 1344–1351, 2008.

[4] T. Alvarez, E. Miller, L. Duska, and E. Oliva, "Molecular profile of grade 3 endometrioid endometrial Carcinoma: is it a type i or type ii endometrial carcinoma?" *American Journal of Surgical Pathology*, vol. 36, no. 5, pp. 753–761, 2012.

[5] P. Mhawech-Fauceglia, D. Wang, J. Kesterson et al., "Gene expression profiles in stage I uterine serous carcinoma in comparison to grade 3 and grade 1 stage i endometrioid

adenocarcinoma," *PLoS ONE*, vol. 6, no. 3, Article ID e18066, 2011.

[6] K. N. Moore and A. Nickles Fader, "Uterine papillary serous carcinoma," *Clinical Obstetrics and Gynecology*, vol. 54, no. 2, pp. 278–291, 2011.

[7] S. Ma, T. K. Lee, B.-J. Zheng, K. W. Chan, and X.-Y. Guan, "CD133$^+$ HCC cancer stem cells confer chemoresistance by preferential expression of the Akt/PKB survival pathway," *Oncogene*, vol. 27, no. 12, pp. 1749–1758, 2008.

[8] B. M. Slomovitz and R. L. Coleman, "The PI3K/AKT/mTOR pathway as a therapeutic target in endometrial cancer," *Clinical Cancer Research*, vol. 18, no. 21, pp. 5856–5864, 2012.

[9] S. Sarmadi, N. Izadi-Mood, K. Sotoudeh, and S. M. Tavangar, "Altered PTEN expression; A diagnostic marker for differentiating normal, hyperplastic and neoplastic endometrium," *Diagnostic Pathology*, vol. 4, no. 1, article 41, 2009.

[10] M. L. Rudd, J. C. Price, S. Fogoros et al., "A unique spectrum of somatic PIK3CA (p110α) mutations within primary endometrial carcinomas," *Clinical Cancer Research*, vol. 17, no. 6, pp. 1331–1340, 2011.

[11] J. V. Lacey Jr., G. L. Mutter, B. M. Ronnett et al., "PTEN expression in endometrial biopsies as a marker of progression to endometrial carcinoma," *Cancer Research*, vol. 68, no. 14, pp. 6014–6020, 2008.

[12] X. Matias-Guiu and J. Prat, "Molecular pathology of endometrial carcinoma," *Histopathology*, vol. 62, no. 1, pp. 111–123, 2013.

[13] S. Acharya, M. L. Hensley, A. C. Montag, and G. F. Fleming, "Rare uterine cancers," *Lancet Oncology*, vol. 6, no. 12, pp. 961–971, 2005.

[14] G. E. Konecny, L. Santos, B. Winterhoff et al., "HER2 gene amplification and EGFR expression in a large cohort of surgically staged patients with nonendometrioid (type II) endometrial cancer," *British Journal of Cancer*, vol. 100, no. 1, pp. 89–95, 2009.

[15] H. Niikura, H. Sasano, G. Matsunaga et al., "Prognostic value of epidermal growth factor receptor expression in endometrioid endometrial carcinoma," *Human Pathology*, vol. 26, no. 8, pp. 892–896, 1995.

[16] C. M. Holland, K. Day, A. Evans, and S. K. Smith, "Expression of the VEGF and angiopoietin genes in endometrial atypical hyperplasia and endometrial cancer," *British Journal of Cancer*, vol. 89, no. 5, pp. 891–898, 2003.

[17] J. Zhang and L. Ma, "MicroRNA control of epithelial-mesenchymal transition and metastasis," *Cancer and Metastasis Reviews*, vol. 31, no. 3-4, pp. 3653–3462, 2012.

[18] K. Kato, "Endometrial cancer stem cells: a new target for cancer therapy," *Anticancer Research*, vol. 32, no. 6, pp. 2283–2293, 2012.

[19] S. A. Hubbard, A. M. Friel, B. Kumar, L. Zhang, B. R. Rueda, and C. E. Gargett, "Evidence for cancer stem cells in human endometrial carcinoma," *Cancer Research*, vol. 69, no. 21, pp. 8241–8248, 2009.

[20] M. Nakamura, S. Kyo, B. Zhang et al., "Prognostic impact of CD133 expression as a tumor-initiating cell marker in endometrial cancer," *Human Pathology*, vol. 41, no. 11, pp. 1516–1529, 2010.

[21] K. J. Dedes, D. Wetterskog, A. Ashworth, S. B. Kaye, and J. S. Reis-Filho, "Emerging therapeutic targets in endometrial cancer," *Nature Reviews Clinical Oncology*, vol. 8, no. 5, pp. 261–271, 2011.

[22] B. Weigelt and S. Banerjee, "Molecular targets and targeted therapeutics in endometrial cancer," *Current Opinion in Oncology*, vol. 24, no. 5, pp. 554–563, 2012.

[23] F. Morgillo, M. A. Bareschino, R. Bianco, G. Tortora, and F. Ciardiello, "Primary and acquired resistance to anti-EGFR targeted drugs in cancer therapy," *Differentiation*, vol. 75, no. 9, pp. 788–799, 2007.

[24] N. C. Turner and J. S. Reis-Filho, "Genetic heterogeneity and cancer drug resistance," *The Lancet Oncology*, vol. 13, no. 4, pp. e178–e185, 2012.

[25] J. A. Engelman, "Targeting PI3K signalling in cancer: opportunities, challenges and limitations," *Nature Reviews Cancer*, vol. 9, no. 8, pp. 550–562, 2009.

[26] C. Widakowich, G. de Castro Jr., E. de Azambuja, P. Dinh, and A. Awada, "Review: side effects of approved molecular targeted therapies in solid cancers," *Oncologist*, vol. 12, no. 12, pp. 1443–1455, 2007.

[27] R. M. Glasspool, J. M. Teodoridis, and R. Brown, "Epigenetics as a mechanism driving polygenic clinical drug resistance," *British Journal of Cancer*, vol. 94, no. 8, pp. 1087–1092, 2006.

[28] T. Ogawa, T. E. Liggett, A. A. Melnikov et al., "Methylation of death-associated protein kinase is associated with cetuximab and erlotinib resistance," *Cell Cycle*, vol. 11, no. 8, pp. 1656–1663, 2012.

[29] G. Ghosh, X. Lian, S. J. Kron, and S. P. Palecek, "Properties of resistant cells generated from lung cancer cell lines treated with EGFR inhibitors," *BMC Cancer*, vol. 12, article 95, 2012.

[30] T. Vu and F. X. Claret, "Trastuzumab: updated mechanisms of action and resistance in breast cancer," *Frontiers in Oncology*, vol. 2, article 62, 2012.

[31] O. Keunen, M. Johansson, A. Oudin et al., "Anti-VEGF treatment reduces blood supply and increases tumor cell invasion in glioblastoma," *Proceedings of the National Academy of Sciences of the United States of America*, vol. 108, no. 9, pp. 3749–3754, 2011.

[32] S. Cannito, E. Novo, A. Compagnone et al., "Redox mechanisms switch on hypoxia-dependent epithelial-mesenchymal transition in cancer cells," *Carcinogenesis*, vol. 29, no. 12, pp. 2267–2278, 2008.

[33] B. Keith and M. C. Simon, "Hypoxia-inducible factors, stem cells, and cancer," *Cell*, vol. 129, no. 3, pp. 465–472, 2007.

[34] M. Pàez-Ribes, E. Allen, J. Hudock et al., "Antiangiogenic therapy elicits malignant progression of tumors to increased local invasion and distant metastasis," *Cancer Cell*, vol. 15, no. 3, pp. 220–231, 2009.

[35] S. Sharma, T. K. Kelly, and P. A. Jones, "Epigenetics in cancer," *Carcinogenesis*, vol. 31, no. 1, pp. 27–36, 2009.

[36] H. B. Salvesen, N. MacDonald, A. Ryan et al., "PTEN methylation is associated with advanced stage and microsatellite instability in endometrial carcinoma," *International Journal of Cancer*, vol. 91, no. 1, pp. 22–26, 2001.

[37] J. Bischoff, A. Ignatov, A. Semczuk et al., "hMLH1 promoter hypermethylation and MSI status in human endometrial carcinomas with and without metastases," *Clinical and Experimental Metastasis*, vol. 29, no. 8, pp. 889–900, 2012.

[38] G. Moreno-Bueno, D. Hardisson, C. Sánchez et al., "Abnormalities of the APC/beta-catenin pathway in endometrial cancer," *Oncogene*, vol. 21, no. 52, pp. 7981–7990, 2002.

[39] X. Liao, M. K.-Y. Siu, K. Y.-K. Chan et al., "Hypermethylation of RAS effector related genes and DNA methyltransferase 1 expression in endometrial carcinogenesis," *International Journal of Cancer*, vol. 123, no. 2, pp. 296–302, 2008.

[40] T.-Z. Yi, J. Guo, L. Zhou et al., "Prognostic value of E-cadherin expression and CDH1 promoter methylation in patients with

endometrial carcinoma," *Cancer Investigation*, vol. 29, no. 1, pp. 86–92, 2011.

[41] S. D. Gore and E. R. Hermes-DeSantis, "Enhancing survival outcomes in the management of patients with higher-risk myelodysplastic syndromes," *Cancer Control*, vol. 16, supplement, pp. 2–10, 2009.

[42] S. A. Kavanaugh, L. A. White, and J. M. Kolesar, "Vorinostat: a novel therapy for the treatment of cutaneous T-cell lymphoma," *American Journal of Health-System Pharmacy*, vol. 67, no. 10, pp. 793–797, 2010.

[43] T.-Z. Yi, J. Li, X. Han et al., "DNMT inhibitors and HDAC inhibitors regulate E-Cadherin and Bcl-2 expression in endometrial carcinoma in vitro and in vivo," *Chemotherapy*, vol. 58, no. 1, pp. 19–29, 2012.

[44] R. Connolly and V. Stearns, "Epigenetics as a therapeutic target in breast cancer," *Journal of Mammary Gland Biology and Neoplasia*, vol. 17, no. 3-4, pp. 3191–4204, 2012.

[45] D. P. Bartel, "MicroRNAs: genomics, biogenesis, mechanism, and function," *Cell*, vol. 116, no. 2, pp. 281–297, 2004.

[46] H. Xia and K. M. Hui, "MicroRNAs involved in regulating epithelial-mesenchymal transition and cancer stem cells as molecular targets for cancer therapeutics," *Cancer Gene Therapy*, vol. 19, no. 11, pp. 723–730, 2012.

[47] T. K. H. Chung, T.-H. Cheung, N.-Y. Huen et al., "Dysregulated microRNAs and their predicted targets associated with endometrioid endometrial adenocarcinoma in Hong Kong women," *International Journal of Cancer*, vol. 124, no. 6, pp. 1358–1365, 2009.

[48] E. Chan, D. E. Prado, and J. B. Weidhaas, "Cancer microRNAs: from subtype profiling to predictors of response to therapy," *Trends in Molecular Medicine*, vol. 17, no. 5, pp. 235–243, 2011.

[49] T. K. H. Chung, T. S. Lau, T. H. Cheung et al., "Dysregulation of microRNA-204 mediates migration and invasion of endometrial cancer by regulating FOXC1," *International Journal of Cancer*, vol. 130, no. 5, pp. 1036–1045, 2012.

[50] H. Zhai, M. Karaayvaz, P. Dong, N. Sakuragi, and J. Ju, "Prognostic significance of miR-194 in endometrial cancer," *Biomarker Research*, vol. 1, article 12, 2013.

[51] P. Dong, M. Kaneuchi, H. Watari et al., "MicroRNA-194 inhibits epithelial to mesenchymal transition of endometrial cancer cells by targeting oncogene BMI-1," *Molecular Cancer*, vol. 10, article 99, 2011.

[52] P. Dong, M. Karaayvaz et al., "Mutant p53 gain-of-function induces epithelial-mesenchymal transition through modulation of the miR-130b-ZEB1 axis," *Oncogene*, 2012.

[53] P. S. Mitchell, R. K. Parkin, E. M. Kroh et al., "Circulating microRNAs as stable blood-based markers for cancer detection," *Proceedings of the National Academy of Sciences of the United States of America*, vol. 105, no. 30, pp. 10513–10518, 2008.

[54] T. Tsuruta, K.-I. Kozaki, A. Uesugi et al., "miR-152 is a tumor suppressor microRNA that is silenced by DNA hypermethylation in endometrial cancer," *Cancer Research*, vol. 71, no. 20, pp. 6450–6462, 2011.

[55] C. S. Gondi and J. S. Rao, "Concepts in in vivo siRNA delivery for cancer therapy," *Journal of Cellular Physiology*, vol. 220, no. 2, pp. 285–291, 2009.

[56] J. A. Broderick and P. D. Zamore, "MicroRNA therapeutics," *Gene Therapy*, vol. 18, no. 12, pp. 1104–1110, 2011.

[57] Y. W. Kong, D. Ferland-McCollough, T. J. Jackson, and M. Bushell, "microRNAs in cancer management," *The Lancet Oncology*, vol. 13, no. 6, pp. e249–e258, 2012.

[58] Y. Chen, X. Zhu, X. Zhang, B. Liu, and L. Huang, "Nanoparticles modified with tumor-targeting scFv deliver siRNA and miRNA for cancer therapy," *Molecular Therapy*, vol. 18, no. 9, pp. 1650–1656, 2010.

[59] E. B. Pasquale, "Eph receptors and ephrins in cancer: bidirectional signalling and beyond," *Nature Reviews Cancer*, vol. 10, no. 3, pp. 165–180, 2010.

[60] M. Tandon, S. V. Vemula, and S. K. Mittal, "Emerging strategies for EphA2 receptor targeting for cancer therapeutics," *Expert Opinion on Therapeutic Targets*, vol. 15, no. 1, pp. 31–51, 2011.

[61] A. A. Kamat, D. Coffey, W. M. Merritt et al., "EphA2 overexpression is associated with lack of hormone receptor expression and poor outcome in endometrial cancer," *Cancer*, vol. 115, no. 12, pp. 2684–2692, 2009.

[62] J.-W. Lee, R. L. Stone, S. J. Lee et al., "EphA2 targeted chemotherapy using an antibody drug conjugate in endometrial carcinoma," *Clinical Cancer Research*, vol. 16, no. 9, pp. 2562–2570, 2010.

[63] G. E. Konecny, R. Agarwal, G. A. Keeney et al., "Claudin-3 and claudin-4 expression in serous papillary, clear-cell, and endometrioid endometrial cancer," *Gynecologic Oncology*, vol. 109, no. 2, pp. 263–269, 2008.

[64] X. Y. Pan, B. Wang, Y. C. Che, Z. P. Weng, H. Y. Dai, and W. Peng, "Expression of claudin-3 and claudin-4 in normal, hyperplastic, and malignant endometrial tissue," *International Journal of Gynecological Cancer*, vol. 17, no. 1, pp. 233–241, 2007.

[65] A. D. Santin, S. Bellone, M. Marizzoni et al., "Overexpression of claudin-3 and claudin-4 receptors in uterine serous papillary carcinoma: novel targets for a type-specific therapy using Clostridium perfringens enterotoxin (CPE)," *Cancer*, vol. 109, no. 7, pp. 1312–1322, 2007.

[66] C. D. Hough, C. A. Sherman-Baust, E. S. Pizer et al., "Large-scale serial analysis of gene expression reveals genes differentially expressed in ovarian cancer," *Cancer Research*, vol. 60, no. 22, pp. 6281–6287, 2000.

[67] L. A. Dainty, J. I. Risinger, C. Morrison et al., "Overexpression of folate binding protein and mesothelin are associated with uterine serous carcinoma," *Gynecologic Oncology*, vol. 105, no. 3, pp. 563–570, 2007.

[68] R. Cubas, M. Li, C. Chen, and Q. Yao, "Trop2: a possible therapeutic target for late stage epithelial carcinomas," *Biochimica et Biophysica Acta*, vol. 1796, no. 2, pp. 309–314, 2009.

[69] J. Varughese, E. Cocco, S. Bellone et al., "Uterine serous papillary carcinomas overexpress human trophoblast-cell-surface marker (trop-2) and are highly sensitive to immunotherapy with hRS7, a humanized anti-trop-2 monoclonal antibody," *Cancer*, vol. 117, no. 14, pp. 3163–3172, 2011.

[70] E. Bignotti, L. Zanotti, S. Calza et al., "Trop-2 protein overexpression is an independent marker for predicting disease recurrence in endometrioid endometrial carcinoma," *BMC Clinical Pathology*, vol. 12, no. 1, article 22, 2012.

[71] S. Imrich, M. Hachmeister, and O. Gires, "EpCAM and its potential role in tumor-initiating cells," *Cell Adhesion and Migration*, vol. 6, no. 1, pp. 30–38, 2012.

[72] K. El-Sahwi, S. Bellone, E. Cocco et al., "Overexpression of EpCAM in uterine serous papillary carcinoma: implications for EpCAM-specific immunotherapy with human monoclonal antibody adecatumumab (MT201)," *Molecular Cancer Therapeutics*, vol. 9, no. 1, pp. 57–66, 2010.

[73] H. Kiefel, S. Bondong, J. Hazin et al., "L1CAM: a major driver for tumor cell invasion and motility," *Cell Adhesion & Migration*, vol. 6, no. 4, pp. 374–384, 2012.

[74] M. Huszar, M. Pfeifer, U. Schirmer et al., "Up-regulation of LICAM is linked to loss of hormone receptors and E-cadherin in aggressive subtypes of endometrial carcinomas," *Journal of Pathology*, vol. 220, no. 5, pp. 551–561, 2010.

[75] H. Schäfer, C. Dieckmann, O. Korniienko et al., "Combined treatment of L1CAM antibodies and cytostatic drugs improve the therapeutic response of pancreatic and ovarian carcinoma," *Cancer Letters*, vol. 319, no. 1, pp. 66–82, 2012.

[76] L. V. Nguyen, R. Vanner, P. Dirks, and C. J. Eaves, "Cancer stem cells: an evolving concept," *Nature Reviews Cancer*, vol. 12, no. 2, pp. 133–143, 2012.

[77] R. J. Gilbertson and T. A. Graham, "Cancer: resolving the stem-cell debate," *Nature*, vol. 488, no. 7412, pp. 462–463, 2012.

[78] S. A. Hubbard, A. M. Friel, B. Kumar, L. Zhang, B. R. Rueda, and C. E. Gargett, "Evidence for cancer stem cells in human endometrial carcinoma," *Cancer Research*, vol. 69, no. 21, pp. 8241–8248, 2009.

[79] M. Nakamura, S. Kyo, B. Zhang et al., "Prognostic impact of CD133 expression as a tumor-initiating cell marker in endometrial cancer," *Human Pathology*, vol. 41, no. 11, pp. 1516–1529, 2010.

[80] K. Kato, "Stem cells in human normal endometrium and endometrial cancer cells: characterization of side population cells," *Kaohsiung Journal of Medical Sciences*, vol. 28, no. 2, pp. 63–71, 2012.

[81] S. A. Mani, W. Guo, M.-J. Liao et al., "The epithelial-mesenchymal transition generates cells with properties of stem cells," *Cell*, vol. 133, no. 4, pp. 704–715, 2008.

[82] A. Singh and J. Settleman, "EMT, cancer stem cells and drug resistance: an emerging axis of evil in the war on cancer," *Oncogene*, vol. 29, no. 34, pp. 4741–4751, 2010.

[83] P. C. Hermann, S. L. Huber, T. Herrler et al., "Distinct populations of cancer stem cells determine tumor growth and metastatic activity in human pancreatic cancer," *Cell Stem Cell*, vol. 1, no. 3, pp. 313–323, 2007.

[84] R. Pang, W. L. Law, A. C. Y. Chu et al., "A subpopulation of CD26$^+$ cancer stem cells with metastatic capacity in human colorectal cancer," *Cell Stem Cell*, vol. 6, no. 6, pp. 603–615, 2010.

[85] B.-B. S. Zhou, H. Zhang, M. Damelin, K. G. Geles, J. C. Grindley, and P. B. Dirks, "Tumour-initiating cells: challenges and opportunities for anticancer drug discovery," *Nature Reviews Drug Discovery*, vol. 8, no. 10, pp. 806–823, 2009.

[86] J. M. Bailey, P. K. Singh, and M. A. Hollingsworth, "Cancer metastasis facilitated by developmental pathways: sonic hedgehog, notch, and bone morphogenic proteins," *Journal of Cellular Biochemistry*, vol. 102, no. 4, pp. 829–839, 2007.

[87] A. Biddle and I. C. Mackenzie, "Cancer stem cells and EMT in carcinoma," *Cancer and Metastasis Reviews*, vol. 31, no. 1-2, pp. 285–293, 2012.

[88] S. Liu, G. Dontu, I. D. Mantle et al., "Hedgehog signaling and Bmi-1 regulate self-renewal of normal and malignant human mammary stem cells," *Cancer Research*, vol. 66, no. 12, pp. 6063–6071, 2006.

[89] Y.-Z. Feng, T. Shiozawa, T. Miyamoto et al., "Overexpression of hedgehog signaling molecules and its involvement in the proliferation of endometrial carcinoma cells," *Clinical Cancer Research*, vol. 13, no. 5, pp. 1389–1398, 2007.

[90] E. Wik, M. B. Ræder, C. Krakstad et al., "Lack of estrogen receptor-α is associated with epithelial-mesenchymal transition and PI3K alterations in endometrial carcinoma," *Clinical Cancer Research*, vol. 19, no. 5, pp. 1094–1105, 2013.

[91] M. Munz, P. A. Baeuerle, and O. Gires, "The emerging role of EpCAM in cancer and stem cell signaling," *Cancer Research*, vol. 69, no. 14, pp. 5627–5629, 2009.

[92] Y. Wang, P. Hanifi-Moghaddam, E. E. Hanekamp et al., "Progesterone inhibition of Wnt/β-catenin signaling in normal endometrium and endometrial cancer," *Clinical Cancer Research*, vol. 15, no. 18, pp. 5784–5793, 2009.

[93] P. B. Gupta, T. T. Onder, G. Jiang et al., "Identification of selective inhibitors of cancer stem cells by high-throughput screening," *Cell*, vol. 138, no. 4, pp. 645–659, 2009.

[94] S. Kusunoki, K. Kato, K. Tabu et al., "The inhibitory effect of salinomycin on the proliferation, migration and invasion of human endometrial cancer stem-like cells," *Gynecologic Oncology*, vol. 129, no. 3, pp. 598–605, 2013.

[95] C. Groth and M. E. Fortini, "Therapeutic approaches to modulating Notch signaling: current challenges and future prospects," *Seminars in Cell and Developmental Biology*, vol. 23, no. 4, pp. 465–472, 2012.

[96] Y. Mitsuhashi, A. Horiuchi, T. Miyamoto, H. Kashima, A. Suzuki, and T. Shiozawa, "Prognostic significance of Notch signalling molecules and their involvement in the invasiveness of endometrial carcinoma cells," *Histopathology*, vol. 60, no. 5, pp. 826–837, 2012.

[97] G. Z. Cheng, W. Zhang, M. Sun et al., "Twist is transcriptionally induced by activation of STAT3 and mediates STAT3 oncogenic function," *Journal of Biological Chemistry*, vol. 283, no. 21, pp. 14665–14673, 2008.

[98] C. H. Chen, S. W. Wang, C. W. Chen et al., "MUC20 overexpression predicts poor prognosis and enhances EGF-induced malignant phenotypes via activation of the EGFR-STAT3 pathway in endometrial cancer," *Gynecologic Oncology*, vol. 128, no. 3, pp. 560–567, 2013.

[99] Ö. H. Yilmaz, R. Valdez, B. K. Theisen et al., "Pten dependence distinguishes haematopoietic stem cells from leukaemia-initiating cells," *Nature*, vol. 441, no. 7092, pp. 475–482, 2006.

[100] L. M. Smith, A. Nesterova, M. C. Ryan et al., "CD133/prominin-1 is a potential therapeutic target for antibody-drug conjugates in hepatocellular and gastric cancers," *British Journal of Cancer*, vol. 99, no. 1, pp. 100–109, 2008.

[101] M. V. Verga Falzacappa, C. Ronchini, L. B. Reavie, and P. G. Pelicci, "Regulation of self-renewal in normal and cancer stem cells," *FEBS Journal*, vol. 279, no. 19, pp. 3559–3572, 2012.

[102] C. Pecqueur, L. Oliver, K. Oizel, L. Lalier, and F. M. Vallette, "Targeting metabolism to induce cell death in cancer cells and cancer stem cells," *International Journal of Cell Biology*, vol. 2013, Article ID 805975, 13 pages, 2013.

[103] M. Mimeault, E. Moore, N. Moniaux et al., "Cytotoxic effects induced by a combination of cydopamine and gefitinib, the selective hedgehog and epidermal growth factor receptor signaling inhibitors, in prostate cancer cells," *International Journal of Cancer*, vol. 118, no. 4, pp. 1022–1031, 2006.

[104] N. Brooks and D. S. Pouniotis, "Immunomodulation in endometrial cancer," *International Journal of Gynecological Cancer*, vol. 19, no. 4, pp. 734–740, 2009.

[105] L. E. Kandalaft, N. Singh, J. B. Liao et al., "The emergence of immunomodulation: combinatorial immunochemotherapy opportunities for the next decade," *Gynecologic Oncology*, vol. 116, no. 2, pp. 222–233, 2010.

[106] M. Vanneman and G. Dranoff, "Combining immunotherapy and targeted therapies in cancer treatment," *Nature Reviews Cancer*, vol. 12, no. 4, pp. 237–251, 2012.

[107] C. Kudo-Saito, H. Shirako, T. Takeuchi, and Y. Kawakami, "Cancer metastasis is accelerated through immunosuppression during snail-induced EMT of cancer cells," *Cancer Cell*, vol. 15, no. 3, pp. 195–206, 2009.

[108] A. Wu, S. Wiesner, J. Xiao et al., "Expression of MHC I and NK ligands on human CD133$^+$ glioma cells: possible targets of immunotherapy," *Journal of Neuro-Oncology*, vol. 83, no. 2, pp. 121–131, 2007.

[109] H. Yu, M. Kortylewski, and D. Pardoll, "Crosstalk between cancer and immune cells: role of STAT3 in the tumour microenvironment," *Nature Reviews Immunology*, vol. 7, no. 1, pp. 41–51, 2007.

[110] F. Nuñez, S. Bravo, F. Cruzat, M. Montecino, and G. V. de Ferrari, "Wnt/β-catenin signaling enhances cyclooxygenase-2 (COX2) transcriptional activity in gastric cancer cells," *PLoS ONE*, vol. 6, no. 4, Article ID e18562, 2011.

[111] R. Fujiwaki, K. Iida, H. Kanasaki, T. Ozaki, K. Hata, and K. Miyazaki, "Cyclooxygenase-2 expression in endometrial cancer: correlation with microvessel count and expression of vascular endothelial growth factor and thymidine phosphorylase," *Human Pathology*, vol. 33, no. 2, pp. 213–219, 2002.

[112] K. Hasegawa, Y. Torii, R. Ishii, S. Oe, R. Kato, and Y. Udagawa, "Effects of a selective COX-2 inhibitor in patients with uterine endometrial cancers," *Archives of Gynecology and Obstetrics*, vol. 284, no. 6, pp. 1515–1521, 2011.

[113] S. Ohno, Y. Ohno, N. Suzuki et al., "Multiple roles of cyclooxygenase-2 in endometrial cancer," *Anticancer Research*, vol. 25, no. 6, pp. 3679–3687, 2005.

5

Treatment of Slaughter House Wastewater in a Sequencing Batch Reactor: Performance Evaluation and Biodegradation Kinetics

Pradyut Kundu, Anupam Debsarkar, and Somnath Mukherjee

Environmental Engineering Division, Civil Engineering Department, Jadavpur University, Kolkata 32, India

Correspondence should be addressed to Somnath Mukherjee; snm ju@yahoo.co.in

Academic Editor: Eldon R. Rene

Slaughterhouse wastewater contains diluted blood, protein, fat, and suspended solids, as a result the organic and nutrient concentration in this wastewater is vary high and the residues are partially solubilized, leading to a highly contaminating effect in riverbeds and other water bodies if the same is let off untreated. The performance of a laboratory-scale Sequencing Batch Reactor (SBR) has been investigated in aerobic-anoxic sequential mode for simultaneous removal of organic carbon and nitrogen from slaughterhouse wastewater. The reactor was operated under three different variations of aerobic-anoxic sequence, namely, (4+4), (5+3), and (3+5) hr. of total react period with two different sets of influent soluble COD (SCOD) and ammonia nitrogen (NH_4^+-N) level 1000 ± 50 mg/L, and 90 ± 10 mg/L, 1000 ± 50 mg/L and 180 ± 10 mg/L, respectively. It was observed that from 86 to 95% of SCOD removal is accomplished at the end of 8.0 hr of total react period. In case of (4+4) aerobic-anoxic operating cycle, a reasonable degree of nitrification 90.12 and 74.75% corresponding to initial NH_4^+-N value of 96.58 and 176.85 mg/L, respectively, were achieved. The biokinetic coefficients (k, K_s, Y, k_d) were also determined for performance evaluation of SBR for scaling full-scale reactor in future operation.

1. Introduction

The continuous drive to increase meat production for the protein needs of the ever increasing world population has some pollution problems attached. Pollution arises from activities in meat production as a result of failure in adhering to Good Manufacturing Practices (GMP) and Good Hygiene Practices (GHP) [1]. Consideration is hardly given to safety practices during animal transport to the abattoir, during slaughter and dressing of hides and flesh [2]. For hygienic reasons abattoirs, use large amount of water in processing operations (slaughtering and cleaning), which produces large amount of wastewater. The major environmental problem associated with this abattoir wastewater is the large amount of suspended solids and liquid waste as well as odor generation [3]. Effluent from slaughterhouses has also been recognized to contaminate both surface and groundwater because during abattoir processing, blood, fat, manure, urine, and meat tissues are lost to the wastewater streams [4]. Leaching into groundwater is

a major part of the concern, especially due to the recalcitrant nature of some contaminants [5]. Blood, one of the major dissolved pollutants in abattoir wastewater, has the highest COD of any effluent from abattoir operations. If the blood from a single cow carcass is allowed to discharge directly into a sewer line, the effluent load would be equivalent to the total sewage produced by 50 people on average day [6]. The major characteristics of abattoir wastes are high organic strength, sufficient organic biological nutrients, adequate alkalinity, relatively high temperature (20 to 30°C) and free of toxic material. Abattoir wastewaters with the previous characteristics are well suited to anaerobic treatment and the efficiency in reducing the BOD_5 ranged between 60 and 90% [7]. The high concentration of nitrates in the abattoir wastewater also exhibits that the wastewater could be treated by biological processes. Nitrogenous wastewater when discharged to receiving water bodies leads to undesirable problems such as algal blooms and eutrophication in addition to oxygen deficit. The dissolved oxygen level further depleted if organic carbon along

TABLE 1: Different available technologies used to treat slaughterhouse wastewater.

Sl no.	Technology adopted	Input characteristics of slaughterhouse wastewater	Observations	References
(1)	Anaerobic treatment of slaughterhouse wastewaters in a UASB (Upflow Anaerobic Sludge Blanket) reactor and in an anaerobic filter (AF).	Slaughterhouse wastewater showed the highest organic content with an average COD of 8000 mg/L, of which 70% was proteins. The suspended solids content represented between 15 and 30% of the COD.	The UASB reactor was run at OLR (Organic Loading Rates) of 1–6.5 kg COD/m³/day. The COD removal was 90% for OLR up to 5 kg COD/m³/day and 60% for an OLR of 6.5 kg COD/m³/day. For similar organic loading rates, the AF showed lower removal efficiencies and lower percentages of methanization.	Ruiz et al. [9]
(2)	Anaerobic sequencing batch reactors.	Influent total chemical oxygen demand (TCOD) ranged from 6908 to 11 500 mg/L, of which approximately 50% were in the form of suspended solids (SS).	Total COD was reduced by 90% to 96% at organic loading rates (OLRs) ranging from 2.07 to 4.93 kg m^{-3} d^{-1} and a hydraulic retention time of 2 days. Soluble COD was reduced by over 95% in most samples.	Massé and Masse [10]
(3)	Moving bed sequencing batch reactor for piggery wastewater treatment.	COD, BOD, and suspended solids in the range of 4700–5900 mg/L, 1500–2300 mg/L, and 4000–8000 mg/L, respectively.	COD and BOD removal efficiency was greater than 80% and 90%, respectively at high organic loads of 1.18–2.36 kg COD/m³·d. The moving-bed SBR gave TKN removal efficiency of 86–93%.	Sombatsompop et al. [11]
(4)	Fixed bed sequencing batch reactor (FBSBR).	The wastewater has COD loadings in the range of 0.5–1.5 Kg COD/m³ per day.	COD, TN, and phosphorus removal efficiencies were at range of 90–96%, 60–88%, and 76–90%, respectively.	Rahimi et al. [12]
(5)	Chemical coagulation and electrocoagulation techniques.	COD and BOD$_5$ of raw wastewater in the range of 5817 ± 473 and 2543 ± 362 mg/L.	Removal of COD and BOD$_5$ more than 99% was obtained by adding 100 mg/L PACl and applied voltage 40 V.	Bazrafshan et al. [13]
(6)	Hybrid upflow anaerobic sludge blanket (HUASB) reactor for treating poultry slaughterhouse wastewater.	Slaughterhouse wastewater showed total COD 3000–4800 mg/L, soluble COD 1030–3000 mg/L, BOD5 750–1890 mg/L, suspended solids 300–950 mg/L, alkalinity (as CaCO$_3$) 600–1340 mg/L, VFA (as acetate) 250–540 mg/L, and pH 7–7.6.	The HUSB reactor was run at OLD of 19 kg COD/m³/day and achieved TCOD and SCOD removal efficiencies of 70–86% and 80–92%, respectively. The biogas was varied between 1.1 and 5.2 m³/m³ d with the maximum methane content of 72%.	Rajakumar et al. [14]
(7)	Anaerobic hybrid reactor was packed with light weight floating media.	COD, BOD and Suspended Solids in the range of 22000–27500 mg/L, 10800–14600 mg/L, and 1280–1500 mg/L, respectively.	COD and BOD reduction was found in the range of 86.0–93.58% and 88.9–95.71%, respectively.	Sunder and Satyanarayan [15]

with nutrient sinks into the water environment. Hence, it is very much necessary to control the discharge of combined organic carbon and nitrogen laden wastewater by means of appropriate treatment. Biological treatment has been proved to be comparatively innocuous and more energy efficient of treating wastewater if good process control could be ensured [8]. Several researchers successfully used different technologies for treatment of slaughterhouse wastewater containing organic carbon and nitrogen (COD and TKN) in laboratory and pilot scale experiment. Table 1 had shown the previous research findings about slaughterhouse wastewater treatment by the different investigators.

Among the various biological treatment processes, sequencing batch reactor (SBR) is considered to be an improved version of activated sludge process, which operates in fill and draw mode for biological treatment of wastewater. An SBR operates in a pseudobatch mode with aeration and sludge settlement both occurring in the same tank. SBRs are operated in fill-react-settle-draw-idle period sequences. The major differences between SBR and conventional continuous-flow, activated sludge system is that the SBR tank carries out the functions of equalization, aeration, and sedimentation in a time sequence rather than in the conventional space sequence of continuous-flow systems. Sequencing batch reactors (SBRs) are advocated as one of the best available techniques (BATs) for slaughterhouse wastewater treatment [16, 17] because they are capable of removing organic carbon, nutrients, and suspended solids from wastewater in a single tank and also have low capital and operational costs.

Biological treatment of wastewater containing organic carbon and nitrogen (COD and TKN) is also carried out in laboratory and pilot scale experiment by several researchers successfully [18–26]. Nutrients in piggery wastewater with high organic matter, nitrogen, and phosphorous content were biological removed by Obaja et al. [27] in a sequencing batch reactor (SBR) with anaerobic, aerobic, and anoxic stages. The SBR was operated with wastewater containing 1500 mg/L ammonium and 144 mg/L phosphate, a removal efficiency of 99.7% for nitrogen and 97.3% for phosphate was obtained. A full-scale SBR system was evaluated by Lo and Liao [28] to remove 82% of BOD and more than 75% of nitrogen after a cycle period of 4.6 hour from swine wastewater. Mahvi et al. [29] carried out a pilot-scale study on removal of nitrogen both from synthetic and domestic wastewater in a continuous flow SBR and obtained a total nitrogen and TKN removal of 70–80% and 85–95%, respectively. An SBR system demonstrated by Lemaire et al. [30] to high degree of biological remove of nitrogen, phosphorus, and COD to very low levels from slaughterhouse wastewater. A high degree removal of total phosphorus (98%), total nitrogen (97%), and total COD (95%) was achieved after a 6-hour cycle period. Moreover, SBRs have been successfully used to treat landfill leachate, tannery wastewater, phenolic wastewater, and various other industrial wastewaters [31–34].

In the present investigation, an attempt has been made to explore the performance efficacy of SBR technology for simultaneous removal of soluble carbonaceous organic matter and ammonia nitrogen from slaughterhouse wastewater and also to determine the biokinetic constants for carbon oxidation, nitrification, and denitrification under different combination of react periods (aerobic/anoxic).

2. Material and Methods

2.1. Seed Acclimatization for Combined Carbon Oxidation and Nitrification.

The active microbial seed was cultured under ambient condition in the laboratory by inoculating 200 mL sludge as collected from an aeration pond of M/S Mokami small-scale slaughterhouse located in the village Nazira, South 24 Parganas district (West Bengal), India, to a growth propagating media composed of 500 mL dextrose solution having concentrations of 1000 mg/L, 250 mL ammonium chloride (NH_4Cl) solution having concentration of 200 mg/L and 250 mL of nutrient solution in 3000 mL capacity cylindrical vessel. The composition of the nutrient solution in 250 mL distilled water was comprised of 60.0 mg K_2HPO_4, 40.0 mg KH_2PO_4, 500.0 mg $MgSO_4 \cdot 7H_2O$, 710.0 mg $FeCl_3 \cdot 6H_2O$, 0.1 mg $ZnSO_4 \cdot 7H_2O$, 0.1 mg $CuSO_4 \cdot 5H_2O$, 8.0 mg $MnCl_2 \cdot 2H_2O$, 0.11 mg $(NH_4)_6Mo_7O_{24}$, 100.0 mg $CaCl_2 \cdot 2H_2O$, 200.0 mg $CoCl_2 \cdot 6H_2O$, 55.0 mg $Al_2(SO_4)_3 \cdot 16H_2O$, 150.0 mg H_3BO_3. Finally 800 mL volume of distilled water was added to liquid mixture to make a volume of 2 L and the mixture was continuously aerated with intermittent feeding with dextrose solution having concentrations of 1000 mg/L and ammonium chloride (NH_4Cl) having concentration of 200 mg/L as a carbon and nitrogen source, respectively. The acclimatization process was continued for an overall period of 90 days. The biomass growth was monitored by the magnitude of sludge volume index (SVI) and mixed liquor suspended solid (MLVSS) concentration

in the reactor. pH in the reactor was maintained in the range 6.8–7.5 by adding required amount of sodium carbonate (Na_2CO_3) and phosphate buffer. The seed acclimatization phase was considered to be over when a steady-state condition was observed in terms of equilibrium COD and NH_4^+-N reduction with respect to a steady level of MLVSS concentration and SVI in the reactor.

Denitrifying seed was cultured separately in 2.0 L capacity aspirator bottle under anoxic condition. 500 gm of digested sludge obtained from the digester of a nearby sewage treatment plant (STP) was added to 1.0 L of distilled water. The solution was filtered and 250 mL of nutrient solution along with 250 mL dextrose solution as carbon source and 100 mL potassium nitrate solution (KNO_3) as the source of nitrate nitrogen (NO_3^--N) was added to it. The resulting solution was acclimatized for denitrification purpose under anoxic condition. Magnetic stirrer was provided for proper mixing of the solution. Denitrifying seed was acclimatized against a nitrate-nitrogen concentration varying from 10–90 mg/L as N, over a period of three months.

2.2. Experimental Procedure.

The experimental work was carried out in a laboratory scale SBR, made of Perspex sheet of 6 mm thickness, having 20.0 L of effective volume. In order to assess the treatability of slaughterhouse wastewater in an SBR, the real-life wastewater samples were collected from two different locations (i) the raw (untreated) wastewater from the main collection pit and (ii) the primary treated effluent from the inlet box of aeration basin. The wastewater samples were collected 4 (four) times over the entire course of the study in 25.0 L plastic containers and stored in a refrigerator at approximately 4.0°C. The effluent quality was examined as per the methods described in "Standard Methods" [35] for determining its initial characteristics which are exhibited in Table 2.

The settled effluent was poured in the reactor of 20.0 L capacity to perform necessary experiments. 2.5 L of preacclimatized mixed seed containing carbonaceous bacteria, nitrifier, and denitrifier was added in the reactor containing 20.0 L of pretreated slaughterhouse wastewater to carry out the necessary experiments. Oxygen was supplied through belt-driven small air compressor. A stirrer of 0.3 KW capacity was installed at the center of the vessel for mixing the content of the reactor. Air was supplied to the reactor during aerobic phase of react period with the help of diffused aeration system. However, during the anoxic phase the stirrer was allowed only to operate for mixing purpose and air supply was cut off. A timer was also connected to compressor for controlling the sequence of different react period (aerobic and anoxic). A schematic diagram of the experimental setup is shown in Figure 1.

The cycle period for the operation of SBR was taken as 10 hour, with a fill period of 0.5 hour, overall react period of 8.0 hours, settle period of 1.0 hour, and idle/decant period of 0.5 hour. The overall react period was divided into aerobic and anoxic react period in the following sequences:

Combination-1: 4-hour aerobic react period and 4-hour anoxic react period.

TABLE 2: Characteristics and composition of slaughterhouse wastewater.

Parameters	Raw wastewater					Pretreated wastewater				
	Set-1	Set-2	Set-3	Set-4	Range	Set-1	Set-2	Set-3	Set-4	Range
pH	8.0	8.2	8.5	8.4	8.0–8.5	7.5	7.2	8.5	7.8	7.5–8.5
TSS (mg/L)	10120	12565	14225	13355	10120–14225	2055	2280	2540	2386	2055–2540
TDS (mg/L)	6345	7056	7840	6865	6345–7840	2800	3065	3230	3185	2800–3230
DO (mg/L)	0.8	1.1	0.9	1.3	0.8–1.3	1.2	1.4	1.5	1.6	1.2–1.6
SCOD (mg/L)	6185	6525	6840	6455	6185–6840	830	945	1045	925	830–1045
BOD$_5$ at 20°C (mg/L)	3000	3200	3500	3350	3000–3500	210	240	265	252	210–265
TKN (mg/L as N)	1050	1130	1200	1165	1050–1200	305	420	525	485	305–525
NH$_4^+$ N (mg/L as N)	650	695	735	710	650–735	95	155	191	125	95–191

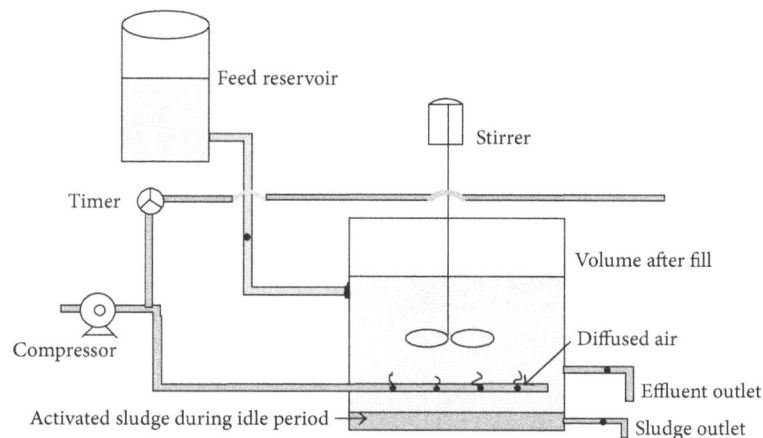

FIGURE 1: A schematic diagram of the experimental setup.

Combination-2: 5-hour aerobic react period and 3-hour anoxic react period.

Combination-3: 3-hour aerobic react period and 5-hour anoxic react period.

The performance study was carried out with pretreated slaughterhouse wastewater with same initial soluble chemical oxygen demand (SCOD) and two different ammonia nitrogen (NH$_4^+$-N) concentration of 1000 ± 50 mg/L and 90 ± 10 mg/L, 1000±50 mg/L and 180±10 mg/L, respectively. During the fill period of 30 min duration, 16.0 L of slaughterhouse wastewater was transferred under gravity from a feeding tank into the reactor. The mechanical mixer was operated continuously with a speed of 400 rpm from the beginning of the fill phase till the end of the total react phase for proper mixing of liquid in the reactor. During the draw phase, the supernatant wastewater was decanted until the liquid volume in the reactor was decreased to 4.0 L. sludge retention time (SRT) was manually controlled by withdrawal of volume of the mixed liquor from the reactor every day at the onset of the commencement of settle phase. The reactor was continuously run for 120 days. The initial pH values in the reactor were kept in between 7.1 and 7.7, whereas the sludge volume index (SVI) has been kept within the range of 75–85 mL/gm, for obtaining good settling property of the biomass. It has been reported that SRT should be longer than 10 days to achieve efficient nitrogen removal [36]. The SRT of 20–25 days as maintained for carbon oxidation and nitrification in the present SBR

system for treatment of wastewater as suggested by Tremblay et al. [37].

During the time course of the study, 100 mL of sample was collected from the outlet of the reactor at every 1.0 hour interval, till completion of the fill period. The samples were analyzed for the following parameters: pH, DO, MLSS, MLVSS, COD, NH$_4^+$-N, NO$_2^-$-N, and NO$_3^-$-N as per the methods described in "Standard Methods" [35]. The pH of the solution was measured by a digital pH meter. NH$_4^+$-N, NO$_2^-$-N, and NO$_3^-$-N were estimated by respective ion selective electrodes in ISE meter. COD was analyzed by closed reflux method using dichromate digestion principle in digester. Dissolved oxygen (DO) was measured electrometrically by digital DO meter. Mixed liquor suspended solids (MLSS) and Mixed liquor volatile suspended solids (MLVSS) were measured by gravimetric method at temperature of 103–105°C and 550 ± 50°C in muffle furnace, respectively.

2.3. Carbon Oxidation and Nitrification Kinetics in SBR. Biokinetic parameters play an important role in designing and optimizing an activated sludge process. The biokinetic constants describe the metabolic performance of the microorganisms when subjected to the substrate and other components of the specific wastewater. These biokinetic coefficients yield a set of realistic design parameters, which can be used in rationalizing the design of the activated sludge process for a specific substrate.

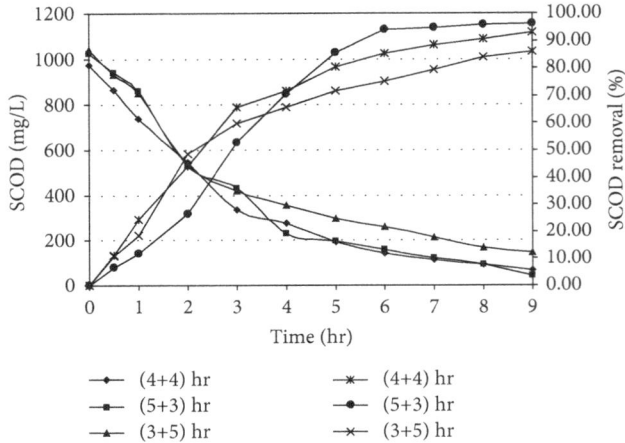

FIGURE 2: Carbon oxidation profile under different react period combination [Initial SCOD = 1000 ± 50 mg/L; Initial NH_4^+-N = 90 ± 10 mg/L as N].

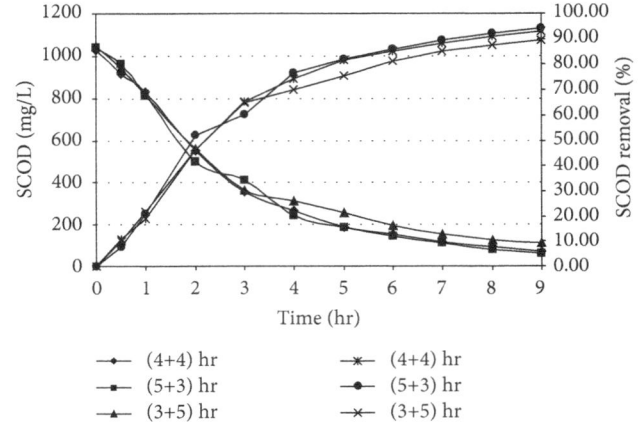

FIGURE 3: Carbon oxidation profile in SBR under different react period combination [Initial SCOD = 1000±50 mg/L; Initial NH_4^+-N = 180 ± 10 mg/L as N].

2.3.1. Substrate Removal Kinetics. The substrate removal constants, namely, half saturation concentration (K_s) and the maximum rate of substrate utilization (k) were determined from the Lawrence and McCarty's modified Monod equation [38] given below:

$$\frac{1}{U} = \left(\frac{K_s}{k}\right)\left(\frac{1}{S}\right) + \frac{1}{k} \qquad (1)$$

S = Substrate (SCOD and NH_4^+-N) concentration at any time in reactor (mg/L), U = Specific substrate utilization rate = $(S_0 - S)/\theta X$ (mg of SCOD or mg of NH_4^+-N/day/mg of MLVSS), θ = Contact time (day), X = MLVSS at any time in the reactor (mg/L), S_0 = Substrate (SCOD and NH_4^+-N) concentration of the influent (mg/L).

The plots made between $1/U$ and $1/S$ develops into a straight line with K_s/k as its slope and $1/k$ as its intercept.

2.3.2. Sludge Growth Kinetics. The sludge growth kinetic constants namely the yield coefficient (Y) and the endogenous decay coefficient (K_d), were determined from the Lawrence and McCarty's modified Monod equation [38] given below:

$$\frac{1}{\theta} = YU - K_d, \qquad (2)$$

where U = Specific substrate utilization rate (mg of SCOD or mg of NH_4^+-N/day/mg of MLVSS), θ = Contact time (day), k_d = Endogenous decay coefficient (day^{-1}), and Y = Yield coefficient (mg of MLVSS produced/mg of SCOD or NH_4^+-N).

A graph drawn between $1/\theta$ and U gives a straight line, with Y as its slope and k_d as its intercept.

2.4. Denitrification Kinetics in SBR. In almost all cases denitrification started occurring at the onset of anoxic period and specific denitrification rate (q_{DN}) was calculated under different initial organic carbon and NH_4^+-N concentrations for different react period combinations, namely, (4+4), (5+3), (3+5) hrs over the respective anoxic environment.

3. Results and Discussion

3.1. Carbon Oxidation Performance. Organic carbon, which is the source of energy for heterogenic and denitrifying microorganism, was estimated as chemical oxygen demand (COD). In the present experiment, in correspondance to an initial SCOD of 975.25 mg/L and initial NH_4^+-N concentration of 87.52 mg/L as N, it has been observed that the major fraction of SCOD removal took place within 4 or 5 hrs of aerobic react period. In anoxic phase, further SCOD removal has been noticed as shown in Figure 2. Li et al. [39] obtained that the maximum removal efficiency of COD (96%) for treatment of slaughterhouse wastewater which was marginally higher than the result of this present study. COD removal profile was also observed in similar pattern in the presence of higher initial NH_4^+-N concentration of 185.24 mg/L as N and initial SCOD of 1028.55 mg/L in a separate set of experiment. The results are plotted in Figure 3. It is revealed from Figures 2 and 3 that the rate of organics utilization by the dominant heterotrophs during initial aerobic react period was more as compared to its rate of removal during anoxic period. The carbon utilization bacteria used up bulk amount COD for energy requirement and growth. The removal efficiency of COD in the suspended growth reactor system depends on COD : TKN ratio. The mean COD : TKN ratio recommended for adequate carbon oxidation and nutrient removal as 10–12 [40]. In the present investigation, COD : TKN ratio was approximately 11.14 which was in agreement in their recommendation. The removal efficiency also depends on react time. The carbon utilizing bacteria obviously and is able to degrade more COD and produce CO_2 with production of new cells due to enhancement of aeration time. A marked improvement has been noticed for higher percentage removal of COD during increase of aeration time. A similar observation was noticed by Kanimozhi and Vasudevan [41]. Due to the increase of time and COD load more cells to be produced eventually higher degree of organic removal. When the react period was changed into 5-hour aerobic followed by a reduced 3.0 hour anoxic, a marginal improvement of SCOD

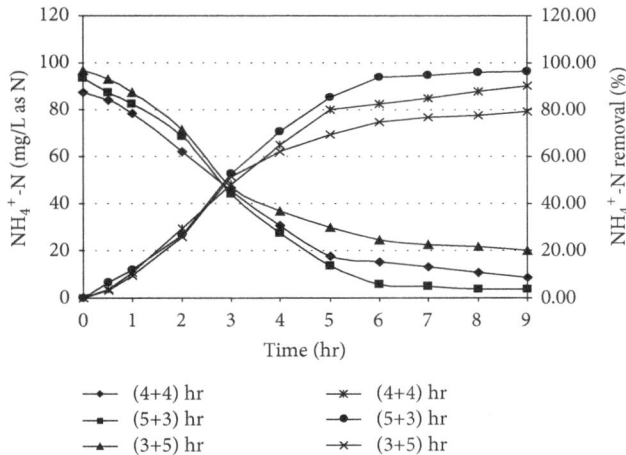

FIGURE 4: Ammonia oxidation profile in SBR under different react period combination [Initial SCOD = 1000±50 mg/L; Initial NH_4^+-N = 90 ± 10 mg/L as N].

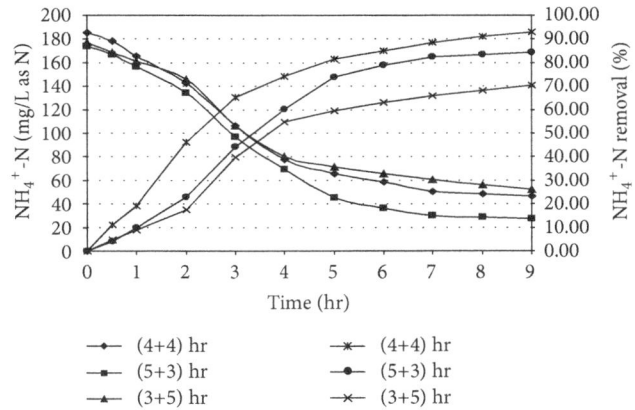

FIGURE 5: Ammonia oxidation profile in SBR under different react period combination [Initial SCOD = 1000±50 mg/L; Initial NH_4^+-N = 180 ± 10 mg/L as N].

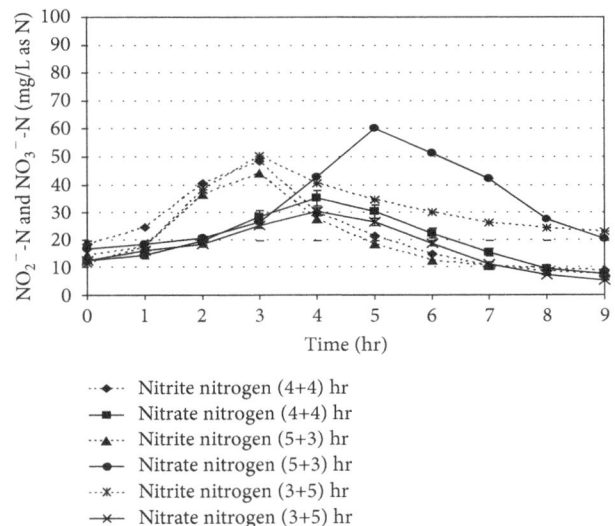

FIGURE 6: Nitrite and nitrate concentration profiles in SBR under different react period combination [initial SCOD = 1000 ± 50 mg/L and initial NH_4^+-N = 90 ± 10 mg/L as N].

removal in aerobic phase (77.27%) and anoxic phase (96.07%) with an initial SCOD of 1023.22 mg/L was observed due to enhanced aeration time. On the other hand, when the react period was subsequently changed to 3.0 hour, aerobic period followed by 5.0-hour anoxic period, a marginal decrease of SCOD removal in aerobic phase (65.64%) and anoxic phase (86.07%) with an initial SCOD of 1042.52 mg/L was obtained due to lag of aeration time.

3.2. Nitrification Performance. Ammonia oxidation took place due to the presence of previously acclimatized nitrifying organisms within the reactor as mixed culture. The nitrification results are shown in Figures 4 and 5. In case of specific cycle period of 4 hr (aerobic) and 4 hr (anoxic), it was observed that at the end of 8 hr react period of reaction, 90.12% nitrification could achieved for an initial NH_4^+-N was approximately 87.52 mg/L as Fongsatitkul et al. [40] obtained maximum 93% removal efficiency of soluble nitrogen for treatment of abattoir wastewater which was slightly higher than our result. The ammonia oxidation occurred in two phases; a fraction of ammonia was assimilated by cell-mass for synthesis of new cell during carbon oxidation and, in the subsequent phase, dissimilatory ammonia removal took place for converting NH_4^+-N into NO_2^--N and NO_3^--N under aerobic period. The dissimilatory removal of ammonia depends on the population of nitrifiers and oxidation time. The descending trend of ammonia removal for higher level of initial concentration of NH_4^+-N was attributed due to limitation of enzymatic metabolism of nitrifiers. When the reactor system was operated in 5 hr (aerobic) and 3 hr (anoxic) mode of react cycle, an overall performance of ammonia oxidation was improved from 90.12 to 96.20% and 84.41% for initial NH_4^+-N of 93.54 mg/L and 173.88 mg/L as N, respectively. Higher oxidation period was also recommended by earlier investigators [42, 43] for describing similar kind of experiment on landfill leachate treatment in SBR. The results reveal the fact that the extension of aeration period helped to enhance the oxidation efficiency for the present system. It was also observed that when aerobic period was reduced to 3.0 hr, ammonia oxidation reduced to 79.18% and 70.53% corresponding to initial NH_4^+-N value of 96.58 mg/L and 176.85 mg/L, respectively, at the end of 8 hr react period.

3.3. Denitrification Performance. The nitrite and nitrate nitrogen (NO_2^--N and NO_3^--N) level in the reactor during the total reaction period is shown in Figure 6. The maximum nitrite level was observed in between 2.5 and 3.0 hr of react period. The peak nitrate (NO_3^-) level was found to be formed close to 4.0 hr of aeration period for (4+4) and (3+5) hr combinations of react period. A time lag of one hour for maximum nitrate formation was also noticed after the attainment of the maximum NO_2^--N level in the reactor. For (5+3) hr react period combination, the formation of NO_3^- showed

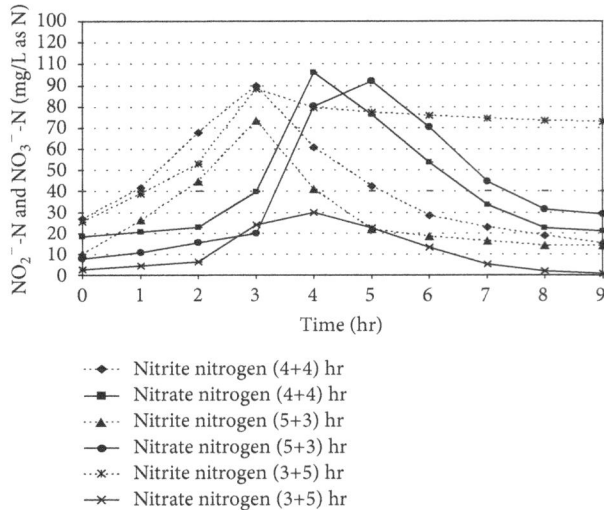

FIGURE 7: Nitrite and nitrate concentration profiles in SBR under different react period combination [initial SCOD = 1000 ± 50 mg/L and initial NH_4^+-N = 180 ± 10 mg/L as N].

a time-dependent factor as the peak was found at the end of 5.0 hrs. In the Figure 6, after 4.0 hr of aeration period, the NO_3^- level was found to be 35.21 mg/L as N corresponding to initial NH_4^+-N level of 87.52 mg/L as N and NO_3^- concentration of 12.35 mg/L as N, respectively. On the other hand, after 5.0 hour of aerated react period, NO_3^--N concentration in the reactor was found to be 60.24 mg/L as N for an initial NH_4^+-N and NO_3^--N concentration of 93.54 and 16.52 mg/L as N, respectively. The maximum NO_3^--N concentration for (3+5) hour react period combination was found to be 25.31 mg/L as N for the initial NH_4^+-N concentration of 96.58 mg/L as N and NO_3^--N level of 12.35 mg/L as N. The experimental results clearly indicate the necessity of longer aeration period for achieving maximum utilization of ammonia by the nitrifiers.

In Figure 7, after 4.0 hour of anoxic react period, nitrate (NO_3^-) was reduced to 22.29 mg/L as N from its peak concentration of 96.22 mg/L as N, which achieved a 76.83% removal of nitrate for initial NH_4^+-N concentration 185.24 mg/L as N. During denitrification phase, the residual soluble COD concentration as available was found to be more than the stoichiometric organic carbon requirement for effective denitrification. When the anoxic react period was reduced to 3.0 hr, it was observed that, nitrate concentration after 5 hr of aerobic period was found to be maximum (92.11 mg/L as N), per cent removal of nitrate descended from 76.83 to 66.16% for initial NH_4^+-N concentration 173.88 mg/L as N, due to insufficient of anoxic period.

3.4. MLVSS, pH, Alkalinity, and DO Profiles in the SBR during Experiment. The pH and alkalinity values of a biological system are vital parameters for microbial denitrification. The value of pH increases for ammonification and denitrification, decreases for organic carbon oxidation and nitrification. Alkalinity is not only important for nitrification and denitrification, but to also be used for indicating the system stability.

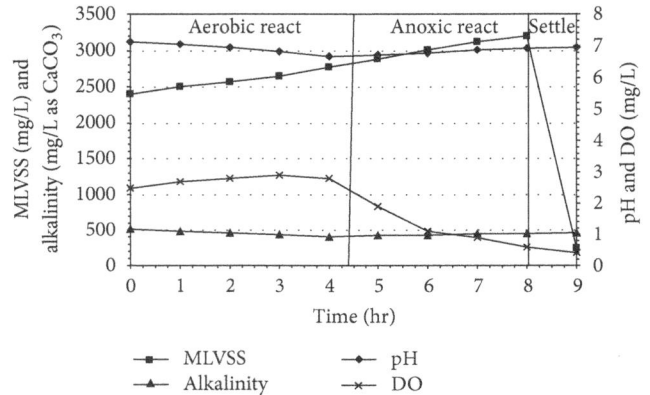

FIGURE 8: MLVSS, pH, alkalinity, and DO profiles for slaughterhouse wastewater treatment in SBR under (4+4) hr react period combination.

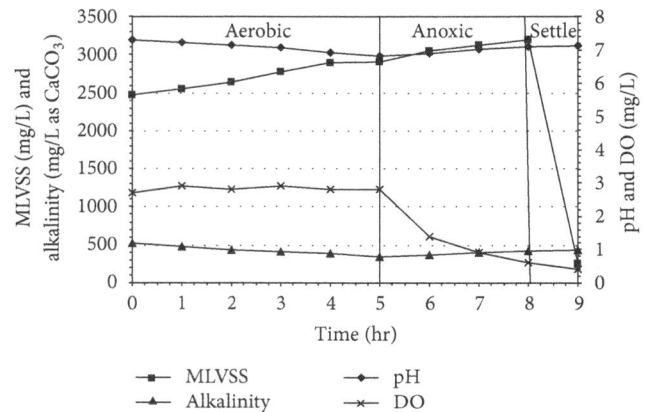

FIGURE 9: MLVSS, pH, alkalinity, and DO profiles for slaughterhouse wastewater treatment in SBR under (5+3) hr react period combination.

Alkalinity was found to have a close correlation with SBR operating conditions, since different extents of nitrification (alkalinity consumption) and denitrification (alkalinity production) contribute to the variation of alkalinity in the system. During the aerobic phase, the minimal value of the pH curve was characterized the end of nitrification (Figures 8, 9, and 10). At the beginning of anoxic react phase, when ammonia nitrogen concentration was reduced considerably, pH starts to increase. This has occurred between 4.0 and 5.0 hr after the starting of aerobic react period in all experimental sets. The DO profile exhibited a sharp fall after which DO concentration decreased markedly at anoxic phase and reached minimum value. In the present, study the DO level remained almost steady during the entire aerobic react period with a marginal increase in DO level, but a marked descending trend was observed during the anoxic period in all the reaction sets irrespective of initial SCOD and ammonia concentrations. Under strict anaerobic condition the DO should be equal to zero, but anoxic environment starts from DO level less than 1.5 mg/L. At the start of anoxic react period most of the cases, DO was found to be less than 1.5 mg/L and at the end of anoxic react period the value becomes less than 1.0 mg/L.

TABLE 3: Evaluation of biokinetic coefficients for carbon oxidation from slaughterhouse wastewater in SBR.

Initial SCOD (mg/L)	(4+4) hr react period combination	(5+3) hr react period combination	(3+5) hr react period combination	Standard values for kinetic constants [44]
1000 ± 50	(i) Substrate utilization- $y = 70.32x + 0.215$ (ii) Microbial growth- $y = 0.522x - 0.051$ *Kinetic constants:* k (day^{-1}) = 4.65 K_s (mg/L SCOD) = 327.06 Y (mg VSS/mg SCOD) = 0.522 k_d (day^{-1}) = 0.051	(i) Substrate utilization- $y = 68.22x + 0.187$ (ii) Microbial growth- $y = 0.622x - 0.057$ *Kinetic constants:* k (day^{-1}) = 5.34 K_s (mg/L SCOD) = 364.81 Y (mg VSS/mg SCOD) = 0.622 k_d (day^{-1}) = 0.057	(i) Substrate utilization- $y = 42.65x + 0.285$ (ii) Microbial growth- $y = 0.485x - 0.047$ *Kinetic constants:* k (day^{-1}) = 3.50 K_s (mg/L SCOD) = 149.64 Y (mg VSS/mg SCOD) = 0.485 k_d (day^{-1}) = 0.047	K (day^{-1}) = (2–10) K_s (mg/L SCOD) = (15–70) Y (mg VSS/mg SCOD) = (0.4–0.8) k_d (day^{-1}) = (0.025–0.075)

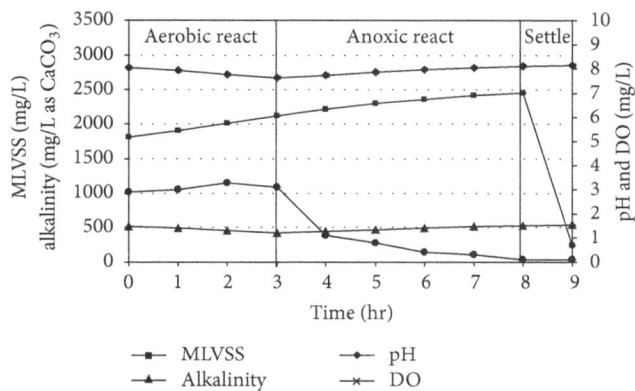

FIGURE 10: MLVSS, pH, alkalinity, and DO profiles for slaughterhouse wastewater treatment in SBR under (3+5) hr react period combination.

3.5. Kinetic Study for Organic Carbon Removal from Slaughterhouse Wastewater in SBR. In the present study, the performance evaluation of the SBR system was also carried out from the view point of reaction kinetics determination for treating slaughterhouse wastewater. The values for the reciprocal of specific substrate utilization rate $(1/U)$ were plotted against the reciprocal of effluent SCOD $(1/S)$ and substrate removal kinetics was evaluated using (1) as stated earlier. The slope of the straight line is (K_s/k) and intercept is $(1/k)$. The reciprocal of the contact time $(1/\theta)$ were plotted against the specific substrate utilization rate (U) and microbial growth kinetics was evaluated using (2). The yield coefficient (Y) was determined from the slope of the line and the endogenous decay coefficient (k_d) was obtained from intercept, $k_d = -C$. The values of biokinetic coefficients (k, K_s, Y, k_d) for combined carbon-oxidation and nitrification are listed in Table 3.

From Table 3, it has been estimated that the value of yield coefficient (Y) for the heterotrophs is in the range from 0.485 to 0.622. The yield coefficient was found to be improved with the increase in aeration period. The half velocity constant (K_s) values were found in the range of 149.64 to 364.81 for different combinations of react period. In the case of (5+3) combination of react cycle, the k and Y values are marginally higher than (4+4) and (3+5) combination. It was attributed to the fact that, after the initial acclimatization; the heterotrophs

converted the carbon content at 5.0 hrs period of time more efficiently. After 5.0 hrs of aerobic react period, the available carbon content was reduced considerably and a fraction of heterotrophs attained endogenous state of condition while the nitrifiers are rejuvenated and started nitrification activity. This metabolism is also supported by the value of endogenous decay rate constant (k_d). In the case of (5+3) combination of react cycle k_d value is found to be 0.057 which is between 0.051 and 0.047 for the cases of (4+4) and (3+5) react period combinations, respectively. The values of biokinetic coefficients, other than K_s, such as k, Y, k_d as obtained from the test results for carbon-oxidation and nitrification are also in congruence with their respective typical values [44].

3.6. Kinetic Study for Ammonium Nitrogen Removal from Slaughterhouse Wastewater in SBR. The nitrification removal kinetics for mixed population (heterotrophs and nitrifiers) followed an identical pattern to organic carbon removal kinetics. A fraction of biological oxidation was attributed to the fact that a mixed population performed in the reactor along with nitrifiers. The linear graphs are plotted between $(1/S)$ and $(1/U)$ for substrate utilization kinetics under three different combinations of react period, namely, (4+4), (5+3), and (3+5), respectively, using (1). Microbial growth kinetics was evaluated using (2), which were determined by plotting straight lines between $(1/\theta)$ and (U) under three different combinations of react period, namely, (4+4), (5+3) and (3+5) hrs, respectively. The kinetic coefficient values for nitrification from the previous plots are given in Table 4. It has been clearly shown earlier that an increasing trend of higher removal efficiency for ammonia oxidation could be observed for extension of the aerobic react period beyond 4 hrs. This previous phenomenon also reflected the magnitudes of biokinetic constants under all experimental combinations of react period. The kinetic coefficients Y, k_d, and K_s were found to be in the range of 0.205 to 0.284, 0.037 to 0.051, and 21.83 to 70.93, respectively. The ammonia concentration found in the slaughterhouse wastewater was very high 180 ± 10 mg/L as N for an inlet SCOD concentration of 1000 ± 50 mg/L, which are not usually present in any municipal wastewater stream. For this reason, the K_s value was found to be higher than the standard values (0.2–5.0 mg/L) considered for nitrification of municipal wastewater stream [44].

TABLE 4: Evaluation of biokinetic coefficients for nitrification from slaughterhouse wastewater in SBR.

Initial NH_4^+-N (mg/L as N)	(4+4) hr react period combination	(5+3) hr react period combination	(3+5) hr react period combination	Standard values for kinetic constants [44]
180 ± 10	(i) Substrate utilization- $y = 2.371x + 0.047$ (ii) Microbial growth- $y = 0.234x - 0.047$ k (day^{-1}) = 21.27 K_s (mg/L NH_4^+-N) = 50.44 Y (mg VSS/mg NH_4^+-N) = 0.234 k_d (day^{-1}) = 0.047	(i) Substrate utilization- $y = 2.412x + 0.034$ (ii) Microbial growth- $y = 0.284x - 0.051$ k (day^{-1}) = 29.41 K_s (mg/L NH_4^+-N) = 70.93 Y (mg VSS/mg NH_4^+-N) = 0.284 k_d (day^{-1}) = 0.051	(i) Substrate utilization- $y = 1.223x + 0.056$ (ii) Microbial growth- $y = 0.205x - 0.037$ k (day^{-1}) = 17.85 K_s (mg/L NH_4^+-N) = 21.83 Y (mg VSS/mg NH_4^+-N) = 0.205 k_d (day^{-1}) = 0.037	k (day^{-1}) = (1–30) K_s (mg/L NH_4^+-N) = (0.2–5.0) Y (mg VSS/mg NH_4^+-N) = (0.1–0.3) k_d (day^{-1}) = (0.03–0.06)

TABLE 5: Denitrification rates during anoxic react phase for treatment of slaughterhouse wastewater in SBR.

Initial NH_4^+-N (mg/L as N)	Initial SCOD (mg/L)	React period combination (Aerobic/Anoxic)	Avg. anoxic SCOD utilization rate (q_{SCOD}) (mg SCOD/gm MLVSS. hr)	Specific denitrification rate (q_{DN}) (mg N/gm MLVSS. hr)							
				1.0 hr	2.0 hr	3.0 hr	4.0 hr	5.0 hr	Avg. (3.0 hrly)	Avg. (4.0 hrly)	Avg. (5.0 hrly)
185.24	1028.55	(4+4)	26.25	4.49	5.57	5.85	3.89	—	5.30	4.95	—
173.88	1023.22	(5+3)	34.87	4.27	5.51	4.16	—	—	4.64	—	—
176.85	1042.52	(3+5)	38.15	4.57	5.55	6.16	7.24	6.23	5.42	5.88	5.95

3.7. Denitrification Rates for Treatment of Slaughterhouse Wastewater in SBR. Specific denitrification rate (q_{DN}) was measured in terms of the rate of NO_3^--N removed per unit mass of denitrifying microorganisms, for three different react period combinations, namely, (4+4), (5+3), and (3+5) under the respective anoxic environment and the results are listed in Table 5. The specific denitrification rate (q_{DN}) is expressed on average basis spanningover respective anoxic periods of 3.0, 4.0, and 5.0 hours. The average specific denitrification rate (q_{DN}), in (5+3), (4+4), and (3+5) cases was found to increase considerably with the increase in average anoxic SCOD utilization rate (q_{SCOD}) when primary treated effluent was considered for treatment in present SBR system. Average specific denitrification rate (q_{DN}) varied from 4.64 to 5.42 mg of N/gm MLVSS. hr for primary treated slaughterhouse wastewater for 3 hr anoxic period. The average 4.0 hourly specific denitrification rate (q_{DN}) varied from 4.95 to 5.88 mg of N/gm MLVSS. hr. The previous rate of specific denitrification rate (q_{DN}) was found to be followed in similar results as reported by Barnes and Bliss [45].

4. Conclusions

The present experimental investigation demonstrated that sequential batch reactor (SBR) is a variable and efficient biological method to treat slaughterhouse wastewater in a single unit. The total react period of 8 hr (4 hr aerobic and 4 hr anoxic) yielded optimum carbon oxidation, nitrification, and denitrification for treatment of carbonaceous and nitrogenous wastewater. The increase in MLVSS level in the reactor exhibited the growth favoring environment of the microorganism. The pH level in the SBR descended initially during aerobic period due to nitrification and carbon oxidation followed by an increasing trend indicating the existence of denitrifiers. This phenomenon has also been established by the variation of alkalinity level during aerobic and anoxic react period. The estimated values of biokinetic coefficients (k, K_s, Y, k_d) showed reasonable agreement with the literature values. The kinetic data and rate reaction constants could be used for the design of a field scale SBR for treating slaughterhouse wastewater. A design rationale can be evaluated on the basis of present experimental data for the purpose of application of this technology in similar plants. The outcome of the present investigation results would be helpful for making a design rationale for SBR treatment of slaughterhouse wastewater and a pilot plant study can be conducted with real-life wastewater sample by application of derived data of present study. In the future scope of the study, microbial genomics study including phosphate removal aspects would be also considered. The influence of solid retention time (SRT) should be explored also. A real-time kinetics profile with automatic data plotting could be derived for explaining the process in more rational way. It is also suggested that optimization of the process and operation variable may be examined with soft computing tools using various statistical approach.

Acknowledgment

This study was supported by the Research Funds of Jadavpur University, Jadavpur, Kolkata, India.

References

[1] A. O. Akinro, I. B. Ologunagba, and O. Yahaya, "Environmental implications of unhygienic operation of a city abattoir in Akure, Western Nigeria," *ARPN Journal of Engineering and Applied Sciences*, vol. 4, no. 9, pp. 311–315, 2009.

[2] V. P. Singh and S. Neelam, "A survey report on impact of abattoir activities management on environments," *Indian Journal of Veterinarians*, vol. 6, pp. 973–978, 2011.

[3] S. M. Gauri, "Treatment of wastewater from abattoirs before land application: a review," *Bioresource Technology*, vol. 97, no. 9, pp. 1119–1135, 2006.

[4] Y. O. Bello and D. T. A. Oyedemi, "Impact of abattoir activities and management in residential neighbourhoods: a case study of Ogbomoso, Nigeria," *Journal of Social Science*, vol. 19, pp. 121–127, 2009.

[5] D. Muhirwa, I. Nhapi, U. Wali, N. Banadda, J. Kashaigili, and R. Kimwaga, "Characterization of wastewater from an abattoir in Rwanda and the impact on downstream water quality," *International Journal of Ecology, Development*, vol. 16, no. 10, pp. 30–46, 2010.

[6] A. O. Aniebo, S. N. Wekhe, and I. C. Okoli, "Abattoir blood waste generation in rivers state and its environmental implications in the Niger Delta," *Toxicological and Environmental Chemistry*, vol. 91, no. 4, pp. 619–625, 2009.

[7] O. Chukwu, "Analysis of groundwater pollution from abattoir Waste in Minna, Nigeria," *Research Journal of Diary Science*, vol. 2, pp. 74–77, 2008.

[8] J. Grady, G. Daigger, and H. Lim, *Biological Wastewater Treatment*, Marcel Dekker, New York, NY, USA, 1999.

[9] I. Ruiz, M. C. Veiga, P. de Santiago, and R. Blfizquez, "Treatment of slaughterhouse wastewater in a UASB reactor and an anaerobic filter," *Bioresource Technology*, vol. 60, no. 3, pp. 251–258, 1997.

[10] D. I. Massé and L. Masse, "Treatment of slaughterhouse wastewater in anaerobic sequencing batch reactors," *Canadian Agricultural Engineering*, vol. 42, no. 3, pp. 131–137, 2000.

[11] K. Sombatsompop, A. Songpim, S. Reabroi, and P. Inkongngam, "A comparative study of sequencing batch reactor and movingbed sequencing batch reactor for piggery wastewater treatment," *Maejo International Journal of Science and Technology*, vol. 5, no. 2, pp. 191–203, 2011.

[12] Y. Rahimi, A. Torabian, N. Mehrdadi, and B. Shahmoradi, "Simultaneous nitrification-denitrification and phosphorus removal in a fixed bed sequencing batch reactor (FBSBR)," *Journal of Hazardous Materials*, vol. 185, no. 2-3, pp. 852–857, 2011.

[13] E. Bazrafshan, F. K. Mostafapour, M. Farzadkia, K. A. Ownagh, and A. H. Mahvi, "Slaughterhouse wastewater treatment by combined chemical coagulation and electrocoagulation process," *PLOS ONE*, vol. 7, no. 6, 2012.

[14] R. Rajakumar, T. Meenambal, P. M. Saravanan, and P. Ananthanarayanan, "Treatment of poultry slaughterhouse wastewater in hybrid upflow anaerobic sludge blanket reactor packed with pleated poly vinyl chloride rings," *Bioresource Technology*, vol. 103, no. 1, pp. 116–122, 2012.

[15] G. C. Sunder and S. Satyanarayan, "Efficient treatment of slaughter house wastewater by anaerobic hybrid reactor packed with special floating media," *International Journal of Chemical and Physical Sciences*, vol. 2, pp. 73–81, 2013.

[16] European Commission, *Integrated Pollution Prevention and Control: Reference Document on Best Available Techniques in the Slaughterhouses and Animal by-Products Industries*, 2005.

[17] A. H. Mahvi, "Sequencing batch reactor: a promising technology in wastewater treatment," *Iranian Journal of Environmental Health Science and Engineering*, vol. 5, no. 2, pp. 79–90, 2008.

[18] N. Z. Al-Mutairi, M. F. Hamoda, and I. A. Al-Ghusain, "Slaughterhouse wastewater treatment using date seeds as adsorbent," *Journal of Environment Science and Health*, vol. 55, pp. 678–710, 2007.

[19] R. Boopathy, C. Bonvillain, Q. Fontenot, and M. Kilgen, "Biological treatment of low-salinity shrimp aquaculture wastewater using sequencing batch reactor," *International Biodeterioration and Biodegradation*, vol. 59, no. 1, pp. 16–19, 2007.

[20] H. S. Kim, Y. K. Choung, S. J. Ahn, and H. S. Oh, "Enhancing nitrogen removal of piggery wastewater by membrane bioreactor combined with nitrification reactor," *Desalination*, vol. 223, no. 1–3, pp. 194–204, 2008.

[21] N. Z. Al-Mutairi, F. A. Al-Sharifi, and S. B. Al-Shammari, "Evaluation study of a slaughterhouse wastewater treatment plant including contact-assisted activated sludge and DAF," *Desalination*, vol. 225, no. 1–3, pp. 167–175, 2008.

[22] D. Roy, K. Hassan, and R. Boopathy, "Effect of carbon to nitrogen (C:N) ratio on nitrogen removal from shrimp production waste water using sequencing batch reactor," *Journal of Industrial Microbiology and Biotechnology*, vol. 37, no. 10, pp. 1105–1110, 2010.

[23] R. Rajagopal, N. Rousseau, N. Bernet, and F. Béline, "Combined anaerobic and activated sludge anoxic/oxic treatment for piggery wastewater," *Bioresource Technology*, vol. 102, no. 3, pp. 2185–2192, 2011.

[24] J. Palatsi, M. Vinas, M. Guivernau, B. Fernandez, and X. Flotats, "Anaerobic digestion of slaughterhouse waste: main process limitations and microbial community interactions," *Bioresource Technology*, vol. 102, no. 3, pp. 2219–2227, 2011.

[25] C. Kern and R. Boopathy, "Use of sequencing batch reactor in the treatment of shrimp aquaculture wastewater," *Journal of Water Sustainability*, vol. 2, no. 4, pp. 221–232, 2012.

[26] F. Wang, Y. Liu, J. Wang, Y. Zhang, and H. Yang, "Influence of growth manner on nitrifying bacterial communities and nitrification kinetics in three lab-scale bioreactors," *Journal of Industrial Microbiology and Biotechnology*, vol. 39, no. 4, pp. 595–604, 2012.

[27] D. Obaja, S. Mac, and J. Mata-Alvarez, "Biological nutrient removal by a sequencing batch reactor (SBR) using an internal organic carbon source in digested piggery wastewater," *Bioresource Technology*, vol. 96, no. 1, pp. 7–14, 2005.

[28] K. V. Lo and P. H. Liao, "A full-scale sequencing batch reactor system for swine wastewater treatment," *Journal of Environmental Science and Health*, vol. 42, no. 2, pp. 237–240, 2007.

[29] A. H. Mahvi, A. R. Mesdaghinia, and F. Karakani, "Nitrogen removal from wastewater in a continuous flow sequencing batch reactor," *Pakistan Journal of Biological Sciences*, vol. 7, no. 11, pp. 1880–1883, 2004.

[30] R. Lemaire, Z. Yuan, B. Nicolas, M. Marcelino, G. Yilmaz, and J. Keller, "A sequencing batch reactor system for high-level biological nitrogen and phosphorus removal from abattoir wastewater," *Biodegradation*, vol. 20, no. 3, pp. 339–350, 2009.

[31] S. Q. Aziz, H. A. Aziz, M. S. Yusoff, and M. J. K. Bashir, "Landfill leachate treatment using powdered activated carbon augmented sequencing batch reactor (SBR) process: optimization by response surface methodology," *Journal of Hazardous Materials*, vol. 189, no. 1-2, pp. 404–413, 2011.

[32] G. Durai, N. Rajamohan, C. Karthikeyan, and M. Rajasimman, "Kinetics studies on biological treatment of tannery wastewater using mixed culture," *International Journal of Chemical and Biological Engineering*, vol. 3, no. 2, pp. 105–109, 2010.

[33] M. Faouzi, M. Merzouki, and M. Benlemlih, "Contribution to optimize the biological treatment of synthetic tannery effluent by the sequencing batch reactor," *Journal of Materials and Environmental Science*, vol. 4, no. 4, pp. 532–541, 2013.

[34] S. Dey and S. Mukherjee, "Performance and kinetic evaluation of phenol biodegradation by mixed microbial culture in a batch reactor," *International Journal of Water Resources and Environmental Engineering*, vol. 2, no. 3, pp. 40–49, 2010.

[35] American Public Health Association, American Water Works Association, Water Pollution Control Federation, *Standard Methods for the Examination of Water and Wastewater*, American Water Works Association, Washington, DC, USA, 20th edition, 1998.

[36] H. Furumai, A. Kazmi, Y. Furuya, and K. Sasaki, "Modeling long term nutrient removal in a sequencing batch reactor," *Water Research*, vol. 33, no. 11, pp. 2708–2714, 1999.

[37] A. Tremblay, R. D. Tyagi, and R. Y. Surampalli, "Effect of SRT on nutrient removal in SBR system," *Practice Periodical of Hazardous, Toxic, and Radioactive Waste Management*, vol. 3, no. 4, pp. 183–190, 1999.

[38] A. W. Lawrence and P. L. McCarty, "Unified basis for biological treatment design and operation," *Journal of the Sanitary Engineering Division*, vol. 96, no. 3, pp. 757–778, 1970.

[39] J. P. Li, M. G. Healy, X. M. Zhan, and M. Rodgers, "Nutrient removal from slaughterhouse wastewater in an intermittently aerated sequencing batch reactor," *Bioresource Technology*, vol. 99, no. 16, pp. 7644–7650, 2008.

[40] P. Fongsatitkul, D. G. Wareham, P. Elefsiniotis, and P. Charoensuk, "Treatment of a slaughterhouse wastewater: effect of internal recycle rate on chemical oxygen demand, total Kjeldahl nitrogen and total phosphorus removal," *Environmental Technology*, vol. 32, no. 15, pp. 1755–1759, 2011.

[41] R. Kanimozhi and N. Vasudevan, "Effect of organic loading rate on the performance of aerobic SBR treating anaerobically digested distillery wastewater," *Clean Technologies and Environmental Policy*, vol. 15, pp. 511–528, 2013.

[42] D. Kulikowska and E. Klimiuk, "Removal of organics and nitrogen from municipal landfill leachate in two-stage SBR reactors," *Polish Journal of Environmental Studies*, vol. 13, no. 4, pp. 389–396, 2004.

[43] J. Doyle, S. Watts, D. Solley, and J. Keller, "Exceptionally high-rate nitrification in sequencing batch reactors treating high ammonia landfill leachate," *Water Science and Technology*, vol. 43, no. 3, pp. 315–322, 2001.

[44] Metcalf and Eddy Inc., *Wastewater Engineering Treatment Disposal Reuse*, Tata McGraw-Hill, 3rd edition, 1995.

[45] D. P. Barnes and P. J. Bliss, *Biological Control of Nitrogen in Wastewater Treatment*, E. & F.N. Spon, London, UK, 1983.

6

Nonviral Gene Targeting at rDNA Locus of Human Mesenchymal Stem Cells

Youjin Hu, Xionghao Liu, Panpan Long, Di Xiao, Jintao Cun, Zhuo Li, Jinfeng Xue, Yong Wu, Sha Luo, Lingqian Wu, and Desheng Liang

State Key Laboratory of Medical Genetics, Central South University, 110 Xiangya Road, Changsha, Hunan 410078, China

Correspondence should be addressed to Desheng Liang; liangdesheng@sklmg.edu.cn

Academic Editor: Ken-ichi Isobe

Background. Genetic modification, such as the addition of exogenous genes to the MSC genome, is crucial to their use as cellular vehicles. Due to the risks associated with viral vectors such as insertional mutagenesis, the safer nonviral vectors have drawn a great deal of attention. *Methods*. VEGF, bFGF, vitamin C, and insulin-transferrin-selenium-X were supplemented in the MSC culture medium. The cells' proliferation and survival capacity was measured by MTT, determination of the cumulative number of cells, and a colony-forming efficiency assay. The plasmid pHr2-NL was constructed and nucleofected into MSCs. The recombinants were selected using G418 and characterized using PCR and Southern blotting. *Results*. BFGF is critical to MSC growth and it acted synergistically with vitamin C, VEGF, and ITS-X, causing the cells to expand significantly. The neomycin gene was targeted to the rDNA locus of human MSCs using a nonviral human ribosomal targeting vector. The recombinant MSCs retained multipotential differentiation capacity, typical levels of hMSC surface marker expression, and a normal karyotype, and none were tumorigenic in nude mice. *Conclusions*. Exogenous genes can be targeted to the rDNA locus of human MSCs while maintaining the characteristics of MSCs. This is the first nonviral gene targeting of hMSCs.

1. Introduction

Human mesenchymal stem cells (hMSCs) are an attractive source of adult stem cells for autologous cell and gene therapy. They have immunosuppressive property and ability to differentiate into multiple cell types present in several tissues [1–3], making them to be a promising cellular vehicle for gene therapy [4, 5]. Current methods most commonly used for genetically modifying hMSCs are based on random transgene integration; however, the uncertainty of the integration site brings problems. Random integration may take place in heterochromatin, leading to silencing [6], or in coding regions, causing disruption of an endogenous gene or interference in the transcription of neighboring sequences [7]. These issues can be addressed by gene targeting, a primary alternative method. Unfortunately, efficient gene targeting in hMSCs has been poorly developed. To our knowledge, only four cases of gene targeting in hMSCs have been reported to date and all of them were based on viral transfer methods. Due to safety concerns related to random integration, nonviral

gene targeting at appropriate transgene harbor deserves great attention, while no actual case based on nonviral delivery has been described in hMSCs.

Human cells have approximately 400 copies of a 45S ribosomal DNA (rDNA) repeat that encodes ribosomal RNA (rRNA) and is distributed over the short arm of the five acrocentric chromosomes 13, 14, 15, 21, and 22. The rRNA gene is transcriptionally active and produces approximately 80% of the total RNA in rapidly dividing cells [8]. It is well known that the human rDNA copy number variations are common among healthy individuals, and a balanced chromosomal translocation involving the rRNA cluster occurs without apparent phenotypic effect. The rDNA cluster exhibited strikingly variable lengths between and within human individuals and showed high intrinsic recombinational instability during both meiosis and mitosis [9]. In addition, the human rRNA gene cluster consists of hundreds copies of tandemly repeated rDNA units. Targeting an exogenous gene into one or a few of the rDNA repeats may not cause loss of function effects of the rRNA genes owing to high copy number of this gene. These properties indicate that the rDNA locus may hold a high intrinsic homologous recombination (HR) activity and this locus is considered to be an ideal safe locus for transgene integration [10, 11]. Several studies have reported that efficient targeted gene addition at the rDNA 28s locus could be achieved based on viral transfer methods. However, because the majority of the integration was randomly located in the genome, the risks of the random integration were unavoidable for these methods [10–12]. Alternatively, based on a nonviral vector, in our previous studies, gene targeting at the rDNA 18S locus had been achieved, and the transgene could be stably expressed ectopically in targeted cells [13, 14]. In this study, a nonviral gene-targeting vector was constructed and targeted gene addition in MSCs was performed at the rDNA locus.

In fact, MSCs are a rare population that comprises only approximately $0.001\% \sim 0.01\%$ of the total bone marrow mononuclear cells. In addition, gene-targeting frequency in MSCs is intrinsically low. The limited number of the targeted MSCs may be a hurdle for their therapeutic use. Therefore, MSCs genetically modified for clinical application typically require extensive expansion *in vitro*. Unlike embryonic stem cells that have an unlimited proliferative lifespan, adult MSCs *in vitro* display a restricted proliferative longevity, a diminishing replication capacity, and an increased loss of differentiation potential [15–17]. The role of growth factors in enhancing the proliferation and survival of MSCs has been widely studied over the past few years. Several suitable factors have been found to improve *ex vivo* expansion in MSCs without altering their stem cell phenotype and multipotent differentiation potentials, including fibroblast-growth-factor- (FGF-) 2, epidermal growth factor (EGF) [18], and platelet-derived-growth-factor- (PDGF-) BB [19–21]. The combined effects of the factors have appeared quite robust [17, 19–21]. The aim of this study was to target an exogenous gene at the rDNA locus of human MSCs using the nonviral rDNA-targeting vector. The proliferation capacity of the MSCs was improved significantly by adding growth factors. The exogenous neomycin (Neo) gene was targeted at the rDNA locus of MSCs. The recombinant MSCs were compared to control MSCs with respect to phenotype, plasticity, multipotency, karyotype, and tumorigenicity. This is the first report of any nonviral gene targeting of human MSCs and this method may be an optimal approach to MSC-based disease modeling and gene therapy.

2. Materials and Methods

2.1. Construction of pHr2-NL. The pHr2-NL plasmid contains a long homologous arm (LHA) corresponding to rDNA +937 to ++6523 and a short homologous arm (SHA) corresponding to rDNA +6523 to +7643. It was generated in five steps. Firstly, a fragment homologous to the rDNA +6523 to +7684 region was amplified from human genomic DNA with the primers (5′-AATCGATTTTGA-TATCTGAGGCAACCCCCTCTCCTCTTGGGC-3′/5′-GTCGCCGCCGGGGACACGCGAA-3′). A fragment homologous to the rDNA +5513 to +6523 was amplified

from human genomic DNA with the primers (5′-GCGGAAGGATCATTAACGGAGCCCGGA-3′/5′-ATAATCGATAGAGGAGAGGGGGTTGCCTCAGGCC-3′). The PCR products were cloned into pGEMT, resulting in pGEM-T-SHA and pGEM-T-LS, respectively. Secondly, the expression cassette ECMV-IRES-Neo-SV40PolyA was amplified from the pHrneo [14] plasmid with the primers (5′-ATAATCGATATAACTTCGTATAATGTATGCTAT-ACGAAGTTATTCTTAAGGAATTCCCCCTCTCCCT-3′/5′-ATAGATATCATAACTTCGTATAATGTATGC-TATACGAAGTTATTAGACGGTCGACCCGTGCGGA-3′). The PCR product (1.8 kb) was cloned into pGEMT, resulting in pGEM-T-INL. Thirdly, the 1.8 kb *ClaI/SacI* fragment from pGEM-T-LS was inserted into the *ClaI* and *SacI* sites of pGEM-T-SHA, generating the plasmid pGEM-T-LS-SHA. Next, the *ClaI/EcoRV* fragment from pGEM-T-INL was inserted into the *ClaI* and *EcoRV* sites of pGEM-T-LS-SHA, generating the plasmid pHr1-NL. Finally, the 5.6 kb *MfeI/SacI* fragment from the T-pHr (constructed previously [14], containing the long homologous arm corresponding to rDNA +937 to +6523) and the 5.5 kb *AAT II/ClaI* fragment from the pHr1-NL, treated with T4 DNA polymerase (FERMENTS) and fastAP thermo sensitive alkaline phosphatase (FERMENTS), respectively, were ligated to generate pHr2-NL.

2.2. Isolation of MSCs from Bone Marrow and Culture Conditions. Informed consent was obtained from all participants according to a protocol approved by the Ethics Committee of State Key Laboratory of Medical Genetics of China (no. 2010-HUMAN-004). Bone marrow was obtained from the iliac bones of two volunteers. MSCs were isolated by using a Histopaque-1077 density gradient (Sigma) as previously described [22, 23]. The cells were cultured in MSC basal medium (L-glucose Dulbecco minimum essential medium (DMEM-L, Sigma-Aldrich, China) supplemented with 10% foetal bovine serum (GIBCO), 100 U of sodium penicillin/mL, and 100 U of streptomycin sulphate/mL). The mononuclear cells were plated at a density of approximately 5×10^5 cells/cm^2. Symmetrical colonies became visible on days 5 to 7, and the cells were subcultured at a seeding density of 1×10^4 cells/cm^2. Growth factors including VEGF (5 ng/μL, PEPRO TECH 100–200), bFGF (10 ng/μL, Invitrogen PHG0263), vitamin C (Vc) (50 μg/mL, Sigma A4544), and ITS-X (insulin-transferrin-selenium-X) (100x Invitrogen 51500056) were added [19, 24, 25].

2.3. Fluorescence-Activated Cell Sorting (FACS) Analysis. Surface markers CD31, CD44, CD45, CD73, CD90, and CD105 (BD Pharmingen, Hunan, China) were analyzed according to the manufacturer's instructions.

2.4. Colony-Forming Unit-Fibroblast (CFU-F) Assay. MSC cells were reseeded at a concentration of 150 cells per 100 mm dish (2.7 cells/cm^2). After 14 days, cultures were stained with 0.5% crystal violet (Sigma). Colonies less than 2 mm in diameter and faintly stained colonies were ignored. Colony-forming efficiency was expressed as the relative number of colonies generated from the number of cells seeded.

2.5. Cell Proliferation Assay. Numbers of cells and cell viability were measured by counting cells on a hemocytometer using the Trypan Blue dye exclusion method. Cells were cultured in 96-well plates (1×10^4 cells/cm^2). Seventy-two hours later, the number of viable cells was determined using an MTT assay. The plates were analyzed using a microplate reader at 570 nm.

2.6. Stable Transfection in MSCs. MSCs were stably transfected with the plasmid pHr2-NL and linearized with *Ahd*I, by nucleofection using the c-17 pulsing program. The DNA/cells ration was 2μg DNA/5×10^5 cells. The transfected MSCs were plated in 100 mm dishes at a density of 1×10^3 cells/cm^2. Twenty-four hours after transfection, the medium was replaced with fresh medium. After culture for another 48 hours, 50μg/mL G418 was added to the culture medium. The medium was refreshed every third day, and the concentrations of G418 on days 3, 6, 9, and 12 were 200μg/mL, 100μg/mL, 100μg/mL, and 15μg/mL, respectively. Medium without G418 was added on day 14, and the drug-resistant cells were cultured without G418 for another 3 weeks. Individual colonies were picked and expanded, and the genomic DNA was extracted using PCR and Southern blotting to detect recombinants.

2.7. PCR Identification of the Site-Integration Colonies. The primer t-up (5'-GTTATCCGCTCACAATTCCACACA-ACATACGA-3') and the primer t-re (5'-GGAGGTCGG-GGGGACGGGTCCGAGGA-3') were used.

2.8. Sequencing the PCR Fragment. PCR products were isolated after migration on 0.8% LMP agarose gels, cloned into the pGEM-T vector, and sequenced with primer-T7 (5'-TAATACGACTCACTATAGGG-3') and primer-SP6 (5'-CATACGATTTAGGTGACACTATAG-3').

2.9. Southern Blotting. After overnight digestion with restriction enzymes *Pvu II, Nco I, EcoR I,* and *Hind III* (New England Biolabs, Ipswich, MA, USA), 3μg of genomic DNA per sample was electrophoresed on a 0.8% agarose gel overnight and then transferred to positively charged nylon membranes (Hybond-N+, Amersham, Piscataway, NJ, USA). DNA molecular weight marker III, digoxigenin-labeled DNA (Roche Diagnostics, Indianapolis, IN, USA), and lambda DNA *Hind III* (TaKaRa, Dalian, China) were used as molecular weight markers. The blots were hybridized with DIG-dUTP-labeled probes overnight at 42°C. After incubation with AP-conjugated DIG antibody (Roche Diagnostics, Indianapolis, IN, USA) and appropriate washing, the signals were detected using CDP-Star (Roche Diagnostics, Indianapolis, IN, USA) as a substrate for chemiluminescence. Probes were generated using DIG-High Prime (Roche Diagnostics, Indianapolis, IN, USA),

and the templates were generated using PCR amplification from pHr2-NL. The primers used for probe 1 (P1) were 5'-CCCGGAAACCTGGCCCTGTCTT-3' and 5'-CTTCGCCCAATAGCAGCCAGTC C-3', and primers for probe 2 (P2) were 5'-AATGGCCGCTTTTCTGGA-3' and 5'-TGTGATGCTATTGCTTTATTTGTA-3'.

2.10. Karyotyping. About 5×10^5 cells from each of the four targeted MSC colonies were treated with 0.08μg/mL colcemid (Sigma, St. Louis, MO, USA) for 2.5 hours. Then cells were trypsinized, centrifuged, and incubated in 0.075 M KCl for 30 minutes at 37°C. After fixing with Carnoy fixative, metaphase chromosome spreads were prepared using the air drying method. Thirty metaphase spreads were evaluated per colony.

2.11. In Vivo Implantation Assay. All animal protocols were approved by the Animal Ethics Committee of the State Key Laboratory of Medical Genetics of China. Twenty-four SCID mice were divided into four groups of six mice each. PBS and a total of 2×10^6 cells of each of the three cell types (heterogenous MSCs derived from the four targeted MSC colonies, wild-type MSCs, and HT1080) were injected subcutaneously over the right ribcage. The skin and underlying soft tissue of the relevant area were dissected, fixed in 4% paraformaldehyde, stained with hematoxylin and eosin, and investigated for possible tumor growth.

2.12. Differentiation Assays. The four MSC colonies subjected to site-specific integration (1-1, 1-2, 2-1, 2-2) were assessed for adipogenic, osteogenic, and chondrogenic potential. Assays of *in vitro* differentiation to osteocytes, chondrocytes, and adipocytes were performed using StemPro Osteogenesis Differentiation Kit, StemPro Chondrogenesis Differentiation Kit, and a StemPro Adipogenesis Differentiation Kit according to the manufacturer's protocol.

2.13. Statistical Analysis. Data sets were expressed as the mean value and standard deviation. The significance of colony-forming efficiency was determined using the Student's t-test. The viable cell numbers derived from media with different additives and the number of oil-red-O-positive cells derived from wild-type and targeted MSCs were analyzed using one-way ANOVA. Differences were considered significant at $P < 0.05$.

3. Results

3.1. Proliferation and Survival of MSCs Treated with Growth Factors. First, the effects of several growth factors, including bFGF, VEGF, Vc, and ITS-X were individually evaluated on the proliferation of MSCs. During a 5-day culture period bFGF significantly increased the number of viable cells relative to cells exposed to plain basal medium, but Vc, ITS-X, and VEGF did not (Figure 1(a)). The cumulative numbers of MSCs cultured in the mediums supplemented with bFGF, Vc, VEGF, and ITS-X were 1.47-, 1.05-, 0.78-, and 1.13-fold higher than those in the basal medium. Next, the effects of

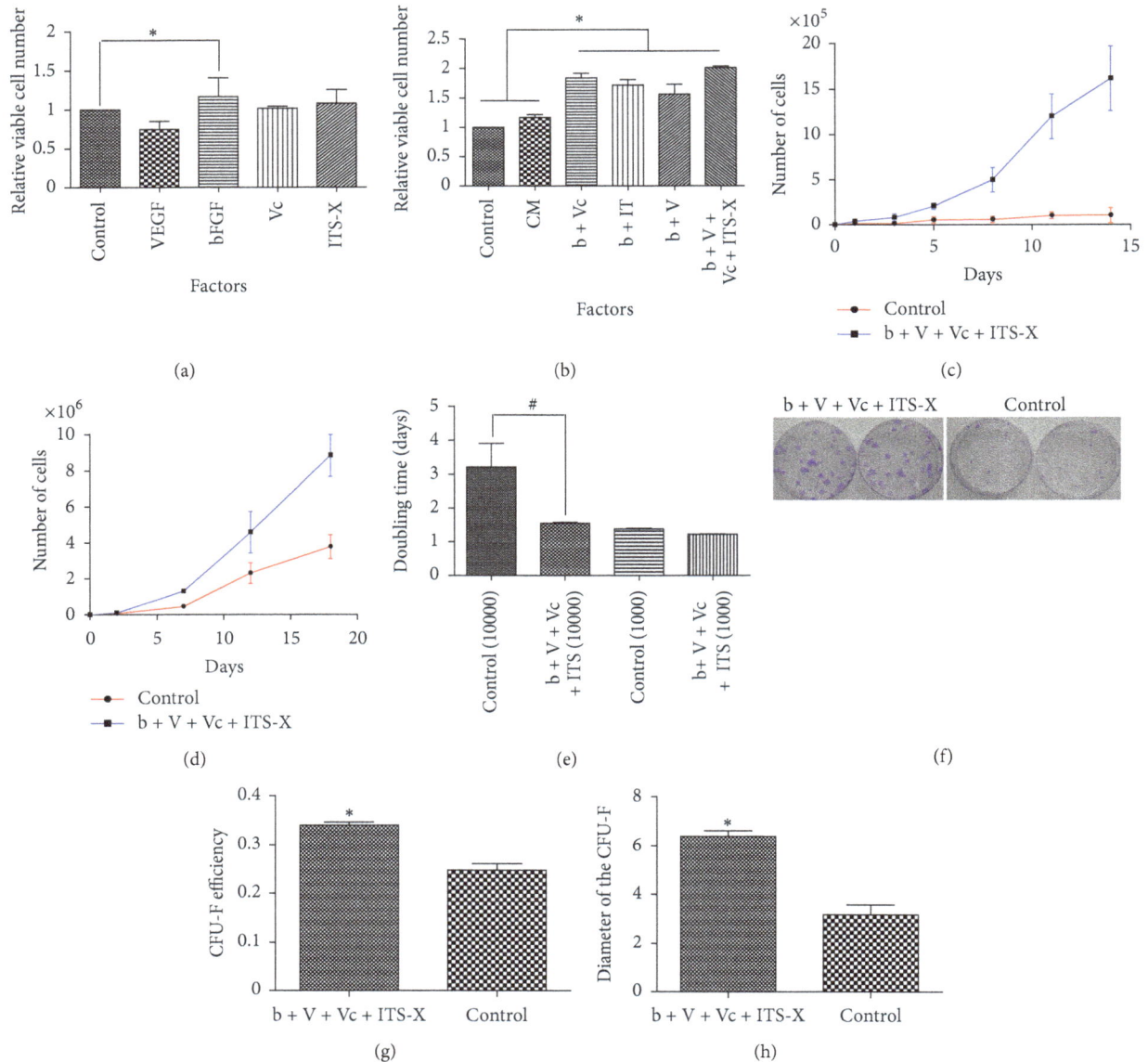

(a)

(b)

(c)

(d)

(e)

(f)

(g)

(h)

FIGURE 1: Effects of VEGF, bFGF, VC, and ITS-X on MSC proliferation. The effects of culture conditions with growth factors alone (a) and (b) in combination were examined on viable cell yield as assayed by MTT. MSC growth curves were generated at plating densities of (c) 1×10^4 cells/cm^2 and (d) 1×10^3 cells/cm^2. (e) Doubling time was calculated ($n = 3$) at these two plating densities. $^{\#}P < 0.05$. (f) CFU-Fs were stained with crystal violet and captured using a camera (Sony). (g) CFU-Fs efficiency ($n = 3$). (h) The average diameter of CFU-Fs from each set of culture conditions. b, bFGF; V, VEGF; Vc, vitamin C; ITS-X, insulin-transferrin-selenium-X; CM, commercial medium from Stem Cell Technologies; control, DMEM with 10% FBS. $^{*}P < 0.05$.

the basal medium supplemented with various combinations of growth factors on the proliferation of MSCs were evaluated. The results showed that all combinations of growth factors increased proliferation. The cumulative numbers of MSCs cultured in the medium supplemented with Vc+bFGF, VEGF+bFGF, ITS-X+bFGF, and VEGF+bFGF+Vc+ITS-X were 2.06-, 1.77-, 1.63-, and 2.39-fold higher than those in the basal medium (Figure 1(b)). At the plating density of 1×10^4 cells/cm^2, the cumulative cell numbers from 3×10^3 cells at passage 6 were on average 15-fold (Figure 1(c)) higher than those in the basal medium after 14 days of incubation. They showed a doubling time of about 1.6 days. At a plating density of 1×10^3 cells/cm^2, the cumulative cell numbers were on average 2.7-fold higher than the basal medium after incubation for 18 days (Figure 1(d)), with a doubling time of about 1.25 days (Figure 1(e)). When the total cell populations were evaluated using a CFU-F assay, the colony-forming efficiency was 34% with the combination of VEGF+bFGF+Vc+ITS-X, which was significantly higher than that in the basal medium (24.7%). The colonies in the VEGF+bFGF+Vc+ITS-X group were clearly larger than the colonies in the basal medium (Figures 1(f)–1(h)).

TABLE 1: Gene targeting in HT1080 cells.

Exp.	N	C	T	S	ATF	RTE
1	1.0	210	6	12	105	50.0%
2	3.0	529	16	30	94	53.3%

Exp.: experiment performed. N: number of cells nucleofected ($\times 10^6$). C: total number of resistant colonies obtained from each experiment. S: number of colonies screened. T: number of colonies screened as targeted recombinants. ATF: absolute targeting frequency ($\times 10^{-6}$) = TC/NS. RTE: relative targeting efficiency = T/S.

TABLE 2: Gene targeting in MSCs.

Don	N	C	S	T	ATF	RTE
1#	0.5	17	9	2	7.6	22.2%
2#	1.0	36	23	3	4.7	13.0%
2#	3.0	98	50	11	7.2	22.0%

Don: donor of bone marrow. N: number of cells nucleofected ($\times 10^6$). C: total number of resistant colonies obtained from each experiment. S: number of colonies screened. T: number of colonies screened as targeted recombinants. ATF: absolute targeting frequency ($\times 10^{-6}$) = TC/NS. RTE: relative targeting efficiency = T/S. 1#, 2#: bone marrow donors.

3.2. Gene Targeting of Human MSCs. We constructed an rDNA-targeting plasmid, pHr2-NL, which introduced a promoterless neomycin (Neo) cassette flanked by two loxP sites into the 45S pre-RNA gene. The cassette was flanked by a long homologous arm (5.6 kb) and a short homologous arm (1.1 kb). The cassette contained an encephalomyocarditis virus internal ribosomal entry site (EMCV-IRES), which enabled resistant gene expression under the control of endogenous RNA polymerase I (Pol I) promoter upstream after HR (Figure 2(a)).

A targeting experiment was first carried out in HT1080 cells. The enrichment efficiency was 50% and the targeting efficiency was 0.01% (Table 1). Then the targeting experiment was performed in triplicate on two groups of MSCs. In the groups exposed to basal medium, a few of drug-resistant cells can be observed but there were no colonies (Figure 2(b)). When exposed to VEGF+bFGF+Vc+ITS-X, many tight colonies were observed (Figure 2(c)). PCR was initially used to detect the positive recombinants; 2 out of 9, 3 out of 23, and 11 out of 50 drug-resistant colonies were found to contain positive recombinants (Table 2) (Figure 2(d)). PCR-positive recombinants were detected by Southern blotting after 5 passages; the results showed only one 8.3 kb band, which indicates that the site-specific integration of the exogenous cassette at the rDNA locus without random integration was present in 4 out of 5 representative PCR-positive colonies. However, an unexpected extra band appeared in one of the PCR-positive colonies, indicating that random integration also took place (Figure 2(e)). Consistent results were produced when the genomic DNA was cut with *Nco I*, *EcoR I*, and *Hind III* (Figure 2(f)).

The MSC colonies that underwent gene targeting were expanded and the cell numbers were counted. On average, 1×10^7 cells were obtained from one targeted MSC colony (Figure 3(a)). The expanded MSCs retained the MSC surface antigene expression (Figures 3(b) and 3(c)) and the ability to differentiate into chondrocytes, adipocytes, and osteocytes *in vitro* (Figure 3(d)). Quantitative analysis indicated that the adipogenic differentiation partially decreased compared with the normal MSCs at passage 6 (Figure 3(e)). They retained a normal karyotype (Figures 4(a) and 4(b)) and failed to develop tumors *in vivo* (Figures 4(c)–4(h)).

4. Discussion

Recent advances have shown that the use of MSCs as therapeutic vehicles may be feasible. The development of gene-targeting methods based on nonviral transfer for hMSCs deserves attention with respect to the advantages of nonviral vectors. The advantages of nonviral gene transfer include low acute toxicity, simplicity, few restrictions on the size of the gene of interest, and feasibility to be produced on a large scale [26–29]. Here, we established a nonviral method and demonstrated that the exogenous Neo gene could be targeted to the rDNA locus of MSCs. The gene-targeted MSCs maintained uniform surface antigen expression and a normal karyotype and did not develop tumors *in vivo*. Nonviral methods based on transposons such as Sleeping Beauty and piggyBac have been reported to be efficient in gene therapy and to be comparable in time-consuming compared with this method. However, safety issues about the transposons are reported. The first safety issue is about the presence of the SB transposase gene and the potential for remobilization of transposons already sited in the recipient genome. The second one is the insertional mutagenesis. The Sleeping Beauty transposon has the most random integration preference of the vectors currently in use for gene therapy [30, 31]. In this study, following antibiotic selection using G418 based on a promoter-strap strategy, the site-specific integration recombinants were selected by PCR and Southern blot assays and expanded *in vitro*. The random recombinants were eliminated.

The low integration efficiency of nonviral gene-targeting addition in mammalian cells has been a major limitation to its application [32]. It is thought that nonhomologous end joining (NHEJ) and HR DNA-repair pathways mediate random integration and site-specific integration, respectively. NHEJ is believed to occur at rates that are three to four orders of magnitude higher than those of HR [33], which makes it relatively easier to obtain colonies that carry a randomly integrated transgene. In this study, the relative gene-targeting frequencies achieved in the rDNA locus were observed to be 13%–22% in hMSCs and 50% in HT1080 cells. By using the nonviral delivery method, the absolute targeting frequency is more than 20-fold higher than that in HT1080 cells at *HPRT*, a most commonly targeted locus [34]. In previous reports, by including 28S rDNA homology arms into the vector design, the integration frequency of a recombinant adeno-associated viral vector in rat hepatocytes was enhanced by 30-fold [10]. The underlying mechanism appears to be the relatively high intrinsic activity of HR at the rDNA locus. Studies on Arabidopsis thaliana and yeast cells have suggested that the rDNA region may have functional components that stimulate HR [35]. Based on the similarity of rDNA structures between different eukaryotic cells, the rDNA region may be

(a)

(b)

(c)

(d)

(e)

(f)

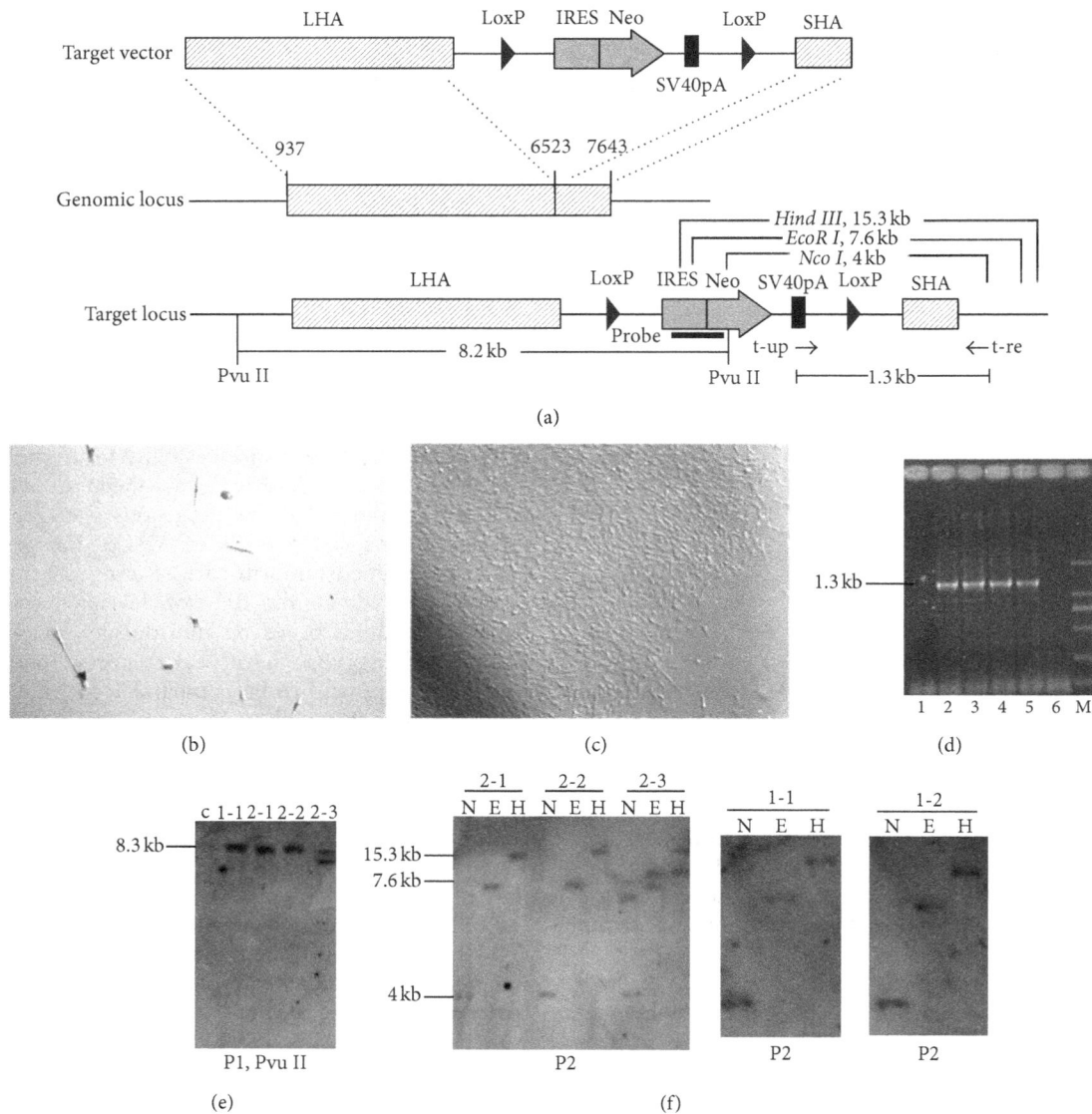

FIGURE 2: Site-specific integration at the rDNA locus of MSCs. (a) Schematic of the construction of pHr2-NL. pHr2-NL contained two inverted expression cassettes, one consisting of an IRES element from the encephalomyocarditis virus, the coding region of the Neo gene, the SV40 polyA signal (SV40pA), and two loxP sites with the same orientation. LoxP sites were recognized by CRE enzyme to remove the Neo cassette after gene targeting. LHA, long homologous arm (U13369:937-6523); SHA, short homologous arm (U13369:6523–7643). The genomic locus indicates the 6.7 kb fragment (U13369:937–7643) required for homologous recombination at the internal transcribed spacer 1 (ITS1) of the rRNA gene. Single cutting sites for restricted enzymes of *Nco I, EcoR I, Hind III*, and *Pvu II* are located at the IRES-Neo frame and outside of the long homologous fragment. The fragment between the two *Pvu II* sites was 8285 bp in size, and it was detected using probe 1 (P1). The expected sizes of the restriction fragments produced by *Nco I, EcoR I, and Hind III* were 4001 bp, 7628 bp, and 15,316 bp, respectively. These were detected using probe 2 (P2). Primer t-up was located at the SV40 polyA. Primer t-re was located outside of the SHA at the hrDNA locus. (b) Drug-resistant cell in basal medium. (c) Drug-resistant colonies in the medium supplemented with VEGF+bFGF+Vc+ITS-X. (d) Identification of colonies with site-specific integration by PCR. The expected fragment, 1.3 kb in size, was amplified from the genomic DNA of colonies using site-specific integration. M, DL200 DNA marker; 1, negative colony; 2–5, positive colonies; 6, wild-type MSCs. (e–f) Southern blotting analysis of the representative recombinants. Genomic DNA digested with *Pvu II, Nco I, EcoR I,* and *Hind III* was analyzed. A specific band was consistently detected in colonies 1-1, 1-2, 2-1, and 2-2. An additional band beside the specific band was detected in colony 2-3. c, control (untransfected MSCs); *N, Nco I; E, EcoR I; H, Hind III*.

a common HR hotspot in the majority of eukaryotic cells [36]. Despite recent success of gene targeting mediated by zinc-finger nuclease (ZFN) [4], we chose not to pursue this strategy because of the laborious design process and the toxicity resulted from the "off-target" effects [37].

To obtain enough cell for clinical use, the MSCs modified at a low efficiency need extensive expansion *in vitro*. Because the gene-targeting efficiency of nonviral vectors was relatively low, there exists a need to expand the targeted cells to get the requisite cell number. Previous reports have shown that by

(a)

(b)

(c)

(d)

(e)

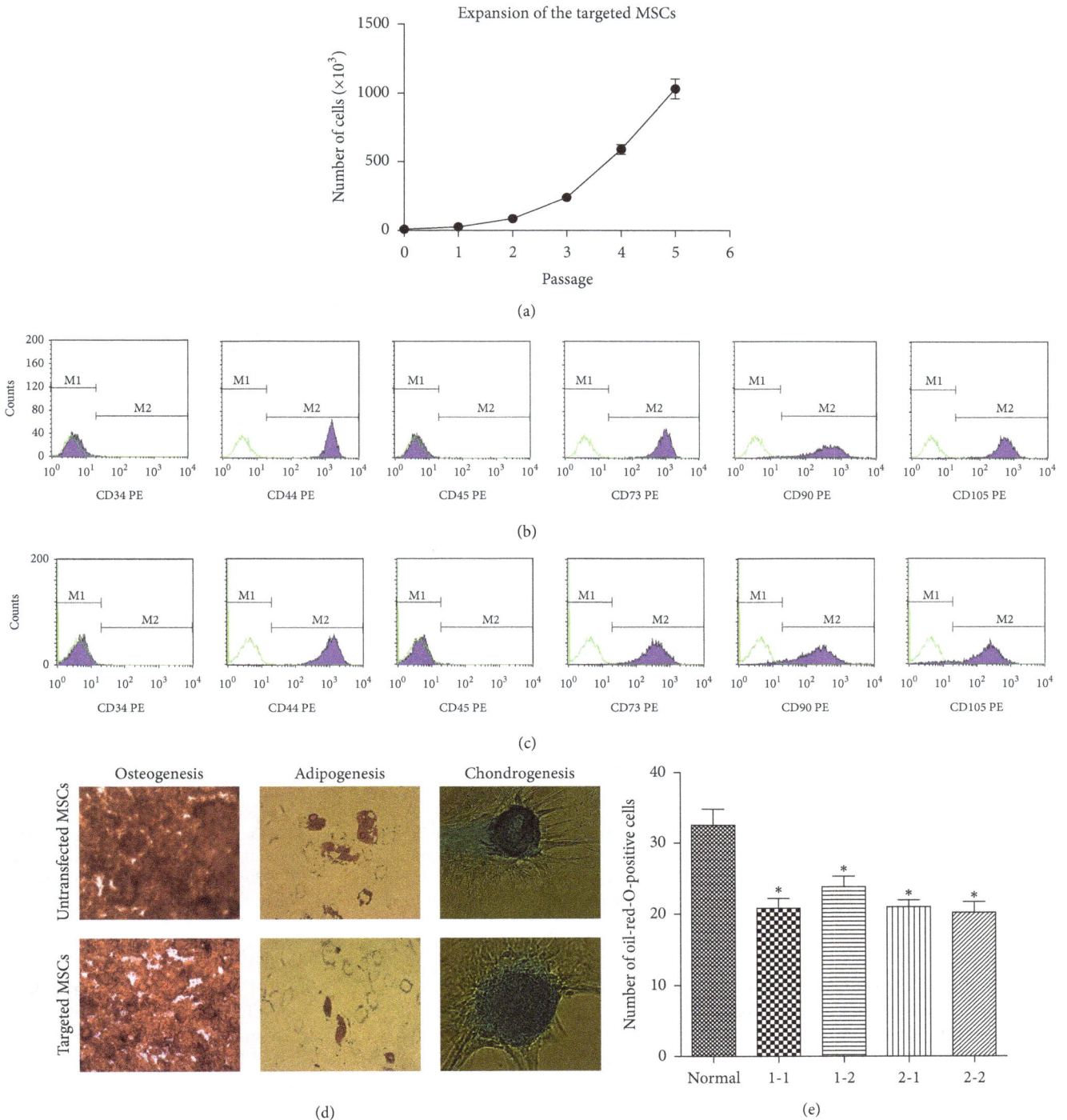

FIGURE 3: MSC surface antigen and differentiation potential detection. (a) Expansion of the targeted MSCs ($n = 3$). About 1×10^4 cells from every targeted colony were expanded. Cell count was performed at every passage. The expanded MSCs were subjected to MSC surface antigen and differentiation potential detection. (b) Flow cytometry analysis of the surface antigen expression of hMSCs untransfected. Green curves represent isotype controls and blue curves represent the specific antibodies. (c) Flow cytometry analysis of the surface antigen of expanded MSCs with gene targeting. (d) Adipogenic, osteogenic, and chondrogenic potential of hMSCs untransfected (upper) or with gene targeting (down). The adipogenic cultures were stained with oil red O to measure the accumulation of intracellular lipids. The osteogenic cultures were stained with alizarin red S to detect calcium deposition. For chondrogenic induction, the pellet sections were stained with alcian blue dye to detect proteoglycans. (e) Quantitative analysis of the adipogenic differentiation. Normal MSCs at passage 6 (normal) and targeted MSCs (1-1, 1-2, 2-1, 2-2) were differentiated into adipocytes. For each colony, the number of the oil-Red-O-positive cells was counted under 6 microscope fields. $^*P < 0.05$. The bar indicates $50\ \mu m$.

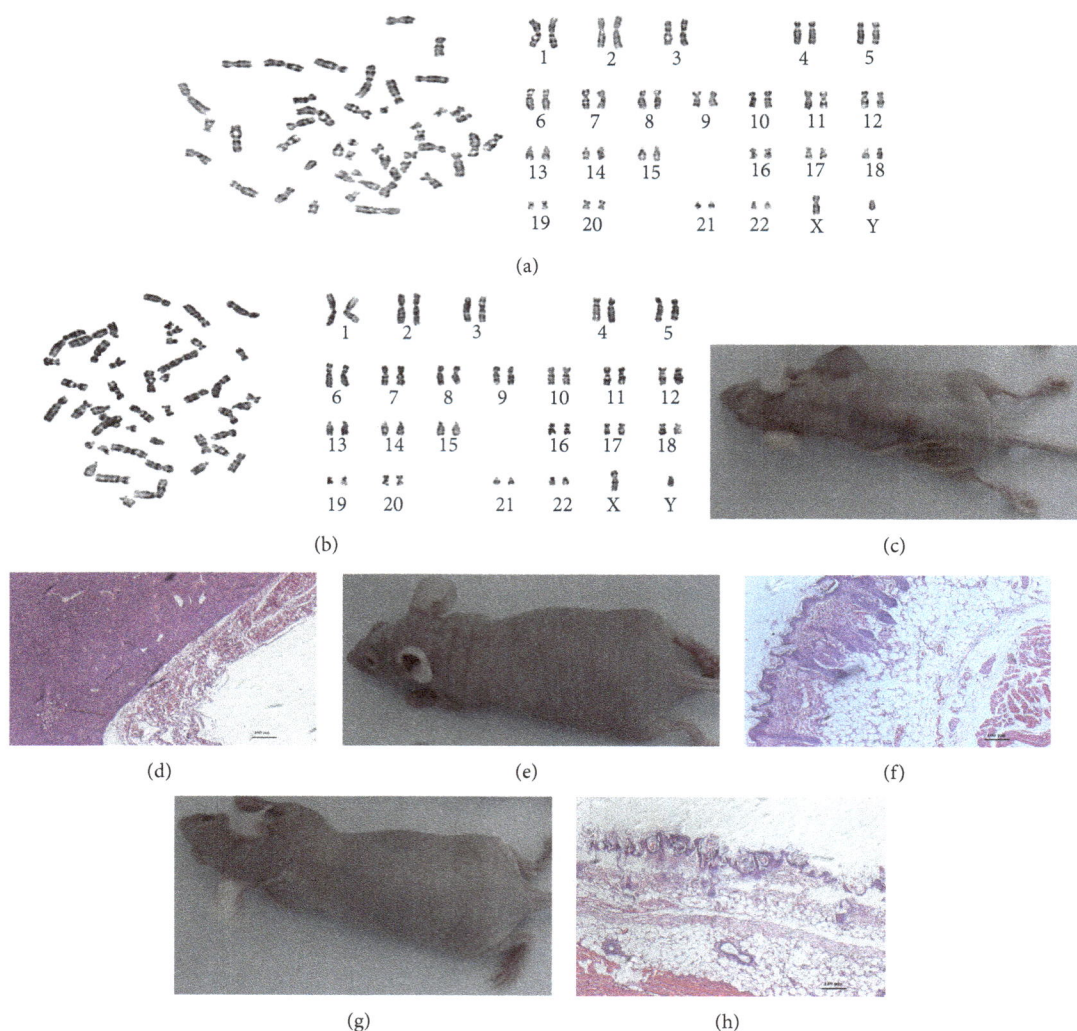

FIGURE 4: Karyotyping and *in vivo* tumor formation. Karyotyping revealed that the (a) untransfected MSCs and (b) MSCs subjected to gene targeting maintained a normal karyotype: 46, xy. (c–h) *In vivo* tumor formation. Untransfected MSCs and those targeted with an exogenous gene did not show (e, g) macroscopic or (f, h) microscopic staining after 8 weeks. (c) In contrast, HT1080 formed macroscopically visible tumors within 4 weeks. (d) Hematoxylin and eosin staining revealed a characteristic tumor growth. The bar indicates 50 μm.

supplementing growth factors the culture conditions could be optimized for inducing proliferation of MSC while maintaining their multipotency. Although combination of two or more supplements have been used [20, 21], the synergistical effects of the factors on the proliferation of MSCs were rarely reported [19]. Our results show that bFGF is critical to MSC growth and that it acted synergistically with vitamin C, VEGF, and ITS-X, causing the cells to expand significantly. Masahiro found continuous FGF stimulation to be necessary for the maintenance of VEGFR2 levels in mice modulating sensitivity to VEGF stimulation [38]. This may explain the synergistic effects of VEGF and bFGF on the proliferation of MSCs. Although the details of the mechanism by which VEGF, bFGF, Vc, and ITS-X synergistically increase cell proliferation are unclear, the most robust growth stimulation was observed with VEGF+bFGF+Vc+ITS-X (Figures 1(c) and 1(d)). In the medium without supplements, after antibiotic

selection, for example, using G418, a few of the resistant MSCs could be observed but they did not form large cell clones and even did not expand to sufficient numbers for characterization (Figure 2(b)). By adding the supplements to the culture medium, the proliferation capacity of the MSCs was obviously improved and at least 1×10^7 cells could be obtained from one targeted recombinant colony. As 11 targeted colonies can be obtained from 3×10^6 MSCs transfected, the total cell number could be calculated as 1.1×10^8. The amount of this level could meet requirement of clinical use ($10^7 \sim 10^8$) [39]. The expanded MSCs retained multipotency, although the adipogenic differentiation partially decreased.

In addition, a second cell behavior critical for the successful use of MSCs is the survival of the cultured cells. The increased survival of MSCs stimulated by growth factors, such as VEGF [25] and Vc [40], may help single targeted MSCs to form colonies. Transferrin and selenite can reduce toxic levels of oxygen radicals and be used as antioxidant in the medium [41]. During the selection process by G418, untargeted MSCs killed may release cytokines to the culture medium. This may increase the survival stress of the targeted MSCs. It was reported that high survival stress such as the oxidative stress could promote cell senescence [42]. This may be why the untargeted MSCs could form colonies but the targeted MSCs cannot form colonies in the basal medium without growth factors. Further improvements proliferation and survival of MSCs by formulation optimization of the different additives and improvement targeting efficiency by optimization of the targeting conditions may help to get more MSCs with targeted modification.

In summary, this study is the first to describe gene targeting of hMSCs using a nonviral delivery system. Exogenous therapeutic genes could be targeted to the rDNA locus of MSCs using the hrDNA vector described herein, and desirable number of the targeted cell could be obtained by improving the proliferation capacity of the MSCs using growth factors. This shows that MSCs have potential as a cellular vehicle for clinical use, and we believe that this method may be useful for autologous therapy of monogenic inheritance disease. Based on the fact that hFVIII integrated at the rDNA locus of several human cell lines expressed efficiently [14] and MSCs can home to sites of ongoing injury/inflammation to release FVIII [43], hFVIII-expressing MSCs generated using the method described herein may bring great hope for the autologous therapy of the hemophilia A, which is the most common inheritable deficiency of coagulation.

Conflict of Interests

No competing financial interests exist.

Acknowledgments

This work was supported by the National Basic Research Program of China (2010CB529903) and the National Natural Science Foundation of China (31071301, 30971298, 81000208, and 81000782).

References

[1] R. E. Bittner, C. Schöfer, K. Weipoltshammer et al., "Recruitment of bone-marrow-derived cells by skeletal and cardiac muscle in adult dystrophic mdx mice," *Anatomy and Embryology*, vol. 199, no. 5, pp. 391–396, 1999.

[2] E. Ferreira, E. Potier, D. Logeart-Avramoglou, S. Salomskaite-Davalgiene, L. M. Mir, and H. Petite, "Optimization of a gene electrotransfer method for mesenchymal stem cell transfection," *Gene Therapy*, vol. 15, no. 7, pp. 537–544, 2008.

[3] D. Baksh, L. Song, and R. S. Tuan, "Adult mesenchymal stem cells: characterization, differentiation, and application in cell and gene therapy," *Journal of Cellular and Molecular Medicine*, vol. 8, no. 3, pp. 301–316, 2004.

[4] B. F. Benabdallah, E. Allard, S. Yao et al., "Targeted gene addition to human mesenchymal stromal cells as a cell-based

[5] L. Kucerova, V. Altanerova, M. Matuskova, S. Tyciakova, and C. Altaner, "Adipose tissue-derived human mesenchymal stem cells mediated prodrug cancer gene therapy," *Cancer Research*, vol. 67, no. 13, pp. 6304–6313, 2007.

[6] J. Ellis, "Silencing and variegation of gammaretrovirus and lentivirus vectors," *Human Gene Therapy*, vol. 16, no. 11, pp. 1241–1246, 2005.

[7] S. Hacein-Bey-Abina, C. Von Kalle, M. Schmidt et al., "*LMO2*-associated clonal T cell proliferation in two patients after gene therapy for SCID-X1," *Science*, vol. 302, no. 5644, pp. 415–419, 2003.

[8] K. Sakai, T. Ohta, S. Minoshima et al., "Human ribosomal RNA gene cluster: identification of the proximal end containing a novel tandem repeat sequence," *Genomics*, vol. 26, no. 3, pp. 521–526, 1995.

[9] D. M. Stults, M. W. Killen, H. H. Pierce, and A. J. Pierce, "Genomic architecture and inheritance of human ribosomal RNA gene clusters," *Genome Research*, vol. 18, no. 1, pp. 13–18, 2008.

[10] L. Lisowski, A. Lau, Z. Wang et al., "Ribosomal DNA integrating rAAV-rDNA vectors allow for stable transgene expression," *Molecular Therapy*, vol. 20, no. 10, pp. 1912–1923, 2012.

[11] D. Schenkwein, V. Turkki, M. K. Ahlroth, O. Timonen, K. J. Airenne, and S. Yla-Herttuala, "rDNA-directed integration by an HIV-1 integrase—I-PpoI fusion protein," *Nucleic Acids Research*, vol. 41, pp. 1–10, 2012.

[12] Z. Wang, L. Lisowski, M. J. Finegold, H. Nakai, M. A. Kay, and M. Grompe, "AAV vectors containing rDNA homology display increased chromosomal integration and transgene persistence," *Molecular Therapy*, vol. 20, no. 10, pp. 1902–1911, 2012.

[13] X. Liu, Y. Wu, Z. Li et al., "Targeting of the human coagulation factor IX gene at rDNA locus of human embryonic stem cells," *PLoS ONE*, vol. 7, no. 5, Article ID e37071, 2012.

[14] X. Liu, M. Liu, Z. Xue et al., "Non-viral ex vivo transduction of human hepatocyte cells to express factor VIII using a human ribosomal DNA-targeting vector," *Journal of Thrombosis and Haemostasis*, vol. 5, no. 2, pp. 347–351, 2007.

[15] E. H. Javazon, K. J. Beggs, and A. W. Flake, "Mesenchymal stem cells: paradoxes of passaging," *Experimental Hematology*, vol. 32, no. 5, pp. 414–425, 2004.

[16] M. M. Bonab, K. Alimoghaddam, F. Talebian, S. H. Ghaffari, A. Ghavamzadeh, and B. Nikbin, "Aging of mesenchymal stem cell in vitro," *BMC Cell Biology*, vol. 7, article 14, 2006.

[17] U. Lindner, J. Kramer, J. Behrends et al., "Improved proliferation and differentiation capacity of human mesenchymal stromal cells cultured with basement-membrane extracellular matrix proteins," *Cytotherapy*, vol. 12, no. 8, pp. 992–1005, 2010.

[18] K. Tamama, H. Kawasaki, and A. Wells, "Epidermal growth factor (EGF) treatment on multipotential stromal cells (MSCs). Possible enhancement of therapeutic potential of MSC," *Journal of Biomedicine & Biotechnology*, vol. 2010, Article ID 795385, 10 pages, 2010.

[19] S. Jung, A. Sen, L. Rosenberg, and L. A. Behie, "Identification of growth and attachment factors for the serum-free isolation and expansion of human mesenchymal stromal cells," *Cytotherapy*, vol. 12, no. 5, pp. 637–657, 2010.

[20] K. Chieregato, S. Castegnaro, D. Madeo, G. Astori, M. Pegoraro, and F. Rodeghiero, "Epidermal growth factor, basic fibroblast

growth factor and platelet-derived growth factor-bb can substitute for fetal bovine serum and compete with human platelet-rich plasma in the ex vivo expansion of mesenchymal stromal cells derived from adipose tissue," *Cytotherapy*, vol. 13, no. 8, pp. 933–943, 2011.

[21] B. Gharibi and F. J. Hughes, "Effects of medium supplements on proliferation, differentiation potential, and in vitro expansion of mesenchymal stem cells," *Stem Cells Translational Medicine*, vol. 1, no. 11, pp. 771–782, 2012.

[22] M. F. Pittenger, A. M. Mackay, S. C. Beck et al., "Multilineage potential of adult human mesenchymal stem cells," *Science*, vol. 284, no. 5411, pp. 143–147, 1999.

[23] A. J. Friedenstein, R. K. Chailakhjan, and K. S. Lalykina, "The development of fibroblast colonies in monolayer cultures of guinea-pig bone marrow and spleen cells," *Cell and Tissue Kinetics*, vol. 3, no. 4, pp. 393–403, 1970.

[24] A. A. Stewart, C. R. Byron, H. Pondenis, and M. C. Stewart, "Effect of fibroblast growth factor-2 on equine mesenchymal stem cell monolayer expansion and chondrogenesis," *American Journal of Veterinary Research*, vol. 68, no. 9, pp. 941–945, 2007.

[25] J. Pons, Y. Huang, J. Arakawa-Hoyt et al., "VEGF improves survival of mesenchymal stem cells in infarcted hearts," *Biochemical and Biophysical Research Communications*, vol. 376, no. 2, pp. 419–422, 2008.

[26] K. L. Douglas, "Toward development of artificial viruses for gene therapy: a comparative evaluation of viral and non-viral transfection," *Biotechnology Progress*, vol. 24, no. 4, pp. 871–883, 2008.

[27] E. H. Chowdhury, "Nuclear targeting of viral and non-viral DNA," *Expert Opinion on Drug Delivery*, vol. 6, no. 7, pp. 697–703, 2009.

[28] K. A. Partridge and R. O. C. Oreffo, "Gene delivery in bone tissue engineering: progress and prospects using viral and nonviral strategies," *Tissue Engineering*, vol. 10, no. 1-2, pp. 295–307, 2004.

[29] C. L. Da Silva, C. Madeira, R. D. Mendes et al., "Nonviral gene delivery to mesenchymal stem cells using cationic liposomes for gene and cell therapy," *Journal of Biomedicine and Biotechnology*, vol. 2010, Article ID 735349, 12 pages, 2010.

[30] M. Di Matteo, E. Belay, M. K. Chuah, and T. Vandendriessche, "Recent developments in transposon-mediated gene therapy," *Expert Opinion on Biological Therapy*, vol. 12, no. 7, pp. 841–858, 2012.

[31] E. L. Aronovich, R. S. McIvor, and P. B. Hackett, "The Sleeping Beauty transposon system: a non-viral vector for gene therapy," *Human Molecular Genetics*, vol. 20, no. 1, pp. R14–R20, 2011.

[32] R. J. Bollag, A. S. Waldman, and R. M. Liskay, "Homologous recombination in mammalian cells," *Annual Review of Genetics*, vol. 23, pp. 199–225, 1989.

[33] S. Iiizumi, A. Kurosawa, S. So et al., "Impact of non-homologous end-joining deficiency on random and targeted DNA integration: implications for gene targeting," *Nucleic Acids Research*, vol. 36, no. 19, pp. 6333–6342, 2008.

[34] R. J. Yáñez and A. C. G. Porter, "Gene targeting is enhanced in human cells overexpressing hRAD51," *Gene Therapy*, vol. 6, no. 7, pp. 1282–1290, 1999.

[35] K. Voelkel-Meiman, R. L. Keil, and G. S. Roeder, "Recombination-stimulating sequences in yeast ribosomal DNA correspond to sequences regulating transcription by RNA polymerase I," *Cell*, vol. 48, no. 6, pp. 1071–1079, 1987.

[36] H. Urawa, M. Hidaka, S. Ishiguro, K. Okada, and T. Horiuchi, "Enhanced homologous recombination caused by the non-transcribed spacer of the rDNA in arabidopsis," *Molecular Genetics and Genomics*, vol. 266, no. 4, pp. 546–555, 2001.

[37] A. Gupta, X. Meng, L. J. Zhu, N. D. Lawson, and S. A. Wolfe, "Zinc finger protein-dependent and -independent contributions to the in vivo off-target activity of zinc finger nucleases," *Nucleic Acids Research*, vol. 39, no. 1, pp. 381–392, 2011.

[38] M. Murakami, L. T. Nguyen, K. Hatanaka et al., "FGF-dependent regulation of VEGF receptor 2 expression in mice," *Journal of Clinical Investigation*, vol. 121, no. 7, pp. 2668–2678, 2011.

[39] M. Rodrigues, L. G. Griffith, and A. Wells, "Growth factor regulation of proliferation and survival of multipotential stromal cells," *Stem Cell Research & Therapy*, vol. 1, no. 4, p. 32, 2010.

[40] F. Wei, C. Qu, T. Song et al., "Vitamin C treatment promotes mesenchymal stem cell sheet formation and tissue regeneration by elevating telomerase activity," *Journal of Cellular Physiology*, vol. 227, no. 9, pp. 3216–3224, 2011.

[41] T. C. Stadtman, "Specific occurrence of selenium in enzymes and amino acid tRNAs," *The FASEB Journal*, vol. 1, no. 5, pp. 375–379, 1987.

[42] H. Alves, A. Mentink, B. Le, C. A. van Blitterswijk, and J. de Boer, "Effect of antioxidant supplementation on the total yield, oxidative stress levels, and multipotency of bone marrow-derived human mesenchymal stromal cells," *Tissue Engineering A*, vol. 19, no. 7-8, pp. 928–937, 2013.

[43] C. D. Porada, C. Sanada, C. J. Kuo et al., "Phenotypic correction of hemophilia A in sheep by postnatal intraperitoneal transplantation of FVIII-expressing MSC," *Experimental Hematology*, vol. 39, no. 12, pp. 1124–1135, 2011.

7

Inhibition of Carrageenan-Induced Acute Inflammation in Mice by Oral Administration of Anthocyanin Mixture from Wild Mulberry and Cyanidin-3-Glucoside

Neuza Mariko Aymoto Hassimotto,[1,2] **Vanessa Moreira,**[3] **Neide Galvão do Nascimento,**[3] **Pollyana Cristina Maggio de Castro Souto,**[3] **Catarina Teixeira,**[3] **and Franco Maria Lajolo**[1,2]

[1] *Laboratório de Química e Bioquímica e Biologia Molecular de Alimentos, Departamento de Alimentos e Nutrição Experimental, FCF, Universidade de São Paulo, Avenida Professor Lineu Prestes 580, Bloco 14, 05508-000 São Paulo, SP, Brazil*
[2] *Núcleo de Apoio à Pesquisa em Alimentos e Nutrição (NAPAN), Universidade de São Paulo, 05508-000 São Paulo, SP, Brazil*
[3] *Laboratório de Farmacologia, Unidade de Inflamação, Instituto Butantan, Avenida Vital Brasil 1500, 05503-000 São Paulo, SP, Brazil*

Correspondence should be addressed to Franco Maria Lajolo; fmlajolo@usp.br

Academic Editor: José Carlos Tavares Carvalho

Anthocyanins are flavonoids which demonstrated biological activities in *in vivo* and *in vitro* models. Here in the anti-inflammatory properties of an anthocyanin-enriched fraction (AF) extracted from wild mulberry and the cyanidin-3-glucoside (C3G), the most abundant anthocyanin in diet, were studied in two acute inflammation experimental models, in the peritonitis and in the paw oedema assays, both of which were induced by carrageenan (cg) in mice. In each trial, AF and C3G (4 mg/100 g/animal) were orally administered in two distinct protocols: 30 min before and 1 h after cg stimulus. The administration of both AF and C3G suppresses the paw oedema in both administration times ($P < 0.05$). In the peritonitis, AF and C3G reduced the polymorphonuclear leukocytes (PMN) influx in the peritoneal exudates when administered 1 h after cg injection. AF was more efficient reducing the PMN when administered 30 min before cg. Both AF and C3G were found to suppress mRNA as well as protein levels of COX-2 upregulated by cg in both protocols, but the inhibitory effect on PGE_2 production in the peritoneal exudates was observed when administered 30 min before cg ($P < 0.05$). Our findings suggest that AF and C3G minimize acute inflammation and they present positive contributions as dietary supplements.

1. Introduction

Anthocyanins, glycosylated polyphydroxy, and polymethoxy derivates of flavilium salt are natural colorants belonging to the flavonoid family and largely intaken from vegetable foods [1]. These pigments are responsible for the pink, red, violet, and blue colours in the flowers, fruits, and vegetables. There is a great variety of anthocyanins spread in the nature but only six are the most common: cyanidin, pelargonidin, malvidin, peonidin, petunidin, and delfinidin [2]. Interest in biological effects of anthocyanins has increased during the last decade because of increasing evidence demonstrating their potential therapeutic effects. Some anthocyanins have demonstrated to inhibit the growth of cancerous cells [3–5], to decrease hyperglycemic levels [6] and to promote antiobesity effects [7, 8]. Furthermore, anthocyanins possess antioxidant [9, 10] and anti-inflammatory [11–13] properties. This group of compounds has been demonstrated to modulate inflammation process dependent on the COX-2 pathway *in vitro* experimental protocols [14–18] and through the inhibition of nitric oxide biosynthesis [10].

Wild black mulberry (*Morus nigra* L.) extracts contains high level of anthocyanins. The identified anthocyanins are mainly cyanidin-3-glucoside (C3G), and in minor level cyanidin-3-rutinoside and pelargonidin derivate [19]. We previously reported that the anthocyanin-enriched extract (AF) obtained from wild black mulberry increased the plasma antioxidant capacity and the plasma catalase activity after oral intake in human [20]. Also, AF demonstrated inhibitory effect on the migration and invasion of a human lung cancer cell [5]. However, there are few studies that use *in vivo* experimental protocols, in order to demonstrate if oral intake of anthocyanins could affect inflammation. Since the anthocyanin is commonly intake daily from vegetable foods, it is important to establish evidence for the effect of anthocyanin consumption on health.

Inflammatory responses are a series of well-coordinated events that depend on the increase in vascular permeability and sequential release of inflammatory mediators, leading to oedema and arrival of inflammatory leukocytes to the site of inflammation, where neutrophils and macrophages are known to recruit and play pivotal roles in acute and chronic inflammation, respectively [21]. Cyclooxygenases (COXs) are the key enzymes in the synthesis of lipid mediators called prostaglandins observed in inflammation events. COXs convert free arachidonic acid, following its release from membrane phospholipids by phospholipases A_2, to prostaglandin H_2, the common precursor for all prostanoids. Nowadays, there are three COX isoforms named COX-1, COX-2, and COX-3 [22, 23]. COX-1 is a housekeeping enzyme, constitutively expressed in most mammalian tissues, and it is responsible for maintaining normal cellular physiologic functions. COX-2 is also present at a basal level in certain tissues, but its expression is induced in inflammatory cells and tissues in response to cellular activation by endotoxin, cytokines, mitogens, and other stimulus [24, 25]. COX-2 is the main enzyme providing a mechanism for the generation of proinflammatory prostanoids, such as prostaglandin E_2 (PGE_2), a potent vasodilator, which enhances oedema formation [26, 27]. COX-3, in turn, has been cloned [28, 29], but its function have yet to be well studied.

Therefore, in this study, we have examined, in mice, the anti-inflammatory activity of oral administration of an anthocyanin-enriched extract obtained from mulberry and its major component, the C3G, in the acute inflammation, peritonitis and paw oedema assays, induced by carrageenan, mainly on COX-2 mRNA and protein expression and PGE_2 production.

2. Material and Methods

2.1. Mulberry Anthocyanin Preparations. The anthocyanin-enriched fraction (AF) was prepared from wild black mulberry according to the previously published method [19]. Briefly, the sample (approximately 5 g) was extracted three times with 100 mL of methanol:water:acetic acid (70 : 30 : 5, v : v : v) (Brinkmann homogeniser, Polytron-Kinematica GmbH, Kriens-Luzern, Sweden) in an ice bath. The homogenate was filtered under reduced pressure

through filter paper (Whatman number 06). The methanol extract obtained was concentrated, under vacuum until methanol content elimination, using a rotary evaporator (Rotavapor RE 120, Buchi, Flawil, Sweden) and made up to 50 mL with distilled water. The extract (25 mL) was passed through polyamide (CC-6, Macherey-Nagel, Germany) column (10 g/60 mL) previously conditioned with 50 mL of methanol and 100 mL of distilled water. Impurities were washed out with distilled water and retained flavonoids were eluted with 120 mL of methanol acidified with 0.1% HCl. The flow rate through the columns was controlled by means of a vacuum manifold Visiprep 24DL (Supelco, Bellefonte, PA). The eluate was evaporated to dryness under reduced pressure at 40°C and dissolved in distilled water prior administration. This fraction corresponds to AF. C3G was further purified from AF according to Chen et al. [5] by passing it through a Bio-Gel P-2 column (40 cm × 2.5 cm) (Bio-Rad Laboratories, Hercules, CG), eluting it with aqueous acetic acid, pH 2.5, and monitoring it by spectrophotometer at 520 nm (Hitachi L-4000 UV-vis detector). The fraction corresponding to C3G, which was confirmed by HPLC-DAD, was collected and lyophilized. C3G was dissolved in distilled water prior to administration.

2.2. Anthocyanin Quantification. For anthocyanin quantification, aliquots of AF and C3G were diluted with methanol: acetic acid (99 : 5, v : v) and filtered through a 0.45 μm PTFE filter (Milipore Ltd., Bedford MA) prior to quantification by HPLC-DAD [19]. The column used was a Prodigy 5 μm ODS3 (250 mm × 4.6 mm i.d., Phenomenex Ltd.) and elution solvents were (A) water : THF : TFA (98 : 2 : 0.1, v : v : v) and (B) acetonitrile. Solvent gradient consisted of 8% B at the beginning, 10% at 5 min, 17% at 10 min, 25% at 15 min, 50% at 25 min, 90% at 30 min, 50% at 32 min, and 8% at 35 min (run time, 35 min). Eluates were monitored at 270 and 525 nm. Flow rate was 1 mL/min; column temperature was 30°C. Peak identification was performed by comparison of retention times and diode array spectral characteristics with the standards and the library spectra. Cochromatography was used when necessary. C3G, C3R, and pelargonidin (Plg) (Extrasynthese, Genay, France) were used as standard. The total anthocyanin content of AF was expressed as C3G equivalent. The anthocyanin composition of AF is 85% C3G, 12% C3R, and 3% Plg derivate and they were previously identified by LC-MS [19]. The anthocyanin profile of AF and the purity of C3G are shown in Figure 1.

2.3. Animals. Male Swiss mice, weighing 18–20 g (approximately four weeks old), were acclimated to housing for at least 1 week prior to investigation. The night before the experiment, food was withdrawn from the cages but water was given *ad libitum*. Animals were randomly assigned to each treatment group and all testing was performed between 8:00 and 9:00 a.m. All animals were handled and experiments were conducted in accordance to the Guidelines for Animal Experimentation of the University of São Paulo, Brazil, after approval by the Ethics Committee of the Pharmacy Faculty of the University of São Paulo (Protocol number 53, FCF-USP).

Inhibition of Carrageenan-Induced Acute Inflammation in Mice by Oral Administration of Anthocyanin
Mixture from Wild Mulberry and Cyanidin-3-Glucoside

57

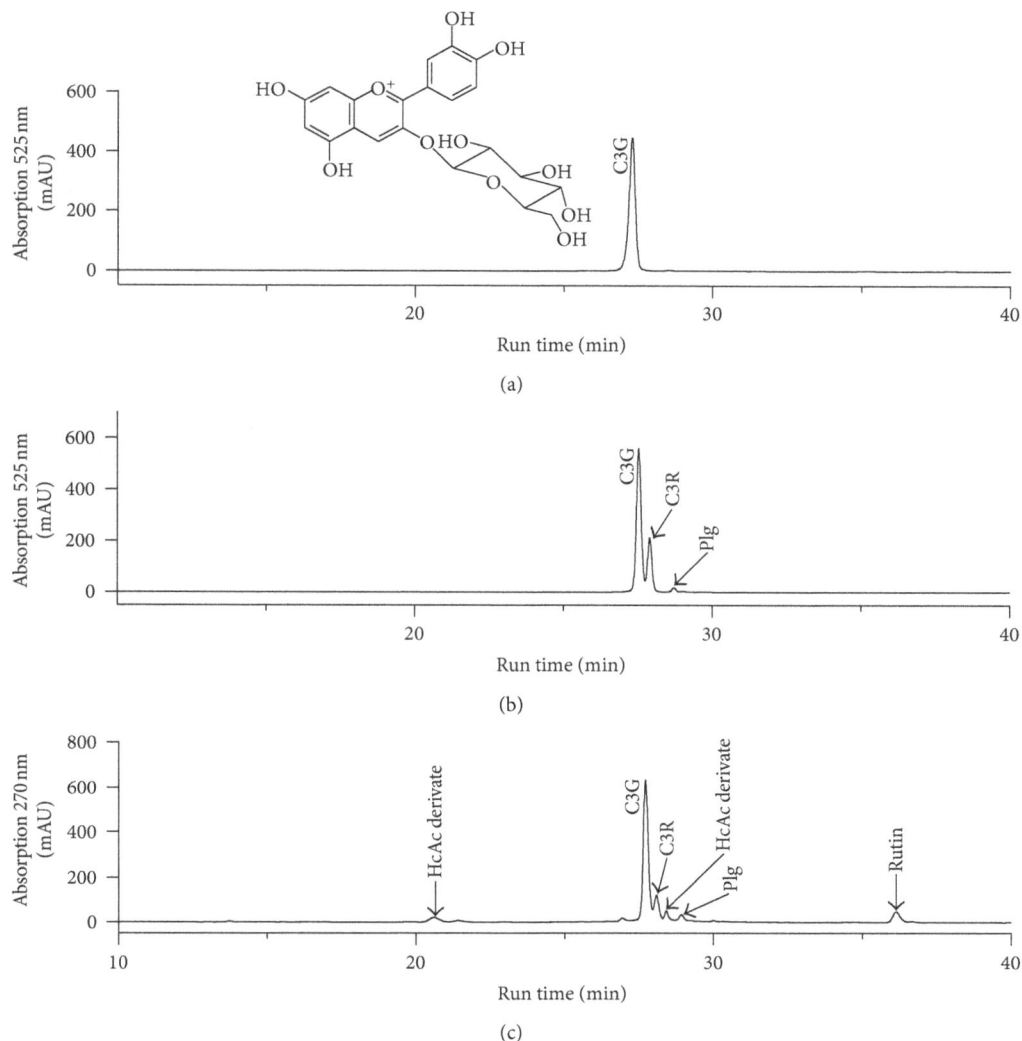

FIGURE 1: HPLC-DAD of cyanidin-3-glucoside (C3G) at 525 nm (a) and anthocyanin profile of AF at 525 nm (b) and 270 nm (c). Peaks were identified by MS/MS as C3G (structure showed), cyanidin-3-rutinoside (C3R), and rutin. Abbreviations: Hydroxycinnamic acid derivate (HcAc derivate) and pelargonidin (Plg).

2.4. Carrageenan-Induced Paw Oedema in Mice.

To assess the effects of the AF and C3G on acute inflammation, the animals were deprived of food overnight and orally administered with an aqueous solution using an intragastric tube as described below.

AF Group: 200 μL of the AF (4 mg C3G equiv/100 g body weight) were administered 30 min before (n = 8) and 1 h after (n = 8) intraplantar (i.pl.) injection of 50 μL cg in saline (0.5% m/v) into the left hind paw.

C3G Group: 200 μL of C3G (4 mg/100 g body weight) were administered 30 min before (n = 8) and 1 h after (n = 8) i.pl. injection of 50 μL carrageenan (cg) in saline (0.5% m/v) into the left hind paw.

Control Group: 200 μL of saline were administered 30 min before (n = 8) and 1 h after (n = 8) i.pl. injection of 50 μL cg in saline (0.5% m/v) into the left hind paw.

Indomethacin Group: indomethacin (1 mg/kg, i.v) was administered 30 min before (n = 8) and 1 h after (n = 8) injection of 50 μL cg in saline (0.5% m/v) into the left hind paw.

The contralateral paw was injected with 50 μL of saline solution and used as a control. The volumes of both hind paws were measured by plethysmometry (model 7140 plethysmometer, Ugo Basile, Italy) 1, 2, 3, 4, and 5 h after the injection of cg. The results were reported as the percent inhibition of the volume increase to be compared with the preinjection paw volume. Mean values of treated groups were compared with mean values of a control group and analyzed using statistical methods.

2.5. Carrageenan-Induced Peritonitis in Mice.

The animals were deprived of food overnight and orally administered with one of the following solutions.

AF Group: 200 μL of the AF (4 mg C3G equiv/100 g body weight) were administered 30 min before ($n = 8$) and 1 h after ($n = 8$) intraperitoneal (i.p.) injection of 1 mL of cg in sterile saline (0.3%, m/v).

C3G Group: 200 μL of the C3G (4 mg/100 g body weight) were administered 30 min before ($n = 8$) and 1 h after ($n = 8$) i.p. injection of 1 mL of cg in sterile saline (0.3%, m/v).

Carrageenan Control Group: 200 μL of saline solution were administered 30 min before ($n = 8$) and 1 h after ($n = 8$) i.p. injection of 1 mL of cg in sterile saline (0.3%, m/v).

Indomethacin Group: 200 μL indomethacin (4 mg/100 g body weight) were administered 30 min before ($n = 8$) and 1 h after ($n = 8$) i.p. injection of 1 mL of cg in sterile saline (0.3%, m/v).

Saline Control Group: 200 μL saline solution were administered 30 min before ($n = 8$) and 1 h after ($n = 8$) i.p. injection of 1 mL of sterile saline solution.

Three hours after cg injections, the animals were killed by overexposure to CO_2 and the peritoneal exudate was withdrawn after washing the peritoneal cavity with 2 mL of saline solution. Aliquots of the washes were used to determine total cell counts. An aliquot of the 1×10^6 and 3×10^6 cells was centrifuged at $800\,g/6\,min/22°C$ and used for COX-2 expression analysis by western blotting and RT-PCR, respectively. The supernatant was used for PGE_2 quantification.

2.6. Leukocyte Harvesting and Counting.
Aliquots of the peritoneal washes were used to determine total cell counts in a Newbauer chamber after dilution (1 : 20, v : v) in Turk's solution (0.2% crystal violet dye in 30% acetic acid). For differential cell counts, cytospin preparations were stained with Hema^3 stain. Differential cell counts were performed by counting at least 100 cells, which were classified as either polymorphonuclear or mononuclear cells, based on conventional morphological criteria.

2.7. Western Blotting.
The precipitate of cells (1×10^6) was lisate with 100 μL of sample buffer [30] and heated for 10 min/100°C. An aliquot of 14 μL of the lisate was separated on SDS-polyacrylamide gels (10%) at 150 V and electrophoretically transferred to nitrocellulose membrane (GE Healthcare, Buckinghamshire, UK). The membrane was blocked with 5% nonfat milk in Tris buffered saline with 0.05% Tween 20 and incubated 1 h at room temperature with the antibody against COX-2 (1 : 1500) (Cayman Chemicals, Ann Arbor, MI) followed by incubation in the same buffer with the appropriate anti-rabbit horseradish peroxidase-conjugated secondary antibody (GE Healthcare, Buckinghamshire, UK) for 1 h at room temperature (1 : 1500). Further, the membrane was also incubated with the antibody against β-actin (1 : 2000) (Sigma, St. Louis, USA) followed by incubation with the anti-mouse secondary horseradish peroxidase-conjugate (1 : 2000) (GE Healthcare, Buckinghamshire, UK). Immunoreactive bands were detected using ECL kit (GE Healthcare, Buckinghamshire, UK). Densities of the bands were determined by a GS 700 Densitometer (Bio-Rad Laboratories, Richmond, CG) using the image analysis software from Molecular Analyst (Bio-Rad Laboratories, Richmond, CG).

2.8. RNA Preparation and Reverse Transcription-Polymerase Chain Reaction (RT-PCR).
Cells (3×10^6) were washed once with sterile saline and mixed with 500 μL of Trizol reagent (Invitrogen, Rockville, MD, EUA) and the RNA was extracted according to the manufacturer's instructions. Complementary DNA was synthesized using an Improm-II Reverse Transcription System (Promega, Madison, WI, USA) according to the manufacturer's instructions and conducted at a thermocycler Gene AMp (PCR System 2400, Applied Biosystems). PCR was performed by denaturing at 94°C for 60 s, annealing at 57°C (COX-2) and 60°C (β-actin) for 1 min and by extension at 72°C for 60 s. Thirty additional cycles for COX-2 and 25 cycles for β-actin were used for amplification. The primer pairs used for analysis were 5$'$-TTTGTTGAGTCGTTCGCCGGACGGA-3$'$ and 5$'$-CGGTATTGAGGAGAAGAGATGGGATT-3$'$ for sense and antisense primers of the COX-2 gene, respectively [31]; 5$'$-TGGAATCCTGTGGCGTCCGTGAAAC-3$'$ and 5$'$-TAAAACGCGGCTCGGTAACGGTCCG-3$'$ for sense and antisense primers of the β-actin gene, respectively [32], used as an inner control.

2.9. PGE_2 Quantification.
Concentrations of PGE_2 were determined by a specific enzyme immunoassay [33] using a commercial kit (Cayman Chemical Company, Ann Arbor, MI). The extraction of PGE_2 was performed on Sep Pak C18 columns (Waters Corporation, Milford, MA) and eluted with ethanol. In brief, 50 μL aliquots of each extracted sample were incubated with the PGE_2 conjugated with acetylcholinesterase and the specific rabbit antiserum in 96-well plates, coated with anti-rabbit IgG mouse monoclonal antibody. After addition of the substrate, the absorbance of the samples was recorded at 405 nm in a microplate reader (Labsystem Multiscan), and concentrations of eicosanoids were estimated from standard curves.

2.10. Statistical Analysis.
Results were presented as mean ± EPM. The statistical analyses were performed by one way analysis of variance (ANOVA) and *Tukey posthoc test* for comparison, using the Statistic software package version 5.0 (StatSoft, Inc.). Results were considered statistically significant for P values <0.05.

3. Results and Discussion

3.1. Effect of C3G and AF on Carrageenan-Induced Paw Oedema.
The oral dose of both extracts and the two protocols applied in this study (30 min before or 1 h after inflammation stimulus) were chosen in order to provide high concentration of C3G in the plasma based in its rapid absorption and excretion [20].

The inflammatory response to subplantar oedema induced by cg in mice was significantly reduced by prior and after oral administration of AF and C3G. Figures 2(a)

Inhibition of Carrageenan-Induced Acute Inflammation in Mice by Oral Administration of Anthocyanin
Mixture from Wild Mulberry and Cyanidin-3-Glucoside

59

and 2(b) show the time course of the paw oedema after i.pl. injection of cg (0.5% m/v). Carrageenan caused progressive increase in the paw oedema 1 h after the injection, presenting the maximum peak at 4 h, decreasing to basal level after 5 h. Before and after treatment of animals with indomethacin significantly reduced cg-induced paw oedema as expected, in comparison with the respective controls (saline). C3G (4 mg/100 g body weight), administered by gavage either 30 min before or 1 h after the cg stimulus significantly decreased ($P < 0.05$) the paw oedema (around 40% and up to 80%, resp.) at the fourth hour after cg injection when compared with the control group (Figures 2(a) and 2(b)). Also, the oral administration of AF decreased the paw oedema approximately 40% in both administration times.

The dose of AF and C3G used in the present study is ten-times lower than that necessary of the anthocyanin mix from tart cherry to suppress the 25% complete Freund's adjuvant and cg-induced paw oedema [13] but closer than ginkgo biloba extract concentration necessary to inhibit the paw oedema induced by cg in rats [34]. This fact suggested that C3G is one of the anthocyanins that presented high anti-inflammatory activity.

It has been established that the paw oedema induced by the subplantar injection of cg is biphasic; the early phase involves the release of the mediators serotonin, histamine, and kinins, while the late phase is characterized by the infiltration of leukocytes and mediated only by prostaglandins [35]. These results suggest that the inhibitory effect of AF or C3G on oedema formation is due to the inhibition of the synthesis and/or release of these mediators, in the early phase of inflammatory effect of cg, especially by inhibiting probably cyclooxygenase products. To support this observation, the data indicate that C3G promoted similar effectiveness in suppressing oedema, when compared to the inhibitory profile of indomethacin, a COX activity inhibitor, on cg-induced inflammation.

3.2. Effect of C3G and AF on Carrageenan-Induced Cellular Influx into Peritoneal Cavity.

Intraperitoneal administration of cg produces a sustained increase in postcapillary *venule* permeability, thereby leading to increased cellular infiltration, particularly of neutrophils [36]. The recruitment of leukocytes from the circulation to sites of inflammation is enhanced by a series of proinflammatory mediators, such as IL-8 and vasoactive amines, ICAM and VCAM, that are produced and released into the tissue by mast cells, macrophages, and activated endothelial cells, as well as transmigrated leukocytes [36].

Figure 3 presents the total leukocyte influx and differential cell into the peritoneal cavities after oral administration of C3G or AF (4 mg/100 g body weight) or indomethacin (4 mg/100 g body weight) or saline (control) 30 min before and 1 h after i.p. injection of cg (0.3% w/v) or saline solution (without stimulus).

The oral administration of AF 30 min before the i.p. injection of cg caused a significant decrease ($P < 0.05$) in the number of total leukocytes (29% decrease) (Figure 3(a)), but not when administered 1 h after the stimulus.

No reduction of total leukocytes in peritoneal exudate was observed when indomethacin was injected 30 min before cg. On the other hand, the C3G decreases the number of total leukocytes when administered 1 h after the cg stimulus (38% decrease) (Figure 3(d)). Similar effects were obtained with indomethacin administration, which promoted reduction of leukocytes (55% decrease) when administered 1 h after i.p. injection of cg.

Differential cell counts showed that leukocytes present in the peritoneal cavity, after i.p. injection of cg, were predominantly polymorphonuclears (PMN), mainly neutrophils, when compared with the group that received saline (without stimulus). The mean values of PMN were $74 \pm 4 \times 10^5$ cells/mL, and $51 \pm 1 \times 10^5$ cells/mL, in the groups that received saline by gavage 30 min before and 1 h after cg injection, respectively (Figures 3(b) and 3(e)). On the other hand, in the group that received saline instead of cg (without stimulus), in both administration times, the mononuclear leukocytes (MN) were predominant ($13 \pm 1 \times 10^5$ cells/mL). In addition, our results showed that cg injection caused a decrease in the number of MN in the peritoneal cavity ($7.1 \pm 0.1 \times 10^5$ cells/mL) (Figures 3(c) and 3(f)).

Like what occurred with the total leukocytes, the number of PMN in peritoneal fluid in mice was significantly reduced when treated with C3G (39% decrease) or indomethacin (40% decrease) 1 h after the i.p cg stimulus, when compared to the control group that received saline orally (Figure 3(e)). On the other hand, AF administered 30 min before cg, promoted a significant decrease in the recruited PMN (24% decrease), compared to the control group (Figure 3(b)).

These results were different from those observed in other tissues, such as air pouch cg inflammation in mice and acute lung inflammation in rats where a decrease in the influx of cell was observed when C3G was previously administered before the cg stimulus [14, 17].

In relation to MN influx, C3G or AF or indomethacin administrated 30 min before cg injection did not change the decrease counts of MN promoted by cg injection (Figures 3(c) and 3(f)), when compared with the group without cg stimulus.

Since C3G was detected intact and in low concentration in plasma of rats after mulberry juice intake [20], the oral intake performed 1 h after cg stimulus probably could provide an ideal concentration of C3G in plasma, resulting in the observed effect. However, this experimental protocol showed that AF is more effective than C3G as a preventive compound against leukocyte migration, suggesting that the complex mixtures of anthocyanins in AF may provide antileukocyte influx effect mainly through a combination of additive and/or synergistic effects.

3.3. Effect of C3G and AF on Carrageenan-Induced Cyclooxygenase-2 Expression in Peritonitis.

The effect of C3G or AF (4 mg/100 g body weight) on cg-induced COX-2 transcription was measured in peritoneal leukocytes by RT-PCR. As shown in Figures 4(a) and 4(c), the i.p. injection of cg (0.3% w/v) drastically increased COX-2 mRNA and protein expression. On the other hand, the oral administration of

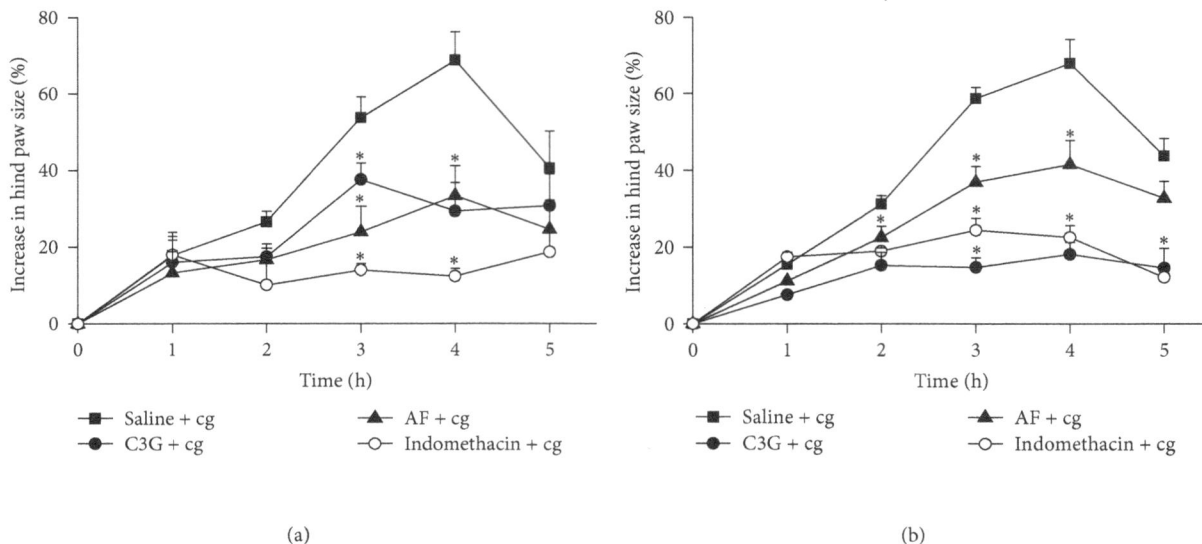

(a)

(b)

FIGURE 2: Effect of C3G and AF on carrageenan-induced paw oedema. Footpad oedema was induced by injection of cg (0.5% w/v in saline, i.pl.) and was evaluated by plethysmometry. C3G or AF (4 mg/100 g body weight) or indomethacin (1 mg/kg, i.v.) or saline (control oedema) was orally administered in two different times: 30 min before (a) and 1 h after (b) i.pl. injection of cg. The increase paw size was measured 1, 2, 3, 4, and 5 h after cg injection. The time zero corresponds to cg injection. The results were expressed as mean ± EPM of 8 mice. Statistically significant difference regarding saline (control group) and C3G and AF and Indomethacin groups is expressed as $^*P < 0.05$.

C3G and AF, either 30 min before and 1 h after cg i.p. injection, clearly downregulated COX-2 mRNA expression (50% reduction) and decreased the levels of COX-2 protein expression, when compared with the control group.

Although some studies have documented that anthocyanins inhibit COX-2 expression in human keratinocyte cell line [15] and cultured macrophages [37, 38] and in asthma model [16], our study provides the first evidence that an anthocyanin mixture or C3G can inhibit, both preventively and therapeutically, the expression of COX-2 protein with a single oral dose. Several lines of evidence clearly established, in *in vitro* models, that the inhibition of some inflammatory cytokines [12, 16] and inhibition of activation of nuclear factor pathway, such as NF-κB [10, 15], could explain the mechanisms of action of anthocyanins on the inhibition of COX-2 expression.

Also, some sources of anthocyanins, such as black soybean anthocyanin and anthocyanins from sweet purple have showed inhibition the COX-2 expression through NF-κB inhibition when administered before the stimulus in inflammation models [11, 12].

3.4. Effect of C3G and AF on Carrageenan-Released PGE₂ in Peritonitis.

Further, this study investigated the effect of C3G and AF (4 mg/100 g body weight) on PGE_2 production, the main inflammatory prostaglandin produced by COX activity, in peritoneal exudates from mice induced by cg. Figures 5(a) and 5(b) showed that i.p. administration of cg induced more than a 25-fold (14.5 ± 2.5 ng/mL) increase in PGE_2 generation compared with the groups without the cg stimulus (0.50 ± 0.05 ng/mL). The PGE_2 concentration

was significantly decreased by the oral pretreatment with C3G, AF, and indomethacin, 30 min before cg injection (4.5 ± 1.0 ng/mL, 5.0 ± 2.0 ng/mL and 2.1 ± 0.1 ng/mL, resp.). In this administration time, the AF and C3G promoted approximately 70% reduction in PGE2 production by cg (Figure 5(a)). On the other hand, the oral treatment of AF or C3G, 1 h after i.p. injection of cg, did not induce any modification in the high levels of PGE_2 release by cg (Figure 5(b)). However, in such experimental condition, the indomethacin suppressed the PGE_2 production by cg stimulus.

Prostaglandin E_2 is a product generated by cyclooxygenases from arachidonic acid, and it is an important mediator in the inflammatory process. In this study, it was observed that after 3 h of administration, cg produced an increase in PGE_2 levels into peritoneal cavity. In parallel, the results showed that C3G produced significant inhibition of PGE_2 production when injected 30 min before cg. However, C3G did not produce such equivalent effectiveness towards cg-induced PGE_2 release when administered 1 h after cg injection. These results are curious because in both administration times used in the present study it was possible to observe that the oral intake of C3G was effective in inhibiting COX-2 expression. Therefore, this suggests that although COX-2 mRNA and protein expression were detected at 3 h after cg injection, this isoform of COX did not present catalytic activity in this period of time. In fact, studies have demonstrated that cg-induced PGE_2 are produced by COX-1 in the first phase, while COX-2-derived PGE_2 turned to be involved in the second phase induced by cg injection [35]. In parallel, our data demonstrated that indomethacin was effective to inhibit PGE_2 production in both administration

Inhibition of Carrageenan-Induced Acute Inflammation in Mice by Oral Administration of Anthocyanin Mixture from Wild Mulberry and Cyanidin-3-Glucoside

61

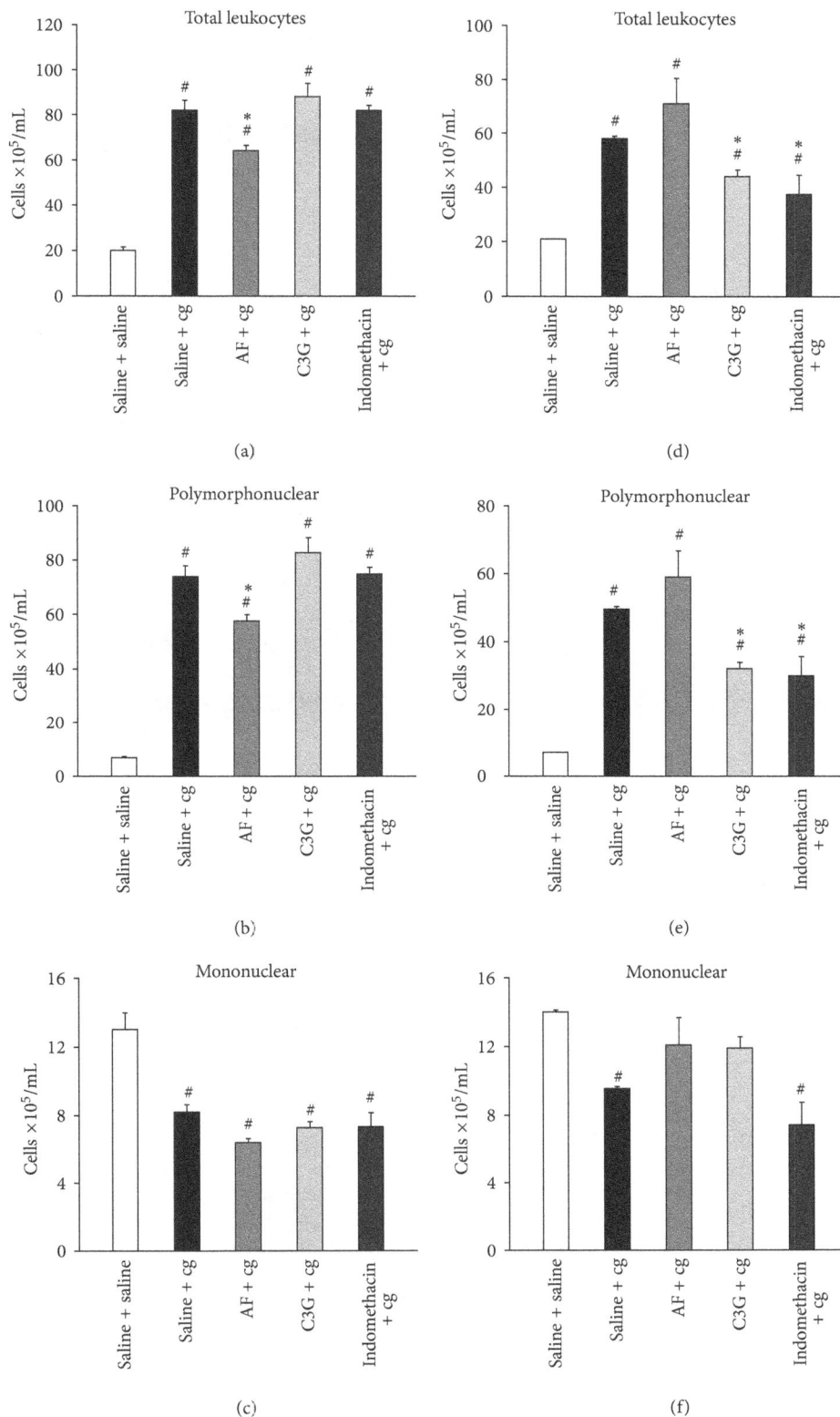

FIGURE 3: Effect of C3G and AF on carrageenan-induced leukocyte influx into peritoneal cavity. Groups of mice received C3G or AF (4 mg/100 g body weight) or indomethacin (4 mg/100 g body weight) or saline (control) by gavage in two different times: 30 min before (a, b, and c) and 1 h after (d, e, and f) cg or saline injection into the peritoneal cavity. Total leukocyte (a, d), PMN (b, e) and MN (c, f) cell counts were determined in peritoneal washes collected 3 h after cg or saline i.p. injection, as described in Section 2. Values are mean ± EPM of 8 animals. $^{\#}P < 0.05$ when compared with the corresponding group without cg stimulus (saline + saline). $^{*}P < 0.05$ when compared with the corresponding control group (saline + cg).

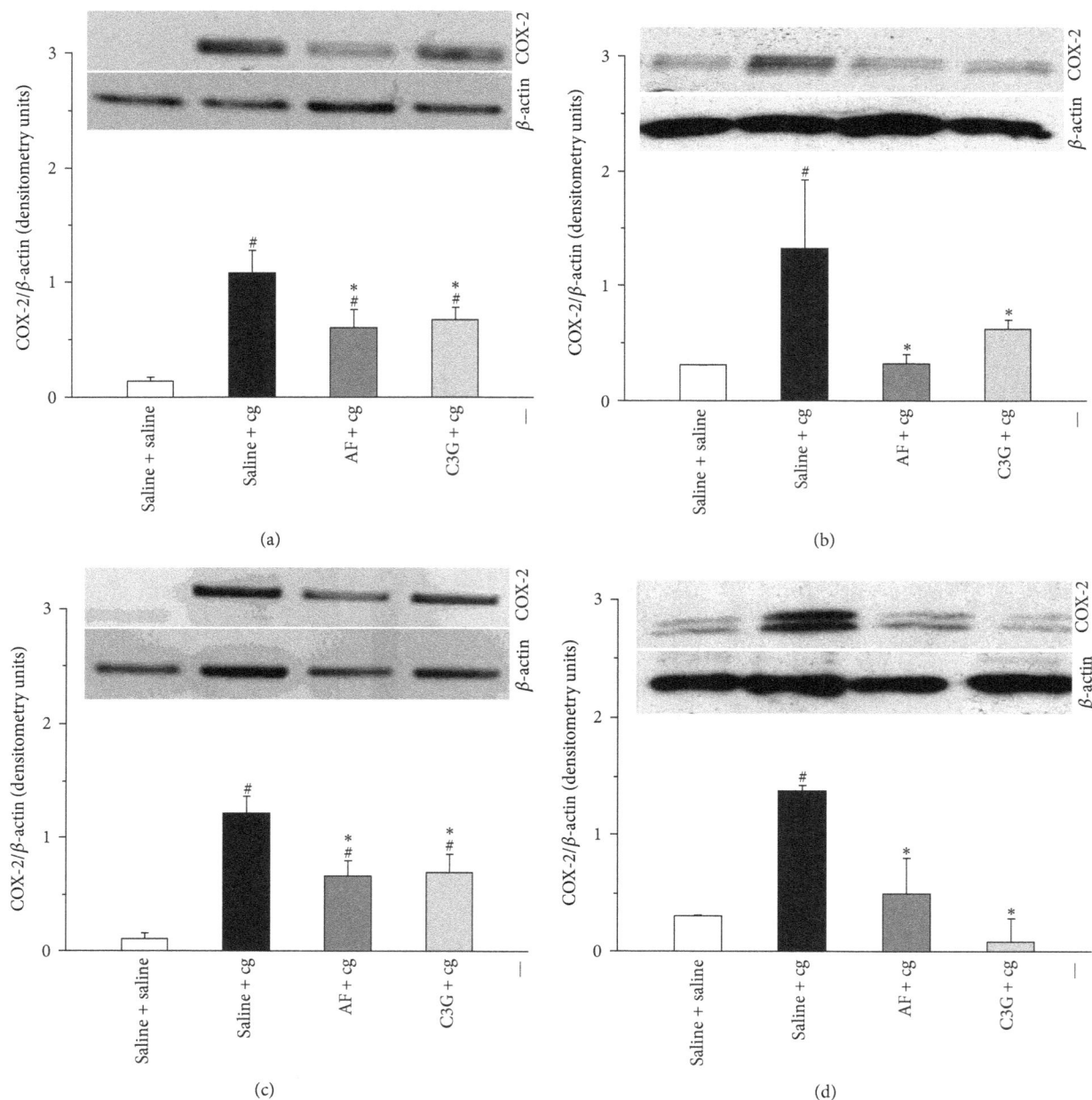

FIGURE 4: Effect of C3G and AF on carrageenan-induced cyclooxygenase-2 expression in peritoneal leukocytes. Groups of mice received C3G or AF (4 mg/100 g body weight) or saline by gavage in two different times: 30 min before or 1 h after cg (0.3% w/v) or saline injection into the peritoneal cavity. Peritoneal leukocytes were collected 3 h after i.p. administration of either cg or saline and whole cells were analyzed for COX-2 expression by RT-PCR and western blotting performed, as described in Section 2. (a and c) RT-PCR of COX-2, and β-actin (loading control); Bar graph shows densitometric analysis of mRNA COX-2. (b and d) Western blotting of COX-2, and β-actin (loading control) of leukocytes present in the inflammatory exudates; bar graph shows densitometric analysis of protein COX-2. The densities (in densitometry units) were normalized with those of β-actin. Results were expressed as mean ± EPM from 8 mice. $^{\#}P < 0.05$ when compared with the corresponding group without cg stimulus (saline + saline). $^{*}P < 0.05$ when compared with the corresponding control group (saline + cg).

times. Although it is generally accepted that nonsteroidal anti-inflammatory drugs such as aspirin and indomethacin are inhibitors of activity of both isoforms of COXs, it is known that these compounds inhibit COX-1 activity more potently than COX-2 in broken cells and in intact cells of mice [39, 40]. In addition, the absence of PGE_2 inhibition when C3G was administered 1 h after cg stimulus compared to the preventive effect obtained by C3G when administered 30 min before the

Inhibition of Carrageenan-Induced Acute Inflammation in Mice by Oral Administration of Anthocyanin
Mixture from Wild Mulberry and Cyanidin-3-Glucoside

63

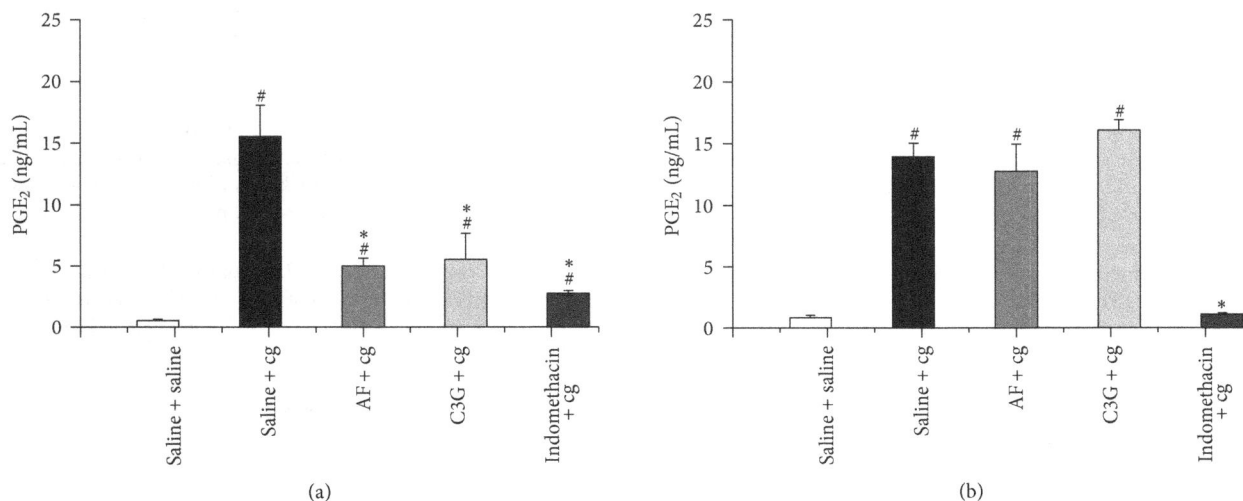

(a)   (b)

FIGURE 5: Effect of C3G and AF on carrageenan-released PGE_2 in peritonitis. Groups of mice received C3G or AF (4 mg/100 g body weight) or indomethacin (4 mg/100 g body weight) or saline (control) by gavage in two different times: 30 min before (a) or 1 h after (b) cg or saline (control) injection into the peritoneal cavity. PGE_2 was quantified in peritoneal exudates collected after 3 h of cg or saline administration. Values are mean ± EPM of 8 mice. $^{\#}P < 0.05$ when compared with the corresponding group without cg stimulus (saline + saline). $^{*}P < 0.05$ when compared with the corresponding control group (saline + cg).

stimulus may be a reflection of the plasma concentrations of this anthocyanin in each administration time.

4. Conclusions

In the present study, AF and C3G have been found to be prophylactic or therapeutically efficient on suppressing cg-induced acute inflammation in mice, like oedema and peritonitis, demonstrating to be an anti-inflammatory component from *Morus nigra*. The results suggest that the anti-inflammatory properties of AF and its major component, the C3G, might be correlated to inhibition of the PMN influx, to downregulation of COX-2 expression, and to inhibition of PGE_2 production.

Acknowledgment

This study was supported by the Fundação de Amparo à Pesquisa do Estado de São Paulo (FAPESP), and Conselho Nacional de Desenvolvimento Científico e Tecnológico (CNPq).

References

[1] A. Castaneda-Ovando, M. D. Pacheco-Hernandez, M. E. Paez-Hernandez, J. A. Rodríguez, and C. A. Galán-Vidal, "Chemical studies of anthocyanins: a review," *Food Chemistry*, vol. 113, no. 4, pp. 859–871, 2009.

[2] M. N. Clifford, "Anthocyanins- nature, occurrence and dietary burden," *Journal Science Food Agriculture*, vol. 80, no. 7, pp. 1063–1072, 2000.

[3] S. Pacifico, B. D'Abrosca, M. Scognamiglio et al., "Metabolic profiling of strawberry grape (Vitis × labruscana cv. "Isabella" components by nuclear magnetic resonance (NMR) and evaluation of their antioxidant and antiproliferative properties," *Journal of Agricultural and Food Chemistry*, vol. 59, no. 14, pp. 7679–7687, 2011.

[4] N. P. Seeram, L. S. Adams, Y. Zhang et al., "Blackberry, black raspberry, blueberry, cranberry, red raspberry, and strawberry extracts inhibit growth and stimulate apoptosis of human cancer cells in vitro," *Journal of Agricultural and Food Chemistry*, vol. 54, no. 25, pp. 9329–9339, 2006.

[5] P. N. Chen, S. C. Chu, H. L. Chiou, W. H. Kuo, C. L. Chiang, and Y. S. Hsieh, "Mulberry anthocyanins, cyanidin 3-rutinoside and cyanidin 3-glucoside, exhibited an inhibitory effect on the migration and invasion of a human lung cancer cell line," *Cancer Letters*, vol. 235, no. 2, pp. 248–259, 2006.

[6] T. Matsui, I. A. Ogunwande, K. J. M. Abesundara, and K. Matsumoto, "Anti-hyperglycemic potential of natural products," *Mini-Reviews in Medicinal Chemistry*, vol. 6, no. 3, pp. 349–356, 2006.

[7] T. Tsuda, F. Horio, K. Uchida, H. Aoki, and T. Osawa, "Dietary cyanidin 3-O-β-D-glucoside-rich purple corn color prevents obesity and ameliorates hyperglycemia in mice," *Journal of Nutrition*, vol. 133, no. 7, pp. 2125–2130, 2003.

[8] T. Tsuda, Y. Ueno, T. Yoshikawa, H. Kojo, and T. Osawa, "Microarray profiling of gene expression in human adipocytes in response to anthocyanins," *Biochemical Pharmacology*, vol. 71, no. 8, pp. 1184–1197, 2006.

[9] K. Wang, P. Jin, S. Cao, H. Shang, Z. Yang, and Y. Zheng, "Methyl jasmonate reduces decay and enhances antioxidant capacity in chinese bayberries," *Journal of Agricultural and Food Chemistry*, vol. 57, no. 13, pp. 5809–5815, 2009.

[10] C. Pergola, A. Rossi, P. Dugo, S. Cuzzocrea, and L. Sautebin, "Inhibition of nitric oxide biosynthesis by anthocyanin fraction of blackberry extract," *Nitric Oxide*, vol. 15, no. 1, pp. 30–39, 2006.

[11] Y. P. Hwang, J. H. Choi, H. J. Yun et al., "Anthocyanins from purple sweet potato attenuate dimethylnitrosamine-induced

liver injury in rats by inducing Nrf2-mediated antioxidant enzymes and reducing COX-2 and iNOS expression," *Food and Chemical Toxicology*, vol. 49, no. 1, pp. 93–99, 2011.

[12] S. L. Yeh, H. M. Wang, P. Y. Chen, and T. C. Wu, "Interactions of β-carotene and flavonoids on the secretion of pro-inflammatory mediators in an in vitro system," *Chemico-Biological Interactions*, vol. 179, no. 2-3, pp. 386–393, 2009.

[13] J. M. Tall, N. P. Seeram, C. Zhao, M. G. Nair, R. A. Meyer, and S. N. Raja, "Tart cherry anthocyanins suppress inflammation-induced pain behavior in rat," *Behavioural Brain Research*, vol. 153, no. 1, pp. 181–188, 2004.

[14] S. W. Min, S. N. Ryu, and D. H. Kim, "Anti-inflammatory effects of black rice, cyanidin-3-O-β-d-glycoside, and its metabolites, cyanidin and protocatechuic acid," *International Immunopharmacology*, vol. 10, no. 8, pp. 959–966, 2010.

[15] K. Tsoyi, H. B. Park, Y. M. Kim et al., "Anthocyanins from black soybean seed coats inhibit UVB-induced inflammatory cylooxygenase-2 gene expression and PGE2 production through regulation of the nuclear factor-κB and phosphatidylinositol 3-kinase/Akt pathway," *Journal of Agricultural and Food Chemistry*, vol. 56, no. 19, pp. 8969–8974, 2008.

[16] S. J. Park, W. H. Shin, J. W. Seo, and E. J. Kim, "Anthocyanins inhibit airway inflammation and hyperresponsiveness in a murine asthma model," *Food and Chemical Toxicology*, vol. 45, no. 8, pp. 1459–1467, 2007.

[17] A. Rossi, I. Serraino, P. Dugo et al., "Protective effects of anthocyanins from blackberry in a rat model of acute lung inflammation," *Free Radical Research*, vol. 37, no. 8, pp. 891–900, 2003.

[18] N. P. Seeram, R. A. Momin, M. G. Nair, and L. D. Bourquin, "Cyclooxygenase inhibitory and antioxidant cyanidin glycosides in cherries and berries," *Phytomedicine*, vol. 8, no. 5, pp. 362–369, 2001.

[19] N. M. A. Hassimotto, M. I. Genovese, and F. M. Lajolo, "Identification and characterisation of anthocyanins from wild mulberry (*Morus nigra* L.) growing in Brazil," *Food Science and Technology International*, vol. 13, no. 1, pp. 17–25, 2007.

[20] N. M. A. Hassimotto, M. I. Genovese, and F. M. Lajolo, "Absorption and metabolism of cyanidin-3-glucoside and cyanidin-3-rutinoside extracted from wild mulberry (*Morus nigra* L.) in rats," *Nutrition Research*, vol. 28, no. 3, pp. 198–207, 2008.

[21] R. Medzhitov, "Origin and physiological roles of inflammation," *Nature*, vol. 454, no. 7203, pp. 428–435, 2008.

[22] F. A. Fitzpatrick, "Cyclooxygenase enzymes: regulation and function," *Current Pharmacology Design*, vol. 10, no. 6, pp. 577–588, 2004.

[23] T. D. Warner and J. A. Mitchell, "Cyclooxygenases: new forms, new inhibitors, and lessons from the clinic," *FASEB Journal*, vol. 18, no. 7, pp. 790–804, 2004.

[24] W. L. Smith, D. L. DeWitt, and R. M. Garavito, "Cyclooxygenases: structural, cellular, and molecular biology," *Annual Review of Biochemistry*, vol. 69, pp. 145–182, 2000.

[25] H. R. Herschman, J. J. Talley, and R. DuBois, "Cyclooxygenase 2 (COX-2) as a target for therapy and noninvasive imaging," *Molecular Imaging and Biology*, vol. 5, no. 5, pp. 286–303, 2003.

[26] J. P. Portanova, Y. Zhang, G. D. Anderson et al., "Selective neutralization of prostaglandin E2 blocks inflammation, hyperalgesia, and interleukin 6 production in vivo," *Journal of Experimental Medicine*, vol. 184, no. 3, pp. 883–891, 1996.

[27] T. G. Brock and M. Peters-Golden, "Activation and regulation of cellular eicosanoid biosynthesis," *TheScientificWorldJournal*, vol. 7, pp. 1273–1284, 2007.

[28] N. V. Chandrasekharan, H. Dai, K. L. T. Roos et al., "COX-3, a cyclooxygenase-1 variant inhibited by acetaminophen and other analgesic/antipyretic drugs: cloning, structure, and expression," *Proceedings of the National Academy of Sciences of the United States of America*, vol. 99, no. 21, pp. 13926–13931, 2002.

[29] B. Kis, J. A. Snipes, T. Isse, K. Nagy, and D. W. Busija, "Putative cyclooxygenase-3 expression in rat brain cells," *Journal of Cerebral Blood Flow and Metabolism*, vol. 23, no. 11, pp. 1287–1292, 2003.

[30] U. K. Laemmli, "Cleavage of structural proteins during the assembly of the head of bacteriophage T4," *Nature*, vol. 227, no. 5259, pp. 680–685, 1970.

[31] J. S. Kang, Y. J. Jeon, S. K. Park, K. H. Yang, and H. M. Kim, "Protection against lipopolysaccharide-induced sepsis and inhibition of interleukin-1β and prostaglandin E2 synthesis by silymarin," *Biochemical Pharmacology*, vol. 67, no. 1, pp. 175–181, 2004.

[32] Y. Bezugla, A. Kolada, S. Kamionka, B. Bernard, R. Scheibe, and P. Dieter, "COX-1 and COX-2 contribute differentially to the LPS-induced release of PGE2 and TxA2 in liver macrophages," *Prostaglandins and Other Lipid Mediators*, vol. 79, no. 1-2, pp. 93–100, 2006.

[33] P. Pradelles, J. Grassi, and J. Maclouf, "Enzyme immunoassays of eicosanoids using acetylcholine esterase as label: an alternative to radioimmunoassay," *Analytical Chemistry*, vol. 57, no. 7, pp. 1170–1173, 1985.

[34] O. M. E. Abdel-Salam, A. R. Baiuomy, S. El-batran, and M. S. Arbid, "Evaluation of the anti-inflammatory, anti-nociceptive and gastric effects of Ginkgo biloba in the rat," *Pharmacological Research*, vol. 49, no. 2, pp. 133–142, 2004.

[35] I. Posadas, M. Bucci, F. Roviezzo et al., "Carrageenan-induced mouse paw oedema is biphasic, age-weight dependent and displays differential nitric oxide cyclooxygenase-2 expression," *British Journal of Pharmacology*, vol. 142, no. 2, pp. 331–338, 2004.

[36] H. L. Malech and J. I. Gallin, "Current concepts: immunology—neutrophils in human diseases," *The New England Journal of Medicine*, vol. 317, no. 11, pp. 687–694, 1987.

[37] X. H. Jin, K. Ohgami, K. Shiratori et al., "Effects of blue honeysuckle (*Lonicera caerulea* L.) extract on lipopolysaccharide-induced inflammation in vitro and in vivo," *Experimental Eye Research*, vol. 82, no. 5, pp. 860–867, 2006.

[38] K. J. Woo, Y. J. Jeong, H. Inoue, J. W. Park, and T. K. Kwon, "Chrysin suppresses lipopolysaccharide-induced cyclooxygenase-2 expression through the inhibition of nuclear factor for IL-6 (NF-IL6) DNA-binding activity," *FEBS Letters*, vol. 579, no. 3, pp. 705–711, 2005.

[39] J. C. Frölich, "A classification of NSAIDs according to the relative inhibition of cyclooxygenase isoenzymes," *Trends in Pharmacological Sciences*, vol. 18, no. 1, pp. 30–34, 1997.

[40] J. A. Mitchell, P. Akarasereenont, C. Thiemermann, R. J. Flower, and J. R. Vane, "Selectivity of nonsteroidal antiinflammatory drugs as inhibitors of constitutive and inducible cyclooxygenase," *Proceedings of the National Academy of Sciences of the United States of America*, vol. 90, no. 24, pp. 11693–11697, 1993.

8

Innovative Strategies to Overcome Biofilm Resistance

Aleksandra Taraszkiewicz, Grzegorz Fila, Mariusz Grinholc, and Joanna Nakonieczna

Laboratory of Molecular Diagnostics, Department of Biotechnology, Intercollegiate Faculty of Biotechnology,
University of Gdansk and Medical University of Gdansk, Kladki 24, 80-822 Gdansk, Poland

Correspondence should be addressed to Joanna Nakonieczna; strzala@biotech.ug.gda.pl

Academic Editor: Tim Maisch

We review the recent literature concerning the efficiency of antimicrobial photodynamic inactivation toward various microbial species in planktonic and biofilm cultures. The review is mainly focused on biofilm-growing microrganisms because this form of growth poses a threat to chronically infected or immunocompromised patients and is difficult to eradicate from medical devices. We discuss the biofilm formation process and mechanisms of its increased resistance to various antimicrobials. We present, based on data in the literature, strategies for overcoming the problem of biofilm resistance. Factors that have potential for use in increasing the efficiency of the killing of biofilm-forming bacteria include plant extracts, enzymes that disturb the biofilm structure, and other nonenzymatic molecules. We propose combining antimicrobial photodynamic therapy with various antimicrobial and antibiofilm approaches to obtain a synergistic effect to permit efficient microbial growth control at low photosensitizer doses.

1. Introduction

Photodynamic therapy dates to the time of the pharaohs and ancient Romans and Greeks, for whom the connection between the sun and health was obvious. Until the 19th century, heliotherapy was the only known form of phototherapy [1]. Heliotherapy was used in thermal stations to cure tuberculosis and to treat ulcers or other skin diseases [2]. The 20th century brought significant developments in phototherapy, particularly in photodynamic therapy (PDT) directed against cancer as well as photodynamic inactivation (PDI) of microorganisms, also known as antimicrobial PDT (APDT). PDT has gained clinical acceptance, and many clinical trials are being conducted, while APDT is in its infancy. As antibiotic therapies become less effective because of increasing microbial resistance to antibiotics, alternative methods such as APDT for fighting infectious diseases are urgently needed. Microbial biofilms cause a large number of chronic infections that are not susceptible to traditional antibiotic treatment [3, 4]. Biofilm-forming microbes are held together by a self-produced matrix that consists of polysaccharides, proteins and extracellular DNA [5, 6].

2. Biofilm: Structure, Biology, and Treatment Problems

A microbial biofilm is defined as a structured community of bacterial cells enclosed in a self-produced polymeric matrix that is adherent to an inert or living surface [4, 7]. The matrix contains polysaccharides, proteins, and extracellular microbial DNA, and the biofilm can consist of one or more microbial (bacterial or fungal) species [5, 8]. The matrix is important because it provides structural stability and protection to the biofilm against adverse environmental conditions, for example, host immunological system and antimicrobial agents [6, 9]. Biofilm-growing microorganisms cause chronic infections which share clinical characteristics, like persistent inflammation and tissue damage [3]. A large number of chronic bacterial infections involve bacterial biofilms, making these infections very difficult to be eradicated by conventional antibiotic therapy [4]. Biofilm formation also causes a multitude of problems in the medical field, particularly in association with prosthetic devices such as indwelling catheters and endotracheal tubes [10]. Biofilms can form on inanimate surface materials such as the inert surfaces of medical devices, catheters, and contact lenses or living tissues, as in endocardium, wounds, and the epithelium of the lungs, particularly in cystic fibrosis patients [8, 11, 12]. Microbial antigens stimulate the production of antibodies, which cannot effectively kill bacteria within the biofilm and may cause immune complex damage to surrounding tissues [13]. Regardless of the presence of excellent cellular and humoral immune reactions, host defense mechanisms are rarely able to resolve biofilm infections [14]. The symptoms caused by the release of planktonic cells from the biofilm can be treated by antibiotic therapy, but the biofilm remains unaffected [15]. Thus, biofilm infection symptoms are recurrent even after several antibiotic therapy cycles, and the only effective means of eradicating the cause of the infection is the removal of the implanted device or the surgical removal of the biofilm that has formed on live tissue [16]. Biofilm-growing bacteria differ from planktonic bacteria with respect to their genetic and biochemical properties. Biofilm-forming bacteria coaggregate with each other and with multiple partners and form coordinated groups attached to an inert or living surface; they surround themselves with polymer matrix, communicate effectively via quorum sensing mechanisms, and express low metabolic activity limiting the impact of conventional antimicrobials acting against actively metabolizing cells [4, 7, 12].

2.1. Biofilm Formation.

Biofilm formation can be divided into three main stages: early, intermediate, and mature [17]. During the early stage, planktonic cells swim along the surface often using their flagella mode of movement or they can be transferred passively with the body fluids (Figure 2). Next, the contact between microorganisms and a surface is made, resulting in the formation of a monolayer of cells [18–20]. At this stage, the bacteria are still susceptible to antibiotics, and perioperative antibiotic prophylaxis can be critical for successful treatment [6, 9]. The importance of the first attachment step was confirmed by experiments with surface attachment-defective (sad) mutant strains of *Pseudomonas aeruginosa*, which are unable to form biofilms [21]. The next step involves irreversible binding to the surface, multiplication of the microorganisms, and the formation of microcolonies [6, 9]. During this stage, the polymer matrix is produced around the microcolonies and generally consists of a mixture of polymeric compounds, primarily polysaccharides (the matrix contributes 50%–90% of the organic matter in biofilms) [22, 23]. Studies on *Candida albicans* have demonstrated that during the third stage (the maturation phase), the amount of extracellular material increases with incubation time until the yeast communities are completely encased within the material [17]. The matrix consists mainly of water, which can be bounded within the capsules of microbial cells or can exist as a solvent [24]. Apart from water and microbial cells, the biofilm matrix is a very complex material. The biofilm matrix consists of polymers secreted by microorganisms within the biofilm, absorbed nutrients and metabolites, and cell lysis products; therefore, all major classes of macromolecules (proteins, polysaccharides, and nucleic acids) are present in addition to peptidoglycan,

lipids, phospholipids, and other cell components [25–27]. The third step of biofilm formation is the formation of a mature community with mushroom-shaped microcolonies [3]. During this stage, the biofilm structure can be disrupted, and microbial cells can be liberated and transferred onto another location/surface, causing expansion of the infection [6, 9].

Biofilm formation is regulated at different stages through diverse mechanisms, among which the best studied is quorum sensing (QS) [28–31]. The QS mechanism involves the production, release, and detection of chemical signaling molecules, which permit communication between microbial cells. The QS process regulates gene expression in a cell-density-dependent manner; for biofilm production, the genes involved in biofilm formation and maturation are activated at a critical population density [32–34]. There are three well-defined groups of signaling QS molecules in bacteria: oligopeptides, acyl homoserine lactones (AHLs), and autoinducer-2 (AI-2) [34]. Gram-positive bacteria predominately use oligopeptides as a communication molecule, and AHLs are specific for Gram-negative bacteria [35, 36]. AI-2 is reported to be a universal signaling molecule that is used for both interspecies and intraspecies communication [34]. Boles and Horswill proposed that the *Staphylococcus aureus agr* quorum sensing system controls not only the switch between planktonic and biofilm growth but also the mechanism of the dispersal of cells from an established biofilm [37]. Moreover, results from our research group indicate that *agr* polymorphism could impact biofilm formation and directly influence bacterial susceptibility to photoinactivation (data not shown).

2.2. Biofilm Resistance.

Infections caused by biofilm-forming bacteria are often difficult to treat. Biofilm formation almost always leads to a large increase in resistance to antimicrobial agents (up to 1000-fold decrease in susceptibility) in comparison with planktonic cultures grown in conventional liquid media [4, 7]. A few mechanisms of biofilm resistance to antibiotics have been proposed. The first proposed mechanism involves the matrix, which represents a physical and chemical barrier to antibiotics. Ciofu et al. [38] demonstrated that the resistance of *P. aeruginosa* biofilms to antimicrobial treatment is related to mucoidy. Mucoid biofilms were up to 1000 times more resistant to tobramycin than nonmucoid biofilms, in spite of similar planktonic MICs [38]. Anderl et al. demonstrated that ciprofloxacin and chloride ion could penetrate a wild-type *Klebsiella pneumoniae* biofilm, while ampicillin could not [39]. By contrast, ampicillin rapidly penetrated a β-lactamase-deficient *K. pneumoniae* biofilm. The authors assumed that the biofilm matrix was not an inherent mechanical barrier to solute mobility and that ampicillin failed to penetrate the biofilm because it was deactivated by the wild-type biofilm at a faster rate than it could diffuse into the film [40]. Jefferson et al. suggested that even though the matrix may not inhibit the penetration of antibiotics, it may retard the rate of penetration enough to induce the expression of genes within the biofilm that mediates resistance [41]. A second hypothesis to explain reduced biofilm susceptibility to antibiotics concerns the metabolic state of microorganisms in a biofilm. Some of the cells located deep inside the biofilm structure experience nutrient limitation and therefore exist in a slow-growing or starved state [42]. Nutrient-depleted zones within the biofilm can result in a stationary phase-like dormancy that may influence the general resistance of biofilms to antibiotics. Walters et al. demonstrated that oxygen penetrated from 50 to 90 μm into colony biofilms formed by *P. aeruginosa* and that the antibiotic action is focused near the air-biofilm interface [43]. This study also showed that oxygen limitation has a role in antibiotic resistance [43]. Slow-growing or nongrowing cells are not very susceptible to many antimicrobial agents because the cells divide infrequently and antibiotics that are active against dividing cells (such as beta-lactams) are not effective. The third hypothesis involves genetic adaptation to different conditions. The mutation frequency of a biofilm-growing microorganism is significantly higher than that of its planktonic form; for *P. aeruginosa*, up to a 105-fold increase in mutability has been observed [44]. A recent study by Ma and Bryers demonstrated that donor populations in biofilms (containing a plasmid with a kanamycin resistance gene) exposed to a sublethal dose of kanamycin exhibited an up to tenfold enhancement in the transfer efficiency of the plasmid [45]. At least some of the cells in a biofilm are likely to adopt a distinct phenotype that is not a response to nutrient limitation but a biologically programmed response to growth on a surface [4, 7]. Several genes are involved in biofilm formation and some of the genes are exclusively expressed in biofilm-growing microorganisms [46, 47].

All published results indicate that a reduction in the efficiency of photodynamic treatment occurs when PDI is applied to biofilm-related experimental models. Thus, it is necessary to identify factors that disrupt biofilm structure or affect biofilm formation.

3. Antimicrobial Photodynamic Therapy

Photodynamic therapy consists of three major components: light, a chemical molecule known as a photosensitizer, and oxygen. The photosensitizer (PS) can be excited by absorbing a certain amount of energy from the light. The excitation occurs when the wavelength range of the light overlaps with absorbance spectrum of the photosensitizer. After excitation, photosensitizers usually form a long-lived triplet-excited state, from which energy can be transferred to biomolecules or directly to molecular oxygen, depending on the reaction type (Figure 1). Type I (Figure 1) reactions involve electron transfer from the triplet state PS to a substrate, for example, unsaturated membrane phospholipids or aminolipids, leading to the production of lipid-derived radicals or hydroxyl radicals (HO$^{\bullet}$) derived from water. These radicals can combine or react with other biomolecules and oxygen to yield hydrogen peroxide, causing lipid peroxidation or leading to the production of reactive oxygen species that can cause cellular damage and cell death [48]. Type II (Figure 1) reactions involve energy transfer from the triplet-state PS to

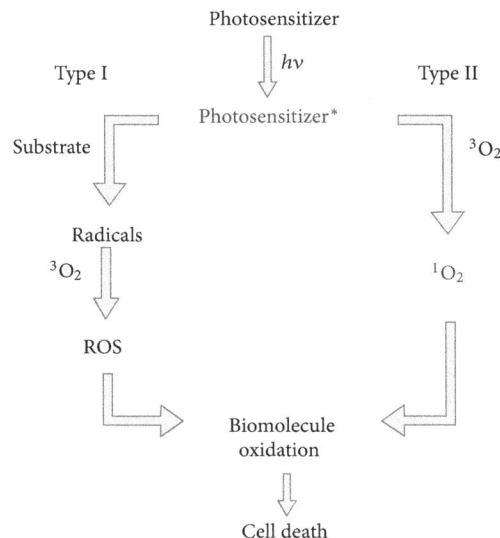

FIGURE 1: Scheme of photodynamic processes. Photosensitizer in excited state forms a long-lived triplet excited state. Type I reactions involve electron transfer from the triplet-state PS to a substrate, leading to production of, for example, lipid-derived radicals which can combine or react with other biomolecules and oxygen, eventually producing reactive oxygen species. In type II reactions, the energy is transferred from the triplet state PS to a ground state (triplet) molecular oxygen to produce excited singlet-state oxygen which can oxidize biomolecules in the cell. Both forms of reactive oxygen can cause cell damage and death.

ground-state (triplet) molecular oxygen to produce excited singlet-state oxygen, which is a very reactive species with the ability to oxidize biomolecules in the cell such as proteins, nucleic acid, and lipids, causing cell damage and death [48]. Both mechanisms can operate in the cell simultaneously, but type II is generally considered the major APDT pathway [49]. There are two major types of cellular damage: DNA damage and the destruction of cellular membranes and organelles. Because the cell is protected by DNA repair systems, DNA damage may not be the main cause of microbial cell death. A large portion of the microbicidal effect of APDT may be due to the disruption of proteins involved in transport and membrane structure and the leakage of cellular contents [49].

Recent studies have shown that the antimicrobial effect can be obtained with the use of photosensitizers belonging to different chemical groups. The most studied PSs are phenothiazine dyes (methylene blue (MB) and toluidine blue O (TBO)), porphyrin and its derivatives, fullerenes, and cyanines and its derivatives (Table 1). More studies have been conducted of forms of microbial growth other than planktonic growth (Table 2). The problem of several chronic microbial infections is now known to be inseparable from biofilm formation by pathogens. Thus, *in vitro* studies have concentrated more on biofilm models as well as *in vivo* models, particularly rat and mouse models of infected wounds (Table 2). As we have discussed previously, when studying the use of photoinactivation in biofilm-related models,

TABLE 1: APDT studies of planktonic microorganisms.

Microorganism	Photosensitizer	References
Staphylococcus aureus, Escherichia coli, Pseudomonas aeruginosa, and *Candida albicans*	Cationic fullerenes	Huang et al., 2010 [60]
Penicillium chrysogenum conidia	Cationic porphyrins	Gomes et al., 2011 [59]
S. aureus	Chlorin e6	Park et al., 2010 [61]
Listeria monocytogenes	MB	Lin et al., 2012 [54]
Candida spp.	MB	Queiroga et al., 2011 [55]
Staphylococcus spp.	MB	Miyabe et al., 2011 [56]
Streptococcus mutans	TBO, MB	Rolim et al., 2012 [57]
Bacillus atrophaeus, Methicillin-resistant *S. aureus Escherichia coli*	TMPyP (5-, 10-, 15-, 20-tetrakis (1-methylpyridinium-4-yl)-porphyrin tetra p-toluenesulfonate)	Maisch et al., 2012 [58]

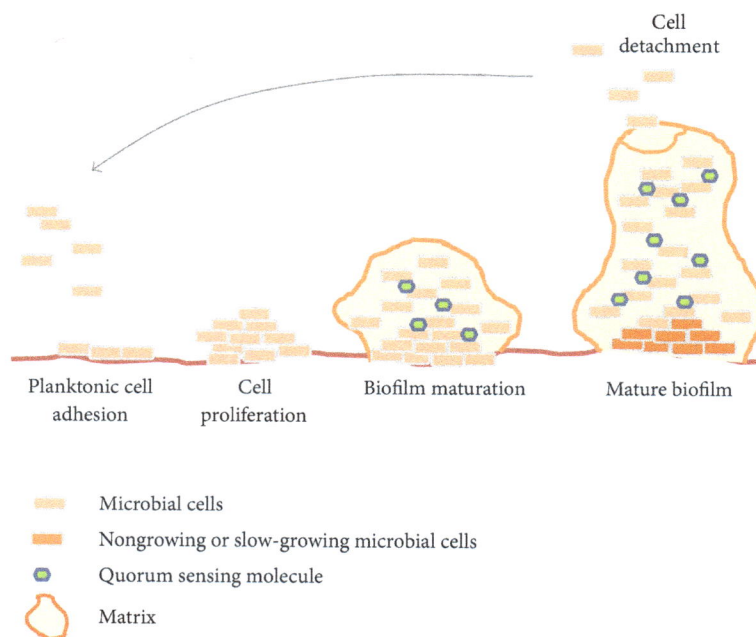

FIGURE 2: Biofilm formation. Planktonic cells adhere to the surface and proliferate. During biofilm maturation, the extracellular matrix and quorum sensing molecules are produced. Mature biofilm is characterized by a large number of matrices, slow-growing microbial cells in the center, and fragmentation which leads to cell detachment and spread of infection.

the mechanism of strain-dependent response to PDI requires further investigation [50–53].

3.1. Recent In Vitro Studies

3.1.1. Planktonic Culture of Microorganisms. In *in vitro* studies of phenothiazine dyes, Lin et al. demonstrated that MB can be successfully used to eradicate *L. monocytogenes* ($3 \log_{10}$ reduction in viability) at a very low concentration of $0.5\,\mu g/mL$ after a 10 min light irradiation (with a tungsten

halogen lamp giving the power output of 165 mW) [54]. Moreover, at higher MB concentrations (up to $1\,\mu g/mL$), the number of viable cells was decreased by up to $7 \log_{10}$ cfu/mL. To inactivate *Candida* species, Queiroga et al. studied much higher concentrations of MB [55]. The PDT effect was strongest in the presence of $150\,\mu g/mL$ MB (78% reduction of CFU/mL) with a light dose of $180\,J/cm^2$ (diode laser InGaAlP, 660 nm). To obtain this light dose, a longer irradiation time was necessary for lower light doses, and the authors suggested that therapy application time should be

TABLE 2: Recent APDT studies of biofilms and animal models.

Microorganism	Photosensitizer	Model	References
P. aeruginosa, Methicillin-resistant S. aureus	MB	Biofilm	Biel et al., 2011 [65]
P. aeruginosa	5-,10-,15-,20-tetrakis(1-methyl-pyridino)-21H, 23H-porphine, tetra-p-tosylate salt (TMP)	Biofilm	Collins et al., 2010 [62]
Candida spp., Trichosporon mucoides, and Kodamaea ohmeri	Cationic nanoemulsion of zinc 2-,9-,16-,23-tetrakis(phenylthio)-29H, 31H-phthalocyanine (ZnPc)	Biofilm	Junqueira et al., 2012 [63]
Enterococcus faecalis	MB	Biofilm	Meire et al., 2012 [66]
Proteus mirabilis P. aeruginosa	Fullerenes B6	Mouse model	Lu et al., 2010 [69]
C. albicans	New MB	Mouse model	Dai et al., 2011 [67]
S. aureus	Chlorin e6	Mouse model	Park et al., 2010 [61]
P. aeruginosa	Hypocrellin B with lanthanide ions (HB:La^{+3})	Mouse model	Hashimoto et al., 2012 [68]

considered as an important factor. Because APDT is related to the production of toxic radicals such as singlet oxygen, the quantity of toxic radicals that are generated should increase as the irradiation time increases [55]. However, our results show no such correlation; for photodynamic inactivation, the light dose is important, not the irradiation time. We obtained the same results for the eradication of S. aureus with a light dose of 12 J/cm^2, whether the irradiation time was 10 or 60 min (data not shown). To inactivate clinical isolates of Staphylococcus species, Miyabe et al. used 3 mM MB and a light fluence of 26.3 J/cm^2 (gallium-aluminum-arsenide laser, 660 nm) to obtain a mean reduction of 6.29 log$_{10}$ cfu/mL [56]. In S. mutans, Rolim et al. did not observe photodynamic activity when MB was used at a concentration of 163.5 μM at 24 J/cm^2 (LED, 640 nm), but a significant reduction (3 log$_{10}$ cfu/mL) was observed when the same concentration of TBO and an equal light dose were used [57]. Maisch et al. reported that incubation of methicillin-sensitive S. aureus (MSSA), methicillin-resistant S. aureus (MRSA), E. coli, and B. atrophaeus with a porphyrin derivative (TMPyP) caused a biologically relevant decrease in CFU/mL upon illumination with multiple light flashes [58]. For MSSA, a TMPyP concentration of 1 μM exhibited a killing efficacy of 2 log$_{10}$ units reduction (at a radiant exposure of 80 J/cm^2), and higher concentrations of TMPyP (10 or 100 μM) caused a further decrease in bacterial survival (more than 5 log$_{10}$ units). E. coli was decreased by 3 log$_{10}$ cfu/mL units after photosensitization with 100 μM TMPyP and a radiant exposure of 20 J/cm^2 and by 5 log$_{10}$ units after a radiant exposure of 40 J/cm^2. However, concentrations of less than 100 μM TMPyP did not induce photodynamic inactivation of E. coli, even with a radiant exposure of up to 80 J/cm^2. MRSA strains that were photosensitized with TMPyP and illuminated under identical conditions exhibited a similar decrease in CFU/mL as that observed for the MSSA strain, indicating that the growth reduction was not dependent on the antibiotic resistance pattern. B. atrophaeus growth was reduced by more than 4 log$_{10}$ by 10 μM TMPyP and a single light flash

of 10 or 20 J/cm^2. For all of the studied strains, higher applied radiant exposures (up to 80 J/cm^2) did not further increase the reduction in growth, and the authors suggested that increasing the radiant exposure appeared to produce a plateau in the killing efficacy [58]. To inactivate P. chrysogenum conidia, Gomes et al. studied porphyrin derivatives based on 5-,10-,15-,20-tetrakis(4-pyridyl) porphyrin and 5-, 10-,15-,20-tetrakis(pentafluorophenyl) porphyrin [59]. A 4 log$_{10}$ unit reduction was observed in the presence of 50 mM 5-,10-,15-,20-tetrakis(N-methylpyridinium-4-yl) porphyrin tetraiodide after 20 min of irradiation (white light at the fluence rate of 200 mW/cm^2). Experiments performed with 100 mM 5-,10-,15-,20-tetrakis(N-methylpyridinium-4-yl) porphyrin tetraiodide and the additional step of removing the PS from the solution by centrifugation did not demonstrate an improvement in the photoinactivation efficiency [59]. Huang et al. studied the effects of PDT on Gram-positive bacteria (S. aureus), two different Gram-negative bacteria (E. coli and P. aeruginosa), and a fungal yeast (C. albicans) [60]. They used fullerene derivatives and white light to illuminate the cells and were able to reduce the growth of all tested microorganisms by 3 to 5 log$_{10}$ units, depending on the microorganism and fullerene derivative. The most efficient was the BF2 derivative [60]. Park et al. demonstrated that chlorin Ce6-mediated PDT significantly reduced the colony formation of S. aureus in a dose-dependent manner [61]. Based on these data, it is clear that APDT can effectively kill various microbial species growing in planktonic culture.

3.1.2. Biofilm Culture of Microorganisms.
It is now well known that infections are mainly associated with biofilm formation. Collins et al. studied the effect of TMP on P. aeruginosa biofilms [62]. A significant decrease in biofilm density was observed, and the majority of the cells within the biofilm were nonviable when 100 μM TMP and 10 min of irradiation (mercury vapor lamp, 220–240 J/cm^2) were used. Moreover, the use of 225 μM TMP and the same light dose resulted in almost complete disruption and clearance of

the studied biofilm [62]. The effect of ZnPc-mediated APDT on yeast biofilms (*C. albicans*, non-*albicans Candida* species and non-*Candida* species) was studied by Junqueira et al. [63]. A gallium-aluminum-arsenide (GaAlAs) laser was used as the light source with the photosensitizer ZnPc at a concentration of 0.25 mg/mL. In all of the studied species, APDT caused reductions in CFU/mL values compared to the control group, but the levels of reduction ranged from 0.33 to 0.85 $\log_{10}$ for the various fungal species. The *Candida* spp. that were most resistant to APDT were *C. albicans, C. glabrata, C. norvegensis, C. krusei,* and *C. lusitaniae* (reduction <0.5 $\log_{10}$). The non-*Candida* pathogens *T. mucoides* and *K. ohmeri* were inactivated by APDT, with reductions of 0.85 $\log_{10}$ and 0.84 $\log_{10}$, respectively [63]. Biel et al. demonstrated that MB-mediated APDT is highly effective in the photoeradication of multispecies bacterial biofilms (multidrug-resistant *P. aeruginosa* and MRSA) [64, 65]. A significant decrease in CFU/mL (>6 $\log_{10}$ units) was achieved when 300 μg/mL MB and a light dose of 60 J/cm^2 (diode laser, 664 nm) were used. The reduction was >7 $\log_{10}$ units when 500 μg/mL MB and two light doses of 55 J/cm^2 separated by a 5-minute break were used [64, 65]. Meire et al. observed a statistically significant 1.9 $\log_{10}$ reduction in the viable counts of *E. faecalis* biofilms treated with 10 mg/mL MB and exposed to a soft laser of an output power of 75 mW (660 nm) for 2 min [66].

For biofilm-based cultures, much higher PS concentrations are required to obtain an APDT killing efficiency comparable to that observed for planktonic cultures. These higher concentrations may be potentially toxic for eukaryotic cells. Thus, it is of great importance to propose a strategy to decrease the PS concentrations used *in vivo* to further facilitate the application of APDT for the treatment of infections in humans and animals. Light parameters such as total light dose, beside the PS concentration, play an important role in APDT efficacy. In general, photoinactivation of microbial cells is dependent on light dose delivered to the sample and its efficacy is increasing with increasing light dose, considering particular light source of specific power density. In fact, lower PS concentration can be substituted by higher light doses, thus giving good opportunity to improve selectivity of APDT in potential clinical applications. The complexity of biological effects of irradiation of microbial cells as well as molecular responses to a PS itself (light-independent effects) demand individual optimization protocols for each reaction.

3.2. Recent In Vivo Studies. Because APDT is an alternative and promising method for treating patients, *in vivo* studies are being conducted. Park et al. performed experiments on biofilms in an *in vivo* mouse model [61]. They demonstrated that Ce6-PDT treatment significantly reduced biofilm formation by *S. aureus* when treated with 10 μM Ce6 and 10 J/cm^2 of laser light. Because the *S. aureus* strain used in the study is bioluminescent, a bioluminescent *in vivo* imaging system (IVIS) was used. The group examined the effect of Ce6-mediated PDT on *in vivo* bacterial growth in a mouse model of skin infection with *S. aureus*, and the reduction

in the intensity of bioluminescence was observed immediately after PDT. Moreover, on the 5th day after infection, the signal was almost undetectable in mice treated with Ce6-mediated PDT [61]. Dai et al. reported that a new MB-mediated APDT effectively treated *C. albicans* skin abrasion infections in mice [67]. In that study, a combination of 400 μM NMB and 78 J/cm^2 red light (Luma-Care lamp) was used to perform APDT 30 min after fungal inoculation, which resulted in a significant decrease in fungal luminescence (only few pixels corresponding to microbes could be observed immediately after APDT). Moreover, no significant reoccurrence of infection was observed at 24 h after APDT [67]. In an *in vivo* study by Hashimoto et al., APDT with 10 μM hypocrellin B with lanthanide ions (HB:La^{+3}) and a light dose of 24 J/cm^2 (blue and red LED) reduced the number of *P. aeruginosa* in burn wounds, delaying bacteremia and decreasing bacterial levels in blood by 2-3 $\log_{10}$ compared to an untreated group [68]. Moreover, mice survival was increased at 24 h [68]. Fullerene-mediated APDT against *P. mirabilis* and *P. aeruginosa* wound infection was investigated by Lu et al. [69]. For *P. mirabilis* infection, 1 mM fullerenes (B6) and illumination with white light yielded a reduction of 96% after 180 J/cm^2, which resulted in a highly significant increase in mouse survival of 82% [69]. For *P. aeruginosa*, the treatment gave a maximum reduction of 95%, but there was no beneficial effect on mouse survival (100% of the mice died within 3 days of infection) [69].

The infectious diseases that can be treated with APDT are mostly found in biofilm form, emphasizing the importance of focusing on the biofilm and its eradication, mass reduction, cell number reduction, and loss of viability. Photodynamic inactivation is a promising treatment option for eradication of microbial infections; however, as a biofilm treatment strategy, it has to overcome the obstacle of exopolymer matrix constituting a physical barrier for the photosensitizers as well as light.

4. Antibiofilm Strategies

Biofilm penetration by biocides or antibiotics is typically strongly hindered. To increase the efficiency of new treatment strategies against bacterial and fungal infections, factors that lead to biofilm growth inhibition, biofilm disruption, or biofilm eradication are being sought. These factors could include enzymes, sodium salts, metal nanoparticles, antibiotics, acids, chitosan derivatives, or plant extracts. All of these factors influence biofilm structure via various mechanisms and with different efficiencies.

4.1. Plant Extracts. Numerous plants are used in folk medicine against various diseases. The increasing antibiotic resistance of pathogenic bacteria has resulted in increased attention by scientists to ethnopharmacology and alternative therapeutic options. Coenye et al. investigated five plant extracts with antibiofilm activity. Sub-MIC concentrations of *Rhodiola crenulata* (arctic root), *Epimedium brevicornum* (rowdy lamb herb), and *Polygonum cuspidatum*

(Japanese knotweed) extracts inhibited *Propionibacterium acnes* biofilm formation by 64.8%, 98.5%, and 99.2%, respectively [70]. Moreover, active compounds within the extracts were identified and tested against three *P. acnes* strains. The most effective compound was resveratrol from *P. cuspidatum*, which reduced biofilm formation by 80% for each strain at a concentration of 0.32% (w/v). Icariin extracted from *E. brevicornum* reduced biofilm formation by 40%–70% at concentrations of 0.01%–0.08% (w/v). The antibiofilm activity of salidroside (0.02%–0.25% concentration) extracted from *R. crenulata* was strain dependent and yielded a biofilm reduction of 40% for *P. acnes* LMG 16711 and less than 20% for other tested strains.

Melia dubia (bead tree) bark extracts were examined by Ravichandiran et al.; at a concentration of 30 mg/mL, these extracts reduced *E. coli* biofilm formation by 84% and inhibited virulence factors such as hemolysins by 20% [71]. Bacterial swarming regulated by quorum sensing mechanisms (QS) was inhibited by 75%, resulting in decreased biofilm expansion [71]. Similar results were reported by Issac Abraham et al. concerning *Capparis spinosa* (caper bush) extract. At a concentration of 2 mg/mL, an inhibition of *E. coli* biofilm formation by 73% was observed [72]. For the pathogens *Serratia marcescens*, *P. aeruginosa*, and *P. mirabilis*, biofilm biomass was reduced by 79%, 75%, and 70%, respectively. Moreover, the mature biofilm structure was disrupted for all of the studied pathogens. Furthermore, the addition of *C. spinosa* extract (100 µg/mL) to a bacterial culture resulted in swimming and swarming inhibition [72]. For *Lagerstroemia speciosa* (giant crape myrtle) extract, 83% biofilm inhibition was achieved at a concentration of 10 mg/mL [73]. Moreover, the anti-QS activity of the *L. speciosa* extract affected tolerance to tobramycin and reduced the expression of virulence factors such as LasA protease, LasB elastase, and pyoverdin [73].

The inhibition of biofilm formation is not the only antibiofilm strategy. Taganna et al. reported that a *Terminalia catappa* (bengal almond) extract at sub-MIC concentrations (500 µg/mL and 1 mg/mL) stimulated biofilm formation; *P. aeruginosa* biofilm formation increased by 220% [74]. Despite increased biofilm formation, the *T. catappa* extract disrupted biofilm structure, and the administration of 1% SDS reduced the biofilm by 46%. Moreover, anti-QS activity and a 50% reduction of LasA expression were observed when the *T. catappa* extract was applied [74].

Highly effective antibiofilm activity was observed for fresh *Allium sativum* extract (fresh garlic extract, FGE). Fourfold treatment of a *P. aeruginosa* biofilm with FGE (24 hrs interval) resulted in biofilm reduction by 6 $\log_{10}$ units. Moreover, *in vivo* prophylactic treatment of a mouse model of kidney infection with FGE (35 mg/mL) for 14 days resulted in a 3 $\log_{10}$ unit decrease in the bacterial load on the fifth day after infection compared to untreated animals. In addition, FGE protected renal tissue from bacterial adherence and resulted in a milder inflammatory response and histopathological changes of infected tissues. Fresh garlic extract inhibited *P. aeruginosa* virulence factors such as pyoverdin, hemolysin, and phospholipase C. Moreover, killing efficacy

and phagocytic uptake of bacteria by peritoneal macrophages were enhanced by garlic extract administration [75].

Extensive studies of the anti-*Staphylococcus epidermidis* biofilm activity of 45 aqueous extracts were published by Trentin et al. [76] At 4 mg/mL, the most effective were extracts derived from *Bauhinia acuruana* branches (orchid tree), *Chamaecrista desvauxii* fruits, *B. acuruana* fruits, and *Pityrocarpa moniliformis* leaves, which decreased biofilm formation by 81.7%, 87.4%, 77.8%, and 77%, respectively. When applied at 10-fold lower concentration, noteworthy biofilm inhibition was observed only in the presence of *Commiphora leptophloeos* stem bark (corkwood) and *Senna macranthera* fruit extracts (reductions of 67.3% and 66.7%, resp.) [76].

Next, Carneiro et al. [77] tested sub-MIC concentrations of casbane diterpene (CS) extracted from *Croton nepetaefolius* bark against two Gram-positive bacteria (*S. aureus* and *S. epidermidis*), five Gram-negative bacteria (*Pseudomonas fluorescens*, *P. aeruginosa*, *Klebsiella oxytoca*, *K. pneumoniae*, and *E. coli*), and three yeasts (*Candida tropicalis*, *C. albicans*, and *C. glabrata*). *S. aureus* and *S. epidermidis* biofilms were significantly disrupted when CS was applied (125 µg/mL and 250 µg/mL, resp.). Among Gram-negative bacteria, *K. oxytoca* biofilms formation were not affected by CS, and *K. pneumoniae* biofilms were reduced by 45%. Administration of CS at a concentration of 125 µg/mL caused complete inhibition of *P. fluorescens* biofilms (by 80%). However, lower concentrations of CS supported *P. aeruginosa* biofilm formation. Similar results were obtained for *E. coli*. The authors explained the observed phenomena by the enhanced production of exopolysaccharides due to the stress induced by the presence of CS in the culture. Casbane diterpene activity against *C. albicans* and *C. tropicalis* was observed, reducing biofilm formation by 50% (at concentrations of 62.5 µg/mL and 15.6 µg/mL, resp.) [77]. *Candida* biofilm formation was inhibited more effectively by *Boesenbergia pandurata* (fingerroot) oil [78]; biofilms were reduced by 63% to 98% when sub-MIC volumes (from 4 µL/mL to 32 µL/mL) were used. Moreover, a significant disruption of mature biofilms was observed when similar volumes of the tested oil were applied [78].

These data confirm that plant extracts have anti-QS, antiseptic, and antivirulence factor properties and can easily inhibit biofilm formation as well as disrupt the mature biofilm structure. Thus, plant extracts in combination with other antimicrobial strategies such as antibiotics or photodynamic inactivation could provide an effective bactericidal tool for the treatment of various bacterial and yeast infections.

4.2. Biofilm-Disrupting Enzymes.

Because the biofilm matrix is composed of DNA, proteins, and extracellular polysaccharides, recent studies have indicated that the disruption of the biofilm structure could be achieved via the degradation of individual biofilm compounds by various enzymes.

4.2.1. Deoxyribonuclease I.

Tetz et al. [79] reported a strong negative impact of deoxyribonuclease I (DNase I) on the structures of biofilms formed by *Acinetobacter baumannii*,

Haemophilus influenzae, K. pneumoniae, E. coli, P. aeruginosa, S. aureus, and *Streptococcus pyogenes*. Using DNase I at a concentration of 10 μg/mL, degradation of mature 24 h formed biofilms by 53.85%, 52.83%, 50.24%, 53.61%, 51.64%, 47.65%, and 49.52%, respectively, was observed. Moreover, bacterial susceptibility to selected antibiotics increased in the presence of DNase I. Azithromycin, rifampin, levofloxacin, ampicillin, and cefotaxime were more effective in the presence of DNase I (5 μg/mL) [79].

Furthermore, Hall-Stoodley et al. [80] reported that DNase I induced biofilm degradation by 66.7%–95% for six clinical isolates of *Streptococcus pneumoniae*, even though the biofilms were grown for six days. The authors revealed that the average biofilm thickness was reduced by 85%–97%, indicating that, within the biofilm, areas composed of lower amounts of extracellular DNA in comparison to adherent cells exist [80].

Moreover, Eckhart et al. [81] investigated the use of DNase I and DNase 1L2 (20 μg/mL) against *S. aureus* and *P. aeruginosa* biofilms. Both enzymes revealed strong antibiofilm activity. After 7 hrs of incubation, *P. aeruginosa* biofilm formation was effectively reduced by DNase 1L2 treatment. However, eighteen hours of incubation in the presence of each enzyme resulted in weak inhibition of biofilm formation. *S. aureus* biofilm formation was significantly reduced by both enzymes, independent of the incubation time [81].

Furthermore, the antibiofilm activity of deoxyribonuclease I (130 μg/mL) in combination with selected antibiotics toward *C. albicans* biofilms was estimated. A reduction of viable counts by 0.5 $\log_{10}$ units was observed for biofilm-growing *C. albicans* incubated with DNase I. Treating *C. albicans* with amphotericin B alone (1 μg/mL) resulted in a 1 $\log_{10}$ unit reduction in cell viability, which increased to 3.5 $\log_{10}$ units in combination with DNase I. At higher concentrations of amphotericin B (>2 μg/mL) and DNase I, cell viability was reduced by 5 $\log_{10}$ units. However, the fungicidal effectiveness of caspofungin and fluconazole decreased when combined with DNase I, indicating that the synergistic effect between the antibiotic and DNase I is dependent on the fungicidal agent used [82].

4.2.2. Lysostaphin. Promising antibiofilm results were also obtained for lysostaphin. Lysostaphin is a natural staphylococcal endopeptidase that can penetrate bacterial biofilms [83, 84]. The antimicrobial properties of lysostaphin were analyzed by Walencka et al. [85], who reported the biofilm inhibitory concentration (BIC) of the enzyme for 13 *S. aureus* and 12 *S. epidermidis* clinical strains. The BIC against 8 *S. aureus* strains was estimated to be between 4 and 32 μg/mL, and for the remaining 5 strains, the BIC value was higher than the maximum tested concentration (>64 μg/mL). The majority of the studied *S. epidermidis* strains were more resistant to lysostaphin activity than were the *S. aureus* strains. Only 2 of the 12 *S. epidermidis* strains exhibited reduced biofilm formation in the presence of 128 μg/mL or 16 μg/mL lysostaphin. For the remaining 10 strains, the BIC value was estimated to be greater than 254 μg/mL. In addition, the combined use of lysostaphin with oxacillin increased the susceptibility of the biofilm-growing bacteria

to the antibiotic. However, no antibiofilm efficiency was observed for hetero-vancomycin-intermediate *S. aureus* and methicillin-resistant *S. epidermidis* strains [85].

High antibiofilm effectiveness of lysostaphin toward *S. aureus* strains was confirmed by Kokai-Kun et al. [86], who used a mouse model to determine the most effective treatment strategy for multiorgan biofilm infection. *S. aureus* biofilms, including methicillin-resistant *S. aureus* (MRSA), were completely eradicated in the presence of lysostaphin when animals were treated with the 15 mg/kg lysostaphin and 50 mg/kg of nafcillin, administered 3 times per day for four days. Moreover, lysostaphin (10 mg/kg) effectively protected indwelling catheters from bacterial infection [86].

In addition, Aguinaga et al. [87] reported that lysostaphin leads to significantly increased antibiotic susceptibility, with strain-dependent activity. The minimal biofilm eradication concentration (MBEC) for MRSA and MSSA strains was estimated for 10 antibiotics in combination with 20 μg/mL lysostaphin. The highest synergistic effect was observed when lysostaphin was combined with doxycycline (MBEC decreased from 4 mg/mL to 0.5 mg/mL) or levofloxacin (MBEC decreased from 2 mg/mL to <1.9 mg/mL) against MRSA and MSSA, respectively [87].

4.2.3. α-Amylases. Craigen et al. [88] analyzed the antibiofilm activity of α-amylases against strains of *S. aureus* and *S. epidermidis*. The tested enzymes effectively reduced formed biofilm and decreased biofilm formation in the case of *S. aureus*. However, no antibiofilm effect of the analyzed enzymes was observed for *S. epidermidis*. Time-course experiments for *S. aureus* showed that biofilms were degraded by 79% within 5 min and by 89% within 30 min of incubation with α-amylases. Amylase at doses of 10, 20, and 100 mg/mL reduced biofilms by 72%, 89%, and 90%, respectively, and inhibited matrix formation by 82%. In fact, *S. aureus* clinical isolates exhibited strain-dependent responses to amylase, but the treatment was successful for each strain. In addition, the antibiofilm activities of amylases from different biological sources were evaluated. The most effective biofilm reduction was reported for α-amylase isolated from *Bacillus subtilis*. Although enzymes derived from human saliva and sweet potato had no effect against preformed biofilms, all of the tested enzymes, regardless of origin, were highly effective in inhibiting biofilm formation [88].

4.2.4. Lyase. Biofilms of two mucoid *P. aeruginosa* strains were treated with gentamycin (64 μg/mL) in combination with alginate lyase (20 U/mL). The studied enzyme caused biofilm matrix liquefaction. Incubation of the biofilm with lyase and gentamycin for 96 h resulted in the complete eradication of the biofilm structure and living bacteria. A reduction of viable counts by 2-3 $\log_{10}$ units was reported for both strains when the combined therapy was applied [89].

4.2.5. Lactonase. Kiran et al. [90] identified lactonase as a potential antibiofilm agent. Biofilms formed by 4 *P. aeruginosa* strains exhibited growth inhibition of 68.8%–76.8% in the presence of enzyme (1 U/mL) compared to the control sample. Moreover, 0.3 U/mL of the enzyme disrupted the biofilm structure and led to increased ciprofloxacin and gentamycin penetration and antimicrobial activity. Additionally, lactonase significantly decreased *P. aeruginosa* virulence factors such as pyocyanin (by 85%–93%), protease activity (by 86%–95%), elastase activity (by 69%–91%), and pyochelin secretion (by 40%–90%) [90].

4.3. Silver Nanoparticles. Silver is a nontoxic antimicrobial metal that can be used in medicine. Kalishwaralal et al. [91] analyzed the antibiofilm activity of silver nanoparticles (AgNPs) against *P. aeruginosa* and *S. epidermidis* strains. Nanoparticles were synthesized with *Bacillus licheniformis* and $AgNO_3$. The mean diameter of the received particles was 50 nm. Incubation with $AgNO_3$-containing nanoparticles (100 nM) inhibited the amount of biofilm formed after 24 h by 98%. Incubation with 50 nM AgNPs reduced exopolysaccharide content, indicating that biofilm formation was inhibited, although bacterial viability was unaffected [91].

Next, Mohanty et al. [92] reported dose-dependent antibiofilm activity of AgNPs against *S. aureus* and *P. aeruginosa*. Silver nanoparticles were prepared in 1% soluble starch with an average particle size of 20 nm. Incubation of biofilms (24 hr incubation) in the presence of 1 μM or 2 μM silver nanoparticles yielded greater than 50% or 85% inhibition of biofilm formation, respectively. Prolonged (48 hr) treatment resulted in 65% and 88% reduction of biofilm formation, respectively. A silver nanoparticle concentration of 0.1 μM did not affect biofilm growth. Moreover, no significant cytotoxic effect was observed at any of the concentrations tested [92].

AgNP activity against *C. albicans* and *C. glabrata* biofilm formation was also estimated. Addition of silver nanoparticles to cultures of *Candida* adherent cells at a concentration of 3.3 μg/mL reduced the percentage of total biomass of adherent *C. glabrata* cells by >90%. Moreover, mature biofilms after the treatment were significantly disrupted (97%) by 54 μg/mL AgNPs. *C. albicans* biofilms exhibited increased resistance in comparison to *C. glabrata* with silver nanoparticle treatment, and an 85% reduction of adherent cell growth was observed at concentration >6.7 μg/mL. No effect on mature biofilms was reported [93].

Chitosan-based silver nanoparticles (CS-AgNPs) reduced *P. aeruginosa* 24 hrs-grown biofilms by >65% at a concentration of 2 μg/mL. *S. aureus* biofilms formation were inhibited by 22% by the same concentration of CS-AgNPs. Treatment with higher dose (5 μg/mL) reduced biofilm formation by 65%. Scanning electron microscopy confirmed the destruction of the *P. aeruginosa* cell membrane by 2 μg/ mL CS-AgNPs. In addition, no cytotoxic effects toward macrophages were observed [94].

4.4. Other Biofilm-Disrupting Factors. As biofilm-related infections have become an increasingly prevalent problem in contemporary medicine, factors that disrupt biofilm structure or exhibit antibiofilm activity have been the subject of intense interest.

The activities of three therapeutic molecules have been evaluated against *E. coli* biofilm formation. At concentrations of 30–125 μg/mL, N-acetyl-L-cysteine reduced biofilm formation by 19.6%–39.7% for 5 of 7 *E. coli* strains. Ibuprofen exhibited greater efficacy, reducing biofilm formation by 37.2% to 44.8% (2–125 μg/mL). Human serum albumin efficiently inhibited biofilm formation at the minimal tested concentration, 8 μg/mL, reducing biofilm formation by 44.9%–79.4% [95].

Arias-Moliz et al. [96] investigated lactic acid at concentrations of 2.5%–20% and demonstrated its antimicrobial activity toward *E. faecalis* and *Enterococcus duran* strains. Complete eradication of biofilms was observed when 15% lactic acid was used for 1 min. In addition, 5% lactic acid reduced the viable cell count by 40.7%–100%. Simultaneous administration with 2% chlorhexidine slightly improved the killing efficacy of lactic acid, while administration with 0.2% cetrimide completely eliminated every tested strain independent of the lactic acid concentration used [96].

Chitosan also exhibits antibiofilm properties [97]. Chitosan nanoparticles were analyzed against 24 hour-formed biofilms of *S. mutans*. The antimicrobial effect of chitosan was tested against the three biofilm layers that could be identified within the mature biofilm structure: the upper (20 μm), middle (15 μm), and lower (2 μm) biofilm layers. High-molecular-weight chitosan displayed biofilm reductions of 21.4% (upper layer), 7.5% (middle layer), and 1.2% (low layer). Low-molecular-weight chitosan reduced 24 hrs-formed biofilms by 93.6%–96.7% in each biofilm layer [97].

Furthermore, Orgaz et al. [98] analyzed the antibiofilm effectiveness of chitosan toward mature biofilms formed by *L. monocytogenes*, *Bacillus cereus*, *S. aureus*, *Salmonella enterica*, and *P. fluorescens*. The *Listeria* biofilm matrix was reduced by >6 $\log_{10}$, 4 $\log_{10}$, and 2.5 $\log_{10}$ units in the presence of 1%, 0.1%, and 0.01% chitosan, respectively. *P. fluorescens* exhibited 5 $\log_{10}$, 1.5 $\log_{10}$, and 1 $\log_{10}$ unit reductions, respectively, in the presence of identical concentrations of chitosan. For *Salmonella* and *Bacillus* species, a greater than 3 $\log_{10}$ unit reduction was not achieved (1% chitosan). The lowest antibiofilm effectiveness (1-2 $\log_{10}$ unit reduction) was obtained for *S. aureus* [98].

Recently, Sun et al. [99] reported the antibiofilm activity of terpinen-4-ol-loaded lipid nanoparticles against *C. albicans* biofilms. The compound used (10 μg/mL) eradicated formed biofilms [99]. Finally, the antibiofilm activity of povidone-iodine (PVP-I) was confirmed by Hosaka et al. [100] against *Porphyromonas gingivalis* and *Fusobacterium nucleatum* biofilms. In the presence (5 min) of 7% PVP-I, 72 hour-formed biofilms of *P. gingivalis* exhibited a 6 $\log_{10}$ unit reduction in viable counts. Lower PVP-I concentrations (2%–5%) reduced biofilms by 2 $\log_{10}$ units. Biofilms formed by *F. nucleatum* were effectively reduced (by >4 $\log_{10}$) after 30 sec of incubation with 5% PVP-I [100].

Recently, numerous antibiofilm researches were published. Considering the fact, that various compounds acting against Gram-positive bacteria, Gram-negative bacteria, or fungi were analyzed, and different stage of biofilm growth (mature biofilm eradication or inhibition of biofilm formation) was assessed, it is difficult to reliably compare all the presented results. Some of the approaches seem, however, to be very promising. Among described plant extracts, fresh garlic showed the highest antibiofilm and antibacterial properties against *P. aeruginosa*. Also Japanese knotweed (*P. cuspidatum*) expresses good efficacy in the treatment of *P. acnes* biofilm formation. What seems, however, to be the most interesting is the ability to search for synergistic effects between different approaches, exemplified by the action of biofilm-disrupting enzymes in combination with antibiotics. *S. aureus* biofilm was completely disrupted by lysostaphin with nafcillin and *P. aeruginosa* biofilm by a combination of lyase with gentamycin, and DNaseI with amphotericin B effectively reduced *C. albicans* biofilm. Chitosan and chitosan-based silver nanoparticles can easily disrupt mature biofilm of *P. aeruginosa* and *S. mutans* and could provide penetration of biofilm structures by antimicrobials. This data suggested that APDT, enzymes, plant extracts, and other compounds can be used in various combinations acting as good antibiofilm and antimicrobial agents. The presented innovative strategies may potentially strongly support classical treatments and cause an increase of their effectiveness.

5. Conclusions

In environments that include the human body, microbial cells form a well-organized structure termed a biofilm. The development of strategies to combat bacteria growing in biofilms is a challenging task; these bacteria are much more resistant to classical antimicrobial therapies and exchange genetic material more easily. Thus, under the pressure of a particular antibiotic, resistant clones are selected. Antimicrobial photodynamic therapy appears to be a very promising therapeutic option to effectively control the growth of microbial biofilms. However, as with other antimicrobial therapies, APDT is generally less effective against microorganisms growing in biofilms than against planktonic cells. Hence, there is a need to develop a therapeutic approach that would (i) increase the sensitivity of the microorganism to already established methods (e.g., antibiotic therapies) by violating the structure of the biofilm or disturbing the communication between a population of microorganisms in the biofilm or (ii) combine several modes of microbicidal action to achieve a synergistic effect. An example of the first approach is to use enzymes that affect the biofilm, while the second approach could be achieved by combining APDT with antibiotics, plant extracts, or biofilm-disrupting enzymes. Moreover, if we combine APDT with the use of enzymes that are specific for microbial structures; the selectivity of the approach will be increased as it potentially will permit the use of lower photosensitizer concentrations. One disadvantage of APDT is the limited amount of data based on animal models. However, the growing number of *in vivo* studies verifying APDT based on various photosensitizers is encouraging and will determine the direction of further research.

Authors' Contribution

A. Taraszkiewicz made substantial contributions to the introduction and antimicrobial-related paragraphs and was also involved in writing and drafting the paper. G. Fila made substantial contributions to the antibiofilm-strategy paragraphs and helped draft the paper. J. Nakonieczna and M. Grinholc made substantial contributions to the conception of the paper and the interpretation of data and were involved in drafting the paper and revising it critically for important intellectual content.

Acknowledgments

This work was supported by Grant no. 1640/B/P01/2010/39 from National Science Centre and Grant no. LIDER/32/36/L-2/10/NCBiR/2011 from the National Centre for Research and Development in Poland.

References

[1] A. F. McDonagh, "Phototherapy: from ancient Egypt to the new millennium," *Journal of Perinatology*, vol. 21, supplement 1, pp. S7–S12, 2001.

[2] R. Roelandts, "The history of phototherapy: something new under the sun?" *Journal of the American Academy of Dermatology*, vol. 46, no. 6, pp. 926–930, 2002.

[3] W. Costerton, R. Veeh, M. Shirtliff, M. Pasmore, C. Post, and G. Ehrlich, "The application of biofilm science to the study and control of chronic bacterial infections," *The Journal of Clinical Investigation*, vol. 112, no. 10, pp. 1466–1477, 2003.

[4] J. W. Costerton, P. S. Stewart, and E. P. Greenberg, "Bacterial biofilms: a common cause of persistent infections," *Science*, vol. 284, no. 5418, pp. 1318–1322, 1999.

[5] C. B. Whitchurch, T. Tolker-Nielsen, P. C. Ragas, and J. S. Mattick, "Extracellular DNA required for bacterial biofilm formation," *Science*, vol. 295, no. 5559, p. 1487, 2002.

[6] N. Høiby, T. Bjarnsholt, M. Givskov, S. Molin, and O. Ciofu, "Antibiotic resistance of bacterial biofilms," *International Journal of Antimicrobial Agents*, vol. 35, no. 4, pp. 322–332, 2010.

[7] J. W. Costerton, "Introduction to biofilm," *International Journal of Antimicrobial Agents*, vol. 11, no. 3-4, pp. 217–221, 1999.

[8] T. Bjarnsholt, P. Ø. Jensen, M. J. Fiandaca et al., "Pseudomonas aeruginosa biofilms in the respiratory tract of cystic fibrosis patients," *Pediatric Pulmonology*, vol. 44, no. 6, pp. 547–558, 2009.

[9] N. Høiby, O. Ciofu, and T. Bjarnsholt, "Pseudomonas aeruginosa biofilms in cystic fibrosis," *Future Microbiology*, vol. 5, no. 11, pp. 1663–1674, 2010.

[10] Y. Taj, F. Essa, F. Aziz, and S. U. Kazmi, "Study on biofilm-forming properties of clinical isolates of Staphylococcus aureus," *Journal of Infection in Developing Countries*, vol. 6, no. 5, pp. 403–409, 2012.

[11] B. W. Ramsey, "Management of pulmonary disease in patients with cystic fibrosis," *The New England Journal of Medicine*, vol. 335, no. 3, pp. 179–188, 1996.

[12] T. F. C. Mah and G. A. O'Toole, "Mechanisms of biofilm resistance to antimicrobial agents," *Trends in Microbiology*, vol. 9, no. 1, pp. 34–39, 2001.

[13] D. M. G. Cochrane, M. R. W. Brown, H. Anwar, P. H. Weller, K. Lam, and J. W. Costeron, "Antibody response to Pseudomonas aeruginosa surface protein antigens in a rat model of chronic lung infection," *Journal of Medical Microbiology*, vol. 27, no. 4, pp. 255–261, 1988.

[14] A. E. Khoury, K. Lam, B. Ellis, and J. W. Costerton, "Prevention and control of bacterial infections associated with medical devices," *ASAIO Journal*, vol. 38, no. 3, pp. M174–M178, 1992.

[15] T. J. Marrie, J. Nelligan, and J. W. Costerton, "A scanning and transmission electron microscopic study of an infected endocardial pacemaker lead," *Circulation*, vol. 66, no. 6, pp. 1339–1341, 1982.

[16] J. W. Costerton, Z. Lewandowski, D. E. Caldwell, D. R. Korber, and H. M. Lappin-Scott, "Microbial biofilms," *Annual Review of Microbiology*, vol. 49, pp. 711–745, 1995.

[17] J. Chandra, D. M. Kuhn, P. K. Mukherjee, L. L. Hoyer, T. McCormick, and M. A. Ghannoum, "Biofilm formation by the fungal pathogen Candida albicans: development, architecture, and drug resistance," *Journal of Bacteriology*, vol. 183, no. 18, pp. 5385–5394, 2001.

[18] E. L. Golovlev, "The mechanism of formation of Pseudomonas aeruginosa biofilm, a type of structured population," *Mikrobiologiya*, vol. 71, no. 3, pp. 293–300, 2002.

[19] A. L. Clutterbuck, C. A. Cochrane, J. Dolman, and S. L. Percival, "Evaluating antibiotics for use in medicine using a poloxamer biofilm model," *Annals of Clinical Microbiology and Antimicrobials*, vol. 6, article 2, 2007.

[20] A. L. Clutterbuck, E. J. Woods, D. C. Knottenbelt, P. D. Clegg, C. A. Cochrane, and S. L. Percival, "Biofilms and their relevance to veterinary medicine," *Veterinary Microbiology*, vol. 121, no. 1-2, pp. 1–17, 2007.

[21] G. A. O'Toole and R. Kolter, "Flagellar and twitching motility are necessary for Pseudomonas aeruginosa biofilm development," *Molecular Microbiology*, vol. 30, no. 2, pp. 295–304, 1998.

[22] H. C. Flemming and J. Wingender, "Relevance of microbial extracellular polymeric substances (EPSs)—part II: technical aspcets," *Water Science and Technology*, vol. 43, no. 6, pp. 9–16, 2001.

[23] H. C. Flemming and J. Wingender, "Relevance of microbial extracellular polymeric substances (EPSs)—part I: structural and ecological aspects," *Water Science and Technology*, vol. 43, no. 6, pp. 1–8, 2001.

[24] C. Mayer, R. Moritz, C. Kirschner et al., "The role of intermolecular interactions: studies on model systems for bacterial biofilms," *International Journal of Biological Macromolecules*, vol. 26, no. 1, pp. 3–16, 1999.

[25] I. W. Sutherland, "Biofilm exopolysaccharides: a strong and sticky framework," *Microbiology*, vol. 147, no. 1, pp. 3–9, 2001.

[26] I. W. Sutherland, "Exopolysaccharides in biofilms, flocs and related structures," *Water Science and Technology*, vol. 43, no. 6, pp. 77–86, 2001.

[27] I. W. Sutherland, "The biofilm matrix—an immobilized but dynamic microbial environment," *Trends in Microbiology*, vol. 9, no. 5, pp. 222–227, 2001.

[28] J. W. Costerton, L. Montanaro, and C. R. Arciola, "Bacterial communications in implant infections: a target for an intelligence war," *International Journal of Artificial Organs*, vol. 30, no. 9, pp. 757–763, 2007.

[29] B. Biradar and P. Devi, "Quorum sensing in plaque biofilms: challenges and future prospects," *Journal of Contemporary Dental Practice*, vol. 12, no. 6, pp. 479–485, 2011.

[30] V. Lazar, "Quorum sensing in biofilms—how to destroy the bacterial citadels or their cohesion/power?" *Anaerobe*, vol. 17, no. 6, pp. 280–285, 2011.

[31] S. Periasamy, H. S. Joo, A. C. Duong et al., "How Staphylococcus aureus biofilms develop their characteristic structure," *Proceedings of the National Academy of Sciences of the United States of America*, vol. 109, no. 4, pp. 1281–1286, 2012.

[32] R. M. Donlan, "Biofilms: microbial life on surfaces," *Emerging Infectious Diseases*, vol. 8, no. 9, pp. 881–890, 2002.

[33] R. M. Donlan and J. W. Costerton, "Biofilms: survival mechanisms of clinically relevant microorganisms," *Clinical Microbiology Reviews*, vol. 15, no. 2, pp. 167–193, 2002.

[34] S. Hooshangi and W. E. Bentley, "From unicellular properties to multicellular behavior: bacteria quorum sensing circuitry and applications," *Current Opinion in Biotechnology*, vol. 19, no. 6, pp. 550–555, 2008.

[35] C. M. Waters and B. L. Bassler, "Quorum sensing: cell-to-cell communication in bacteria," *Annual Review of Cell and Developmental Biology*, vol. 21, pp. 319–346, 2005.

[36] N. C. Reading and V. Sperandio, "Quorum sensing: the many languages of bacteria," *FEMS Microbiology Letters*, vol. 254, no. 1, pp. 1–11, 2006.

[37] B. R. Boles and A. R. Horswill, "Agr-mediated dispersal of Staphylococcus aureus biofilms," *PLoS Pathogens*, vol. 4, no. 4, Article ID e1000052, 2008.

[38] O. Ciofu, L. F. Mandsberg, H. Wang, and N. Hoiby, "Phenotypes selected during chronic lung infection in cystic fibrosis patients: implications for the treatment of Pseudomonas aeruginosa biofilm infections," *FEMS Immunology & Medical Microbiology*, vol. 65, no. 2, pp. 215–225, 2012.

[39] J. N. Anderl, M. J. Franklin, and P. S. Stewart, "Role of antibiotic penetration limitation in Klebsiella pneumoniae biofilm resistance to ampicillin and ciprofloxacin," *Antimicrobial Agents and Chemotherapy*, vol. 44, no. 7, pp. 1818–1824, 2000.

[40] J. N. Anderl, J. Zahller, F. Roe, and P. S. Stewart, "Role of nutrient limitation and stationary-phase existence in Klebsiella pneumoniae biofilm resistance to ampicillin and ciprofloxacin," *Antimicrobial Agents and Chemotherapy*, vol. 47, no. 4, pp. 1251–1256, 2003.

[41] K. K. Jefferson, D. A. Goldmann, and G. B. Pier, "Use of confocal microscopy to analyze the rate of vancomycin penetration through Staphylococcus aureus biofilms," *Antimicrobial Agents and Chemotherapy*, vol. 49, no. 6, pp. 2467–2473, 2005.

[42] M. R. W. Brown, D. G. Allison, and P. Gilbert, "Resistance of bacterial biofilms to antibiotics: a growth-rate related effect?" *Journal of Antimicrobial Chemotherapy*, vol. 22, no. 6, pp. 777–780, 1988.

[43] M. C. Walters III, F. Roe, A. Bugnicourt, M. J. Franklin, and P. S. Stewart, "Contributions of antibiotic penetration, oxygen limitation, and low metabolic activity to tolerance of Pseudomonas aeruginosa biofilms to ciprofloxacin and tobramycin," *Antimicrobial Agents and Chemotherapy*, vol. 47, no. 1, pp. 317–323, 2003.

[44] K. Driffield, K. Miller, J. M. Bostock, A. J. O'neill, and I. Chopra, "Increased mutability of Pseudomonas aeruginosa in biofilms," *Journal of Antimicrobial Chemotherapy*, vol. 61, no. 5, pp. 1053–1056, 2008.

[45] H. Ma and J. D. Bryers, "Non-invasive determination of conjugative transfer of plasmids bearing antibiotic-resistance genes in biofilm-bound bacteria: effects of substrate loading and antibiotic selection," *Applied Microbiology and Biotechnology*. In press.

[46] M. K. Yadav, S. K. Kwon, C. G. Cho, S. W. Park, S. W. Chae, and J. J. Song, "Gene expression profile of early in vitro biofilms of Streptococcus pneumoniae," *Microbiology and Immunology*, vol. 56, no. 9, pp. 621–629, 2012.

[47] E. Szczuka, K. Urbanska, M. Pietryka, and A. Kaznowski, "Biofilm density and detection of biofilm-producing genes in methicillin-resistantStaphylococcus aureus strains," *Folia Microbiol (Praha)*. In press.

[48] M. Wainwright, "Photodynamic antimicrobial chemotherapy (PACT)," *Journal of Antimicrobial Chemotherapy*, vol. 42, no. 1, pp. 13–28, 1998.

[49] M. R. Hamblin and T. Hasan, "Photodynamic therapy: a new antimicrobial approach to infectious disease?" *Photochemical and Photobiological Sciences*, vol. 3, no. 5, pp. 436–450, 2004.

[50] M. Grinholc, B. Szramka, J. Kurlenda, A. Graczyk, and K. P. Bielawski, "Bactericidal effect of photodynamic inactivation against methicillin-resistant and methicillin-susceptible Staphylococcus aureus is strain-dependent," *Journal of Photochemistry and Photobiology B*, vol. 90, no. 1, pp. 57–63, 2008.

[51] M. Grinholc, M. Richter, J. Nakonieczna, G. Fila, and K. P. Bielawski, "The connection between agr and SCCmec elements of staphylococcus aureus strains and their response to photodynamic inactivation," *Photomedicine and Laser Surgery*, vol. 29, no. 6, pp. 413–419, 2011.

[52] M. Grinholc, J. Zawacka-Pankau, A. Gwizdek-Wiśniewska, and K. P. Bielawski, "Evaluation of the role of the pharmacological inhibition of staphylococcus aureus multidrug resistance pumps and the variable levels of the uptake of the sensitizer in the strain-dependent response of staphylococcus aureus to PPArg2-based photodynamic inactivation," *Photochemistry and Photobiology*, vol. 86, no. 5, pp. 1118–1126, 2010.

[53] J. Nakonieczna, E. Michta, M. Rybicka, M. Grinholc, A. Gwizdek-Wisniewska, and K. P. Bielawski, "Superoxide dismutase is upregulated in Staphylococcus aureus following protoporphyrin-mediated photodynamic inactivation and does not directly influence the response to photodynamic treatment," *BMC Microbiology*, vol. 10, article 323, 2010.

[54] S. L. Lin, J. M. Hu, S. S. Tang, X. Y. Wu, Z. Q. Chen, and S. Z. Tang, "Photodynamic inactivation of methylene blue and tungsten-halogen lamp light against food pathogen Listeria monocytogenes," *Photochemistry and Photobiology*, vol. 88, no. 4, pp. 985–991, 2012.

[55] A. S. Queiroga, V. N. Trajano, E. O. Lima, A. F. Ferreira, A. S. Queiroga, and F. A. Limeira Jr., "In vitro photodynamic inactivation of Candida spp. by different doses of low power laser light," *Photodiagnosis and Photodynamic Therapy*, vol. 8, no. 4, pp. 332–336, 2011.

[56] M. Miyabe, J. C. Junqueira, A. C. B. P. da Costa, A. O. C. Jorge, M. S. Ribeiro, and I. S. Feist, "Effect of photodynamic therapy on clinical isolates of Staphylococcus spp," *Brazilian Oral Research*, vol. 25, no. 3, pp. 230–234, 2011.

[57] J. P. Rolim, M. A. de-Melo, S. F. Guedes et al., "The antimicrobial activity of photodynamic therapy against Streptococcus mutans using different photosensitizers," *Journal of Photochemistry and Photobiology B*, vol. 106, pp. 40–46, 2012.

[58] T. Maisch, F. Spannberger, J. Regensburger, A. Felgentrager, and W. Baumler, "Fast and effective: intense pulse light photodynamic inactivation of bacteria," *Journal of Industrial Microbiology and Biotechnology*, vol. 39, no. 7, pp. 1013–1021, 2012.

[59] M. C. Gomes, S. M. Woranovicz-Barreira, M. A. Faustino et al., "Photodynamic inactivation of Penicillium chrysogenum conidia by cationic porphyrins," *Photochemical & Photobiological Sciences*, vol. 10, no. 11, pp. 1735–1743, 2011.

[60] L. Huang, M. Terakawa, T. Zhiyentayev et al., "Innovative cationic fullerenes as broad-spectrum light-activated antimicrobials," *Nanomedicine*, vol. 6, no. 3, pp. 442–452, 2010.

[61] J. H. Park, Y. H. Moon, I. S. Bang et al., "Antimicrobial effect of photodynamic therapy using a highly pure chlorin e6," *Lasers in Medical Science*, vol. 25, no. 5, pp. 705–710, 2010.

[62] T. L. Collins, E. A. Markus, D. J. Hassett, and J. B. Robinson, "The effect of a cationic porphyrin on pseudomonas aeruginosa biofilms," *Current Microbiology*, vol. 61, no. 5, pp. 411–416, 2010.

[63] J. C. Junqueira, A. O. Jorge, J. O. Barbosa et al., "Photodynamic inactivation of biofilms formed by Candida spp., Trichosporon mucoides, and Kodamaea ohmeri by cationic nanoemulsion of zinc 2, 9, 16, 23-tetrakis(phenylthio)-29H, 31H-phthalocyanine (ZnPc)," *Lasers in Medical Science*. In press.

[64] M. A. Biel, C. Sievert, M. Usacheva et al., "Reduction of endotracheal tube biofilms using antimicrobial photodynamic therapy," *Lasers in Surgery and Medicine*, vol. 43, no. 7, pp. 586–590, 2011.

[65] M. A. Biel, C. Sievert, M. Usacheva, M. Teichert, and J. Balcom, "Antimicrobial photodynamic therapy treatment of chronic recurrent sinusitis biofilms," *International Forum of Allergy & Rhinology*, vol. 1, no. 5, pp. 329–334, 2011.

[66] M. A. Meire, T. Coenye, H. J. Nelis, and R. J. De Moor, "Evaluation of Nd:YAG and Er:YAG irradiation, antibacterial photodynamic therapy and sodium hypochlorite treatment on Enterococcus faecalis biofilms," *International Endodontic Journal*, vol. 45, no. 5, pp. 482–491, 2012.

[67] T. Dai, d. A. Bil, G. P. Tegos, and M. R. Hamblin, "Blue dye and red light, a dynamic combination for prophylaxis and treatment of cutaneous Candida albicans infections in mice," *Antimicrobial Agents and Chemotherapy*, vol. 55, no. 12, pp. 5710–5717, 2011.

[68] M. C. Hashimoto, R. A. Prates, I. T. Kato, S. C. Nunez, L. C. Courrol, and M. S. Ribeiro, "Antimicrobial photodynamic therapy on drug-resistant Pseudomonas aeruginosa-induced infection. An in vivo study," *Photochemistry and Photobiology*, vol. 88, no. 3, pp. 590–595, 2012.

[69] Z. Lu, T. Dai, L. Huang et al., "Photodynamic therapy with a cationic functionalized fullerene rescues mice from fatal wound infections," *Nanomedicine*, vol. 5, no. 10, pp. 1525–1533, 2010.

[70] T. Coenye, G. Brackman, P. Rigole et al., "Eradication of Propionibacterium acnes biofilms by plant extracts and putative identification of icariin, resveratrol and salidroside as active compounds," *Phytomedicine*, vol. 19, no. 5, pp. 409–412, 2012.

[71] V. Ravichandiran, K. Shanmugam, K. Anupama, S. Thomas, and A. Princy, "Structure-based virtual screening for plant-derived SdiA-selective ligands as potential antivirulent agents against uropathogenic Escherichia coli," *European Journal of Medicinal Chemistry*, vol. 48, pp. 200–205, 2012.

[72] S. V. Issac Abraham, A. Palani, B. R. Ramaswamy, K. P. Shunmugiah, and V. R. Arumugam, "Antiquorum sensing and antibiofilm potential of Capparis spinosa," *Archives of Medical Research*, vol. 42, no. 8, pp. 658–668, 2011.

[73] B. N. Singh, H. B. Singh, A. Singh, B. R. Singh, A. Mishra, and C. S. Nautiyal, "Lagerstroemia speciosa fruit extract modulates quorum sensing-controlled virulence factor production and biofilm formation in Pseudomonas aeruginosa," *Microbiology*, vol. 158, part 2, pp. 529–538, 2012.

[74] J. C. Taganna, J. P. Quanico, R. M. G. Perono, E. C. Amor, and W. L. Rivera, "Tannin-rich fraction from Terminalia catappa inhibits quorum sensing (QS) in Chromobacterium violaceum and the QS-controlled biofilm maturation and LasA staphylolytic activity in Pseudomonas aeruginosa," *Journal of Ethnopharmacology*, vol. 134, no. 3, pp. 865–871, 2011.

[75] K. Harjai, R. Kumar, and S. Singh, "Garlic blocks quorum sensing and attenuates the virulence of Pseudomonas aeruginosa," *FEMS Immunology and Medical Microbiology*, vol. 58, no. 2, pp. 161–168, 2010.

[76] D. D. S. Trentin, R. B. Giordani, K. R. Zimmer et al., "Potential of medicinal plants from the Brazilian semi-arid region (Caatinga) against Staphylococcus epidermidis planktonic and biofilm lifestyles," *Journal of Ethnopharmacology*, vol. 137, no. 1, pp. 327–335, 2011.

[77] V. A. Carneiro, H. S. Dos Santos, F. V. S. Arruda et al., "Casbane diterpene as a promising natural antimicrobial agent against biofilm-associated infections," *Molecules*, vol. 16, no. 1, pp. 190–201, 2011.

[78] S. Taweechaisupapong, S. Singhara, P. Lertsatitthanakorn, and W. Khunkitti, "Antimicrobial effects of Boesenbergia pandurata and Piper sarmentosum leaf extracts on planktonic cells and biofilm of oral pathogens," *Pakistan Journal of Pharmaceutical Sciences*, vol. 23, no. 2, pp. 224–231, 2010.

[79] G. V. Tetz, N. K. Artemenko, and V. V. Tetz, "Effect of DNase and antibiotics on biofilm characteristics," *Antimicrobial Agents and Chemotherapy*, vol. 53, no. 3, pp. 1204–1209, 2009.

[80] L. Hall-Stoodley, L. Nistico, K. Sambanthamoorthy et al., "Characterization of biofilm matrix, degradation by DNase treatment and evidence of capsule downregulation in Streptococcus pneumoniae clinical isolates," *BMC Microbiology*, vol. 8, article 173, 2008.

[81] L. Eckhart, H. Fischer, K. B. Barken, T. Tolker-Nielsen, and E. Tschachler, "DNase1L2 suppresses biofilm formation by Pseudomonas aeruginosa and Staphylococcus aureus," *British Journal of Dermatology*, vol. 156, no. 6, pp. 1342–1345, 2007.

[82] M. Martins, M. Henriques, J. L. Lopez-Ribot, and R. Oliveira, "Addition of DNase improves the in vitro activity of antifungal drugs against Candida albicans biofilms," *Mycoses*, vol. 55, no. 1, pp. 80–85, 2012.

[83] I. Belyansky, V. B. Tsirline, T. R. Martin et al., "The addition of lysostaphin dramatically improves survival, protects porcine biomesh from infection, and improves graft tensile shear strength," *Journal of Surgical Research*, vol. 171, no. 2, pp. 409–415, 2011.

[84] I. Belyansky, V. B. Tsirline, P. N. Montero et al., "Lysostaphin-coated mesh prevents staphylococcal infection and significantly improves survival in a contaminated surgical field," *The American Journal of Surgery*, vol. 77, no. 8, pp. 1025–1031, 2011.

[85] E. Walencka, B. Sadowska, S. Rózalska, W. Hryniewicz, and B. Rózalska, "Lysostaphin as a potential therapeutic agent for staphylococcal biofilm eradication," *Polish Journal of Microbiology*, vol. 54, no. 3, pp. 191–200, 2005.

[86] J. F. Kokai-Kun, T. Chanturiya, and J. J. Mond, "Lysostaphin eradicates established Staphylococcus aureus biofilms in jugular vein catheterized mice," *Journal of Antimicrobial Chemotherapy*, vol. 64, no. 1, pp. 94–100, 2009.

[87] A. Aguinaga, M. L. Francés, J. L. Del Pozo et al., "Lysostaphin and clarithromycin: a promising combination for the eradication of Staphylococcus aureus biofilms," *International Journal of Antimicrobial Agents*, vol. 37, no. 6, pp. 585–587, 2011.

[88] B. Craigen, A. Dashiff, and D. E. Kadouri, "The use of commercially available alpha-amylase compounds to inhibit and remove Staphylococcus aureus biofilms," *Open Microbiology Journal*, vol. 5, pp. 21–31, 2011.

[89] M. A. Alkawash, J. S. Soothill, and N. L. Schiller, "Alginate lyase enhances antibiotic killing of mucoid Pseudomonas aeruginosa in biofilms," *Acta Pathologica, Microbiologica et Immunologica Scandinavica*, vol. 114, no. 2, pp. 131–138, 2006.

[90] S. Kiran, P. Sharma, K. Harjai, and N. Capalash, "Enzymatic quorum quenching increases antibiotic susceptibility of multidrug resistant Pseudomonas aeruginosa," *Iranian Journal of Microbiology*, vol. 3, no. 1, pp. 1–12, 2011.

[91] K. Kalishwaralal, S. BarathManiKanth, S. R. K. Pandian, V. Deepak, and S. Gurunathan, "Silver nanoparticles impede the biofilm formation by Pseudomonas aeruginosa and Staphylococcus epidermidis," *Colloids and Surfaces B*, vol. 79, no. 2, pp. 340–344, 2010.

[92] S. Mohanty, S. Mishra, P. Jena, B. Jacob, B. Sarkar, and A. Sonawane, "An investigation on the antibacterial, cytotoxic, and antibiofilm efficacy of starch-stabilized silver nanoparticles," *Nanomedicinem*, vol. 8, no. 6, pp. 916–924, 2012.

[93] D. R. Monteiro, L. F. Gorup, S. Silva et al., "Silver colloidal nanoparticles: antifungal effect against adhered cells and biofilms of Candida albicans and Candida glabrata," *Biofouling*, vol. 27, no. 7, pp. 711–719, 2011.

[94] P. Jena, S. Mohanty, R. Mallick, B. Jacob, and A. Sonawane, "Toxicity and antibacterial assessment of chitosancoated silver nanoparticles on human pathogens and macrophage cells," *International Journal of Nanomedicine*, vol. 7, pp. 1805–1818, 2012.

[95] P. Naves, G. del Prado, L. Huelves et al., "Effects of human serum albumin, ibuprofen and N-acetyl-l-cysteine against biofilm formation by pathogenic Escherichia coli strains," *Journal of Hospital Infection*, vol. 76, no. 2, pp. 165–170, 2010.

[96] M. T. Arias-Moliz, P. Baca, S. Ordonez-Becerra, M. P. Gonzalez-Rodriguez, and C. M. Ferrer-Luque, "Eradication of enterococci biofilms by lactic acid alone and combined with chlorhexidine and cetrimide," *Medicina Oral Patologia Oral y Cirugia Bucal*, vol. 17, no. 5, pp. 902–906, 2012.

[97] L. E. C. de Paz, A. Resin, K. A. Howard, D. S. Sutherland, and P. L. Wejse, "Antimicrobial effect of chitosan nanoparticles on Streptococcus mutans biofilms," *Applied and Environmental Microbiology*, vol. 77, no. 11, pp. 3892–3895, 2011.

[98] B. Orgaz, M. M. Lobete, C. H. Puga, and C. S. Jose, "Effectiveness of chitosan against mature biofilms formed by food related bacteria," *International Journal of Molecular Sciences*, vol. 12, no. 1, pp. 817–828, 2011.

[99] L. M. Sun, C. L. Zhang, and P. Li, "Characterization, antibiofilm, and mechanism of action of novel PEG-stabilized lipid nanoparticles loaded with terpinen-4-ol," *Journal of Agricultural and Food Chemistry*, vol. 60, no. 24, pp. 6150–6156, 2012.

[100] Y. Hosaka, A. Saito, R. Maeda et al., "Antibacterial activity of povidone-iodine against an artificial biofilm of Porphyromonas gingivalis and Fusobacterium nucleatum," *Archives of Oral Biology*, vol. 57, no. 4, pp. 364–368, 2012.

9

Strategies to Characterize Fungal Lipases for Applications in Medicine and Dairy Industry

Subash C. B. Gopinath,[1,2] **Periasamy Anbu,**[3] **Thangavel Lakshmipriya,**[4] **and Azariah Hilda**[1]

[1] Center for Advanced Studies in Botany, University of Madras, Guindy Campus, Chennai, Tamil Nadu 600025, India
[2] Electronics and Photonics Research Institute, National Institute of Advanced Industrial Science and Technology, Central 5, 1-1-1 Higashi, Tsukuba, Ibaraki 305-8565, Japan
[3] Department of Biological Engineering, College of Engineering, Inha University, Incheon 402-751, Republic of Korea
[4] Department of Mathematics, SBK College, Madurai Kamaraj University, Aruppukottai, Tamil Nadu 626101, India

Correspondence should be addressed to Subash C. B. Gopinath; gopi-subashchandrabose@aist.go.jp
and Periasamy Anbu; anbu25@yahoo.com

Academic Editor: Bidur Prasad Chaulagain

Lipases are water-soluble enzymes that act on insoluble substrates and catalyze the hydrolysis of long-chain triglycerides. Lipases play a vital role in the food, detergent, chemical, and pharmaceutical industries. In the past, fungal lipases gained significant attention in the industries due to their substrate specificity and stability under varied chemical and physical conditions. Fungal enzymes are extracellular in nature, and they can be extracted easily, which significantly reduces the cost and makes this source preferable over bacteria. Soil contaminated with spillage from the products of oil and dairy harbors fungal species, which have the potential to secrete lipases to degrade fats and oils. Herein, the strategies involved in the characterization of fungal lipases, capable of degrading fatty substances, are narrated with a focus on further applications.

1. Introduction

Lipases are defined as triacylglycerol acyl hydrolases (EC 3.1.1.3) and are involved in the hydrolysis of fats and oils to yield glycerol and free fatty acids [1] (Figure 1(a)). Lipases belong to the class of serine hydrolases and do not require any cofactor. Under natural conditions, lipases catalyze the hydrolysis of ester bonds at the interface between an insoluble substrate phase and the aqueous phase where the enzyme remains dissolved [2] (Figure 1(b)). Lipases are involved in conversion reactions, such as esterification, interesterification, transesterification, alcoholysis, acidolysis, and aminolysis [3]. Many microorganisms such as bacteria, yeasts, molds, and a few protozoa are known to secrete lipases for the digestion of lipid materials [1, 4–12]. Microbes, being ubiquitous in distribution, are highly successful at surviving in a wide range of environmental conditions owing to their great plasticity and physiological versatility and have been the subject of several reviews [13, 14]. Due to efficient enzyme

systems, microbes thrive well in inhospitable habitats [15]. With mechanisms for adapting to environmental extremes and for the utilization of their trophic niche, the ability of microorganisms to produce extracellular enzymes is of great survival value [16]. Among different microbial enzymes, lipases are widely documented among bacteria, fungi, plants, and animals [17, 18].

Extracellular secretion has been well studied for a number of fungi, primarily zygomycetes [19], hyphomycetes [20], and yeasts [21, 22]. Lipase production has also been reported for some ascomycetes [23] and coelomycetes [24]. Lipolytic activity has been observed in *Mucor* spp. [25, 26], *Lipomyces starkeyi* [27], *Rhizopus* spp. [26, 28–30], *Geotrichum candidum* [25, 31–34], *Pencillium* spp. [9, 28, 35, 36], *Acremonium strictum* [37], *Candida rugosa* [38], *Humicola lanuginosa* [39], *Cunninghamella verticillata* [40], and *Aspergillus* spp. [11, 41]. Considering the importance of fungal lipases, their applications are discussed and the techniques involved in lipase generation have been gleaned recently [1].

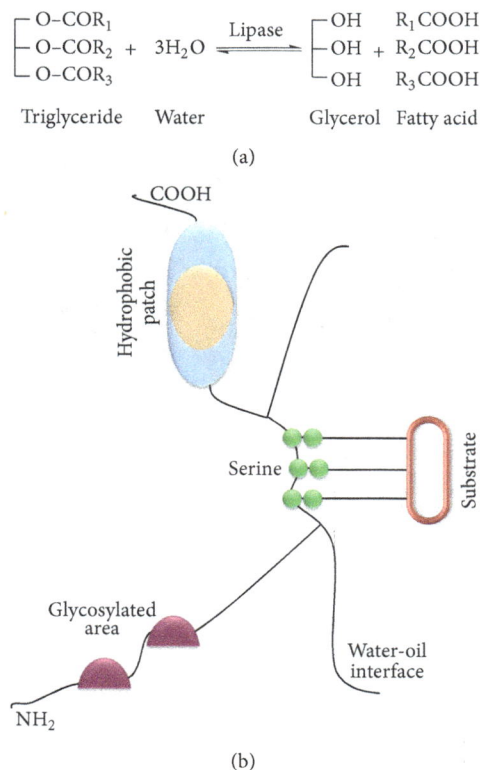

FIGURE 1: (a) Hydrolysis of triglyceride by lipase [1]. Upon hydrolysis triglyceride converts into glycerol and fatty acid. (b) Representation of a molecule of lipase with its features. The substrate can be any triglyceride [2]. Substrate interactive regions are displayed.

Fungi are involved in the degradation of undesirable materials or compounds converting them into harmless, tolerable, or useful products. The undesirable materials include sewage waste from domestic and industrial complexes and plant, animal, and agricultural wastes, oil spills, and dairy waste. The role of fungi in bioremediation processes in varied environments has been well documented [16, 30, 42, 43]. There has been an increasing awareness of potentially harmful effects of the worldwide spillage of oil and fatty substances in both saline and fresh waters. Domestic waste is also considered as the pollutants as it has a high amount of fatty and oil substances. Industrial and domestic wastes harbor fungal species of greater potential in degrading fats and oils. Besides waste disposal, bioconversion by fungal activity results in the production of a vast number of useful substances. Thus, waste can be converted into a resource. Bearing in mind the importance of lipolytic fungal enzymes from different disposal sources, this overview focuses on strategies to characterize the fungal lipases with an emphasis on a wide range of applications.

2. Lipolytic Fungal Species from Oil-Spill Wastes

Due to usage of vegetable oils for cooking, these oils are released into the open environment both at the production

level and by domestic users. To keep the environment clean, these oils should be degraded by using environmentally-friendly technology. Several oil-industries have been established at both small scale (Figure 2(a)) and large scale (Figure 2(b)). Oil spillages from these production points (Figure 2(c)) cause a hindrance in ensuring environmental hygiene due to the formation of clogs in drain pipes [44]. Cleanup and recovery of oil wastes is difficult and depends upon many factors, including the type of oil spilled, the temperature of water affecting evaporation, and biodegradation. Microbial degradation is one of the most important events to ameliorate oil pollution in the environment. Fungi that produce lipases are found in diverse habitats including oil-contaminated soils, wastes around oil processing factories, domestic waste points, and dairy products [27]. Gopinath et al. [16] have isolated 34 fungal species from oil-spill contaminated soils, collected in major cities of India. These species were tested for their survival with the changes in seasons. Twelve fungal species from oil-mill effluent composts at Nsukka have been studied and it was found that *Aspergillus* spp. are more common; however, the higher lipase producers are *Trichoderma* sp. followed by *Aspergillus* spp. [9]. D'Annibale et al. [45] used olive mill waste water as the substrate to determine lipase production. Lipase producing fungal species were also recovered from compost heaps, coal tips, and industrial wastes [43]. Cihangir and Sarikaya [42] have isolated *Aspergillus* sp. from the soil samples collected in Turkey. Extracellular lipase of *Rhizopus* sp. isolated from oil-contaminated soil was recently characterized [30].

3. Screening Lipase Production on Agar Solid Surface

Studies on mycoflora are significant as they could harbor species of the highest potential for degradation. The industrial demand for new lipase sources continues to stimulate the isolation and screening of new lipolytic microorganisms. In view of the interesting applications of microbial lipases, it could be of tremendous value to screen and identify microorganisms of highest potential for the biodegradation of oils and fats. Although, different screening strategies have been proposed for the determination of lipase activity, assays using agar plates are highly recommended, because it is an easier method with lower cost. Assay using agar plates are performed due to the fact that activities for lipases are hard to determine because of the water-soluble enzyme acting on substrates which are insoluble [46, 47].

To isolate fungal species from the oil-spill contaminated soils, screening studies were performed by Gopinath et al. [16, 40] using different substrates on agar plates. These methods with different substrates include Tween-20 (Figure 3(a)), tributyrin (Figure 3(b)), and vegetable oil in the presence of Rhodamine (Figure 3(c)). Due to the oil rich environments of the substrates, special attention was given to screening of lipolytic enzymes. On the Tween-20 substrate, a visible precipitate appeared due to the deposition of calcium salt crystals formed by the liberated fatty acid by the action of lipase or by clearing of such a precipitate due to complete

FIGURE 2: Oil production and spillage. (a) Crushing system in small-scale oil production. (b) Crushing system in large-scale oil production. (c) Spillage from oil production points. These releasing points are potential sources of environmental issues.

degradation of the fatty acid salts. Brockman [48] suggested that the primary role of calcium ions is to remove fatty acids formed during hydrolysis as insoluble calcium soaps, and thus changing the substrate-water interface relationship to favorable conditions for enzyme action. The development of a clear crystal zone of Tween-20 around the fungus was also an indication of lipolytic activity, and this zone can be measured. Using Tween-20 as the substrate, Salihu et al. [36] screened different fungal species for the production of lipases. Another substrate, tributyrin, is convenient because it is easily dispersed in water by shaking or stirring without the addition of any emulsifiers. Tributyrin is a very strong surface-active substance, and its hydrolysis can be followed by measuring the increase in the diameter of the clear zone. Nevertheless, the observed zones of clearing could be the activity response of nonspecific esterases, which may have little or no activity against the long-chain triglycerides [49]. Hence, it is imperative to use another method to confirm true

lipase activity. Tributyrin agar plates were used to investigate lipase production by new strains and 18 strains were found to be positive [42]. Using tributyrin formation of the clear zone around the fungal colony showed different mutant strains that produced extracellular lipases [50]. By using Tween-20 and tributyrin substrates, lipolytic activity (high and moderate activity) was evidenced by 19 and 32 species, respectively [51]. The lipolytic potential of this fungus was also confirmed by the Rhodamine method because the enzyme will fluoresce with orange compound (Figure 3(c)) as reported by Kouker and Jaeger [52]. Furthermore, Hou and Johnston [53] as well as Lee and Rhee [54] proved that this method is highly sensitive and reliable as a lipase assay. In a recent study, the Rhodamine method with olive mill wastewater was used to determine the production of lipases by *Aspergillus ibericus* [11]. Savitha et al. [3] used Rhodamine fluorescence-based assay to screen 32 fungal species from different sources. Our previous results provide very useful information about

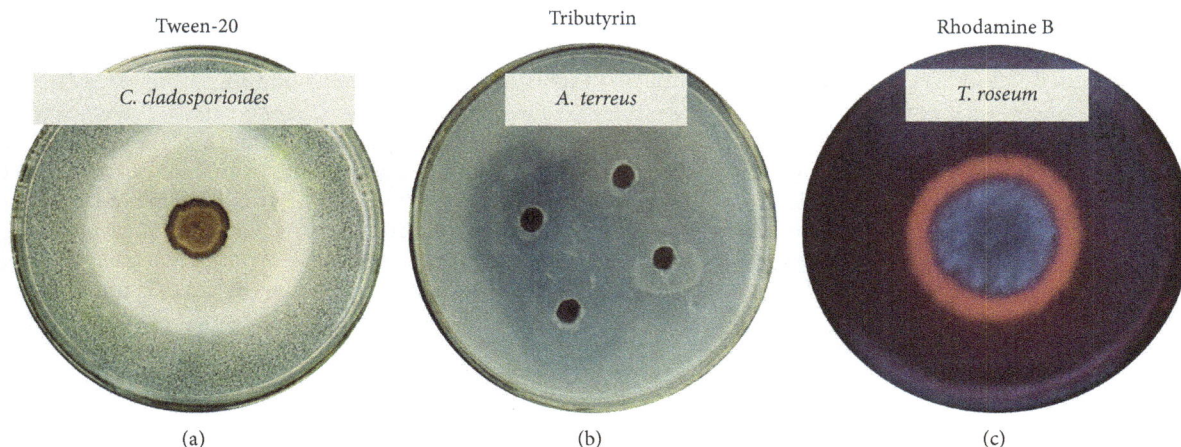

FIGURE 3: Agar plate screening for lipases. Using the substrates (a) Tween, (b) tributyrin, and (c) vegetable oil. In the Tween method, formation of calcium crystals was observed. The tributyrin method shows a clear zone, whereas in the Rhodamine method, formation of fluorescence with fatty substrate was observed under UV illumination. Active zones are increasing with a period of incubation time and these zones can be measured.

the degradation of vegetable oils by *Cunninghamella verticillata* in the presence of Rhodamine [51]. Based on the above screening strategies, Gopinath et al. [16] revealed the following fungi as potential candidates that secrete enzymes lipases, *Absidia corymbifera, Aspergillus fumigatus, Aspergillus japonicus, Aspergillus nidulans, Aspergillus terreus, Cunninghamella verticillata, Curvularia pallescens, Fusarium oxysporum, Geotrichum candidum, Mucor racemosus, Penicillium citrinum, Penicillium frequentans, Rhizopus stolonifer,* and *Trichoderma viride.* They also conducted screening studies for other enzymes and confirmed that some of the isolated fungal species could also secrete amylases, proteases, and cellulases in addition to lipases, representing the ability of these species to survive in a wide range of environmental substrates [16]. These observations provided interesting perspectives, demonstrating that fungi isolated from oil-rich environments represent a source of several enzymes potentially exploitable for biotechnological purposes. In another study, 59 fungi were screened by measuring the formation of halos on the agar plate used for lipase screening [55]. Preliminary screening studies for lipase production by fungi were also carried out on agar plates using olive oil or emulsified tributyrin by gum arabic [56]. Kumar et al. [57] screened fungi with bromophenol blue dye supplemented agar plates with olive oil as the substrate.

4. Degradation of Oils by Lipases

The use of specific microbial lipases to catalyze interesterification reactions became considerable interest because of its advantages over chemical catalysts. Traditionally, fatty acids are manufactured by the hydrolysis of oils at high temperature and pressure. However, lipase hydrolysis is an energy saving process because oil degradation of fatty acids (the reaction) can be carried out at room temperature and pressure [58]. Industrially important chemicals manufactured from fats and oils by chemical processes could be produced by lipases

with rapidity and better specificity under mild conditions [59, 60]. Lipases are one among several kinds of extracellular enzymes that perform the function of recycling large quantities of insoluble organic material in nature [61]. Apart from numerous applications such as transesterification [62, 63], ester synthesis, production of biosurfactants [64], and application in food and dairy industry [35, 65, 66], the enzyme lipase has a proven role as a useful interesterification catalyst. Interesterification is a technology by which fatty acids within a triacylglycerol molecule can be interchanged with regard to their positional distribution. The process of fat splitting, along with interesterification, is an essential tool in the manufacture of new tailor made fats and oils. Enzyme-catalyzed reactions of lipids are of considerable interest in view of their possible applications in the biotechnology of fats and oils. The technique of fat splitting plays an important role in the manufacture of soaps and other industrial products like candles from conventional minor oils. Different isolates from oil mill effluent have been tested for their ability to degrade the different oils and the potential of individual species varied with the type of fatty acid residues in the oil (Table 1). From this study, it was revealed that the behavior of lipases from different fungal species is different in terms of their biochemical characteristics. Teng and Xu [67] analyzed the production of lipase from *Rhizopus chinensis* under experimental conditions and Bapiraju et al. [68] performed a similar study with mutants of *Rhizopus* sp. Studies have also been documented with lipase from *Penicillium* spp. [69, 70]. Extensive review on the production of lipases from different microbes has been published [71].

5. Purification Strategies for Lipases

Knowledge of the purified lipase activities to be used for biotechnological purposes is mandatory, and it can be the basis or other applications. Various purification strategies have been reviewed for the lipase enzyme [72–74]. In the case

TABLE 1: Degree of utilization (%) of vegetable oils by fungal species.

Organism	Olive	Soybean	Groundnut	Cottonseed	Sunflower
A. strictum	20	60	40	90	90
C. verticillata	80	80	90	40	90
G. candidum	90	10	90	40	80
M. racemosus	50	60	60	10	50
R. miehei	30	90	90	80	90
R. stolonifer	60	70	80	20	60
T. roseum	90	60	80	80	80
T. viride	50	20	30	50	30

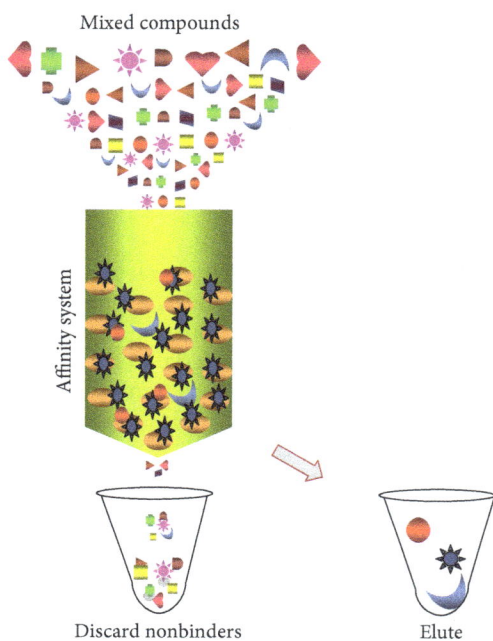

FIGURE 4: Purification of lipase using an affinity system. Separation of lipase from mixed compounds is indicated. Bound lipase can be eluted by creating stringent conditions.

of extracellular lipases, it is primarily important to remove other contaminants from the compound mixture containing lipase by suitable strategy (Figure 4). The conventional purification strategies give a low yield due to a large hydrophobic surface near the active site. Novel purification steps are mandatory to increase overall enzyme yields and it could be achieved by opting an appropriate chromatography system. One of the choices is hydrophobic interaction chromatography and it is considered as a common strategy [75–77]. In addition to this, ion exchange and gel filtration chromatography are commonly preferred methods [75, 78–80]. A reversed micellar system, membrane processes, immunopurification, hydrophobic interaction chromatography with an epoxy-activated spacer arm (ligand), column chromatography using polyethylene glycol (PEG)/sepharose gel, and aqueous two-phase systems are also recommended [81]. Kumarevel et al. [82] reported a stepwise purification strategy for fungal lipases to remove other components released from

the fungus *Cunninghamella verticillata* extracellularly, using acetone precipitation as the important step. To avoid many steps in this study and to minimize the impurities as much as possible the experiment was repeated with 50% acetone saturation with a gradual increments of 5% acetone. Using the above methods, many lipases from different microorganisms have been reported, and molecular masses of 31 and 19 kDa have been reported for the lipase of *Aspergillus niger* by Hofelmann et al. [83], 21.4 kDa for *Rhodotorula pilimena* [84], 30 kDa for *Rhizopus japonicus* [85], 51 kDa for *Pichia burtonii* [86], 25 kDa for *Aspergillus oryzae* [87], 49 kDa for *Mucor hiemalis* [88], 35.5 kDa for *Aspergillus niger* [89], 49 kDa for *Cunninghamella verticillata* [40], and 32 kDa for *Geotrichum candidum* [34]. Different strategies for lipase purification with the varied sources were recently described in detail by Singh and Mukhopadhyay [1], and it seems that the production of lipases from fungal species results in different molecular sizes, due to variations in the number of amino acid residues. Saxena et al. [73] summarized the purification strategies for microbial lipases. Overall, traditional purification strategies are considered time consuming with lower yields and the trends are moving towards aqueous two-phase extraction, and purification in ionic liquids and purification based on lipase-lipase interaction [10].

6. Statistical Calculations

Statistical calculations were focused in the past, due to their reliable prediction for the experimental conditions for enzyme studies to be optimized [34, 40, 90–92]. In statistics, response surface methodology (RSM) has referred the relationships between several explanatory variables and one or more response variables. This method was introduced initially by Box and Wilson [93]. The main idea of RSM is to get an optimal response by using a sequence of designing experiments, and it was suggested to use a second-degree polynomial model to perform RSM. Box-Behnken design experiments are one of the most common, and this is an independent quadratic design without an embedded factorial design. Different combinations of midpoints are used for experiments; for example, with 3 experimental parameters, 17 experiments can be run and it yields a predicted result (Figure 5). The Box-Behnken design is where the outcome unit (Y) is related to experimental variables by a response equation,

$$Y = f(X_1, X_2, X_3, \ldots, X_k). \quad (1)$$

As mentioned above a second-degree quadratic polynomial is used to represent the function in the range of interest,

$$Y = R_0 + \sum_{i=1}^{k} R_i X_i + \sum_{i=1}^{k} R_{ii} X_i^2 + \sum_{i=1, i<j}^{k-1} \sum_{j=2}^{k} R_{ii} X_i X_j + \varepsilon, \quad (2)$$

where $X_1, X_2, X_3, \ldots, X_k$ are the input variables which effect the response Y, R_0, R_i, R_{ii}, and R_{ij} ($i = 1 - k$, $j = 1 - k$) are the known parameters, and ε is the random error. A second-order model is designed such that the variance of Y is constant for all points equidistant from the center of the design

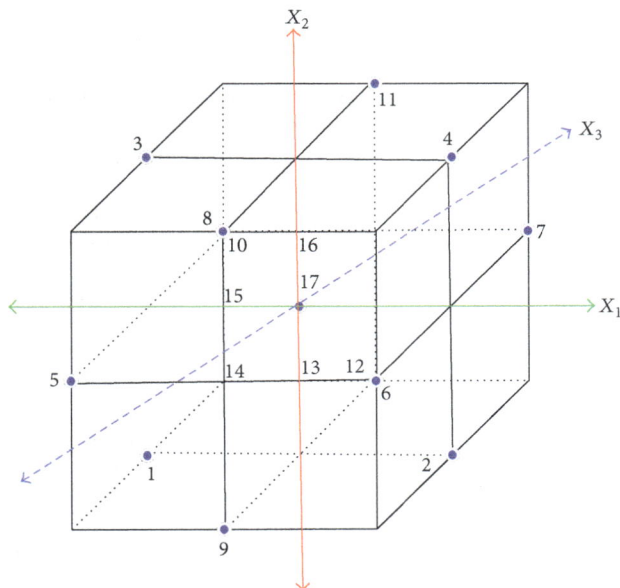

FIGURE 5: Box-Behnken design for experiments. Different combinations at the midpoints used for experiments are shown with 3 experimental parameters and 17 experiments run. A number of experiments vary with the number of experimental parameters.

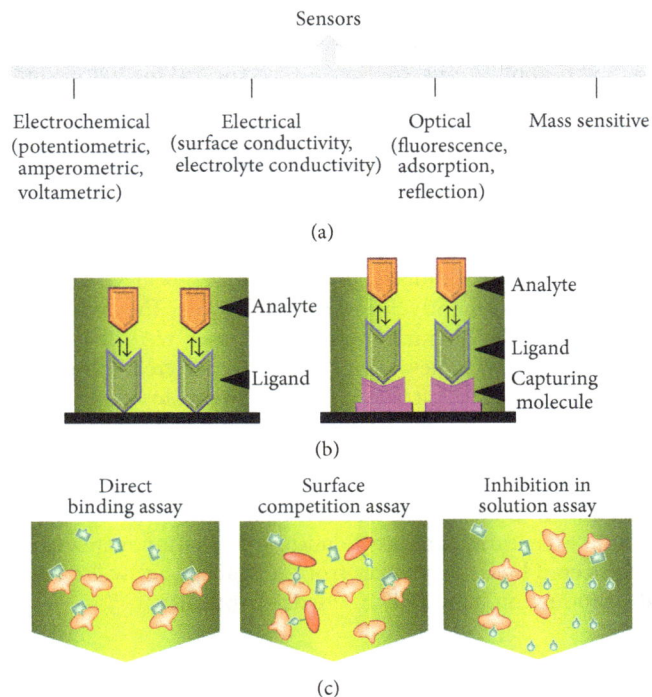

FIGURE 6: Sensing systems. (a) Types of sensors. (b) Strategy of immobilizing ligand and analyte. (c) Methods involved with ligand and analyte binding. Sensitivity depends on the interactive molecules.

(Figure 5). The validity of the model can be determined based on Student's t-test. The Fisher-test, P value, t-test, and R^2, and so forth can be used to evaluate the model as well as to determine the optimal processing conditions. The Fisher-test with a very low probability value ($P_{model} > F = 0.0001$) showed that the regression model had a very high significance. The model reliability of fit was checked by means of the determination coefficient (R^2). This model fits the experimental range studied perfectly when the value of R^2 is adjusted to nearly one. Using the Box-Behnken design, conditions were optimized for lipase production by *Geotrichum candidum* [34] and optimization of purified lipase from *Cunninghamella verticillata* for physical parameters was also shown [40]. Using these optimized conditions, the purification steps were reduced and purified lipase was used for crystallization studies [82]. Other than Box-Behnken, the Plackett-Burman design and the Luedeking-Piret model can also be used for different optimization studies. Statistical design experiments by Plackett-Burman were used to evaluate the production of lipase by *Candida rugosa* [1].

7. Biosensors for Lipase

A biosensor is a combination of a biological component with a physicochemical detector and it assists with analysis of biomolecular interactions. Development of an analytical system will help us to find a minute amount of biological agents within mixed compounds. Different types of sensing systems have been proposed for determining biomolecular interactions and environmental monitoring [94–98]. In general biosensors may be classified as electrochemical, electrical, optical, or mass sensitive (Figure 6(a)). The core design

for sensors mainly includes three components, probe-target recognition, signal transduction, and physical readout [99]. On the sensor surfaces the lipase can be either ligand or analyte. Ligand can be immobilized directly on the sensor surface or indirectly via an immobilized surface chemical linkage (Figure 6(b)). The direct adsorption of the molecules on the sensor surface leads to rapid, simple, and cheaper strategies compared to immobilization by chemical means. Using these strategies the interactions of lipase with oil or molecules for the purpose of interactive analysis and environmental monitoring can be done by different assay formats (Figure 6(c)). Various detection and measurement methods or strategies are discussed on microbial lipases [100]. Phospholipases are potential markers for diagnosing diseases in the pancreas and coronary arteries [101, 102]. In all, free and phosphatidyl-bound choline in milk and a dietary supplement can be determined quantitatively, using a phospholipase D packed bioreactor [103]. A surface acoustic wave sensing system was generated to measure pancreatic lipase [104]. Lipase activity based on glycerol dehydrogenase/NADH oxidase was reported based on amperometric sensor [105]. Lipases can be immobilized on the sensing surface and can function as lipid biosensors for blood cholesterol determinations [106].

8. Conclusions

Fungi are capable of producing several enzymes for their survival within a wide range of substrates. Among those enzymes, lipases are predominantly used in several applications. These fat-splitting enzymes are attractive because of their applications in fields relevant to medicine and dairy industry. Lipases play a major role as the biocatalysts and microbial lipases can be produced in large scale by overexpression. The disadvantage lipase enzyme is that it continues to be active due to turnover reaction and may need to optimize the reaction condition and specificity with different sources of lipase [107]. Indeed, various methods have been proposed for the different lipases to survive under variant physical and chemical conditions. The strategies involved in the characterization of lipases were discussed here, suitable for large-scale production. The great advantage of fungal lipases is that they are easily amenable to extraction due to their extracellular nature, which will significantly reduce the cost and makes these lipases more attractive than those bacteria. Furthermore, with available sources such as LIPABASE (database for true lipases), which provides taxonomic, structural, and biochemical information, genetically engineered lipase sequences from fungal species will hasten the production, especially in the dairy industry.

Acknowledgment

P. Anbu would like to thank Inha University, Republic of Korea, for the financial support from Inha University Research Grant.

References

[1] A. K. Singh and M. Mukhopadhyay, "Overview of fungal lipase: a review," *Applied Biochemistry and Biotechnology*, vol. 166, no. 2, pp. 486–520, 2012.

[2] P. K. Ghosh, R. K. Saxena, R. Gupta, R. P. Yadav, and S. Davidson, "Microbial lipases: production and applications," *Science progress*, vol. 79, pp. 119–157, 1996.

[3] J. Savitha, S. Srividya, R. Jagat et al., "Identification of potential fungal strain(s) for the production of inducible, extracellular and alkalophilic lipase," *African Journal of Biotechnology*, vol. 6, no. 5, pp. 564–568, 2007.

[4] D. V. Vadehra, "Staphylococcal lipases," *Lipids*, vol. 9, no. 3, pp. 158–165, 1974.

[5] A. R. Macrae, "Lipase-catalyzed interesterification of oils and fats," *Journal of the American Oil Chemists' Society*, vol. 60, no. 2, pp. 291–294, 1983.

[6] G. M. Frost and D. A. Moss, "Production of enzymes by fermentation," in *Biotechnology*, H. J. Rehm and G. Reed, Eds., vol. 7a, pp. 65–211, Verlag Chemie, Weinheim, Germany, 1987.

[7] G. Ginalska, R. Bancerz, and T. Korniłłowicz-Kowalska, "A thermostable lipase produced by a newly isolates *Geotrichum*-like strain, R59," *Journal of Industrial Microbiology and Biotechnology*, vol. 31, no. 4, pp. 177–182, 2004.

[8] H. M. Saeed, T. I. Zaghloul, A. I. Khalil, and M. T. Abdelbaeth, "Purification and characterization of two extracellular lipases from *Pseudomonas aeruginosa* Ps-x," *Polish Journal of Microbiology*, vol. 54, no. 3, pp. 233–240, 2005.

[9] C. O. Nwuche and J. C. Ogbonna, "Isolation of lipase producing fungi from palm oil mill effluent (POME) dump sites at Nsukka," *Brazilian Archives of Biology and Technology*, vol. 54, no. 1, pp. 113–116, 2011.

[10] S. Nagarajan, "New tools for exploring, 'old friends-microbial lipases," *Applied Biochemistry and Biotechnology*, vol. 168, pp. 1163–1196, 2012.

[11] L. Abrunhosa, F. Oliveira, D. Dantas, C. Goncalves, and I. Belo, "Lipase production by *Aspergillus ibericus* using olive mill wastewater," *Bioprocess and Biosystem Engineering*, vol. 36, pp. 285–291, 2013.

[12] P. Anbu, M. Noh, D. Kim, J. Seo, B. Hur, and K. H. Min, "Screening and optimization of extracellular lipases by *Acinetobacter* species isolated from oil-contaminated soil in South Korea," *African Journal of Biotechnology*, vol. 10, no. 20, pp. 4147–4156, 2011.

[13] W. Stuer, K. E. Jaeger, and U. K. Winkler, "Purification of extracellular lipase from *Pseudomonas aeruginosa*," *Journal of Bacteriology*, vol. 168, no. 3, pp. 1070–1074, 1986.

[14] B. N. Johri, J. D. Alurralde, and J. Klein, "Lipase production by free and immobilized protoplasts of *Sporotrichum (Chrysosporium) thermophile apinis*," *Applied Microbiology and Biotechnology*, vol. 33, no. 4, pp. 367–371, 1990.

[15] W. B. Cooke, *The Ecology of Fungi*, CRC Press, Boca laton, Fla, USA, 1979.

[16] S. C. B. Gopinath, P. Anbu, and A. Hilda, "Extracellular enzymatic activity profiles in fungi isolated from oil-rich environments," *Mycoscience*, vol. 46, no. 2, pp. 119–126, 2005.

[17] U. T. Bornscheuer, "Microbial carboxyl esterases: classification, properties and application in biocatalysis," *FEMS Microbiology Reviews*, vol. 26, no. 1, pp. 73–81, 2002.

[18] F. Hasan, A. A. Shah, and A. Hameed, "Industrial applications of microbial lipases," *Enzyme and Microbial Technology*, vol. 39, no. 2, pp. 235–251, 2006.

[19] M. W. Akhtar, A. Q. Mirza, M. N. Nawazish, and M. I. Chughtai, "Effect of triglycerides on the production of lipids and lipase by *Mucor hiemalis*," *Canadian Journal of Microbiology*, vol. 29, no. 6, pp. 664–669, 1983.

[20] H. Chander, V. K. Batish, S. S. Sannabhadti, and R. A. Srinivasan, "Factors affecting lipase production in *Aspergillus wentii*," *Journal of Food Science*, vol. 45, pp. 598–600, 1980.

[21] G. B. Ksandopulo, "Effects of some fats and surfactants on lipase activity of *Geotrichum* fungi," *Mikrobiologiya*, vol. 43, no. 6, pp. 850–853, 1974.

[22] Y. Tsujisaka, S. Okumura, and M. Iwai, "Glyceride synthesis by four kinds of microbial lipase," *Biochimica et Biophysica Acta*, vol. 489, no. 3, pp. 415–422, 1977.

[23] B. A. Oso, "The lipase activity of *Talaromyces emersonii*," *Canadian Journal of Botany*, vol. 56, pp. 1840–1843, 1978.

[24] A. S. Reddy and S. M. Reddy, "Lipase activity of two seed-borne fungi of sesamum (*Sesamum indicum* Linn.)," *Folia Microbiologica*, vol. 28, no. 6, pp. 463–466, 1983.

[25] J. T. M. Wouters, "The effect of tweens on the lipolytic activity of *Geotrichum candidum*," *Antonie van Leeuwenhoek*, vol. 33, no. 1, pp. 365–380, 1967.

[26] R. G. Jensen, "Detection and determination of lipase (acylglycerol hydrolase) activity from various sources," *Lipids*, vol. 18, no. 9, pp. 650–657, 1983.

[27] H. Sztajer, I. Maliszewska, and J. Wieczorek, "Production of exogenous lipases by bacteria, fungi, and actinomycetes," *Enzyme and Microbial Technology*, vol. 10, no. 8, pp. 492–497, 1988.

[28] H. Sztajer and I. Maliszewska, "The effect of culture conditions on lipolytic productivity of *Penicillium citrinum*," *Biotechnology Letters*, vol. 11, no. 12, pp. 895–898, 1989.

[29] T. Hoshino, T. Yamane, and S. Shimizu, "Selective hydrolysis of fish oil by lipase to concentrate n-3 polyunsaturated fatty acid," *Agricultural and Biological Chemistry*, vol. 54, pp. 1459–1467, 1990.

[30] P. Thota, P. K. Bhogavalli, P. R. Vallem, and V. Sreerangam, "Biochemical characterization of an extracellular lipase from new strain of *Rhizopus* sp. isolated from oil contaminated soil," *International Journal of Plant, Animal and Environmental Sciences*, vol. 2, pp. 41–45, 2012.

[31] Y. Tsujisaka, M. Iwai, and Y. Tominaga, "Purification, crystallization and some properties of lipase from *Geotrichum candidum* link," *Agricultural and Biological Chemistry*, vol. 37, no. 6, pp. 1457–1464, 1973.

[32] M. K. Tahoun, E. Mostafa, R. Mashaly, and S. Abou-Donia, "Lipase induction in *Geotrichum candidum*," *Milchwiss*, vol. 37, pp. 86–88, 1982.

[33] T. Jacobsen, J. Olsen, and K. Allermann, "Substrate specificity of *Geotrichum candidum* lipase preparations," *Biotechnology Letters*, vol. 12, no. 2, pp. 121–126, 1990.

[34] S. C. B. Gopinath, A. Hilda, T. L. Priya, G. Annadurai, and P. Anbu, "Purification of lipase from *Geotrichum candidum*: conditions optimized for enzyme production using Box-Behnken design," *World Journal of Microbiology and Biotechnology*, vol. 19, no. 7, pp. 681–689, 2003.

[35] S. E. Petrovic, M. Skrinjar, A. Becarevic, I. F. Vujicic, and L. Banka, "Effect of various carbon sources on microbial lipases biosynthesis," *Biotechnology Letters*, vol. 12, no. 4, pp. 299–304, 1990.

[36] A. Salihu, M. Z. Alam, M. I. AbdulKarim, and H. M. Salleh, "Suitability of using palm oil mill effluent as a medium for lipase production," *African Journal of Biotechnology*, vol. 10, no. 11, pp. 2044–2052, 2011.

[37] C. N. Okeke and B. N. Okolo, "The effect of cultural conditions on the production of lipase by *Acremonium strictum*," *Biotechnology Letters*, vol. 12, no. 10, pp. 747–750, 1990.

[38] S.-H. Wu, Z.-W. Guo, and C. J. Sih, "Enhancing the enantioselectivity of *Candida* lipase catalyzed ester hydrolysis via noncovalent enzyme modification," *Journal of the American Chemical Society*, vol. 112, no. 5, pp. 1990–1995, 1990.

[39] T. Iizumi, K. Nakamura, and T. Fukase, "Purification and characterization of a thermostable lipase from newly isolated *Pseudomonas* sp. KWI-56," *Agricultural and Biological Chemistry*, vol. 54, pp. 1253–1258, 1990.

[40] S. C. B. Gopinath, A. Hilda, T. L. Priya, and G. Annadurai, "Purification of lipase from *Cunninghamella verticillata* and optimization of enzyme activity using response surface methodology," *World Journal of Microbiology and Biotechnology*, vol. 18, no. 5, pp. 449–458, 2002.

[41] S. Gopinath, A. Hilda, and P. Anbu, "Screening methods for detecting lipolytic enzymes by *Aspergillus* species," *Acta Botanica Indica*, vol. 28, pp. 41–44, 2000.

[42] N. Cihangir and E. Sarikaya, "Investigation of lipase production by a new isolate of *Aspergillus* sp.," *World Journal of Microbiology and Biotechnology*, vol. 20, no. 2, pp. 193–197, 2004.

[43] V. Gunasekaran and D. Das, "Lipase fermentation: progress and prospects," *Indian Journal of Biotechnology*, vol. 4, no. 4, pp. 437–445, 2005.

[44] G. R. Lemus and A. K. Lau, "Biodegradation of lipidic compounds in synthetic food wastes during composting," *Canadian Biosystems Engineering*, vol. 44, pp. 33–39, 2002.

[45] A. D'Annibale, G. G. Sermanni, F. Federici, and M. Petruccioli, "Olive-mill wastewaters: a promising substrate for microbial lipase production," *Bioresource Technology*, vol. 97, no. 15, pp. 1823–1833, 2006.

[46] O. Redondo, A. Herrero, J. F. Bello et al., "Comparative kinetic study of lipases A and B from *Candida rugosa* in the hydrolysis of lipid p-nitrophenyl esters in mixed micelles with Triton X-100," *Biochimica et Biophysica Acta*, vol. 1243, no. 1, pp. 15–24, 1995.

[47] R. Kanchana, U. D. Muraleedharan, and S. Raghukumar, "Alkaline lipase activity from the marine protists, thraustochytrids," *World Journal of Microbiology and Biotechnology*, vol. 27, no. 9, pp. 2125–2131, 2011.

[48] H. L. Brockman, "General features of lipolysis reaction scheme interfacial structure and experimental approaches," in *Lipaseseds*, B. Borgstrom and H. L. Brockman, Eds., pp. 3–46, Elsevier, Amsterdam, The Netherlands, 1984.

[49] R. G. Roberts, W. H. Morrison III, and J. A. Robertson, "Extracellular lipase production by fungi from sunflower seed," *Mycologia*, vol. 79, pp. 265–273, 1987.

[50] L. Toscano, V. Gochev, G. Montero, and M. Stoytcheva, "Enhanced production of extracellular lipase by novel mutant strain of *Aspergillus niger*," *Biotechnology and Biotechnological Equipment*, vol. 25, pp. 2243–2247, 2011.

[51] S. C. B. Gopinath, *Studies on oil-mill effluent mycoflora and their lipolytic acitivity [Ph.D. thesis]*, University of Madras, Chennai, India, 1998.

[52] G. Kouker and K.-E. Jaeger, "Specific and sensitive plate assay for bacterial lipases," *Applied and Environmental Microbiology*, vol. 53, no. 1, pp. 211–213, 1987.

[53] C. T. Hou and T. M. Johnston, "Screening of lipase activities with cultures from the agricultural research service culture collection," *Journal of the American Oil Chemists' Society*, vol. 69, no. 11, pp. 1088–1097, 1992.

[54] S. Y. Lee and J. S. Rhee, "Production and partial purification of a lipase from *Pseudomonas putida* 3SK," *Enzyme and Microbial Technology*, vol. 15, pp. 617–623, 1993.

[55] G. Colen, R. G. Junqueira, and T. Moraes-Santos, "Isolation and screening of alkaline lipase-producing fungi from Brazilian Savanna soil," *World Journal of Microbiology and Biotechnology*, vol. 22, pp. 881–885, 2006.

[56] K. H. Domsch, W. Gams, and T. H. Anderson, *Compendium of Soil Fungi*, IHW, Eching, Germany, 2nd edition, 1993.

[57] S. Kumar, N. Katiyar, P. Ingle, and S. Negi, "Use of evolutionary operation (EVOP) factorial design technique to develop a bioprocess using grease waste as a substrate for lipase production," *Bioresource Technology*, vol. 102, no. 7, pp. 4909–4912, 2011.

[58] V. R. Murty, J. Bhat, and P. K. A. Muniswaran, "Hydrolysis of oils by using immobilized lipase enzyme: a review," *Biotechnology and Bioprocess Engineering*, vol. 7, no. 2, pp. 57–66, 2002.

[59] H. G. Davis, R. H. Green, D. R. Kelly, and S. M. Roberts, *Biotransformations in Preparative Organic Synthesis*, Academic Press, London, UK, 1990.

[60] E. N. Vulfson, "Industrial applications of lipases," in *Lipases*, P. Wooley and S. B. Petersen, Eds., p. 271, Cambridge University Press, Cambridge, UK, 1994.

[61] P. Gowland, M. Kernick, and T. K. Sundaram, "Thermophilic bacterial isolates producing lipase," *FEMS Microbiology Letters*, vol. 48, no. 3, pp. 339–343, 1987.

[62] J. Harwood, "The versatility of lipases for industrial uses," *Trends in Biochemical Sciences*, vol. 14, no. 4, pp. 125–126, 1989.

[63] S. Bloomer, P. Adlercreutz, and B. Mattiasson, "Triglyceride interesterification by lipases. 1. Cocoa butter equivalents from a fraction of palm oil," *Journal of the American Oil Chemists' Society*, vol. 67, no. 8, pp. 519–524, 1990.

[64] J. Chopineau, F. D. McCafferty, M. Therisod, and M. Klibanov, "Production of biosurfactants from sugar alcohols and vegetable oils catalyzed by lipases in a non-aqueous medium," *Biotechnology and Bioengineering*, vol. 31, pp. 208–214, 1988.

[65] T. Jacobsen, B. Jensen, J. Olsen, and K. Allermann, "Extracellular and cell-bound lipase activity in relation to growth of *Geotrichum candidum*," *Applied Microbiology and Biotechnology*, vol. 32, no. 3, pp. 256–261, 1989.

[66] E. Espinosa, S. Sanchez, and A. Farres, "Nutritional factors affecting lipase production by *Rhizopus delemar* CDBB H313," *Biotechnology Letters*, vol. 12, no. 3, pp. 209–214, 1990.

[67] Y. Teng and Y. Xu, "Culture condition improvement for whole-cell lipase production in submerged fermentation by *Rhizopus chinensis* using statistical method," *Bioresource Technology*, vol. 99, no. 9, pp. 3900–3907, 2008.

[68] K. V. V. S. N. Bapiraju, P. Sujatha, P. Ellaiah, and T. Ramana, "Sequential parametric optimization of lipase production by a mutant strain *Rhizopus* sp. BTNT-2," *Journal of Basic Microbiology*, vol. 45, no. 4, pp. 257–273, 2005.

[69] A. P. Kempka, N. L. Lipke, T. Da Luz Fontoura Pinheiro et al., "Response surface method to optimize the production and characterization of lipase from *Penicillium verrucosum* in solid-state fermentation," *Bioprocess and Biosystems Engineering*, vol. 31, no. 2, pp. 119–125, 2008.

[70] G. D. L. P. Vargas, H. Treichel, D. de Oliveira, S. C. Beneti, D. M. G. Freire, and M. Di Luccio, "Optimization of lipase production by *Penicillium simplicissimum* in soybean meal," *Journal of Chemical Technology and Biotechnology*, vol. 83, no. 1, pp. 47–54, 2008.

[71] H. Treichel, D. de Oliveira, M. A. Mazutti, M. Di Luccio, and J. V. Oliveira, "A review on microbial lipases production," *Food and Bioprocess Technology*, vol. 3, no. 2, pp. 182–196, 2010.

[72] A. A. Palekar, P. T. Vasudevan, and S. Yan, "Purification of lipase: a review," *Biocatalysis and Biotransformation*, vol. 18, no. 3, pp. 177–200, 2000.

[73] R. K. Saxena, A. Sheoran, B. Giri, and W. S. Davidson, "Purification strategies for microbial lipases," *Journal of Microbiological Methods*, vol. 52, no. 1, pp. 1–18, 2003.

[74] R. Sharma, Y. Chisti, and U. C. Banerjee, "Production, purification, characterization, and applications of lipases," *Biotechnology Advances*, vol. 19, no. 8, pp. 627–662, 2001.

[75] S. Imamura and S. Kitaura, "Purification and characterization of a monoacylglycerol lipase from the moderately thermophilic *Bacillus* sp. H-257," *Journal of Biochemistry*, vol. 127, no. 3, pp. 419–425, 2000.

[76] J. A. Queiroz, C. T. Tomaz, and J. M. S. Cabral, "Hydrophobic interaction chromatography of proteins," *Journal of Biotechnology*, vol. 87, no. 2, pp. 143–159, 2001.

[77] P. Fuciños, L. Pastrana, A. Sanromán, M. A. Longo, J. A. Hermoso, and M. L. Rúa, "An esterase from *Thermus thermophilus* HB27 with hyper-thermoalkalophilic properties: purification, characterisation and structural modelling," *Journal of Molecular Catalysis B*, vol. 70, no. 3-4, pp. 127–137, 2011.

[78] A. M. Abdou, "Purification and partial characterization of psychrotrophic *Serratia marcescens* lipase," *Journal of Dairy Science*, vol. 86, no. 1, pp. 127–132, 2003.

[79] D. Litthauer, A. Ginster, and E. V. Skein, "Pseudomonas luteola lipase: a new member of the 320-residue *Pseudomonas* lipase family," *Enzyme and Microbial Technology*, vol. 30, no. 2, pp. 209–215, 2002.

[80] E. A. Snellman, E. R. Sullivan, and R. R. Colwell, "Purification and properties of the extracellular lipase, LipA, of *Acinetobacter* sp. RAG-1," *European Journal of Biochemistry*, vol. 269, no. 23, pp. 5771–5779, 2002.

[81] R. Gupta, N. Gupta, and P. Rathi, "Bacterial lipases: an overview of production, purification and biochemical properties," *Applied Microbiology and Biotechnology*, vol. 64, no. 6, pp. 763–781, 2004.

[82] T. S. Kumarevel, S. C. B. Gopinath, A. Hilda, N. Gautham, and M. N. Ponnusamy, "Purification of lipase from *Cunninghamella verticillata* by stepwise precipitation and optimized conditions for crystallization," *World Journal of Microbiology and Biotechnology*, vol. 21, no. 1, pp. 23–26, 2005.

[83] M. Hofelmann, J. Hartmann, A. Zink, and P. Schreier, "Isolation, purification and characterization of lipase isozymes from a technical *Aspergillus niger* enzyme," *Journal of Food Science*, vol. 50, pp. 1721–1726, 1985.

[84] J. M. Muderhwa, R. Ratomahenina, and M. Pina, "Purification and properties of the lipases from *Rhodotorula pilimanae* Hedrick and Burke," *Applied Microbiology and Biotechnology*, vol. 23, no. 5, pp. 348–354, 1986.

[85] M. Suzuki, H. Yamamoto, and M. Mizugaki, "Purification and general properties of a metal-insensitive lipase from *Rhizopus japonicus* NR 400," *Journal of Biochemistry*, vol. 100, no. 5, pp. 1207–1213, 1986.

[86] A. Sugihara, T. Senoo, A. Enoki, Y. Shimada, T. Nagao, and Y. Tominaga, "Purification and characterization of a lipase from *Pichia burtonii*," *Applied Microbiology and Biotechnology*, vol. 43, no. 2, pp. 277–281, 1995.

[87] J. Toida, Y. Arikawa, K. Kondou, M. Fukuzawa, and J. Sekiguchi, "Purification and characterization of Triacylglycerol lipase from *Aspergillus oryzae*," *Bioscience, Biotechnology and Biochemistry*, vol. 62, no. 4, pp. 759–763, 1998.

[88] A. Hiol, M. D. Jonzo, D. Druet, and L. Comeau, "Production, purification and characterization of an extracellular lipase from *Mucor hiemalis f. hiemalis*," *Enzyme and Microbial Technology*, vol. 25, no. 1-2, pp. 80–87, 1999.

[89] V. M. H. Namboodiri and R. Chattopadhyaya, "Purification and biochemical characterization of a novel thermostable lipase from *Aspergillus niger*," *Lipids*, vol. 35, no. 5, pp. 495–502, 2000.

[90] P. Anbu, S. C. B. Gopinath, A. Hilda, T. L. Priya, and G. Annadurai, "Purification of keratinase from poultry farm isolate-*Scopulariopsis brevicaulis* and statistical optimization of enzyme activity," *Enzyme and Microbial Technology*, vol. 36, no. 5-6, pp. 639–647, 2005.

[91] P. Anbu, S. C. B. Gopinath, A. Hilda, T. L. Priya, and G. Annadurai, "Optimization of extracellular keratinase production by poultry farm isolate *Scopulariopsis brevicaulis*," *Bioresource Technology*, vol. 98, no. 6, pp. 1298–1303, 2007.

[92] P. Anbu, G. Annadurai, J. Lee, and B. Hur, "Optimization of alkaline protease production from *Shewanella oneidensis* MR-1 by response surface methodology," *Journal of Chemical Technology and Biotechnology*, vol. 84, no. 1, pp. 54–62, 2009.

[93] G. E. P. Box and K. B. Wilson, "On the experimental attainment of optimum condition," *Journal of the Royal Statistical Society Series*, vol. 13, pp. 1–45, 1951.

[94] E. Baldrich, A. Restrepo, and C. K. O'Sullivan, "Aptasensor development: elucidation of critical parameters for optimal

aptamer performance," *Analytical Chemistry*, vol. 76, no. 23, pp. 7053–7063, 2004.

[95] T. M. A. Gronewold, S. Glass, E. Quandt, and M. Famulok, "Monitoring complex formation in the blood-coagulation cascade using aptamer-coated SAW sensors," *Biosensors and Bioelectronics*, vol. 20, no. 10, pp. 2044–2052, 2005.

[96] K. J. Odenthal and J. J. Gooding, "An introduction to electrochemical DNA biosensors," *Analyst*, vol. 132, no. 7, pp. 603–610, 2007.

[97] C. A. Marquette and L. J. Blum, "Electro-chemiluminescent biosensing," *Analytical and Bioanalytical Chemistry*, vol. 390, no. 1, pp. 155–168, 2008.

[98] S. C. B. Gopinath, K. Awazu, and M. Fujimaki, "Waveguide-mode sensors as aptasensors," *Sensors*, vol. 12, no. 2, pp. 2136–2151, 2012.

[99] D. Li, S. Song, and C. Fan, "Target-responsive structural switching for nucleic acid-based sensors," *Accounts of Chemical Research*, vol. 43, no. 5, pp. 631–641, 2010.

[100] C. A. Thomson, P. J. Delaquis, and G. Mazza, "Detection and measurement of microbial lipase activity: a review," *Critical Reviews in Food Science and Nutrition*, vol. 39, no. 2, pp. 165–187, 1999.

[101] N. Agarwal and C. S. Pitchumoni, "Assessment of severity in acute pancreatitis," *American Journal of Gastroenterology*, vol. 86, no. 10, pp. 1385–1391, 1991.

[102] H. S. Oei, I. M. Van der Meer, A. Hofman et al., "Lipoprotein-associated phospholipase A2 activity is associated with risk of coronary heart disease and ischemic stroke: the Rotterdam Study," *Circulation*, vol. 111, no. 5, pp. 570–575, 2005.

[103] S. Pati, F. Palmisano, M. Quinto, and P. G. Zambonin, "Quantitation of major choline fractions in milk and dietary supplements using a phospholipase D bioreactor coupled to a choline amperometric biosensor," *Journal of Agricultural and Food Chemistry*, vol. 53, no. 18, pp. 6974–6979, 2005.

[104] K. Ge, D. Liu, K. Chen, L. Nie, and S. Yao, "Assay of pancreatic lipase with the surface acoustic wave sensor system," *Analytical Biochemistry*, vol. 226, no. 2, pp. 207–211, 1995.

[105] I. B. Rejeb, F. Arduini, A. Amine, M. Gargouri, and G. Palleschi, "Amperometric biosensor based on Prussian Blue-modified screen-printed electrode for lipase activity and triacylglycerol determination," *Analytica Chimica Acta*, vol. 594, no. 1, pp. 1–8, 2007.

[106] E. J. Herrera-López, "Lipase and phospholipase biosensors: a review," *Methods in Molecular Biology*, vol. 861, pp. 525–543, 2012.

[107] R. Aravindan, P. Anbumathi, and T. Viruthagiri, "Lipase applications in food industry," *Indian Journal of Biotechnology*, vol. 6, no. 2, pp. 141–158, 2007.

10

Antihyperglycemic Effect of *Ginkgo biloba* Extract in Streptozotocin-Induced Diabetes in Rats

Daye Cheng,[1] Bin Liang,[2] and Yunhui Li[3]

[1] *Department of Transfusion, The First Hospital of China Medical University, North Nanjing Street No. 155, Shenyang 110001, China*
[2] *High Vocational Technological College, China Medical University, Shenyang 11001, China*
[3] *Department of Clinical Laboratory, No. 202 Hospital, Shenyang 110003, China*

Correspondence should be addressed to Daye Cheng; dayecheng_cmu@yahoo.cn

Academic Editor: Fabio Ferreira Perazzo

The *Ginkgo biloba* extract (GBE) has been reported to have a wide range of health benefits in traditional Chinese medicine. The aim of this study was to evaluate the antihyperglycemic effects of GBE on streptozotocin- (STZ-) induced diabetes in rats. Diabetes was induced in male *Wistar* rats by the administration of STZ (60 mg/kg b.w.) intraperitoneally. GBE (100, 200, and 300 mg/kg b.w.) was administered orally once a day for a period of 30 days. Body weight and blood glucose levels were determined in different experimental days. Serum lipid profile and antioxidant enzymes in hepatic and pancreatic tissue were measured at the end of the experimental period. Significant decreases in body weight and antioxidant ability and increases in blood glucose, lipid profile, and lipid peroxidation were observed in STZ-induced diabetic rats. The administration of GBE and glibenclamide daily for 30 days in STZ-induced diabetic rats reversed the above parameters significantly. GBE possesses antihyperglycemic, antioxidant, and antihyperlipidemia activities in STZ-induced chronic diabetic rats, which promisingly support the use of GBE as a food supplement or an adjunct treatment for diabetics.

1. Introduction

Diabetes is a global epidemic with an estimated worldwide prevalence of 246 million people in 2007 and forecasts to rise to 300 million by 2025 [1]; consequently, diabetes presents a major challenge to healthcare systems around the world. Diabetes is a metabolic disorder of multiple etiologies characterized by chronic hyperglycemia with disturbance of carbohydrate, fat, and protein metabolism resulting from defects in insulin secretion, insulin action, or both [2]. Many oral antihyperglycemic agents, such as sulfonylurea and biguanides, are available along with insulin for the treatment of diabetes, but these agents have significant side effects, and some are ineffective in chronic diabetes patients [3]. Thus, there is an increasing need of new natural antihyperglycemic products especially nutraceuticals with less side effects, safe, and high antihyperglycemic potential.

Previous studies have demonstrated that diabetes exhibits enhanced oxidative stress and high reactive oxygen species (ROS) in pancreatic islets due to persistent and chronic hyperglycemia, thereby depletes the activity of antioxidative defense system, and thus promotes free radical generation [4]. A number of mechanisms or pathways by which hyperglycemia, the major contributing factor of increased ROS production, causes tissue damage or diabetic complications have been identified [5]. Also, reduced antioxidant levels as a result of increased free radical production in experimental diabetes have been reported [6]. The rise in free radical activity is suggested to play an important role in lipid peroxidation and protein oxidation of cellular structures resulting in cell injury and implicated in the pathogenesis of vascular disease which are the mainly cause of morbidity and mortality in diabetes [7]. Streptozotocin (STZ) is frequently used to induce diabetes in experimental animals through its toxic effects on pancreatic β-cells [8, 9] and as a potential inducer of oxidative stress. It has been reported that diabetes induced by STZ is the best characterized system of xenobiotic-induced diabetes and the commonly used model for the screening of antihyperglycemic activities.

Traditional medicines derived mainly from plants played an important role in the management of diabetes. *Ginkgo biloba* is a dioecious tree with a history of use in traditional Chinese medicine and has many pharmacologic effects. The mechanism of action of *Ginkgo* is believed to be linked with its functions as a neuroprotective agent, an antioxidant, a free-radical scavenger, a membrane stabilizer, and an inhibitor of the platelet-activating factor, and so on [10–13]. *Ginkgo biloba* extract (GBE) from *Ginkgo biloba* leaves is commonly used in dietary supplements for ailments and has showed excellent clinical effects in many cases. The goal of this study was to evaluate the effects of GBE on the antihyperglycemic ability in STZ-induced diabetes rats. Furthermore, the positive roles of natural products in the correction of oxidative stress and hyperlipidaemia, which are diabetes-related complication, were also assessed.

2. Materials and Methods

2.1. Materials. The powder form of GBE was purchased from Hangzhou Greensky Biological Tech (Hangzhou, China). All reagents used in this research were of analytical grade and obtained from Shenyang Biotechnology Co. Ltd.

2.2. Induction of Diabetes to Experimental Rats. Diabetes was induced by a single intraperitoneal injection of a freshly buffered (0.1 M citrate, pH 4.5) solution of STZ at a dosage of 60 mg/kg body weight (b.w.). After 72 h of STZ administration, the tail vein blood was collected to determine fasting blood glucose level. Only rats with fasting blood glucose over 250 mg/dL were considered diabetic and included in the experiments. Treatment with GBE started after the last STZ injection. Blood samples were drawn at 48 hours, 15 days, and 30 days till the end of the study (30 days).

2.3. Experimental Design. A total of 70 matured normoglycemic male *Wistar* rats (12–14 weeks of age, weighing about 180 ± 10 g) were collected for this experiment. Animals were acclimated for a period of 7 days in our laboratory condition prior to the experiment. The rats were fed with standard laboratory diet and allowed to drink water *ad libitum*. Animal experiments were carried out in accordance with institutional ethical guidelines for the care of laboratory animals of the China Medical University.

Rats randomly selected were divided into seven groups, comprising ten rats each. The treatment schedule was as follows: (1) normal control group (N group), (2) STZ control group (D group), (3) GBE- (200 mg/kg b.w.) treated control group (N + G group), (4) GBE- (100 mg/kg b.w.) treated STZ group (D+LG group), (5) GBE- (200 mg/kg b.w.) treated STZ group (D+MG group), (6) GBE- (300 mg/kg b.w.) treated STZ group (D + HG group), and (7) glibenclamide- (5 mg/kg b.w.) treated STZ group (D + GLI group). The treatment with GBE and glibenclamide started after the last STZ injection, where the vehicle, GBE, and glibenclamide were administrated orally to the respective group rats. After 48 hours, 15 days,

and 30 days of treatment, the rats were fasted overnight, and blood glucose and body weight were measured in the morning. The serum lipid profile was determined after 30 days of treatment. At the end of 30-day experiment, all rats were anesthetized with pentobarbital sodium (35 mg/kg) and euthanized by cervical decapitation. The liver and pancreas were excised immediately from the animals, washed with ice-chilled physiological saline, and stored at −80°C.

2.4. Determination of Serum Glucose and Lipid Profile. The serum concentrations of glucose, triglyceride (TG), total cholesterol (TC), high-density lipoprotein cholesterol (HDL-C), and low-density lipoprotein cholesterol (LDL-C) were determined using commercially available kits (BIOSINO Biotechnology and Science INC, China).

2.5. Determination of Oxidative Stress Markers. The superoxide dismutase (SOD), glutathione peroxidase (GSH-Px), catalase (CAT) activities, and glutathione (GSH) level in hepatic and pancreatic tissue were measured using commercially available kits (Jiancheng Bioengineering Institute, Nanjing, China). Lipid peroxidation was measured as malondialdehyde (MDA) level in hepatic and pancreatic tissue according to Jain's method [14].

2.6. Statistical Analysis. Results are expressed as the means ± SD. Statistical analysis was performed by ANOVA for multiple comparisons (SPSS, Version 15.0). A value of $P < 0.05$ was considered significant.

3. Results

3.1. Effect of GBE on Body Weight in Experimental Groups. Body weights of rats in the seven groups were monitored during the experimental period. As shown in Table 1, there is no difference in the different groups before treatment and after 48 h ($P > 0.05$). Body weights of rats in D group were lower than those in other groups after 15 days and 30 days ($P < 0.01$). STZ caused a significant weight loss of rats in D group while treatment with GBE at different concentrations or glibenclamide suppressed the decrease in the body weight.

3.2. Effect of GBE on Blood Glucose in Experimental Groups. The STZ-induced diabetic rats exhibited a significant increase in fasting blood glucose (299 ± 10 mg/dL) as compared to non-STZ-treated rats (49 ± 6 mg/dL) before the initial treatment of GBE or glibenclamide ($P < 0.01$). After GBE and glibenclamide treatment, the changes of blood glucose levels in different experimental groups were shown in Figure 1. At 48 hours, the levels of blood glucose in D group (298.3 ± 11.2 mg/dL), D+LG group (291.6±8.8 mg/dL), D + MG group (286.4 ± 9.2 mg/dL), D + HG group (283.5 ± 5.6 mg/dL), and D + GLI group (252.9 ± 9.1 mg/dL) were higher than N group (46.8 ± 4.6 mg/dL) and N + G group (45.3 ± 3.3 mg/dL) ($P < 0.01$). The administration of GBE or glibenclamide for 30 days in D+LG group (155.5±11.6 mg/dL), D+MG group (143.1 ± 9.6 mg/dL), D + HG group (85.0 ± 6.7 mg/dL), and D + GLI group (40.9 ± 5.4 mg/dL) caused a significant

TABLE 1: Changes of the body weight (g) of rats in the seven groups during the experimental period of 30 days.

	Before treatment	48 h	15 days	30 days
N group	178.3 ± 5.6	183.4 ± 6.5	214.3 ± 7.7	220.3 ± 6.3
D group	180.1 ± 6.9	177.3 ± 5.9	174.1 ± 10.2[a]	169.3 ± 11.6[a]
N + G group	176.9 ± 5.8	186.1 ± 8.8	209.6 ± 9.6[b]	220.8 ± 10.0[b]
N + LG group	179.3 ± 7.6	183.0 ± 7.2	187.6 ± 5.5[a,b]	206.3 ± 5.9[a,b]
N + MG group	177.4 ± 5.1	180.6 ± 4.6	190.6 ± 4.9[a,b]	213.6 ± 9.9[b]
N + HG group	178.6 ± 7.7	186.9 ± 5.8	199.5 ± 7.1[a,b]	216.2 ± 7.8[b]
N + GLI group	180.2 ± 6.3	187.9 ± 5.0	196.3 ± 6.1[a,b]	212.1 ± 11.8[b]

Values are means ± SD for 10 rats in each group. N group: normal control; D group: diabetes group; N + G group: normal control plus GBE; D + LG group: diabetes plus low GBE treatment; D + MG group: diabetes plus middle GBE treatment; D + HG group: diabetes plus high GBE treatment; D + GLI group: diabetes plus glibenclamide treatment.
[a]Indicates statistical significance of $P < 0.01$ compared to N group; [b]$P < 0.01$ compared to D group.

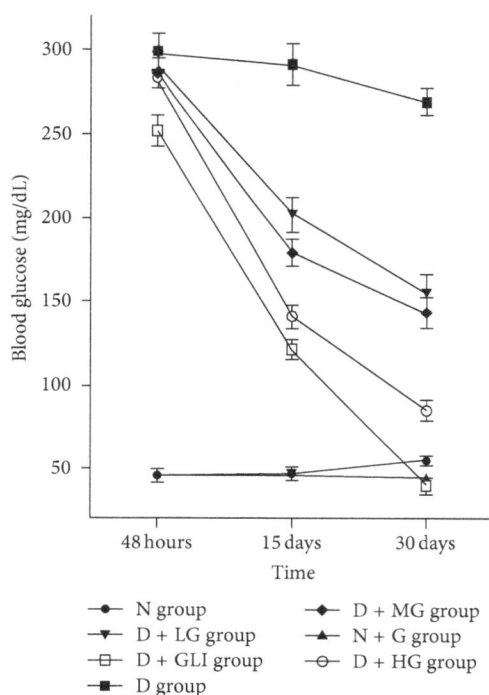

FIGURE 1: Changes of blood glucose of rats in the seven groups during the experimental period of 30 days. Values are mean ± SD ($n = 10$ animals).

growing body of evidence showing that STZ-induced diabetes can also induce anomaly of serum TC, TG, HDL-C, and LDL-C. Table 2 showed the levels of serum TG, TC, LDL-C, and HDL-C of rats in different experimental groups. Rats in D group displayed a significant increase in the levels of TG, TC, and LDL-C in comparison with N group ($P < 0.01$). However, serum HDL-C level of rats in D group was significantly lower than that of rats in N group ($P < 0.01$). Similar with the glibenclamide-treated STZ group, GBE administration showed a significant decrease in the levels of serum TG, TC, and LDL-C or a significant increase in the level of HDL-C after 30-day treatment when compared with D group.

3.4. Determination of Oxidative Stress Marker in Hepatic and Pancreatic Tissue. As shown in Table 3, a marked increase of MDA production and decrease of antioxidant level (GSH) and antioxidant enzyme activity (SOD, CAT, GSH-Px) were observed in the hepatic and pancreatic tissues of rats in D group when compared with N group ($P < 0.01$). GBE and glibenclamide treatment significantly inhibited the formation of MDA and raised antioxidant level (GSH) and antioxidant enzyme activity (SOD, CAT, GSH-Px) in a dose-dependent manner. Likewise, GBE exhibited the same antioxidation effects as glibenclamide at the dose of 5 mg/kg b.w.

4. Discussion

Diabetes is increasing at an alarming rate worldwide, which can mainly be attributed to the sedentary life style and calorie-rich diet. Diabetes is often linked with abnormal lipid metabolism and is considered as a major factor for the development of atherosclerosis and cardiovascular complication [15]. Recently, the WHO Expert Committee recommended the importance to investigate and explore hypoglycemic agents from plant origin because plants used in the traditional medicine have fewer side effects than synthetic drugs [16]. Currently, GBE is widely used for medicine in China and showed excellent clinical effects in many aspects, and the pharmacological mechanisms including modification of Ca^{2+} signaling [17], clearing oxygen free radical [18], decreasing lipid peroxidation, and promoting the synthesis and release

decrease in blood glucose levels when compared with D group ($269.7 ± 8.4$ mg/dL) ($P < 0.05$), but all GBE-treated group retained high blood glucose (>100 mg/dL) after 15 days. GBE caused a significant dose- ($P < 0.01$) and time-dependent reduction ($P < 0.01$) in blood glucose levels of diabetic rats. The blood glucose values of diabetic rates showed a tendency to normal levels after administration of GBE at 300 mg/kg b.w. and glibenclamide, 5 mg/kg b.w. in 30 days.

3.3. Effect of GBE on Serum Lipid Profiles in Experimental Groups. Our observation provides further support to the

TABLE 2: Effect of GBE treatment on serum lipid profile of rats in experimental groups.

	N group	D group	N + G group	D + LG group	D + MG group	D + HG group	D + GLI group
TG (mmol/L)	1.48 ± 0.35	2.32 ± 0.51[a]	1.46 ± 0.38[b]	1.75 ± 0.40[c]	1.65 ± 0.41[b]	1.52 ± 0.28[b]	1.36 ± 0.40[b]
TC (mmol/L)	1.69 ± 0.26	2.71 ± 0.48[a]	1.61 ± 0.30[b]	2.02 ± 0.51[b]	1.83 ± 0.44[b]	1.77 ± 0.33[b]	1.65 ± 0.31[b]
HDL-C (mmol/L)	1.35 ± 0.23	0.92 ± 0.30[a]	1.21 ± 0.28[b]	0.98 ± 0.25	1.19 ± 0.23[c]	1.28 ± 0.30[c]	1.20 ± 0.22[b]
LDL-C (mmol/L)	0.56 ± 0.09	0.73 ± 0.11[a]	0.51 ± 0.08[b]	0.68 ± 0.09	0.67 ± 0.03	0.62 ± 0.06[c]	0.57 ± 0.10[b]

Values are means ± SD for 10 rats in each group. N group: normal control; D group: diabetes group; N + G group: normal control plus GBE; D + LG group: diabetes plus low GBE treatment; D + MG group: diabetes plus middle GBE treatment; D + HG group: diabetes plus high GBE treatment; D + GLI group: diabetes plus glibenclamide treatment.
[a]Indicates statistical significance of $P < 0.01$ compared to N group; [b]$P < 0.01$ compared to D group.
[c]$P < 0.05$ compared to D group.

of epoprostenol [6]. Increasing evidence has demonstrated that GBE has a potential efficacy in glucose metabolism and lipid metabolism. Therefore, the present study was aimed to assess the effect of GBE on hyperglycemia, lipid profile, and enzymatic and nonenzymatic antioxidants in STZ-induced diabetic rats.

In our study, Figure 1 shows the changes in body weight of normal and streptozotocin-induced diabetic rats. STZ-injection-induced diabetes is associated with the characteristic loss of body weight which is due to increased muscle wasting and due to loss of tissue proteins [19]. As expected in D group, the body weight of rats was progressively reduced, and the treatment of diabetic rats with glibenclamide and GBE improved body weight significantly, which indicates the prevention of muscle tissue damage due to hyperglycemic condition. In addition, STZ injection caused diabetes, which may be due to destruction of β-cells of the islets of Langerhans [20]. G. B. Kudolo has already reported that ingestion of 120 mg of GBE as a single dose for 3 months for individuals leads to an increase in pancreatic β-cell function [21, 22]. Zhou et al. proposed that GBE improved insulin sensitivity mainly by enhancing insulin receptor substrate 2 transcription and preventing insulin resistance. Similarly, our data showed that the daily administration of GBE for 30 consecutive days abolished the blood glucose increase in the STZ-induced diabetic rats. This effect was dose dependent and time dependent. The reduction of blood glucose may be either due to the increased level of plasma insulin in diabetic rats, which may influence the stimulation of pancreatic insulin secretion from β-cells in islets of Langerhans, or due to the enhanced transport of blood glucose to peripheral tissue. Our results supported the reported evidence that GBE has the potential to prevent insulin resistance and is a promising antidiabetic drug [23].

Previous studies have demonstrated that GBE exerts multidirectional lipid-lowering effects on the rat metabonome, including limitation of the absorption of cholesterol, inactivation of HMG-CoA, and favorable regulation of profiles of essential polyunsaturated fatty acid [24]. In diabetes, hyperglycemia is accompanied with dyslipidemia [25] characterized by increase in TC, LDL, VLDL, and TG and fall in HDL. This altered serum lipid profile was reversed towards normal after treatment with GBE. The possible mechanism through which GBE exerts its anti-hyperlipidemic effect might include the changed activity of cholesterol biosynthesis

enzymes and/or the changed level of lipolysis which are under the control of insulin [26]. It is reported that GBE treatment could decrease the capacity of LDL to carry free cholesterol to various tissues without affecting the capacity of HDL to carry cholesterol back to the liver in rats [27]. In addition, GBE treatment can partially reverse ethanol-induced dyslipidemia at dose levels of 48 and 96 mg/kg b.w. in rats by reducing the lipid peroxidation induced by ethanol [28]. Our results indicated that the lipid-lowering effect of GBE could be an indirect consequence of amelioration of insulin resistance or direct hypolipidemic effect mediated through other mechanisms [29].

Hyperglycemia is a main cause for elevated free radical levels, followed by production of ROS, which can lead to increased lipid peroxidation and altered antioxidant defense and further impair glucose metabolism in biological system [30]. An imbalance between oxidation and antioxidant status has been shown to play an important role in mediating insulin resistance [29]. Overwhelming free radicals generated due to oxidative stress may develop several adverse effects commonly seen in diabetes such as neuropathy, nephropathy, retinopathy, and vascular disorders [31]. The major antioxidant enzymes, including SOD, CAT, and GSH-Px, are regarded as the first line of the antioxidant defense system against ROS generated in vivo during oxidative stress and act cooperatively at different sites in the metabolic pathway of free radicals [32]. In our study, reduced activities of SOD, CAT, and GSH-Px in the liver and pancreas have been observed in diabetic rats. The administration of GBE for 30 days increased the SOD, CAT, and GSH-Px activity and GSH level in the liver and pancreas of diabetic rats. Robertson et al. demonstrated that antioxidants have been shown to brake the worsening of diabetes by improving β-cells function in animal models and suggested that enhancing antioxidant defense mechanisms in pancreatic islets may be a valuable pharmacologic approach to managing diabetes [33]. Modak MA et al. reported that the control of hyperglycemia leads to improvement in oxidative stress profile, and enhancing antioxidant defense mechanisms in pancreatic islets helps them to cope better with oxidative stress. Since GBE is a complex mixture of ingredients with a unique broad spectrum of pharmacological activities, it probably acts through several different mechanisms covering ROS scavenger and/or enhancing antioxidant ability. Moreover, Naik et al. have demonstrated that GBE is scavengers of

TABLE 3: Effect of GBE treatment on hepatic and pancreatic oxidative stress markers of rats in experimental groups.

	N group	D group	N + G group	D + LG group	D + MG group	D + HG group	D + GLI group
Liver							
GSH (μg/mg protein)	512.36 ± 91.25	365.50 ± 84.77[a]	529.36 ± 101.23[b]	400.28 ± 79.46	451.33 ± 89.44[c]	486.36 ± 98.31[b]	482.09 ± 106.30[c]
CAT (U/mg protein)	362.63 ± 18.52	302.05 ± 22.82[a]	374.47 ± 26.36[b]	313.25 ± 24.47	322.52 ± 17.52[c]	349.96 ± 25.01[b]	344.80 ± 20.91[b]
SOD (U/mg protein)	542.35 ± 24.21	235.27 ± 34.22[a]	602.0 ± 40.50[b]	322.21 ± 32.65[b]	414.11 ± 35.33[b]	505.20 ± 41.52[b]	523.36 ± 36.66[b]
GSH-Px (U/mg protein)	3023 ± 217	2236 ± 259[a]	3111 ± 285[b]	2567 ± 198[b]	2758 ± 252[b]	2904 ± 220[b]	2808 ± 244[b]
MDA (nmol/mg protein)	9.03 ± 1.25	16.91 ± 1.81[a]	8.85 ± 1.33[b]	14.28 ± 1.57[b]	10.05 ± 1.01[b]	9.66 ± 1.41[b]	10.36 ± 1.55[b]
Pancreas							
GSH (μg/mg protein)	444.08 ± 71.44	294.14 ± 69.27[a]	408.50 ± 66.85[b]	312.58 ± 75.55	360.00 ± 57.19[c]	405.81 ± 63.36[b]	406.16 ± 56.39[b]
CAT (U/mg protein)	298.02 ± 21.30	174.22 ± 14.52[a]	287.34 ± 22.71[b]	196.35 ± 18.90[b]	209.42 ± 23.00[b]	212.56 ± 20.66[b]	274.30 ± 21.94[b]
SOD (U/mg protein)	415.00 ± 33.20	197.38 ± 36.55[a]	385.52 ± 32.11[b]	222.44 ± 36.40	356.78 ± 37.80[b]	379.91 ± 38.92[b]	364.82 ± 40.24[b]
GSH-Px (U/mg protein)	658 ± 89	462 ± 92[a]	711 ± 101[b]	555 ± 74[c]	596 ± 94[b]	623 ± 78[b]	539 ± 79[b]
MDA (nmol/mg protein)	8.77 ± 1.03	14.91 ± 1.44[a]	8.05 ± 1.11[b]	12.50 ± 1.48[b]	10.87 ± 1.61[b]	9.22 ± 1.31[b]	8.99 ± 1.40[b]

Values are means ± SD for 10 rats in each group. N group: normal control; D group: diabetes group; N + G group: normal control plus GBE; D + LG group: diabetes plus low GBE treatment; D + MG group: diabetes plus middle GBE treatment; D + HG group: diabetes plus high GBE treatment; D + GLI group: diabetes plus glibenclamide treatment.
[a] Indicates statistical significance of $P < 0.01$ compared to N group; [b] $P < 0.01$ compared to D group.
[c] $P < 0.05$ compared to D group.

free radicals by increasing levels of free radical scavenging enzymes [34]. It is also reported that GBE may reduce the oxidative stress in the reperfused myocardium and increase the antioxidant activity in ischemia reperfusion rats. Further, lipid peroxidation measurement is a more practical and safer method to evaluate the factors causing cellular injury and the activation of the common pathway. Tissue MDA content, the final product of lipid breakdown caused by oxidative stress, is an important indicator of free radical-induced lipid peroxidation [35–37]. GBE-treated rats showed decreased level of MDA, suggesting that GBE has antioxidant capacity.

5. Conclusion

In summary, GBE possesses antihyperglycemic, antioxidant, and antihyperlipidemic activities in STZ-induced chronic diabetic rats, which promisingly support the use of GBE as a food supplement or an adjunct treatment for diabetics. Moreover, further work is necessary to elucidate in detail the mechanism of action of the GBE at the cellular and molecular levels.

Conflict of Interests

The authors declare that they have no conflict of interests.

Acknowledgment

The authors thank Professors Guo Zhengdong and Li Hong in the China Medical University Animal Research Laboratory for their help with experimental techniques.

References

[1] D. R. Whiting, L. Guariguata, C. Weil, and J. Shaw, "IDF diabetes atlas: global estimates of the prevalence of diabetes for 2011 and 2030," *Diabetes Research and Clinical Practice*, vol. 94, pp. 311–321, 2011.

[2] B. S. Nayak and L. Roberts, "Relationship between inflammatory markers, metabolic and anthropometric variables in the Caribbean type 2 diabetic patients with and without microvascular complications," *Journal of Inflammation*, vol. 3, article no. 17, 2006.

[3] L. Pari and R. Saravanan, "Antidiabetic effect of diasulin, a herbal drug, on blood glucose, plasma insulin and hepatic enzymes of glucose mretabolism hyperglycaemic rats," *Diabetes, Obesity and Metabolism*, vol. 6, no. 4, pp. 286–292, 2004.

[4] O. Savu, C. Ionescu-Tirgoviste, V. Atanasiu, L. Gaman, R. Papacocea, and I. Stoian, "Increase in total antioxidant capacity of plasma despite high levels of oxidative stress in uncomplicated type 2 diabetes mellitus," *The Journal of International Medical*, vol. 40, pp. 709–716, 2012.

[5] F. Giacco and M. Brownlee, "Oxidative stress and diabetic complications," *Circulation Research*, vol. 107, no. 9, pp. 1058–1070, 2010.

[6] M. M. Ali and F. G. Agha, "Amelioration of streptozotocin-induced diabetes mellitus, oxidative stress and dyslipidemia in rats by tomato extract lycopene," *Scandinavian Journal of Clinical and Laboratory Investigation*, vol. 69, no. 3, pp. 371–379, 2009.

[7] S. M. Son, "Reactive oxygen and nitrogen species in pathogenesis of vascular complications of diabetes," *Journal of Diabetes & Metabolism*, vol. 36, pp. 190–198, 2012.

[8] A. N. Shafik, "Effects of topiramate on diabetes mellitus induced by streptozotocin in rats," *European Journal of Pharmacology*, vol. 684, pp. 161–167, 2012.

[9] F. Hu, W. Zhu, M. Chen, Q. Shou, and Y. Li, "Biological activities of Chinese propolis and Brazilian propolis on streptozotocin-induced type 1 diabetes mellitus in rats," *Evidence-based Complementary and Alternative Medicine*, vol. 2011, Article ID 468529, 8 pages, 2011.

[10] I. Ilieva, K. Ohgami, K. Shiratori et al., "The effects of Ginkgo biloba extract on lipopolysaccharide-induced inflammation in vitro and in vivo," *Experimental Eye Research*, vol. 79, no. 2, pp. 181–187, 2004.

[11] A. Louajri, S. Harraga, V. Godot, G. Toubin, J. P. Kantelip, and P. Magninc, "The effect of Ginkgo biloba extract on free radical production in hypoxic rats," *Biological and Pharmaceutical Bulletin*, vol. 24, no. 6, pp. 710–712, 2001.

[12] K. Ashok Shenoy, S. N. Somayaji, and K. L. Bairy, "Evaluation of hepatoprotective activity of Ginkgo biloba in rats," *Indian Journal of Physiology and Pharmacology*, vol. 46, no. 2, pp. 167–174, 2002.

[13] A. Eckert, U. Keil, S. Kressmann et al., "Effects of EGb 761® Ginkgo biloba extract on mitochondrial function and oxidative stress," *Pharmacopsychiatry*, vol. 36, supplement 1, pp. S15–S23, 2003.

[14] S. K. Jain, S. N. Levine, J. Duett, and B. Hollier, "Elevated lipid peroxidation levels in red blood cells of streptozotocin-treated diabetics rats," *Metabolism: Clinical and Experimental*, vol. 39, no. 9, pp. 971–975, 1990.

[15] S. B. Biddinger and C. R. Kahn, "From mice to men: insights into the insulin resistance syndromes," *Annual Review of Physiology*, vol. 68, pp. 123–158, 2006.

[16] F. J. Alarcon-Aguilara, R. Roman-Ramos, S. Perez-Gutierrez, A. Aguilar-Contreras, C. C. Contreras-Weber, and J. L. Flores-Saenz, "Study of the anti-hyperglycemic effect of plants used as antidiabetics," *Journal of Ethnopharmacology*, vol. 61, no. 2, pp. 101–110, 1998.

[17] C. Shi, F. Wu, and J. Xu, "H2O2 and PAF mediate Aβ1-42-induced Ca2+ dyshomeostasis that is blocked by EGb761," *Neurochemistry International*, vol. 56, no. 8, pp. 893–905, 2010.

[18] Y. H. Niu, X. Y. Yang, and W. S. Bao, "Protective effects of Ginkgo biloba extract on cultured rat cardiomyocytes damaged by H_2O_2," *Acta Pharmacologica Sinica*, vol. 20, no. 7, pp. 635–638, 1999.

[19] S. K. Swanston-Flatt, C. Day, C. J. Bailey, and P. R. Flatt, "Traditional plant treatments for diabetes. Studies in normal and streptozotocin diabetic mice," *Diabetologia*, vol. 33, no. 8, pp. 462–464, 1990.

[20] G. Kavalali, H. Tuncel, S. Göksel, and H. H. Hatemi, "Hypoglycemic activity of Urtica pilulifera in streptozotocin-diabetic rats," *Journal of Ethnopharmacology*, vol. 84, no. 2-3, pp. 241–245, 2003.

[21] G. B. Kudolo, "The effect of 3-month ingestion of Ginkgo biloba extract on pancreatic β-cell function in response to glucose loading in normal glucose tolerant individuals," *Journal of Clinical Pharmacology*, vol. 40, no. 6, pp. 647–654, 2000.

[22] G. B. Kudolo, "The effect of 3-month ingestion of Ginkgo biloba extract (EGb 761) on pancreatic β-cell function in response to glucose loading in individuals with non-insulin-dependent diabetes mellitus," *Journal of Clinical Pharmacology*, vol. 41, no. 6, pp. 600–611, 2001.

[23] L. Zhou, Q. Meng, T. Qian, and Z. Yang, "*Ginkgo biloba* extract enhances glucose tolerance in hyperinsulinism-induced hepatic cells," *Journal of Natural Medicines*, vol. 65, no. 1, pp. 50–56, 2011.

[24] Q. Zhang, G. J. Wang, J. Y. A et al., "Application of GC/MS-based metabonomic profiling in studying the lipid-regulating effects of *Ginkgo biloba* extract on diet-induced hyperlipidemia in rats," *Acta Pharmacologica Sinica*, vol. 30, no. 12, pp. 1674–1687, 2009.

[25] M. S. De Sereday, C. Gonzalez, D. Giorgini et al., "Prevalence of diabetes, obesity, hypertension and hyperlipidemia in the central area of Argentina," *Diabetes and Metabolism*, vol. 30, no. 4, pp. 335–339, 2004.

[26] S. B. Sharma, A. Nasir, K. M. Prabhu, P. S. Murthy, and G. Dev, "Hypoglycaemic and hypolipidemic effect of ethanolic extract of seeds of Eugenia jambolana in alloxan-induced diabetic rabbits," *Journal of Ethnopharmacology*, vol. 85, no. 2-3, pp. 201–206, 2003.

[27] Z. X. Yao, Z. Han, K. Drieu, and V. Papadopoulos, "*Ginkgo biloba* extract (Egb 761) inhibits β-amyloid production by lowering free cholesterol levels," *Journal of Nutritional Biochemistry*, vol. 15, no. 12, pp. 749–756, 2004.

[28] P. Yao, F. Song, K. Li et al., "*Ginkgo biloba* extract prevents ethanol induced dyslipidemia," *American Journal of Chinese Medicine*, vol. 35, no. 4, pp. 643–652, 2007.

[29] B. Ramesh and D. Saralakumari, "Antihyperglycemic, hypolipidemic and antioxidant activities of ethanolic extract of Commiphora mukul gum resin in fructose-fed male Wistar rats," *Journal of Physiology and Biochemistry*, vol. 68, no. 4, pp. 573–582, 2012.

[30] M. S. Balasubashini, R. Rukkumani, P. Viswanathan, and V. P. Menon, "Ferulic acid alleviates lipid peroxidation in diabetic rats," *Phytotherapy Research*, vol. 18, no. 4, pp. 310–314, 2004.

[31] H. F. Al-Azzawie and M. S. S. Alhamdani, "Hypoglycemic and antioxidant effect of oleuropein in alloxan-diabetic rabbits," *Life Sciences*, vol. 78, no. 12, pp. 1371–1377, 2006.

[32] D. Cheng and H. Kong, "The effect of lycium barbarum polysaccharide on alcohol-induced oxidative stress in rats," *Molecules*, vol. 16, no. 3, pp. 2542–2550, 2011.

[33] R. P. Robertson, Y. Tanaka, H. Takahashi, P. O. T. Tran, and J. S. Harmon, "Prevention of oxidative stress by adenoviral overexpression of glutathione-related enzymes in pancreatic islets," *Annals of the New York Academy of Sciences*, vol. 1043, pp. 513–520, 2005.

[34] S. R. Naik, V. W. Pilgaonkar, and V. S. Panda, "Evaluation of antioxidant activity of *Ginkgo biloba* phytosomes in rat brain," *Phytotherapy Research*, vol. 20, no. 11, pp. 1013–1016, 2006.

[35] H. Okur, M. Kucukaydin, K. Kose, O. Kontas, P. Dogan, and A. Kazez, "Hypoxia-induced necrotizing enterocolitis in the immature rat: the role of lipid peroxidation and management by vitamin E," *Journal of Pediatric Surgery*, vol. 30, no. 10, pp. 1416–1419, 1995.

[36] C. Kabaroglu, M. Akisu, S. Habif et al., "Effects of L-arginine and L-carnitine in hypoxia/reoxygenation-induced intestinal injury," *Pediatrics International*, vol. 47, no. 1, pp. 10–14, 2005.

[37] U. Bicakci, B. Tander, E. Aritürk et al., "Effects of omeprazole and gentamicin on the biochemical and histopathological alterations of the hypoxia/reoxygenation induced intestinal injury in newborn rats," *Pediatric Surgery International*, vol. 21, no. 10, pp. 800–805, 2005.

11

Rhizobium pongamiae sp. nov. from Root Nodules of Pongamia pinnata

Vigya Kesari, Aadi Moolam Ramesh, and Latha Rangan

Department of Biotechnology, Indian Institute of Technology Guwahati, Assam 781 039, India

Correspondence should be addressed to Latha Rangan; latha_rangan@yahoo.com

Academic Editor: Eldon R. Rene

Pongamia pinnata has an added advantage of N2-fixing ability and tolerance to stress conditions as compared with other biodiesel crops. It harbours "rhizobia" as an endophytic bacterial community on its root nodules. A gram-negative, nonmotile, fast-growing, rod-shaped, bacterial strain VKLR-01^T was isolated from root nodules of *Pongamia* that grew optimal at 28°C, pH 7.0 in presence of 2% NaCl. Isolate VKLR-01 exhibits higher tolerance to the prevailing adverse conditions, for example, salt stress, elevated temperatures and alkalinity. Strain VKLR-01^T has the major cellular fatty acid as $C_{18:1}$ ω7c (65.92%). Strain VKLR-01^T was found to be a nitrogen fixer using the acetylene reduction assay and PCR detection of a *nif*H gene. On the basis of phenotypic, phylogenetic distinctiveness and molecular data (16S rRNA, *rec*A, and *atp*D gene sequences, G + C content, DNA-DNA hybridization etc.), strain VKLR-01^T = (MTCC 10513^T = MSCL 1015^T) is considered to represent a novel species of the genus *Rhizobium* for which the name *Rhizobium pongamiae* sp. nov. is proposed. *Rhizobium pongamiae* may possess specific traits that can be transferred to other rhizobia through biotechnological tools and can be directly used as inoculants for reclamation of wasteland; hence, they are very important from both economic and environmental prospects.

1. Introduction

Pongamia pinnata (L.) Pierre is a nonedible "pioneer" biodiesel and medicinal tree species of the family Leguminosae that grows in multiple geoclimatic conditions, ranging from humid, tropical, subtropical regions to cooler and semiarid zones [1–3]. Nitrogen is an important nutrient for plant growth and yield; however, its availability in soils is limited. Modern agriculture depends on chemically synthesized N fertilizers which are expensive and require fossil fuels for production, adding to greenhouse gas emissions. Biological nitrogen fixation is a useful and important alternative [4], especially in biofuel production [2, 5]. *Pongamia* can grow on low-fertility land due to its nodulation properties and good N_2-fixing symbiotic associations with "rhizobia" (a polyphyletic assemblage of alphaproteobacteria family: *Rhizobiaceae*), thus minimizing competition with food crops or related fertilizer, water, and land resources needed for food and fodder production [6]. The sustainable production of plant oils for biodiesel production from a tree crop such as

P. pinnata, which can be cultivated on marginal lands, has the potential to not only provide a renewable energy resource but in addition alleviate the competitive situation that exists with food crops as biofuels and associated arable land and water use. It is also used in agriculture and environmental management, due to its insecticidal and nematicidal properties [7]. Finally, *Pongamia* has been identified as a resource for agroforestry, urban landscaping to suspend the pollutants and the bioameloriation of degraded lands.

Isolation and identification of authentic and effective rhizobia isolates are required to support *P. pinnata* plantations in nitrogen-poor soils. *Pongamia* trees are purportedly able to grow in a wide range of environments: in the tropics, with temperatures from 13–45°C, saline soils, and in soils with a range of pH including sodic soils, and they are an ideal candidate for reforestation of marginal lands [3, 5]. The ability of rhizobia to grow in these diverse environmental conditions will be important for the establishment and success of *Pongamia* plantations on these unfertile lands. It, thus, satisfies all "sustainability criteria" expected from modern second- and third-generation biofuel crops. However, little attention has been paid to the occurrence of nitrogen-fixing endophytic bacteria in the rhizospheres of this biodiesel tree which is important for their diverse applicabilities as well as agronomic and ecological significances. In preliminary studies, the effective nodulations of *P. pinnata* with three strains of rhizobia (*Bradyrhizobium japonicum* strain CB1809, *Bradyrhizobium sp.* strain CB564, and *Rhizobia sp.* strain NGR234) were demonstrated [1].

The importance of characterizing indigenous rhizobia of *Pongamia* cannot be overemphasized. Still there has not been any detailed study of phenotypic characteristics and symbiotic effectiveness of rhizobia isolates which naturally nodulate *Pongamia* considering its potential value in sustainable agriculture and role in agroforestry. Therefore, in this work, we attempt to isolate the nitrogen-fixing rhizobial symbiont strain from nodules of *P. pinnata* occurring in North Guwahati, Assam, India. The objectives were to determine the exact taxonomic position of isolated and identified strain by using a polyphasic characterization that included determination of phenotypic and biochemical properties, phylogenetic investigations based on 16S rRNA, *atp*D, and *rec*A gene sequences, and genetic analysis. Further investigations were also performed in order to verify the nodulation and nitrogen-fixing property of the isolated bacterium strain.

2. Materials and Methods

2.1. Soil Sampling and Isolation of Rhizobia. *P. pinnata* saplings approximately 2-3 months old found in North Guwahati (26°14′6″ N; 91°41′28″ E), Assam, India, were uprooted during April 2010 containing distinct nodules. Nodules excised from the roots were surface sterilized with 70% (v/v) ethanol for 1 min. Subsequently, nodules were treated with 10% (w/v) sodium hypochlorite for 15 min and washed with sterile distilled water (3x). Single surface-sterilized nodule (approximately 2 mm) was opened into two halves with a sterile blade, and the central parts of the nodule were scooped with blunt needle, macerated, and diluted in 500 μL of saline water (0.9%). Roughly 100 μL of the inoculum was spread on yeast extract-mannitol (YEM) and tryptone-yeast extract (TYE) agar plates and incubated at 28°C for 1–3 days. The purity of the culture was verified by repeated streaking of single colony onto YEM agar [8] with 25 mg Kg^{-1} (w/v) congo red. Single purified isolate was maintained in YEM broth containing 20% (v/v) glycerol at −80°C.

2.2. Growth and Phenotypic Characteristics. Cell size and morphology of the root nodule isolate were determined using scanning electron microscopy (LEO 1430 VP; Leo Electron Microscopy, Ltd., Cambridge, the UK) at 10 kV. For the micromorphology study, cells from the exponential growth phase (grown in YEM broth at 28°C) were harvested by centrifugation and fixed in 2.5% (w/v) glutaraldehyde for 45 min. The cells were washed with phosphate-buffered saline (PBS) and applied to ethanol dehydration series (at 50%, 70%, 90%, and 100% for 10 min each) (v/v) followed by critical-point drying with CO_2 and sputter-coating with gold as

described by Boyde and Wood [15]. A growth characteristic of isolate was recorded at different temperatures (4, 25, 28, 30, 37, 42, and 50°C) in YEM broth and agar until 48 h of culture. The ability to grow in acid and alkaline media was also tested by inoculating the isolate onto YEM broth and YEM agar plates adjusted to various pH values (pH 4.0–11.0 at intervals of 1 pH units) using 1 N HCl/1 N NaOH. The NaCl tolerance of the isolate was tested by growing in YEM broth and YEM agar plates containing 0%, 1%, 2%, 3%, 4%, and 5% (w/v) NaCl.

2.3. Gram Staining: Biochemical and Physiological Characteristic. Gram reaction was determined with the bioMérieux Gram-stain kit according to the manufacturer instructions. Acid production from different carbohydrates was determined by employing the API 50 CH system (bioMérieux) according to the manufacturers instructions. The ability of the isolate to utilize various carbon and amino acids compounds as sole carbon and nitrogen source was investigated using the method described by Lindström and Lehtomäki [16]. Def 9 (carbon source) and Def 8 (nitrogen source) agar mediums were used, and appropriate controls were maintained. Results were noted after 3 days of incubation at 28°C. A set of physiological characteristics including catalase and oxidase tests and nitrate reduction were assessed using protocols described by Shieh et al. [17]. Gelatin hydrolysis and methyl red test were also performed using the methods of Smibert and Krieg [18]. The intrinsic antibiotic resistance tests for the isolate were performed by disc-diffusion assay in YEM agar against 16 different antibiotics (Discs, HiMedia) of concentrations ranging from 2 to 100 μg. Cell biomass for the analysis of isoprenoid quinones was obtained from the isolate grown on yeast extract-mannitol broth (YEM) (12 h, 28°C, 180 rpm).

2.4. FAME Analysis. Fatty acid methyl esters (FAMEs) of the isolate were extracted and prepared according to the standard protocol of the MIDI Sherlock/Hewlett Packard Microbial Identification System as described by Sasser [19]. For cellular FAME analysis, isolate was grown on trypticase soy broth agar (TSBA), which consists of 30 g trypticase soy broth and 15 g of agar (BBL) for 12 h at 28°C under aerobic conditions (180 rpm). The fatty acid methyl esters extracts were analyzed by Gas Chromatography MODEL 6850 (Agilent Gas Chromatography) equipped with an Agilent ultra 2 capillary column.

2.5. DNA Extraction, PCR Amplification, and Sequencing. Cell biomass for DNA extraction was obtained from the isolate grown on yeast extract-mannitol (YEM) broth (12 h, 28°C, 180 rpm). Chromosomal DNA was isolated and purified using Sigma's GenElute Bacterial Genomic Kit. The 16S rRNA gene was amplified by PCR using two consensus primer fD1 and rD1 [20]. Polymerase chain reaction (PCR) of 16S rRNA gene was performed in 25 μL volume mixing the template DNA (10 ng) with 1x PCR buffer (Bioline), 2.5 mM MgCl$_2$ (Bioline), 0.5 U *Taq* DNA polymerase (Bioline), 2.5 mM dNTPs each, 0.4 μM (each) primers fD1 and rD1 using a DNA thermal cycler (Applied Biosystems, USA). The following temperature profile was used for DNA amplification: an initial denaturation at 95°C for 5 min followed by 35 cycles of 94°C for 1 min, 55°C for 1 min, and 72°C for 2 min and a final extension step of 72°C for 10 min. Reaction products were electrophoresed on a 1.3% (w/v) agarose gel and were purified using a Qiaquick PCR purification kit (Qiagen) before sequencing. Sequencing reactions were performed using the ABI PRISM Dye Terminator Cycle Sequencing Kit (Applied Biosystems, USA) with the primers IRF1, 1050R, 800F, and 800R [21], and they were analyzed in an automatic sequencer ABI PRISM 3730 sequencer (Applied Biosystems). PCR amplifications of other housekeeping genes *recA* and *atpD* were performed under the conditions described by Yoon et al. [9]. The primer sequences that were used for amplification and sequencing of 16S rRNA, *recA*, and *atpD* genes are listed in (Supplementary Table 1; see Supplementary Material available online at http://dx.doi.org/10.1155/2013/165198.) The sequences of these genes were compared with the sequences available from GenBank using BLASTN program [22] and were aligned using ClustalW2 multiple sequence alignment [23]. Phylogenetic trees were inferred using the neighbor-joining method [24], and distances were determined according to the Kimura-2 model [25]. Bootstrap analysis was based on 1000 resamplings. The MEGA 4.0 version [26] was used for all analyses.

2.6. Nitrogen Fixation and Nodulation Assessment Test. The acetylene reduction assay (ARA) was used to test the isolate for potential nitrogen fixations. The amount of ethylene produced was measured using 10% (v/v) acetylene according to the method of Li and MacRae [27] using a Hewlett Packard 4890 GC equipped with a Porapak N column. Isolate was subjected to *nif*H-specific PCR amplification using the primers (Supplementary Table 1) of Poly et al. [28]. Nodulation test for isolate was performed by PCR amplification of *nodD* gene (Supplementary Table 1) as described by Yoon et al. [9].

2.7. DNA Hybridization and G + C Content. The DNA G + C content was determined as described by Tamaoka and Komagata [29]. Isolate was disrupted using a French pressure cell (ThermoSpectronic), and the DNA in the crude lysate was purified by chromatography on hydroxyapatite as described by Cashion et al. [30]. DNA was hydrolyzed, dephosphorylated, and analyzed for its G + C content by HPLC [31]. Nonmethylated Lambda DNA (Sigma) was used as a reference. DNA-DNA hybridization was assessed for isolate against reference strain *Rhizobium radiobacter* DSM 30147^T (= AB247615) that showed 97% sequence similarity. DNA-DNA hybridization was carried out as described by De Ley et al. [32] under consideration of the modifications described by Huss et al. [33] using a model Cary 100 Bio UV/VIS spectrophotometer equipped with Peltier-thermostatted 6 × 6 multicell changer and a temperature controller with *in situ* temperature probe (Varian). The highest and lowest values obtained were excluded, and the means of the duplicates were quoted as DNA-DNA relatedness values. Analysis of respiratory quinones was carried out by the Identification Service and Dr. Brian Tindall, DSMZ, Braunschweig, Germany, according to the method of Komagata and Suzuki [34].

3. Results and Discussion

Since *P. pinnata* is introduced as the most important multipurpose tree for biodiesel production, it has become the most widespread legume in India and other parts of the world. This predominance has resulted from the massive implantation of the species for multipurpose use in a broad edaphic range including urban and social forestries to alleviate the environmental imbalance. The community structure of *Pongamia* root-nodule bacteria has been addressed by a few studies that assessed nodulation ability by endogenous rhizobia and also few strains commonly associated with *Glycine max* and their stimulatory effect on nodule number and plant growth [1, 35]. In this present study, we extend the work on *Pongamia* root-nodulating bacteria to isolate and characterize the novel *Rhizobium* species with traits that make it competitive in stress environments from the root nodules of biodiesel crop *P. pinnata* from North Guwahati, Assam, India.

3.1. Trapping and Isolating Root-Nodulating Bacteria. Root nodules were observed in all of the *Pongamia* saplings uprooted from the sampling site, indicating that root nodules occur widely in this legume crop growing naturally in North Guwahati, Assam, India. The plants, however, varied in the extent to which they were nodulated as nodules varied in shapes and sizes formed on primary as well as secondary roots (Supplementary Figure 1). Different shapes of root nodules of *P. pinnata* observed in the present study may be related to different developmental phases of the nodule ontogeny, and it is, therefore, not surprising that the nodule morphology has been used as a taxonomic marker [36]. Phenotypic traits of *Pongamia* saplings, namely, mean shoot length, mean root length, and the number of nodules observed in *P. pinnata*, were 28.25 ± 3.24, 12.17 ± 2.19, and 9.90 ± 2.31, respectively. Transverse section of single nodule (approximately 20 μm thick) in SEM image revealed the absence of any visible bacteria in the outer wall portion of the nodule. But when the middle portion was focused, each cell was fully filled with rod-shaped bacteria (Supplementary Figures 2A and 2B). Therefore, the middle portion of the root nodule was further used for isolation of single pure bacterial colony specific to *P. pinnata* and named as VKLR-01.

3.2. Growth, Phenotypic, Biochemical and Physiological Characteristics of Isolate. The first visible growth of the bacterium was observed as a small white shiny dot-like structure which increased in size from 1.5–3.5 mm (24 h) to 4.0–5.5 mm (48 h) in both YEM and TYE plates at 28°C (Supplementary Figure 3A). The generation time noted was 0.67 h in YEM medium. Pure culture was obtained from individual colony designated as VKLR-01. Isolate VKLR-01 showed creamy or white opaque, round or convex, and gummy colonies, with little or moderate extracellular polysaccharide production (EPS) having a diameter of 1.5–3.5 mm after growth for 24 h at 28°C. Metabolism is strictly aerobic. Isolate VKLR-01 appears to be a fast-growing rhizobial strain, forming colonies of 4–5.5 mm in diameter in 2-day time. Existing reports show that trees are as often nodulated by fast-growing as by slow-growing rhizobia [37].

TABLE 1: Phenotypic characteristics of *Rhizobium pongamiae* VKLR-01[T] and type strains of phylogenetically related *Rhizobium* species.

Characteristic	1	2	3	4
Origin	Root nodule of *P. pinnata* (North Guwahati, Assam)	*Galega orientalis* (Finland)	ND	*Ficus benjamina* (Florida)
Cell morphology (μm)	Rods (0.4–0.5 × 1.4–1.6)	Rods (0.9–1.0 × 1.5–1.8)[a]	Rods (0.6–1 × 1.5–3.0)	Rods
Flagella	ND	1-2	Several, peritrichous	Several, peritrichous
Nodulation	+	+	−	−
pH range	6–11	5.0–9.5[a]	ND	ND
Growth at/in				
40°C	+	−	ND	−
1% NaCl	+	+	+	+
2% NaCl	+	−	+	+
4% NaCl	+	−	−	−
Utilization as carbon source				
Sucrose	+	+	ND	ND
Arabinose	−	+	+	ND
Mannitol	+	+	+	ND
Lactose	+	−	ND	ND
Glucose	−	+	ND	ND
Maltose	+	ND	+	+
Melibiose	+	+	+	ND

Strains: 1, *R. pongamiae* VKLR-01[T] (present study); 2, *R. galegae* LMG 6214[T] 3; 3, *A. tumefaciens* (biovar 1); 4, *R. larrymoorei* AF3-10[T] [9–12]. [a]Unpublished data of Wang et al. (personal communication) [13]. +: positive; −: negative; ND: data not available. All strains are Gram negative, aerobic, rod shaped, and nonspore forming. All 7 strains are positive for oxidase, catalase, and nitrate reductase tests.

The SEM image of the purified isolate VKLR-01 from an exponential phase revealed the bacterium to be rod shaped, nonmotile having a cell dimension of 0.4-0.5 μm width and 1.4–1.6 μm length, respectively (Supplementary Figure 3B). Isolate VKLR-01 occurs at temperature range of 25–30°C (optimal at 28°C) and can tolerate up to 42°C, but no growth at 4 and 50°C. Isolate VKLR-01 can tolerate the salt concentration varied in the range of 1%–4% NaCl, but no visible growth was observed at 5% NaCl concentration in the YEM medium. The isolate also grew at pH 6.0 to pH 11.0, but no visible growth was observed at and below pH 5.0. Optimum growth conditions for the isolate VKLR-01 were temperature of 28–30°C, pH of 7.0–8.0, and 2% (w/v) NaCl. Temperature is known to influence survival, growth, and nitrogen fixation of *Rhizobium* [38]. Isolate VKLR-01 can tolerate extreme environmental conditions such as temperature up to 42°C and 4% NaCl, which differentiates this strain from other species. Similar results were found with rhizobia that nodulate *Lotus corniculatus* [39]. Generally, rhizobia collected from high temperature areas are resistant to high temperatures, and their tolerance is probably due to their adaptation to the extreme air temperatures inherent of tropical climate [40]. Salinity tolerance of the host is often the limiting factor in determining effective symbiosis of compatible rhizobia under saline conditions [41]. As the isolate VKLR-01 is able to adapt even in the presence of such unfavorable environmental conditions, this strain may be used for generation of genetically modified rhizobia by using genetic engineering tools. There is a report available that

showed transferring of a 10 kb DNA fragment constructed from a wild-type strain of *Sinorhizobium* to *Rhizobium etli* (a sensitive strain) having resistance to several antibiotics, 4% NaCl, low and high pHs, heavy metals, and a temperature as high as 43°C [42]. Another application of this strain would be as an inoculum for a number of crop legumes that will significantly improve nodulation and nitrogen fixation and may lead to increase in plant dry matter under a low-level N fertilizer in low-fertility land.

Isolate VKLR-01 is Gram negative and nonspore forming. Gas is not produced from raffinose, sucrose, arabinose, mannitol, lactose, and glucose. Acid was produced from fermentation of sucrose, mannitol, and lactose but not from raffinose, arabinose, and glucose. Report showed that slow growth of rhizobia associated with woody tree species is related to alkali production, and fast growth is related to acid producers [43]. In the absence of specific taxonomic information, very fast-, fast-, and intermediate-growing (all acid-producing) will be referred to here as *Rhizobium*. In this study, mean generation times of the acid-producing isolate VKLR-01 were within the ranges reported in the literature for *Rhizobium* [38].

Isolate VKLR-01 did not grow in media without a carbon source (control). Isolate VKLR-01 utilizes lactose, sucrose, mannitol, D-ribose, maltose, D-galactose, glycerol, sorbitol, sodium citrate, inositol, D-fructose, D-mannose, N-acetyl glucosamine, pyruvate, dextran, α-ketoglutarate, and melibiose as the sole carbon sources and L-histidine, L-arginine, and L-proline as the sole nitrogen sources. Isolate VKLR-01 was positive for oxidase, catalase, and nitrate reduction test; however, it showed negative result for gelatin hydrolysis and methyl red tests. The intrinsic antibiotic resistance test for different antibiotics revealed that the isolate VKLR-01 is highly sensitive to antibiotics such as gentamicin, streptomycin, and tetracycline and is resistant to antibiotics like penicillin G, whereas for antibiotics like ampicillin and chloramphenicol it either mildly sensitive or moderately sensitive. Phenotypic characteristics of the isolate is VKLR-01 are shown in Table 1. Isolate VKLR-01 contained ubiquinone-10 (Q-10), at a peak ratio of approximately 100% as the predominant isoprenoid quinone.

3.3. FAME Analysis.

The most abundant fatty acids are summed feature 8 (65.92%; comprising $C_{18:1}$ $\omega 7c$ and/or $C_{18:1}$ $\omega 6c$) followed by $C_{16:0}$ iso (10.43%), summed feature 2 (7.42%, comprising $C_{14:0}$ 3OH/$C_{16:1}$ iso I and an unidentified fatty acid with an equivalent chain length of 10.9525) followed by $C_{16:0}$ 3OH (4.19%), $C_{13:1}$ at 12-13 (2.80%), and $C_{19:0}$ cyclo $\omega 8c$ (2.76%), and summed feature 3 (2.42%; comprising $C_{16:1}$ $\omega 7c$/$C_{16:1}$ $\omega 6c$). Studies also showed that CFA profile of 5 Rhizobium species (R. soli DS-42[T], R. huautlense LMG 18254[T], R. galegae LMG 6214[T], R. loessense CIP 108030[T], and R. cellulosilyticum DSM 18291[T]) contains $C_{18:1}$ $\omega 7c$ as the major fatty acid, although there were differences in the proportions of some other fatty acids [9]. The cellular fatty acid profile for isolate VKLR-01 is shown in Table 2.

3.4. Genotyping by 16S rRNA, recA, and atpD Gene Sequences.

Ribosomal RNA is consider the most useful of the highly conserved sequences available for the measurement of phylogenetic relationships [10]. The almost complete 16S rRNA gene sequence of isolate VKLR-01 determined in this study comprised 1428 nucleotides (approximately 95% of the Escherichia coli 16S rRNA sequences). In the neighbor-joining tree based on 16S rRNA gene sequences, isolate VKLR-01 fell within the clade comprising Rhizobium species (Figure 1). The gene sequence similarities between isolate VKLR-01 and Rhizobium radiobacter LMG 383[T] were 97% and were 94% with Rhizobium rubi LMG 156[T], Rhizobium alkalisoli CCBAU 01393[T], and Rhizobium vignae CCBAU 05176[T], respectively. Gene sequence similarity values of not more than 93% were found when isolate VKLR-01 was compared with other species in the genus Rhizobium. The node to which isolate VKLR-01 belonged was also supported in phylogenetic trees generated with the maximum-likelihood and maximum parsimony algorithms (data not shown). In the neighbor-joining tree based on recA gene and atpD gene sequences, isolate VKLR-01 formed distinct phylogenetic lineages within the clade comprising Rhizobium species (Supplementary Figures 4A and 4B). Isolate VKLR-01 exhibited 81 to 92% recA gene sequence similarity and 80% to 94% atpD gene sequence similarity to Rhizobium species used in this study, respectively. In the present study, phylogenetic analysis of 16S rRNA housekeeping gene, other housekeeping genes like atpD and recA, and other methods of genomic investigations revealed that the isolate VKLR-01 from the root nodules of P. pinnata occurring in North Guwahati, Assam, India,

TABLE 2: Cellular fatty acid composition (%) of Rhizobium pongamiae VKLR-01[T] and type strains of phylogenetically related Rhizobium species.

Fatty acid	1	2	3
Straight-chain fatty acid			
12:0	00.55	—	—
14:0	00.75	—	0.11
15:0	—	—	0.29
16:0	10.43	7.7	9.03
17:0			0.13
18:0	00.66	0.9	0.17
Unsaturated fatty acid			
13:1 at 12-13	02.80	—	0.49
16:1 $\omega 5c$	—	—	—
17:1 $\omega 8c$	00.52	—	—
15:1 $\omega 8c$	—	0.8	—
17:1 $\omega 6c$	—	—	—
17:1 $\omega 8c$	—	—	0.22
18:1 $\omega 5c$	—	—	—
18:1 $\omega 7c$	—	76.2	—
Hydroxy fatty acid	—	—	—
12:0 3-OH	—	—	—
13:0 2-OH	—	—	—
15:0 3-OH	—	—	0.03
15:1 3-OH iso	—	—	—
16:0 3-OH	04.19	2.0	4.76
17:0 3-OH	—	—	0.18
18:0 2-OH	—	—	—
18:0 3-OH	00.58	0.7	—
15:1 G iso	—	—	—
16:0 iso	00.67	—	—
10-Methyl 18:0 TBSA	00.34	—	—
10-Methyl 19:0	—	2.1	1.09
11-Methyl 18:1 $\omega 7c$	—	—	0.23
17:0 cyclo	—	—	1.60
19:0 cyclo $\omega 8c$	02.76	3.9	18.78
20:2 $\omega 6, 9c$	—	—	0.01
20:3 $\omega 6, 9, 12c$	—	—	0.33
Unknown (ECL 18.794)	—	—	—
Summed features*	—	—	—
2	7.41	4.6	8.18
3	02.42	0.52	—
4	—	—	1.62
7	—	—	52.41
8	65.92	—	—

Strains: 1, R. pongamiae sp. nov.; 2, R. galegae LMG 6214[T] 12 [9]; 3, A. tumefaciens [14]. Values are percentages of the total amount of fatty acid compounds present for those 14 species. ECL: equivalent chain length. —: not detected.

*Summed features 2 (12:0 aldehyde? and/or 16:1 iso I and/or 14:0 3-OH and/or unknown ECL 10.928 and/or unknown ECL 10.9525 and/or 15:1 iso H/I, 13:0 3-OH).

*Summed features 3 (16:1 $\omega 7c$ and/or 16:1 $\omega 6c$ and/or 15:0 iso 2-OH).

*Summed features 4 (iso-17:1 I and/or anteiso-17:1 B and/or 15:0 iso 2-OH, 16:1 $\omega 7c$).

*Summed features 7 (18:1 $\omega 7c$ and/or $\omega 9$ trans and/or $\omega 12$ trans and/or 18:1 $\omega 7c$ and/or $\omega 9c$ and/or $\omega 12$trans).

*Summed features 8 (18:1 $\omega 7c$ and/or 18:1 $\omega 6c$).

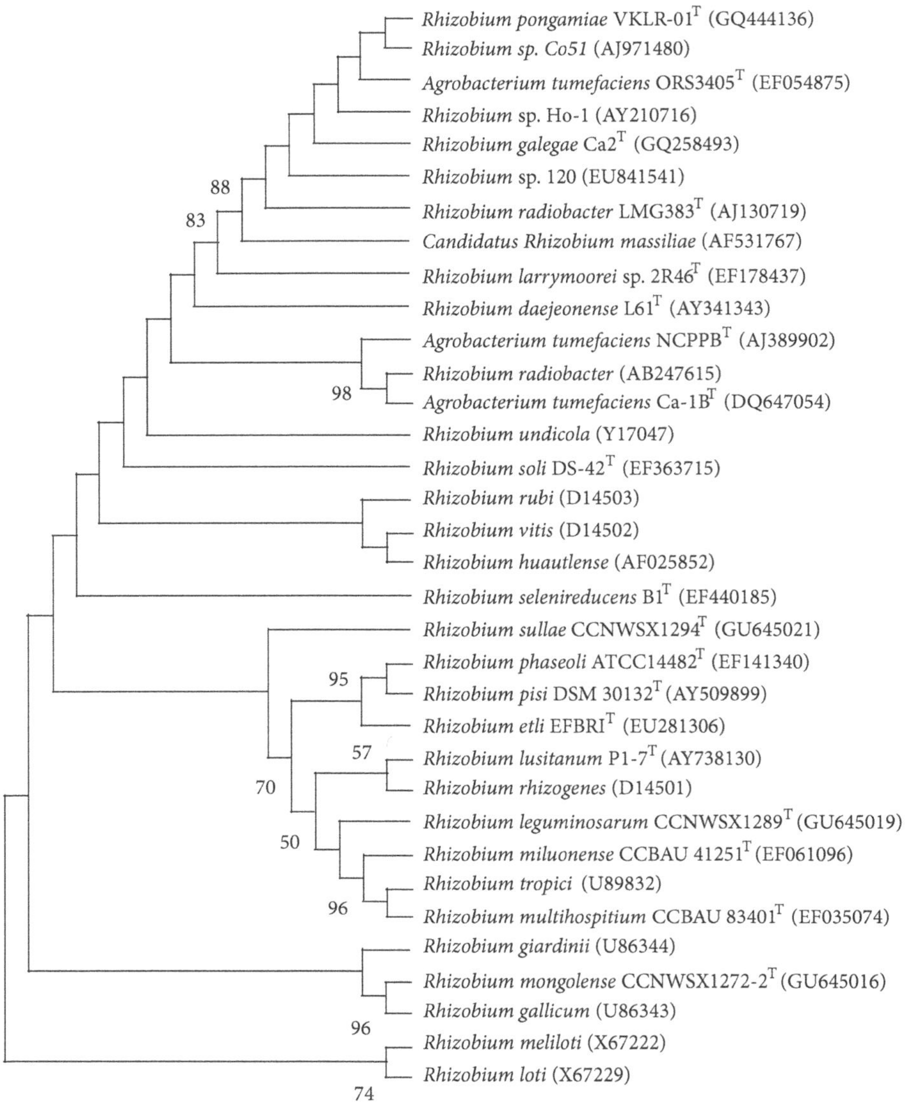

FIGURE 1: Dendrogram depicting the phylogenetic relationships of *Rhizobium pongamiae* VKLR-01[T] within the family *Rhizobiaceae* determined using 16S rRNA gene sequence analysis and generated with the MEGA 4.0 software as described in text. Bootstrap values based on 1000 replications are listed as percentages at branching points. Bootstrap values below 50 were omitted from the dendrogram. Bar, 0.002 substitutions per nucleotide position.

represented distinct genotype. These gene sequence similarity values are below the cut-off value of 97%, the level normally judged sufficient to justify the proposal of a novel bacterial species [44].

3.5. Nitrogen Fixation and Nodulation Assessment Test. In addition to 16S, multilocus sequence analysis (MLSA) is recommended for better resolution of phylogenetic relationships and species identification of novel bacterial strains [45]. In this study, we choose the *nod*D and *nif*H genes and corresponding primers from *R. leguminosarum biovar trifolii, R. leguminosarum biovar viciae,* and *S. meliloti,* which works well with fast-growing rhizobia. Isolate VKLR-01 was able to reduce acetylene to ethylene, and when subjected to *nif*H-specific amplification, it amplified an expected product

of 620 bp (Supplementary Figure 5A). The amplified 620 bp fragments of isolate VKLR-01 were sequenced and were found to show 83.0% to 90.0% sequence similarity with other *nif*H sequences from the NCBI database. Nodulation test was performed by PCR amplification of *nod*D gene. Isolate VKLR-01 amplified an expected product of 540 bp (Supplementary Figure 5B). The *nif*H and *nod*D genes amplification results confirm that the isolate VKLR-01 is a nitrogen fixer and plays a role in nodule formation. This will have important implications for biofuel production where reducing inputs (urea-based fertilizers) is highly desirable for production on nutrient-exhausted land [5].

3.6. G + C and DNA Hybridization Tests. The DNA G + C content of the isolate VKLR-01 is 59.1 mol% well within the range of values for the genus *Rhizobium* [46]. However, DNA G + C content of the isolate VKLR-01 was lower than that of the type strains *R. soli* DS-42^T (60.8 mol%), *R. galegae* LMG 6214^T (63.0 mol%), *R. loessense* CIP 108030^T (59.5 mol%), *R. tianshanense* 6 (63 mol%), and *R. tianshanense* A-1BS^T (61 mol%), but higher than that of *R. huautlense* LMG 18254^T (57 mol%) and *R. cellulosilyticum* DSM 18291^T (57 mol%), respectively [9, 47], and it exhibited mean DNA-DNA relatedness values of 51.9% to the type strain of phylogenetically related *Rhizobium* species (*Rhizobium radiobacter* DSM 30147^T).

DNA-DNA hybridization provides a useful strategy to establish the taxonomic place and identity of novel strain [48]. Isolate VKLR-01 exhibited mean DNA-DNA relatedness values of 51.9% to the type strain of phylogenetically related *Rhizobium* species (*Rhizobium radiobacter* DSM 30147^T). Since isolate VKLR-01 shares DNA-DNA hybridization value of less than 70% with reference strain DSM 30147^T, the isolate is regarded as a distinct *Rhizobium* species [47]. The phylogenetic distinctiveness, together with the DNA-DNA relatedness data and differential phenotypic properties, is sufficient to allocate isolate VKLR-01 to a species that is separate from the recognized *Rhizobium* species and named as *Rhizobium pongamiae* [44].

3.7. Rhizobium pongamiae sp. nov. On the basis of characterization of phenotypic features, cellular fatty acid profile, cluster analysis, PCR amplification of *nif*H and *nod*D genes, and DNA base composition, DNA-DNA hybridization, the isolate VKLR-01 (= MTCC 10513^T = MSCL 1015^T) from root nodules of *P. pinnata* is considered to represent a novel species within the genus *Rhizobium*. The name for isolate VKLR-01^T proposed is *Rhizobium pongamiae* sp. nov. (pon.ga'mi.ae. N.L. gen. n. pongamiae of Pongamia).

4. Conclusion

In conclusion, the results based on diverse phenotypic, physiological, biochemical, and molecular studies confirmed the novelty as well as abiotic stress-tolerance potential of the isolated bacterium *Rhizobium pongamiae* (strain VKLR-01^T) obtained from root nodules of *P. pinnata*, a legume biodiesel crop growing in North Guwahati, Assam, India. Metabolism of the isolate is strictly aerobic and able to fix atmospheric nitrogen and could be defined as a novel species according to the current standards for definition of bacterial (rhizobial) species. The ecological success of the *R. pongamiae* (strain VKLR-01^T) is that it has specific traits for abiotic stresses, for example, salt, drought, and alkaline tolerance as revealed from the results discussed above which may reflect its advantages for wasteland reclamation, reforestation, and native ecosystem restoration of low-fertility soil. These specific traits of *R. pongamiae* may also be transferred to other rhizobia through biotechnological tools to generate genetic engineered rhizobia beneficial for agricultural point of view. *R. pongamiae* may also be used in several other biotechnological applications such as the production of polysaccharides,

enzymes, and antibiotics, which will be the focus of research in future investigations for biotechnological purposes.

Disclosure

The GenBank/EMBL/DDBJ accession numbers for the 16S rRNA, *rec*A, and *atp*D gene sequences of strain VKLR-01^T are GQ444136, HM626171, and HM626172, respectively. The isolated novel bacterium *R. pongamiae* VKLR-01^T has assigned culture collection numbers (= MTCC 10513^T = MSCL 1015^T).

Conflict of Interests

There is no conflict of interests.

Authors' Contribution

All of the authors contributed to a similar extent overall, and all authors have seen and agreed to the submitted paper.

Funding

Vigya Kesari and Aadi Moolam Ramesh acknowledge the Council of Scientific and the Industrial Research (CSIR) and the Ministry of Human Resources Development (MHRD), Government of India, for the award of the Senior Research Fellowship (SRF). Latha Rangan acknowledges support from the Department of Biotechnology (DBT), the Government of India, for funding the research.

Acknowledgments

The authors would like to thank Professor H. G. Truper for his suggestions on Latin nomenclature for the novel isolate. Thanks are expressed to the Royal Life Sciences Pvt., Ltd., Secunderabad and DSMZ GmbH, Germany, for their excellent technical assistance related to FAME and molecular analyses. The support extended by the Central Instrument Facility (CIF), the Indian Institute of Technology Guwahati (IITG), related to scanning electron microscopy studies is greatly acknowledged. Authors are thankful to the Forest Department Officials of Sila Forest, North Guwahati, India, for the kind supply of study material.

References

[1] P. T. Scott, L. Pregelj, N. Chen, J. S. Hadler, M. A. Djordjevic, and P. M. Gresshoff, "*Pongamia pinnata*: an untapped resource for the biofuels: industry of the future," *Bioenergy Research*, vol. 1, pp. 2–11, 2008.

[2] V. Kesari, A. Das, and L. Rangan, "Physico-chemical characterization and antimicrobial activity from seed oil of *Pongamia pinnata*, a potential biofuel crop," *Biomass and Bioenergy*, vol. 34, no. 1, pp. 108–115, 2010.

[3] N. Mukta and Y. Sreevalli, "Propagation techniques, evaluation and improvement of the biodiesel plant, *Pongamia pinnata* (L.)

Pierre-A review," *Industrial Crops and Products*, vol. 31, no. 1, pp. 1–12, 2010.

[4] M. Sahgal and B. N. Johri, "The changing face of rhizobial systematics," *Current Science*, vol. 84, no. 1, pp. 43–48, 2003.

[5] H. T. Murphy, D. A. O'Connell, G. Seaton et al., "A common view of the opportunities, challenges and research actions for *Pongamia* in Australia," *Bioenergy Research*, vol. 5, no. 3, pp. 773–800, 2012.

[6] I. O. A. Odeh, D. K. Y. Tan, and T. Ancev, "Potential suitability and viability of selected biodiesel crops in Australian marginal agricultural lands under current and future climates," *Bioenergy Research*, vol. 4, no. 3, pp. 165–179, 2011.

[7] K. E. Kabir, F. Islam, and A. R. Khan, "Insecticidal effect of the petroleum ether fraction obtained from the leaf extract of *Pongamia glabra* Vent. on the American cockroach, *Periplaneta americana* (L.) (Dictyoptera: Blattidae)," *International Pest Control*, vol. 43, pp. 152–154, 2001.

[8] J. M. Vincent, *A Manual for the Practical Study of the Root-Nodule Bacteria: IBP Handbook no. 15*, Blackwell Scientific, Oxford, UK, 1970.

[9] J.-H. Yoon, S.-J. Kang, H.-S. Yi, T.-K. Oh, and C.-M. Ryu, "*Rhizobium soli* sp. nov., isolated from soil," *International Journal of Systematic and Evolutionary Microbiology*, vol. 60, no. 6, pp. 1387–1393, 2010.

[10] C. R. Woese, "Bacterial evolution," *Microbiological Reviews*, vol. 51, no. 2, pp. 221–271, 1987.

[11] H. Bouzar and J. B. Jones, "Agrobacterium larrymoorei sp. nov., a pathogen isolated from aerial tumours of Ficus benjamina," *International Journal of Systematic and Evolutionary Microbiology*, vol. 51, no. 3, pp. 1023–1026, 2001.

[12] B. De Las Rivas, Á. Marcobal, and R. Muñoz, "Allelic diversity and population structure in Oenococcus oeni as determined from sequence analysis of housekeeping genes," *Applied and Environmental Microbiology*, vol. 70, no. 12, pp. 7210–7219, 2004.

[13] E. T. Wang, P. Van Berkum, D. Beyene et al., "*Rhizobium huautlense* sp. nov., a symbiont of *Sesbania herbacea* that has a close phylogenetic relationship with *Rhizobium galegae*," *International Journal of Systematic Bacteriology*, vol. 48, no. 3, pp. 687–699, 1998.

[14] S. W. Tighe, P. De Lajudie, K. Dipietro, K. Lindström, G. Nick, and B. D. W. Jarvis, "Analysis of cellular fatty acids and phenotypic relationships of *Agrobacterium*, *Bradyrhizobium*, *Mesorhizobium*, *Rhizobium* and *Sinorhizobium* species using the Sherlock Microbial Identification System," *International Journal of Systematic and Evolutionary Microbiology*, vol. 50, no. 2, pp. 787–801, 2000.

[15] A. Boyde and C. Wood, "Preparation of animal tissues for surface-scanning electron microscopy," *Journal of Microscopy*, vol. 90, no. 3, pp. 221–249, 1969.

[16] K. Lindström and S. Lehtomäki, "Metabolic properties, maximum growth temperature and phage sensitivity of *Rhizobium* sp. (Galega) compared with other fast-growing rhizobia," *FEMS Microbiology Letters*, vol. 50, no. 2-3, pp. 277–287, 1988.

[17] W. Y. Shieh, A.-L. Chen, and H.-H. Chiu, "*Vibrio aerogenes* sp. nov., a facultatively anaerobic marine bacterium that ferments glucose with gas production," *International Journal of Systematic and Evolutionary Microbiology*, vol. 50, no. 1, pp. 321–329, 2000.

[18] R. M. Smibert and N. R. Krieg, "Phenotypic characterization," in *Methods for General and Molecular Bacteriology*, P. Gerhardt, R. G. E. Murray, W. A. Wood, and N. R. Krieg, Eds., pp. 607–654, American Society for Microbiology, Washington, DC, USA, 1994.

[19] M. Sasser, "Identification of bacteria through fatty acid analysis," in *Methods in Phytobacteriology*, Z. Klement, K. Rudolph, and D. C. Sands, Eds., pp. 199–204, Akademiai Kiado, Budapest, Hungary, 1990.

[20] G. Laguerre, M.-R. Allard, F. Revoy, and N. Amarger, "Rapid identification of rhizobia by restriction fragment length polymorphism analysis of PCR-amplified 16S rRNA genes," *Applied and Environmental Microbiology*, vol. 60, no. 1, pp. 56–63, 1994.

[21] D. J. Lane, "16S/23S rRNA sequencing," in *Nucleic Acid Techniques in Bacterial Systematic*, E. Strackebrandt and M. Goodfellow, Eds., pp. 115–175, Wiley, Chischester, UK, 1991.

[22] S. F. Altschul, W. Gish, W. Miller, E. W. Myers, and D. J. Lipman, "Basic local alignment search tool," *Journal of Molecular Biology*, vol. 215, no. 3, pp. 403–410, 1990.

[23] R. Chenna, H. Sugawara, T. Koike et al., "Multiple sequence alignment with the Clustal series of programs," *Nucleic Acids Research*, vol. 31, no. 13, pp. 3497–3500, 2003.

[24] N. Saitou and M. Nei, "The neighbor-joining method: a new method for reconstructing phylogenetic trees," *Molecular biology and evolution*, vol. 4, no. 4, pp. 406–425, 1987.

[25] M. Kimura, "A simple method for estimating evolutionary rates of base substitutions through comparative studies of nucleotide sequences," *Journal of Molecular Evolution*, vol. 16, no. 2, pp. 111–120, 1980.

[26] K. Tamura, J. Dudley, M. Nei, and S. Kumar, "MEGA4: Molecular Evolutionary Genetics Analysis (MEGA) software version 4.0," *Molecular Biology and Evolution*, vol. 24, no. 8, pp. 1596–1599, 2007.

[27] R. Li and I. C. MacRae, "Specific identification and enumeration of *Acetobacter diazotrophicus* in sugarcane," *Soil Biology and Biochemistry*, vol. 24, no. 5, pp. 413–419, 1992.

[28] F. Poly, L. J. Monrozier, and R. Bally, "Improvement in the RFLP procedure for studying the diversity of nifH genes in communities of nitrogen fixers in soil," *Research in Microbiology*, vol. 152, no. 1, pp. 95–103, 2001.

[29] J. Tamaoka and K. Komagata, "Determination of DNA base composition by reversed-phase high-performance liquid chromatography," *FEMS Microbiology Letters*, vol. 25, no. 1, pp. 125–128, 1984.

[30] P. Cashion, M. A. Holder Franklin, J. McCully, and M. Franklin, "A rapid method for the base ratio determination of bacterial DNA," *Analytical Biochemistry*, vol. 81, no. 2, pp. 461–466, 1977.

[31] M. Mesbah, U. Premachandran, and W. B. Whitman, "Precise measurement of the G+C content of deoxyribonucleic acid by High-Performance Liquid Chromatography," *International Journal of Systematic Bacteriology*, vol. 39, no. 2, pp. 159–167, 1989.

[32] J. De Ley, H. Cattoir, and A. Reynaerts, "The quantitative measurement of DNA hybridization from renaturation rates," *European Journal of Biochemistry*, vol. 12, no. 1, pp. 133–142, 1970.

[33] V. A. R. Huss, H. Festl, and K. H. Schleifer, "Studies on the spectrophotometric determination of DNA hybridization from renaturation rates," *Systematic and Applied Microbiology*, vol. 4, no. 2, pp. 184–192, 1983.

[34] K. Komagata and K. Suzuki, "Lipid and cell wall analysis in bacterial systematic," *Methods in Microbiology*, vol. 19, pp. 161–206, 1987.

[35] M. H. Siddiqui, "Nodulation study of a few legume tree species during seedling stage," *Nitrogen Fixing Tree Research Reports*, vol. 7, p. 6, 1989.

[36] P. Felker and P. R. Clark, "Nitrogen fixation (acetylene reduction) and cross inoculation in 12 *Prosopis* (mesquite) species," *Plant and Soil*, vol. 57, no. 2-3, pp. 177–186, 1980.

[37] F. M. S. Moreira, K. Haukka, and J. P. W. Young, "Biodiversity of rhizobia isolated from a wide range of forest legumes in Brazil," *Molecular Ecology*, vol. 7, no. 7, pp. 889–895, 1998.

[38] P. H. Graham, M. J. Sadowsky, H. H. Keyser et al., "Proposed minimal standards for the description of new genera and species of root- and stem-nodulating bacteria," *International Journal of Systematic Bacteriology*, vol. 41, no. 4, pp. 582–587, 1991.

[39] A. Baraibar, L. Frioni, M. E. Guedes, and H. Ljunggren, "Symbiotic effectiveness and ecological characterization of indigenous *Rhizobium loti* populations in Uruguay," *Pesquisa Agropecuaria Brasileira*, vol. 34, no. 6, pp. 1011–1017, 1999.

[40] A. Fterich, M. Mahdhi, and M. Mars, "TThe effects of Acacia tortilis sub. raddiana, soil texture and soil depth on soil microbial and biochemical characteristics in arid zones of Tunisia," *Land Degradation and Development*, 2011.

[41] G. F. Craig, C. A. Atkins, and D. T. Bell, "Effect of salinity on growth of four strains of *Rhizobium* and their infectivity and effectiveness on two species of Acacia," *Plant and Soil*, vol. 133, no. 2, pp. 253–262, 1991.

[42] R. Defez, B. Senatore, and D. Camerini, "Genetically modified rhizobia as a tool to improve legume growth in semi-arid conditions," in *Mediterranean Conference of Rhizobiology Workshop on Symbiotic Nitrogen Fixation for Mediterranean Area*, Montpellier, France, July 2000.

[43] D. W. Odee, J. M. Sutherland, E. T. Makatiani, S. G. McInroy, and J. I. Sprent, "Phenotypic characteristics and composition of rhizobia associated with woody legumes growing in diverse Kenyan conditions," *Plant and Soil*, vol. 188, no. 1, pp. 65–75, 1997.

[44] E. Stackebrandt and B. M. Goebel, "Taxonomic note: a place for DNA-DNA reassociation and 16S rRNA sequence analysis in the present species definition in bacteriology," *International Journal of Systematic Bacteriology*, vol. 44, no. 4, pp. 846–849, 1994.

[45] R. Rivas, M. Martens, P. de Lajudie, and A. Willems, "Multilocus sequence analysis of the genus *Bradyrhizobium*," *Systematic and Applied Microbiology*, vol. 32, no. 2, pp. 101–110, 2009.

[46] W. Chen, E. Wang, S. Wang, Y. Li, X. Chen, and Y. Li, "Characteristics of *Rhizobium tianshanense* sp. nov., a moderately and slowly growing root nodule bacterium isolated from an arid saline environment in Xinjiang, People's Republic of China," *International Journal of Systematic Bacteriology*, vol. 45, no. 1, pp. 153–159, 1995.

[47] L. G. Wayne, D. J. Brenner, R. R. Colwell et al., "Report of the ad hoc committee on reconciliation of approaches to bacterial systematic," *International Journal of Systematic Bacteriology*, vol. 37, pp. 463–464, 1987.

[48] A. Willems, F. Doignon-Bourcier, J. Goris et al., "DNA-DNA hybridization study of *Bradyrhizobium* strains," *International Journal of Systematic and Evolutionary Microbiology*, vol. 51, no. 4, pp. 1315–1322, 2001.

12

Analysis of Structures and Epitopes of Surface Antigen Glycoproteins Expressed in Bradyzoites of *Toxoplasma gondii*

Hua Cong,[1] Min Zhang,[1] Qingli Zhang,[2] Jing Gong,[3] Haizi Cong,[4] Qing Xin,[5] and Shenyi He[1]

[1] *Department of Human Parasitology, School of Medicine, Shandong University, No. 44 Wenhuaxi Road, Jinan, Shandong 250012, China*

[2] *Laboratory of Morphology, School of Medicine, Shandong University, No. 44 Wenhuaxi Road, Jinan, Shandong 250012, China*

[3] *Cancer Research Center, School of Medicine, Shandong University, No. 44 Wenhuaxi Road, Jinan, Shandong 250012, China*

[4] *Department of College English, Shandong University, No. 44 Wenhuaxi Road, Jinan, Shandong 250012, China*

[5] *School Hospital of Shandong University, No. 73 Jingshi Road, Jinan, Shandong 250012, China*

Correspondence should be addressed to Hua Cong; conghua@sdu.edu.cn

Academic Editor: María Sol Arias Vázquez

Toxoplasma gondii is a protozoan parasite capable of infecting humans and animals. Surface antigen glycoproteins, SAG2C, -2D, -2X, and -2Y, are expressed on the surface of bradyzoites. These antigens have been shown to protect bradyzoites against immune responses during chronic infections. We studied structures of SAG2C, -2D, -2X, and -2Y proteins using bioinformatics methods. The protein sequence alignment was performed by T-Coffee method. Secondary structural and functional domains were predicted using software PSIPRED v3.0 and SMART software, and 3D models of proteins were constructed and compared using the I-TASSER server, VMD, and SWISS-spdbv. Our results showed that SAG2C, -2D, -2X, and -2Y are highly homologous proteins. They share the same conserved peptides and HLA-I restricted epitopes. The similarity in structure and domains indicated putative common functions that might stimulate similar immune response in hosts. The conserved peptides and HLA-restricted epitopes could provide important insights on vaccine study and the diagnosis of this disease.

1. Introduction

Toxoplasma gondii (*T. gondii*) is a species of parasitic protozoa in the genus *Toxoplasma* that can be carried by many warm-blooded animals including humans [1]. There are three infectious stages in a complex life cycle of *T. gondii*: the tachyzoites, the bradyzoites, and the sporozoites [2]. A bradyzoite is a slowly replicating version of the parasite, which is responsible for chronic infection of *T. gondii* [3]. In chronic toxoplasmosis, the parasitophorous vacuoles containing the reproductive bradyzoites form cysts in the tissues of the muscles and brain [4].

The surface antigen of *T. gondii* that plays roles in the processes of host cell attachment and host immune evasion is dominated by a SRS (SAG1-related sequence) family of proteins which includes the SAG1-like sequence branch

and the SAG2-like sequence branch [5]. SRS proteins are expressed in a stage-specific manner. SAG1, SAG2A, SAG2B, SAG3, SRS1, SRS2, and SRS3 are mainly expressed on the tachyzoite surface [6]. Studies have indicated that SAG2 members participate in the process of parasite's invasion to the host, and their antibodies could block the further attachment of *T. gondii* on host cells [7, 8]. Previous studies have demonstrated that *T. gondii* parasites with a deletion of SAG2C, -2D, -2X, and -2Y gene cluster are less capable of maintaining a chronic infection in the brain [9]. It revealed that SAG2CDXY are important for persistence of cysts in the brain and these antigens might protect bradyzoites against an immune response. Contrary to SAG2A and SAG2B, which are expressed in tachyzoites, SAG2C, -2D, -2X, and -2Y appeared to be expressed exclusively on the surface of bradyzoites [9, 10]. However, among 160 members of the

TABLE 1: The original resources for SAG2C, D, X, and Y.

Protein	Size[a]	Coding gene location in types I, II, and III parasites[b]	Derivation[c]
SAG2C	365 aa	TGGT1_chrX: 7,358,316–7,360,225 (−); TGME49_chrX: 7353954–7,365860 (−); TGVEG_chrX: 7,440,957–7,442,866 (−)	SRS49D
SAG2D	196 aa	TGGT1_chrX: 7,352,615–7,353,874 (−); TGME49_chrX: 7,353,328–7,354,587 (−); TGVEG_chrX: 7,435,249–7,436,508 (−)	SRS49C
SAG2X	367 aa	TGGT1_chrX: 7,352,615–7,353,874 (−); TGME49_chrX: 7,353,328–7,354,587 (−); TGVEG_chrX: 7,435,249–7,436,508 (−)	SRS49B
SAG2Y	316 aa	TGGT1_chrX: 7,358,316–7,360,225 (−); TGME49_chrX: 7,355,817–7,356,572 (−); TGVEG_chrX: 7,440,957–7,442,866 (−)	SRS49A

[a] Size is the amino acid number that the protein has.
[b] Coding gene is the location of the gene that coded the protein.
[c] SRS domain-containing protein number.

SRS family, only three proteins' structures were reported. They are (i) the tachyzoite-expressed SAG1 [11], (ii) the bradyzoite-expressed BSR4 [12], and (iii) the sporoSAG [13]. The structure and function domains of SAG2C, -2D, -2X, and -2Y are still not very clear.

In this study, we sought to predict the structure and function domains of SAG2C, -2D, -2X, and -2Y by bioinformatics methods. The protein sequence alignments were performed by the T-Coffee method. Secondary structural and functional domains were predicted using the software PSIPRED v3.0 and SMART software. The 3D structure model of each protein was mapped using the I-TASSER server. The structural similarities of these proteins were summarized and possible functions of some key amino acids were predicted using the space confrontation by VMD and SWISS-spdbv. Furthermore, HLA-restricted epitopes of SAG2C, -2D, -2X, and -2Y proteins were predicted via algorithms.

2. Methods

2.1. Data Resources. The protein sequences were derived from ToxoDB 5.1 (http://toxodb.org/toxo/). *Toxoplasma gondii* has three common types: type I, *T. gondii* GT1 (TGGT1_chrX 7,429,598); type II, *T. gondii* ME49 (TGME49_chrX 7,419,075); type III, *T. gondii* VEG (TGVEG_chrX 7,553,721). The original resources are listed in Table 1.

2.2. Modular Architecture Identification. Multiple sequence alignment tool, T-Coffee (http://www.tcoffee.org/) [14, 15], was used to obtain the alignment analysis among SAG2C, SAG2D, SAG2X, and SAG2Y. The secondary structures were constructed using the software PSIPREDv3.0 (http://bioinf.cs.ucl.ac.uk/psipred/) [16, 17]. Simple modular architecture research identification and annotation of signaling domain sequences were analyzed via a web-based tool, SMART (http://smart.embl-heidelberg.de/) [18].

The 3D models of proteins were constructed by I-TASSER, a protein structure server on the website

http://zhanglab.ccmb.med.umich.edu/I-TASSER/, which is considered to predict protein 3D structures that have more than 100 amino acids [19–21]. VMD is a molecular visualization software for displaying, animating, and analyzing large biomolecular systems using 3D graphics and built-in scripts (http://www.ks.uiuc.edu/Research/vmd/). VMD was used to read standard Protein Data Bank (PDB) files and display the contained structure [22–25]. Swiss-Pdb Viewer (http://www.expasy.org/spdbv/) is an application that provides a user friendly interface allowing analyses of several proteins at the same time. The proteins can be superimposed in order to obtain structural alignments and compare their active domains. We deduced amino acid mutations, H bonds, angles, and distances between atoms from the intuitive graphic and menu interface. 3D protein molecular fitness analysis was performed for SAG2C, -2D and SAG2X, -2Y [22, 23].

2.3. Conserved HLA-Restricted Epitopes Prediction. Consensus methods including ANN, SMM, and CombLib-Sidney in immune epitope database IEDB (http://www.immuneepitope.org/) were used to predict HLA-restricted epitopes [26–28]. We used this tool to determine each peptide sequence's ability to bind to the specific HLA class I molecule.

3. Results and Discussion

3.1. Amino Acid Sequence Alignment Analysis. SAG2C, SAG2D, SAG2X, and SAG2Y are positioned next to each other on chromosome X. The molecular masses of SAG2C, -2D, -2X, and -2Y are 32–38 kDa, 18–20 kDa 31–34 kDa, and 28–30 kDa, respectively [9]. Multiple sequence alignment for SAG2C, -2D, -2X, and -2Y shows that the four proteins sequences have 97% similarity (Figure 1). In Particular, SAG2C (184 to 364) has a 98% sequence identity to SAG2D (14 to 196) and SAG2X (184 to 367) has a 99% sequence identity to SAG2Y (128 to 300). The protein sequence alignment analysis indicated that SAG2C, -2D, -2X, and -2Y have high

T-Coffee, version 9.02.r1228 (2012-02-16 18:15:12, revision 1228, build 336)
Cedric Notredame
CPU time: 0 s
Score = 97
*
BAD AVG GOOD
*

SAG2C	: 96
SAG2D	: 98
SAG2X	: 96
SAG2Y	: 96
Cons	: 97

```
SAG2C    1   MAAAHSAAACRYSTFWPCLLRERQSGTSSVSFVYPSQFLSLSLLVILTGSFAQQSAGNQANSQSV-TCESNASPL   74
SAG2D    1   M--------------------------------------------------------------------------   1
SAG2X    1   METAHSTVANRYSTFRPALFQGIRSDHSSFSVLRSCRCLPLYLFVIFACNFVQHSQGDSVPQQPVPTCSPSASPL   75
SAG2Y    1   MQQPV-----------------------------------------------------AK---ASEVQSCSGSATPL   21

Cons     1   *                                                                            75

SAG2C    75  VLRITSKTNEVKFKCGTDLQLRENPAGSNKFWGNAACTKEVDASSVTFTSSPSPAKVAGNKGTEYSLALKNSSLP   149
SAG2D    2   ---------------------------------------------------------------------------   1
SAG2X    76  SLNLTTTKKVVKFKCDEGLELHKKPTEDGKLCGNIACTKEIDASAFTFTPQSQGAKNNSTPDTEYSLGVKG-NLP   149
SAG2Y    22  FLRLSSTNQEVKFKCGAGLQLRAKPENDEMLWGNAACTKEVDASSLIFTPPTSKTKS----ESGYSLKVKSDALL   92

Cons     76                                                                               150

SAG2C    150 LSPFTVYFSCDPPSTTGV-GETGKAKVPASATTCIVQVSVFSQTAVTVPETNKCKNGQVTVAVTSKSKSVTFGCS   223
SAG2D    2   --------------SGRVEASDPKKSCLVQVSVFSQQPTPVPDSNKCKNGQVTVAVTSKSKSVTFGCS           55
SAG2X    150 TTPLTVYFSCDPKSAAGG-GGSARVEASESNKPCVVQVSVFSQKPTPVPDANKCKDGQVALAITSTTKSVTFGCN   223
SAG2Y    93  KTPFTVYFLCDPTSSSGGDSGSKQVTGSGSSTTCLIQVSVFSQKPTPVPDANKCKDGQVALAITSTTKSVTFGCN   167

Cons     151                    :  :  .  ..::  *:*******  **.:****.***.:*.** .*******       225

SAG2C    224 EGATLKPALLDHVFIEKATEKSGGASTGREEEVVLQDLVPNSSLVENAANTGNDTVGYTLSCPDLPSSPQNIFYK   298
SAG2D    56  EGATLKPALLEHVFIEKATEKSGGASTGREEEVVLQDLVPNSSLVENAANTGNDTVGYTLSCPDLPSSPQNIFYK   130
SAG2X    224 ESATLEPRFFEHVFIEEQSQSSRAAATLEAKEVVLQALVPNSSLVENAAGTTATAVGYTLSCPNLPSSAQTFFYK   298
SAG2Y    168 ESATLEPRFFEHVFIEEQSQSSRAAATLEVKEVVLQALVPNSSLVENAAGTTATAVGYTLSCPNLPSSAQTFFYK   242

Cons     226 *.***:* :::*****: ::.* .*:*  ;:***** ***********.*   :*******:****.*.:*      300

SAG2C    299 CVSPA---SAREQVGTQTECKVLINIEEKPEAETPATPEPSRGEQGVVLGSAFMIAFISCFALVAGNM-----F   364
SAG2D    131 CVSSA---SAREQVGTQTECKVLINIEEKPEAETPATPEPSRGEQGVVLGSAFMIAFISCFALVTGNM-----F   196
SAG2X    299 CVPPNPPSGGSGRTGTPEGCKVLISVEKRPDSDATATPAPSGGERGIVLSSSFMIVSISLVTTATGKL-----F   367
SAG2Y    243 CVPPNPPSGGSGRTGTPEGCKVLISVEKRPDSDATATPAPSGGEQGVVLSSSFMIVSMSFAAMVAGNGFSAICC   316

Cons     301 **..       ..  :.** *****.*::*:::. *** ** **:*.** *.***. :*   : .*:          374
```

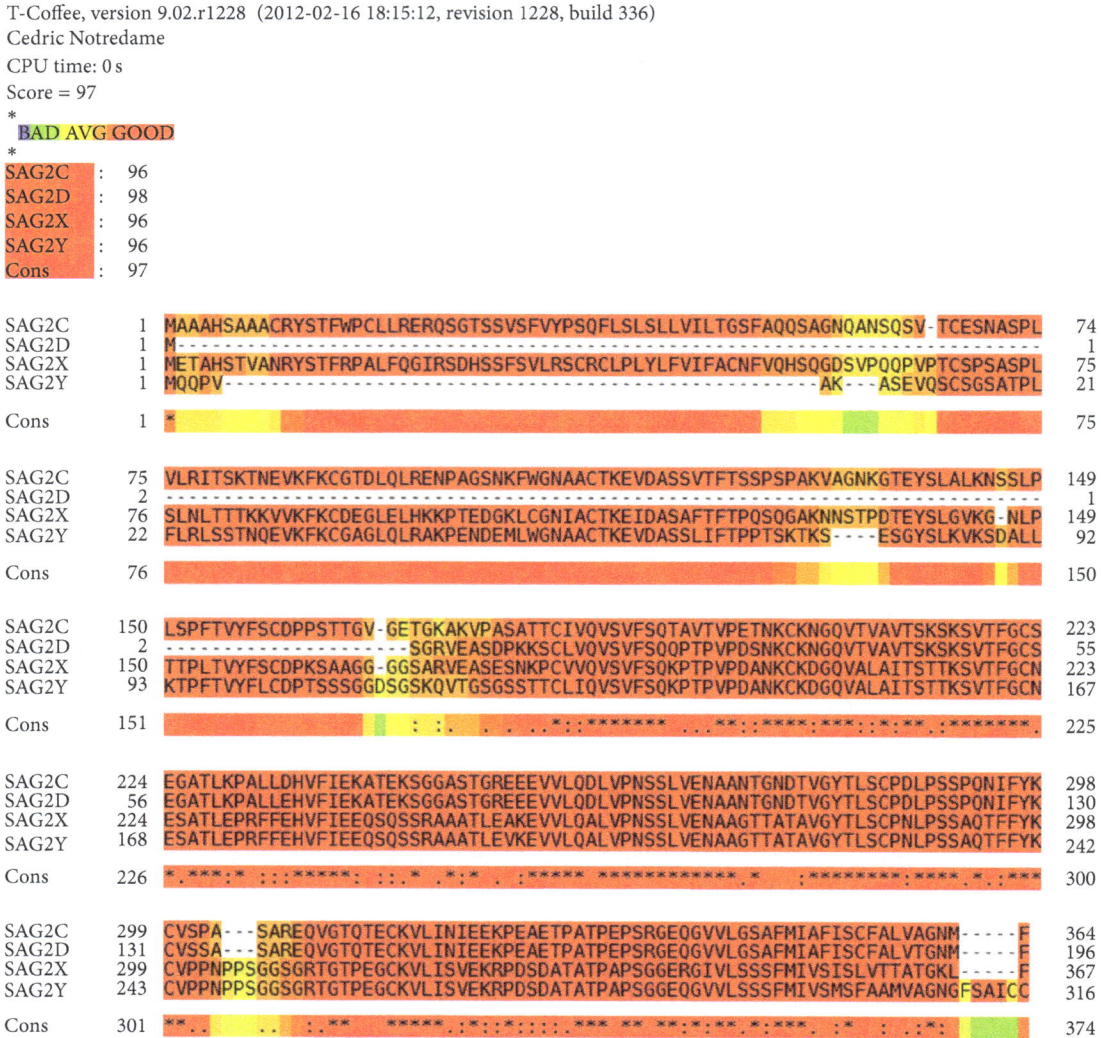

FIGURE 1: Alignment analysis for SAG2C, -2D, -2X, -2Y proteins. T-Coffee: multiple sequence alignment tools were used to obtain the alignment analysis result for SAG2C, -2D, -2X, and -2Y. Color bar indicated the identity, from bad identity to good identity.

homologous sequences. However, when including SAG2A and SAG2B in the alignment analysis, the consensus dropped to 73%, even though the consensus between SAG2A and SAG2B has very good score 84%. It indicates that a great difference exists among SAG2A, -2B and SAG2C, -2D, -2X, -2Y.

3.2. 2-D Structure Alignment for SAG2C, -2D, -2X, and -2Y Proteins.

PSIPRED v. 3.0 was used to predict the secondary structures of SAG2C, -2D, -2X, and -2Y proteins. Figure 2 showed that SAG2C protein has two α-helixes, 19 β-strands, and 20 coils; SAG2D protein has one α-helix, 9 β-strands, and 10 coils; SAG2X protein has 3 α-helixes, 14 β-strands, and 18 coils; SAG2Y protein has two α-helixes, 15 β-strands, and 18 coils. Obviously, there was a long α-helix on the C-terminal of all the proteins. SAG2D protein has similar secondary structure elements as SAG2C protein resides from 169 to 364. SAG2X and SAG2Y also have quite similar secondary structures except for a little discrepancy: SAG2X have one more helix than SAG2Y and one strand less than SAG2Y.

Furthermore, we used SMART to identify domains of these proteins (Figure 3). SAG2C, SAG2X, and SAG2Y all have two domains, while SAG2D only has one domain. SAG2D has an insertion of an adenosine, causing a frame shift and a premature stop codon, presumably leading to a truncated protein. SAG2C and SAG2D have transmembrane segments, while no transmembrane segments were identified on SAG2X and SAG2Y. From Figure 3, we could see that these proteins have no signal peptides, indicating that they are mature proteins. Members of the SAG2 family also differ in terms of open reading frame size, with the smaller SAG2D protein consisting of only one SAG domain, whereas SAG2C, SAG2X, and SAG2Y contain two SAG domains interrupted by a single intron. This indicates that SAG2C, SAG2X, and SAG2Y proteins have similar structure domains except SAG2D protein, which only has one domain.

```
SAG2C  1    CCCCEECCCCCCCCCCCCCCCCCEECCCCEEEEHHHHHHHHHHHHHHHHCCCCCCCCCC   60
SAG2D  1    C
SAG2X  1    CCCCCCCCCCCCCCCCCCCCCCCCCCCCCCCEEECCCCCHHHHHHHHHHCCCCCCCCCCC  60
SAG2Y  1    CCCCC                                                 CC      7

SAG2C  61   CCCCE ECCCCCCCEEEEEECCCCEEEEECCCCCCCCCCCCCCCEEECCCCCCCCCCCCC  119
SAG2D  2
SAG2X  61   CCCCCCCCCCCCCCEEEEEECCCCEEEEECCCCCCCCCCCCCCCCEECCCCCCCCCCCCC  120
SAG2Y  8    CCCCCCCCCCCCCEEEEEECCCCEEEEECCCCCCCCCCCCCCCCEEECCCCCCCCCCCCC   67

SAG2C  120  CCCCCCCCCCEECCCCCCEEEEEECCCCCCCCCCEEEEECCCCCCCC CCCCCCCCCCCC  178
SAG2D  2                                               CCCCCCCCC       11
SAG2X  121  CCCCCCCCCCCCCCCCCEEEEEECC CCCCCCCEEEEECCCCCCCCC CCCCCCCCCCC  178
SAG2Y  68   EEECCCCCCCCC      CCEEEEEECCCCCCCCCEEEEECCCCCCCCCCCCCCCCCCC  123

SAG2C  179  CCCCEEEEEEECCCCCCCCCCCCCCCCCEEEEEECCCCEEEEECCCCCCCCCCCCCCCCC  238
SAG2D  12   CCEEEEEEEECCCCCCCCCCCCCCCCCEEEEEECCCCEEEEECCCCCCCCCCCCCCEEEC   71
SAG2X  179  CCCCEEEEEEECCCCCCCCCCCCCCCCCEEEECCCCEEEEECCCCCCCCCCCCCCCCCEE  238
SAG2Y  124  CCCCEEEEEEECCCCCCCCCCCCCCCCCEEEECCCCCEEEEECCCCCCCCCCCCCCCCEE  183

SAG2C  239  CCCCCCCCCCCCCCCEEEECCCCCCCCEEEECCCCCCCCCCEEEEECCCCCCCCCEEEEE  298
SAG2D  72   CCCCCCCCCCCCCCCEEEECCCCCCCEEEECCCCCCCCCCEEEEECCCCCCCCCEEEEEE  131
SAG2X  239  CCCCHHHHHHCCEEEECCCCCCCCEEEECCCCCCCCCCCEEEECCCCCCCCCEEEEEE    298
SAG2Y  184  CCCCCCHHHHHHCCCEEEEEECCCCCEEEECCCCCCCCCCEEEEECCCCCCCCCEEEEE   243

SAG2C  299  EECCC    CCCCCCCCCCEEEEEEECCCCCCCCCCCCCCCCCCHHHHHHHHHHHHHHH  356
SAG2D  132  EECCC    CCCCCCCCCCEEEEEECCCCCCCCCCCCCCCCCCCCHHHHHHHHHHHHHH  188
SAG2X  299  CCCCCCCCCCCCCCCCCCCCEEEEEEECCCCCCCCCCCCCCCCCCHHHHHHHHHHHHHH   358
SAG2Y  244  EECCCCCCCCCCCCCCCCCCEEEEEEECCCCCCCCCCCCCCCCCCHHHHHHHHHHHHHH   303

SAG2C  357  HHHHCCC    C                                                 364
SAG2D  189  HHHHHCC    C                                                 196
SAG2X  359  HHHHHCCC     C                                               367
SAG2Y  304  HHHHHHHHHHCCC                                                316
```

FIGURE 2: The 2D Structures of SAG2C, -2D, -2X, and -2Y proteins. PSIPREDv3.0 was used to predict the secondary structure for SAG2C, -2D, -2X, and -2Y proteins (C stands for coil, H stands for α-helix, and E stands for β-strand).

3.3. Construction of 3D Model for SAG2C, -2D, -2X, and -2Y Proteins.

3D model of SAG2C, -2D, -2X, and -2Y proteins were constructed by I-TASSER server. Five models were set up for each protein by Dr. Zhang's lab [19]. We selected the model with highest confidence C-score, which estimates the quality of predicted models by I-TASSER. It was calculated based on the significance of threading template alignments and the convergence parameters of the structure assembly simulations [20]. C-score is typically in the range of $[-5, 2]$, and model with a C-score above 2 suggested a high confidence.

Low temperature replicas (decoys) generated during the simulation were clustered by SPICKER and top five cluster centroids were selected to generate full atomic protein models. The cluster density was defined as the number of structure decoys at each unit of space in the SPICKER cluster. A higher cluster density meant that the structure occurs more often in the simulation trajectory and therefore a better quality model. Table 2 showed the parameters for construction D model of each protein.

The best model of each protein was selected and viewed via VMD program (Figure 4). SAG2C, -2X, and -2Y have obvious two domains, D1 and D2, which are formed by two β-strands separated by one α-helix; SAG2D has one domain which is formed by one β-strand separated by one α-helix. The β-strands rotate to form a sheet tube that is a common character of these proteins. Furthermore, the binding sites of residues in the model were predicted and showed in Table 3.

Previous analysis of SAG2C, -2D, -2X, and -2Y structures revealed that the five on three β sandwich fold of SAG2 was most similar to the _T. gondii_ bradyzoite-expressed BSR4 with TM-scores of 0.583, 0.661, 0.672, and 0.670, respectively (Table 4). BSR4 is a prototypical bradyzoite surface antigen encoded in a cluster of SRS genes on chromosome IV, including the closely related paralogs SRS6 and SRS9 [8, 9]. Sequence alignment shows that SAG2C, -2D, -2X, and -2Y share 71% sequence identity with the tachyzoite-expressed BSR4. This observation is consistent with the prediction that stage-specific structural features might play an important role in the process of infection, dissemination, and pathogenesis

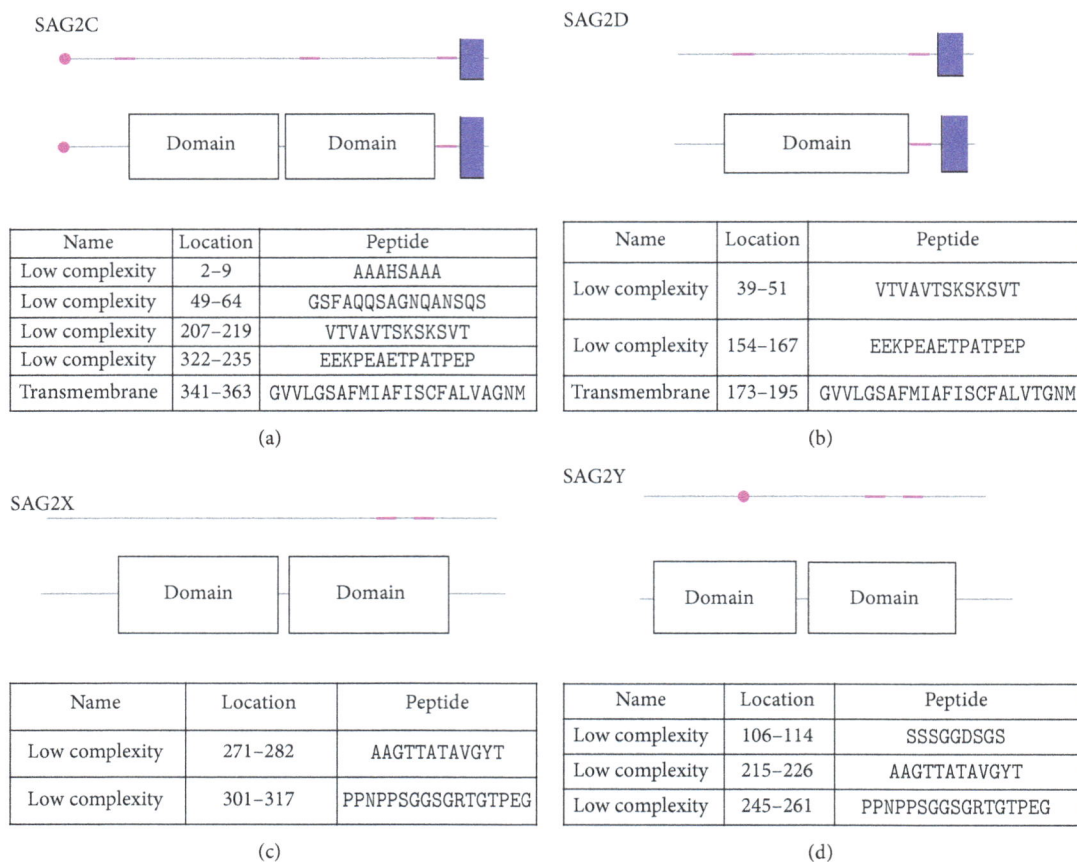

Name	Location	Peptide
Low complexity	2–9	AAAHSAAA
Low complexity	49–64	GSFAQQSAGNQANSQS
Low complexity	207–219	VTVAVTSKSKSVT
Low complexity	322–235	EEKPEAETPATPEP
Transmembrane	341–363	GVVLGSAFMIAFISCFALVAGNM

(a)

Name	Location	Peptide
Low complexity	39–51	VTVAVTSKSKSVT
Low complexity	154–167	EEKPEAETPATPEP
Transmembrane	173–195	GVVLGSAFMIAFISCFALVTGNM

(b)

Name	Location	Peptide
Low complexity	271–282	AAGTTATAVGYT
Low complexity	301–317	PPNPPSGGSGRTGTPEG

(c)

Name	Location	Peptide
Low complexity	106–114	SSSGGDSGS
Low complexity	215–226	AAGTTATAVGYT
Low complexity	245–261	PPNPPSGGSGRTGTPEG

(d)

FIGURE 3: Prediction for protein domain. A web-based tool—SMARTM was used to figure out the domains of these proteins: transmembrane segments predicted by the TMHMM2 program (segments in blue color), segments of low compositional complexity determined by the SEG program (segments in purple color), signal peptides determined by the SignalP program (segments in red color), and domain (segments in gray color).

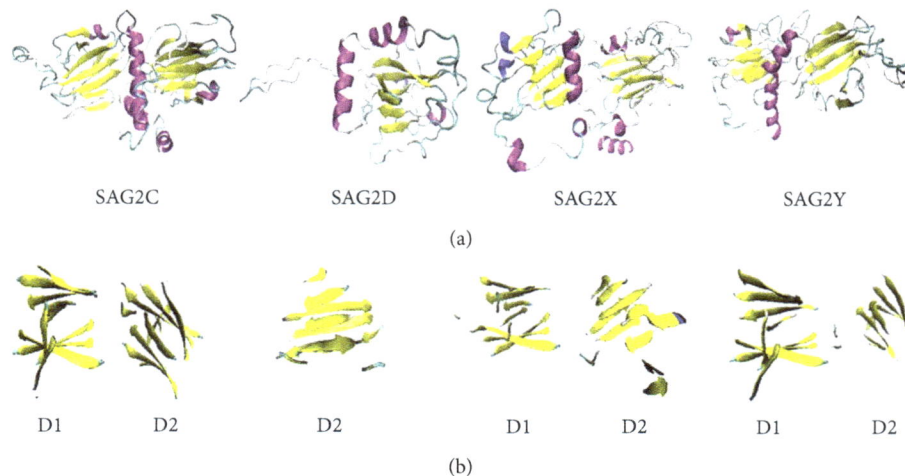

FIGURE 4: The 3D models of SAG2C, -2D, -2X, and -2Y. The sequences of proteins were sent to Dr. Zhang's lab from the website http://zhanglab.ccmb.med.umich.edu/I-TASSER/. The 3D models with the highest score for each protein were selected. The models were viewed by VMD software, color method was secondary structure (yellow: β-strands, purple: α-helix, gray: coil), and draw method was new cartoon. The domain of each model was shown out in sheet form.

TABLE 2: Parameters for predicted best 3-D model.

Name	C-score[a]	TM-score[b]	RMSD[b]	No. of decoys[c]	Cluster density[d]
SAG2C	−2.18	0.46 ± 0.15	11.8 ± 4.5	460	0.0285
SAG2D	−1.45	0.54 ± 0.15	8.5 ± 4.5	4662	0.0842
SAG2X	−2.22	0.45 ± 0.15	11.9 ± 4.4	455	0.0280
SAG2Y	−1.59	0.52 ± 0.15	9.9 ± 4.6	2877	0.0525

[a]C-score is a confidence score for estimating the quality of predicted models by I-TASSER. C-score is typically in the range of [−5, 2], where a C-score of higher value signifies a model with a high confidence and vice versa.
[b]TM-score and RMSD are known standards for measuring structural similarity between two structures which are usually used to measure the accuracy of structure modeling when the native structure is known.
[c]Number of decoys represents the number of structural decoys that are used in generating each model.
[d]Cluster density represents the density of cluster.

TABLE 3: Prediction binding site residues in the model.

Protein model	C-score^{LB}[a]	TM-score[b]	RMSD[c]	IDEN[d]	Cov.[e]	BS-score[f]	Lig. name	Predicted binding site residues in the model
SAG2C	0.09	0.350	7.07	0.031	0.604	0.84	MAL	222, 223, 224, 225, 226, 227, 273, 279
SAG2D	0.11	0.435	4.90	0.043	0.668	0.75	ANP	96, 153, 155
SAG2X	0.06	0.388	5.94	0.053	0.597	0.79	NA	261, 292, 326
SAG2Y	0.07	0.412	6.36	0.057	0.687	0.77	FES	144, 145, 146, 147, 149, 150, 151, 153, 243

[a]C-score^{LB} is the confidence score of predicted binding site. C-score^{LB} values range between [0-1], where a higher score indicates a more reliable ligand-binding site prediction.
[b]TM-score is a measure of global structural similarity between query and template protein.
[c]RMSD is the RMSD between residues that are structurally aligned by TM-align.
[d]IDEN is the percentage sequence identity in the structurally aligned region.
[e]Cov. represents the coverage of global structural alignment and is equal to the number of structurally aligned residues divided by length of the query protein.
[f]BS-score is a measure of local similarity (sequence and structure) between template binding site and predicted binding site in the query structure.

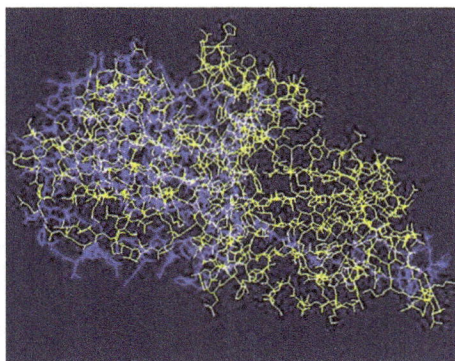

Fit: SAG2C and SAG2D

(a)

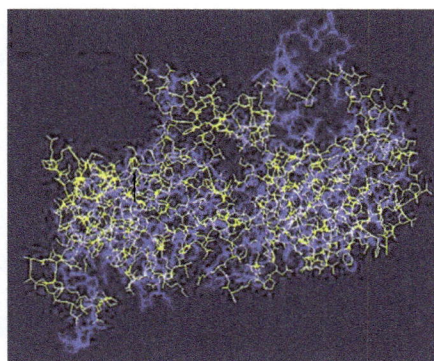

Fit: SAG2X and SAG2Y

(b)

FIGURE 5: Fit analyses of SAG2C, -2D, -2X, and -2Y. Swiss-Pdb Viewer was used to show the fitness. SAG2C and SAG2X are in yellow color; SAG2D and SAG2Y are in blue color.

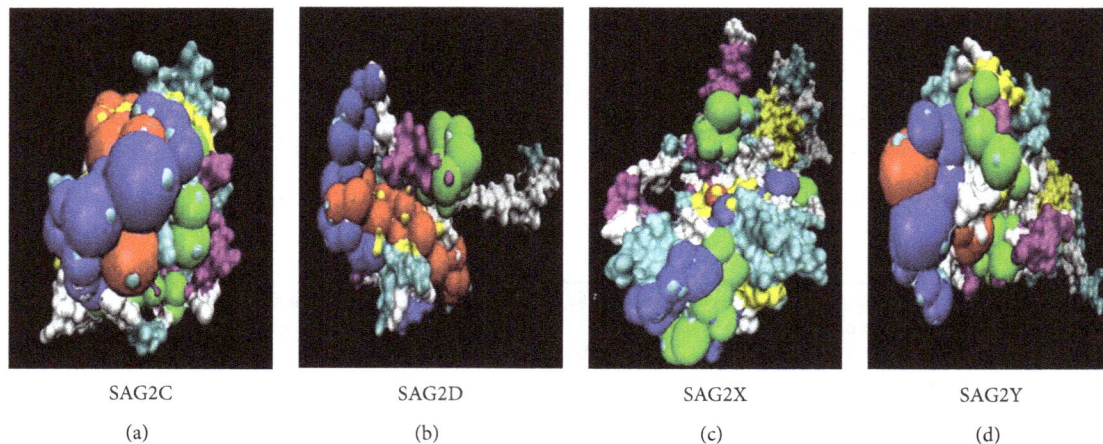

FIGURE 6: Identifying HLA-restricted epitopes on the surface of 3D models. The predicted HLA-restricted epitopes sequences shown in Tables 2 and 3 were marked out on the surface of 3D models of SAG2C, -2D, -2X, and -2Y. The 3D structures of proteins were shown using Surf method. Red color balls stand for epitopes restricted by HLA-A*1101, green color balls stand for epitopes restricted by HLA-A*0201, and blue color balls stand for epitopes restricted by HLA-B*0702.

TABLE 4: Similarity between SAG2C, -2D, -2X, -2Y and top identified structural analogs BSR4 protein.

Protein model	TM-score[a]	RMSD[b]	IDEN[c]	Cov.[d]
SAG2A	0.583	1.20	0.244	0.596
SAG2B	0.661	2.42	0.167	0.755
SAG2C	0.673	1.68	0.179	0.698
SAG2D	0.670	2.85	0.194	0.763

[a]TM-score of the structural alignment between the query structure and known structures in the PDB library.
[b]RMSD is the RMSD between residues that are structurally aligned by TM-align.
[c]IDEN is the percentage sequence identity in the structurally aligned region.
[d]Cov. represents the coverage of the alignment by TM-align and is equal to the number of structurally aligned residues divided by length of the query protein.

in *T. gondii*. In BSR4, two strands are organized in an antiparallel fashion, followed by another strand on the lower face of the β sandwich. The dimeric structure of SAG1 showed a β sandwich, two parallel outside strands with an opposite one in between [29]. The overall topology of the five on three β sandwich D2 domain is conserved between SAG2C, -2D, -2X, -2Y and BSR4. A detailed comparison of SAG2C, -2D, -2X, -2Y and BSR4 reveals a similarity in topology of the D1 and D2 domain consistent with the lower Z-score from the Dali search.

By comparison, the next most similar structure is SproSAG (surface antigen glycoprotein) with a substantially reduced TM-score [30, 31]. SproSAG is a dominant surface coat protein expressed on the surface of sporozoites. SproSAG crystallized as a monomer and displayed unique features of the SRS β sandwich fold compared to SAG1 and BSR4 [9]. Intriguingly, the structural diversity is localized to the upper sheets of the β sandwich fold and may have important implications for multimerization and host cell

ligand recognition. By fit analysis, SAG2D fits well on the C-terminal of the protein SAG2C. SAG2X and SAG2Y fit pretty well from C-terminal to N-terminal (Figure 5).

3.4. Conserved HLA-Restricted CD8⁺ T Cells Epitope Prediction.

Epitope prediction algorithm consensus was used to predict peptides that could stimulate human to induce effective and protective immune response against *T. gondii*. We want to see if they have similar epitopes scattered on the surface of their protein. The epitopes from SAG2C, -2D, -2X, and -2Y were predicted using the software from IEDB (http://www.immuneepitope.org/) which could identify novel HLA-class I restricted CD8⁺ T cell epitopes derived from *T. gondii*. 16 peptides were selected based on a high HLA allele binding score (percentile rank < 3).

From Table 5, we can see three HLA-A*0201-restricted peptides: VVLGSAFMI, FMIAFISCF, AFISCFALV; four HLA-A*1101-restricted peptides: QVTVAVTSK, SSPQNIFYK, QVGTQTECK, KVLINIEEK; and two HLA-B*0702-restricted peptides: LPSSPQNIF, KPEAETPAT shared by SAG2C and SAG2D. From Table 6, we can see two HLA-A*0201-restricted peptides: ALVPNSSLV, VLSSSFMIV; three HLA-A*1101-restricted peptides: ALAITSTTK, SSAQTFFYK, KVLISVEKR; and two HLA-B*0702-restricted peptides: LPSSAQTFF, RPDSDATAT shared by SAG2X and SAG2Y.

More interestingly, when we marked the HLA-restricted epitopes on the alignment sequences of the proteins, we found that the epitopes restricted by the same type of HLA allele are located at the same domains of the proteins (Figure 6). Our results indicated that the epitopes from SAG2C, -2D, -2X, and -2Y can be recognized by the proper MHC-I molecular and present on the cell surface to induce immune response in the host CD8⁺ T cells which might be helpful on vaccine study and diagnosis for this parasitic disease. Some identified peptides from these proteins have been

TABLE 5: Predicted HLA restricted CD8$^+$ T cell epitopes for SAG2C, -2D of *T. gondii*.

Allele	Sequence	Pecentile rank	Method used	Location	
				SAG2C	SAG2D
HLA-A*0201	VVLGSAFMI	3.4	Consensus (*ANN, SMM, CombLib Sidenev* 2008)	342–350	174–182
HLA-A*0201	FMIAFISCF	1.4	Consensus (*ANN, SMM, CombLib Sidenev* 2008)	348–356	180–188
HLA-A*0201	AFISCFALV	2.8	Consensus (*ANN, SMM*)	351–359	183–191
HLA-A*1101	QVTVAVTSK	1.45	Consensus (*ANN, SMM*)	206–214	38–46
HLA-A*1101	SSPQNIFYK	0.2	Consensus (*ANN, SMM*)	290–298	122–130
HLA-A*1101	QVGTQTECK	2.35	Consensus (*ANN, SMM*)	308–316	140–148
HLA-A*1101	KVLINIEEK	1	Consensus (*ANN, SMM*)	316–324	148–156
HLA-B*0702	LPSSPQNIF	1.1	Consensus (*ANN, SMM, CombLib Sidenev* 2008)	288–296	120–128
HLA-B*0702	KPEAETPAT	2.6	Consensus (*ANN, SMM, CombLib Sidenev* 2008)	324–332	156–164

TABLE 6: Predicted HLA-restricted CD8$^+$ T cell epitopes for SAG2X, -2Y of *T. gondii*.

Allele	Sequence	Pecentile rank	Method used	Location	
				SAG2C	SAG2D
HLA-A*0201	ALVPNSSLV	1.6	Consensus (*ANN, SMM, CombLib Sidenev* 2008)	260–268	204–212
HLA-A*0201	VLSSSFMIV	1.2	Consensus (*ANN, SMM, CombLib Sidenev* 2008)	346–354	290–298
HLA-A*1101	ALAITSTTK	1.6	Consensus (ANN, SMM)	208–216	152–160
HLA-A*1101	SSAQTFFYK	0.1	Consensus (*ANN, SMM*)	289–297	234–242
HLA-A*1101	KVLISVEKR	2.75	Consensus (*ANN, SMM*)	311–319	263–271
HLA-B*0702	LPSSAQTFF	3	Consensus (*ANN, SMM, CombLib Sidenev* 2008)	285–293	232–240
HLA-B*0702	RPDSDATAT	2.2	Consensus (*ANN, SMM, CombLib Sidenev* 2008)	314–322	271–279

proven to be recognized by PBMC cells from proper HLA-restricted *T. gondii* seropositive individuals and significantly induced IFN-γ production in T cells from immunized mice [32, 33] and therefore confirmed our predictions.

4. Conclusions

In this study, we have conducted a detailed bioinformatic and structural characterization analysis of the bradyzoite proteins SAG2C, -2D, -2X, and -2Y. The characterization of SAG2C, -2D, -2X, and -2Y provided structural view of the *T. gondii* SRS family members at chronic bradyzoite stage. Our bioinformatic analysis clearly showed that SAG2C, -2D, -2X, and -2Y are homologous protein members of the SAG2 subfamily. Consistently, our structural analysis demonstrated that SAG2C, -2D, -2X, and -2Y are similar to two other bradyzoite SAG2 members, BSR4 and SPOROSAG, rather than tachyzoite SAG1. This result indicated that SAG2 family has conserved structure at bradyzoite stage but a great difference from SAG1 at tachyzoite stage. Furthermore, the predicted conserved peptides and HLA-restricted epitopes shed interesting light on vaccine study and diagnosis for this parasitic disease.

Conflict of Interests

The authors wish to declare that there is no known conflict of interests associated with this publication and there has been no other significant financial support for this work that could have influenced its outcome.

Acknowledgments

This study was supported by a Grant from the National Natural Science Foundation Project of China (no. 81171604) and no. 49 China postdoc foundation.

References

[1] R. McLeod, F. Kieffer, M. Sautter, T. Hosten, and H. Pelloux, "Why prevent, diagnose and treat congenital toxoplasmosis?" *Memorias do Instituto Oswaldo Cruz*, vol. 104, no. 2, pp. 320–344, 2009.

[2] A. M. Tenter, A. R. Heckeroth, and L. M. Weiss, "*Toxoplasma gondii*: from animals to humans," *International Journal for Parasitology*, vol. 30, no. 12, pp. 1217–1258, 2000.

[3] W. K. Chew, M. J. Wah, S. Ambu, and I. Segarra, "*Toxoplasma gondii*: determination of the onset of chronic infection in mice and the in vitro reactivation of brain cysts," *Experimental Parasitology*, vol. 130, no. 1, pp. 22–25, 2012.

[4] D. Soldati and J. C. Boothroyd, "Transient transfection and expression in the obligate intracellular parasite *Toxoplasma gondii*," *Science*, vol. 260, no. 5106, pp. 349–352, 1993.

[5] C. Jung, C. Y. Lee, and M. E. Grigg, "The SRS superfamily of Toxoplasma surface proteins," *International Journal for Parasitology*, vol. 34, no. 3, pp. 285–296, 2004.

[6] I. D. Manger, A. B. Hehl, and J. C. Boothroyd, "The surface of Toxoplasma tachyzoites is dominated by a family of glycosylphosphatidylinositol-anchored antigens related to SAG1," *Infection and Immunity*, vol. 66, no. 5, pp. 2237–2244, 1998.

[7] C. Lekutis, D. J. P. Ferguson, and J. C. Boothroyd, "*Toxoplasma gondii*: identification of a developmentally regulated family of

genes related to SAG2," *Experimental Parasitology*, vol. 96, no. 2, pp. 89–96, 2000.

[8] A. V. Machado, B. C. Caetano, R. P. Barbosa et al., "Prime and boost immunization with influenza and adenovirus encoding the *Toxoplasma gondii* surface antigen 2 (SAG2) induces strong protective immunity," *Vaccine*, vol. 28, no. 18, pp. 3247–3256, 2010.

[9] J. P. J. Saeij, G. Arrizabalaga, and J. C. Boothroyd, "A cluster of four surface antigen genes specifically expressed in bradyzoites, SAG2CDXY, plays an important role in *Toxoplasma gondii* persistence," *Infection and Immunity*, vol. 76, no. 6, pp. 2402–2410, 2008.

[10] S. K. Kim, A. Karasov, and J. C. Boothroyd, "Bradyzoite-specific surface antigen SRS9 plays a role in maintaining *Toxoplasma gondii* persistence in the brain and in host control of parasite replication in the intestine," *Infection and Immunity*, vol. 75, no. 4, pp. 1626–1634, 2007.

[11] X. L. He, M. E. Grigg, J. C. Boothroyd, and K. C. Garcia, "Structure of the immunodominant surface antigen from the *Toxoplasma gondii* SRS superfamily," *Nature Structural Biology*, vol. 9, no. 8, pp. 606–611, 2002.

[12] J. Crawford, O. Grujic, E. Bruic, M. Czjzek, M. E. Grigg, and M. J. Boulanger, "Structural characterization of the bradyzoite surface antigen (BSR4) from *Toxoplasma gondii*, a unique addition to the surface antigen glycoprotein 1-related superfamily," *Journal of Biological Chemistry*, vol. 284, no. 14, pp. 9192–9198, 2009.

[13] J. Crawford, E. Lamb, J. Wasmuth, O. Grujic, M. E. Grigg, and M. J. Boulanger, "Structural and functional characterization of SporoSAG: a SAG2-related surface antigen from *Toxoplasma gondii*," *Journal of Biological Chemistry*, vol. 285, no. 16, pp. 12063–12070, 2010.

[14] C. Notredame, D. G. Higgins, and J. Heringa, "T-coffee: a novel method for fast and accurate multiple sequence alignment," *Journal of Molecular Biology*, vol. 302, no. 1, pp. 205–217, 2000.

[15] P. Di Tommaso, S. Moretti, I. Xenarios et al., "T-Coffee: a web server for the multiple sequence alignment of protein and RNA sequences using structural information and homology extension," *Nucleic Acids Research*, vol. 39, no. 2, pp. W13–W17, 2011.

[16] D. W. A. Buchan, S. M. Ward, A. E. Lobley, T. C. O. Nugent, K. Bryson, and D. T. Jones, "Protein annotation and modelling servers at University College London," *Nucleic Acids Research*, vol. 38, no. 2, Article ID gkq427, pp. W563–W568, 2010.

[17] M. I. Sadowski and D. T. Jones, "The sequence-structure relationship and protein function prediction," *Current Opinion in Structural Biology*, vol. 19, no. 3, pp. 357–362, 2009.

[18] J. Schultz, R. R. Copley, T. Doerks, C. P. Ponting, and P. Bork, "SMART: a web-based tool for the study of genetically mobile domains," *Nucleic Acids Research*, vol. 28, no. 1, pp. 231–234, 2000.

[19] A. Roy, D. Xu, J. Poisson, and Y. Zhang, "A protocol for computer-based protein structure and function prediction," *Journal of Visualized Experiments*, vol. 3, no. 57, p. e3259, 2011.

[20] Y. Zhang, "I-TASSER server for protein 3D structure prediction," *BMC Bioinformatics*, vol. 9, article 40, 2008.

[21] A. Roy, A. Kucukural, and Y. Zhang, "I-TASSER: a unified platform for automated protein structure and function prediction.," *Nature protocols*, vol. 5, no. 4, pp. 725–738, 2010.

[22] W. Humphrey, A. Dalke, and K. Schulten, "VMD: visual molecular dynamics," *Journal of Molecular Graphics*, vol. 14, no. 1, pp. 33–38, 1996.

[23] T. Schwede, J. Kopp, N. Guex, and M. C. Peitsch, "SWISS-MODEL: an automated protein homology-modeling server," *Nucleic Acids Research*, vol. 31, no. 13, pp. 3381–3385, 2003.

[24] N. Guex and M. C. Peitsch, "SWISS-MODEL and the Swiss-PdbViewer: an environment for comparative protein modeling," *Electrophoresis*, vol. 18, no. 15, pp. 2714–2723, 1997.

[25] N. Guex, M. C. Peitsch, and T. Schwede, "Automated comparative protein structure modeling with SWISS-MODEL and Swiss-PdbViewer: a historical perspective," *Electrophoresis*, vol. 30, no. 1, pp. S162–S173, 2009.

[26] M. Moutaftsi, B. Peters, V. Pasquetto et al., "A consensus epitope prediction approach identifies the breadth of murine T(CD8+)-cell responses to vaccinia virus," *Nature Biotechnology*, vol. 24, no. 7, pp. 817–819, 2006.

[27] J. Sidney, E. Assarsson, C. Moore et al., "Quantitative peptide binding motifs for 19 human and mouse MHC class i molecules derived using positional scanning combinatorial peptide libraries," *Immunome Research*, vol. 4, no. 1, article 2, 2008.

[28] C. Lundegaard, M. Nielsen, and O. Lund, "The validity of predicted T-cell epitopes," *Trends in Biotechnology*, vol. 24, no. 12, pp. 537–538, 2006.

[29] A. Barragan and L. David Sibley, "Transepithelial migration of *Toxoplasma gondii* is linked to parasite motility and virulence," *Journal of Experimental Medicine*, vol. 195, no. 12, pp. 1625–1633, 2002.

[30] B. Wallner and A. Elofsson, "Can correct protein models be identified?" *Protein Science*, vol. 12, no. 5, pp. 1073–1086, 2003.

[31] M. Graille, E. A. Stura, M. Bossus et al., "Crystal structure of the complex between the monomeric form of *Toxoplasma gondii* surface antigen 1 (SAG1) and a monoclonal antibody that mimics the human immune response," *Journal of Molecular Biology*, vol. 354, no. 2, pp. 447–458, 2005.

[32] H. Cong, E. J. Mui, W. H. Witola et al., "Towards an immunosense vaccine to prevent toxoplasmosis: protective *Toxoplasma gondii* epitopes restricted by HLA-A*0201," *Vaccine*, vol. 29, no. 4, pp. 754–762, 2011.

[33] H. Cong, E. J. Mui, W. H. Witola et al., "Human immunome, bioinformatic analyses using HLA supermotifs and the parasite genome, binding assays, studies of human T cell responses, and immunization of HLA-A*1101 transgenic mice including novel adjuvants provide a foundation for HLA-A03 restricted CD8+T cell epitope based, adjuvanted vaccine protective against *Toxoplasma gondii*," *Immunome Research*, vol. 6, no. 1, article 12, 2010.

13

Biomedical Applications of Fermenticin HV6b Isolated from *Lactobacillus fermentum* HV6b MTCC10770

Baljinder Kaur, Praveen P. Balgir, Bharti Mittu, Balvir Kumar, and Neena Garg

Department of Biotechnology, Punjabi University, Patiala, Punjab 147002, India

Correspondence should be addressed to Baljinder Kaur; baljinderbt@hotmail.com

Academic Editor: Hakan Bermek

Fermenticin HV6b is a class IIa antimicrobial peptide produced by *Lactobacillus fermentum* HV6b MTCC 10770 isolated from human vaginal ecosystem. It shows growth inhibition of a wide range of opportunistic pathogens of humans, for example, *Bacteroides, Gardnerella vaginalis, Mobiluncus, Staphylococci,* and *Streptococci,* associated with bacterial vaginosis in humans. It does possess an impressive sperm immobilization and spermicidal activity tested against human sperms which makes it an attractive proposition for formulating antibacterial vaginosis and contraceptive products. Apart from this, *in vitro* studies conducted against four different tissue models have indicated its potential to be used as a component of anticancerous drug therapy as it is reported to induce apoptosis in cancerous cells. This information could be integrated in future studies focusing on *in vivo* assessment of anticancerous activity of lactic acid bacterial toxins or bacteriocins.

1. Introduction

Bacterial vaginosis (BV) is diseased vaginal state where natural balance of vaginal microflora gets disturbed due to growth of anaerobic bacteria such as *Bacteroides* spp., *Escherichia coli, G. vaginalis, Mobiluncus* spp., *Mycoplasma hominis, Peptostreptococcus* spp., *Staphylococci, Streptococci,* and/or viruses [1]. Disease is characterized by a milky or gray vaginal discharge with foul odor, presence of "clue cells," and an increased vaginal pH of >4.5 [2, 3]. Bacterial vaginosis can induce complications in normal pregnancy [4] and also shows inflammation in pelvic regions. Toxins released by BV pathogens cross placenta and cause permanent brain damage that results in development of neurodegenerative disorders such as Parkinson's disease and Schizophrenia in new borne [5]. Disease also raises risk of sexually transmitted diseases (STDs) including HIV/AIDs, chlamydis, herpes, gonorrhea, and trichomoniasis. Presently, antibiotics such as metronidazole used to treat BV give initial cure rates of approximately 90% or better [6]. It undergoes oxidative metabolism in the liver, which results in formation of several metabolites and becomes widely distributed in the body [7]. Several side effects were observed to be associated with metronidazole antibiotic therapy such as suppression of healthy microflora of vagina, diarrhea, dizziness, headache, loss of appetite, nausea or vomiting, stomach pain, or cramps [8, 9]. Therefore, it is necessary to have an alternative form of treatment that could help in complete eradication of BV.

Bacteriocins produced by LAB have potential biomedical applications where they may provide valuable alternatives to antibiotics for the treatment of human and animal infections. Antimicrobially active lactobacilli were commonly used to develop products for prevention and treatment of genital infections [10–12]. Bacteriocins that are active against vaginal pathogens are also reported as having spermicidal activity [13, 14]. This feature makes them attractive for formulation of feminine health care and contraceptive products. A number of studies claim that bacteriocins are effective in the prevention of tooth decay and gingivitis and therefore could be included in mouth washes [15–19]. Bacteriocins have also been suggested as a potent antineoplastic agent. They have shown distinct promise as a diagnostic agent for some cancers; bacteriocins have also tested as AIDS drugs [20] but have not progressed beyond *in vitro* tests on cell lines. Fermenticin HV6b is an antimicrobial peptide produced by recently isolated and characterized *Lactobacillus fermentum* HV6b MTCC10770 from human vaginal ecosystem. Bacteriocin is proteinaceous in nature which possesses antimicrobial activity against bacterial vaginosis associated pathogens such as *Bacteroides* species, *Candida albicans*, *Gardnerella vaginalis*, *Listeria monocytogenes*, *Micrococcus flavus*, *Neisseria mucosa*, *Pediococcus acidilactici*, *Proteus mirabilis*, and *Staphylococcus albus* [21]. Keeping in view, the health benefits extended by GRAS bacteriocin of lactic acid bacteria, present investigation was carried out with the aim to explore potential biomedical applications of fermenticin HV6b *in vitro*.

2. Materials and Methods

2.1. Procurement and Maintenance of Culture. *Lactobacillus fermentum* HV6b MTCC10770 was grown in MRS medium, pH 6.5 at 37°C. Standard indicator strains, namely, *Enterococcus faecalis* and *Pediococcus acidilactici* LB42, were procured from Professor R. K. Malik (NDRI, Karnal, India) and cultivated in MRS medium at 37°C. *G. vaginalis* ATCC 14018 was revived and maintained in Casman's medium containing *Gardnerella* active supplement (constituting gentamycin sulphate, nalidixic acid, and amphotericin B) and 5% w/v defibrinated human blood [21]. Bacterial strains were maintained as 20% w/v glycerol stocks stored at −20°C.

2.2. Production and Purification of Fermenticin HV6b. Pure bacteriocin preparation was prepared from an overnight grown culture of *Lactobacillus fermentum* HV6b MTCC 10770 in MRS (supplemented with 0.1% w/v Tween 80, pH 6.5) by conventional adsorption-desorption method [22].

2.3. Bacteriocin Activity Assay. The antimicrobial activity of the bacteriocin preparation was tested against an array of opportunistic human pathogens (as indicated afterwards) using well-diffusion assay as described by Cintas et al. [23]. Bacteriocin activity was calculated as arbitrary unit (AU) and expressed as AU/mL according to standard protocol of Pucci et al. [24].

2.4. Determination of Antimicrobial Activity of Bacteriocins. Antimicrobial activity of bacteriocin fermenticin HV6b (crude as well as purified) was characterized using well-diffusion method [23] against *Bacteroides fragilis* MTCC1045, MTCC3298, and MTCC1350, *Candida albicans* ATCC10231 and MTCC183, *Gardnerella vaginalis* ATCC14018, *Micrococcus flavus* ATCC10240, *Neisseria gonorrhoeae* ATCC19424, *N. mucosa* MTCC1772, *Proteus mirabilis* NCIM2387, *Staphylococcus albus* ATCC11631, *S. aureus* MTCC737 and NCTC7447, *Streptococcus agalactiae* NCIM2401, *S. faecalis* MTCC459, *S. pyogenes* NCTC10869, and *S. thermophilus* MTCC1928. General human pathogens used are *Bacillus subtilis* ATCC6633, *Clostridium perfringens* MTCC450, *Escherichia coli* BL21 (DE3) MTCC1679, MTCC1652, and MTCC1650, *Enterococcus faecalis* (Laboratory isolate), ATCC29212, *Klebsiella pneumoniae* NCIM2883 and NCIM2401, *Leuconostoc mesenteroides* MTCC107, *Listeria monocytogenes* MTCC657, *Pseudomonas aeruginosa* ATCC10662, *Salmonella typhi* NCTC5760, *Vibrio cholera* ATCC14104, and *Yersinia enterocolitica* MTCC861.

Nonpathogenic microorganisms assayed in the study are *Lactobacillus brevis* MTCC1750, *L. bulgaricus* NCDC253, *L. casei* NCIM2651, *L. helveticus* NCIM2126, *L. leichmanni* NCIM2027, *L. pentosus* NCIM2669, *L. plantarum* NCIM2912, *Lactococcus lactis* subsp. *cremoris* MTCC1484, and *Pediococcus acidilactici* LB42.

2.5. Semen Sample Collection and Analysis. Partially purified fermenticin HV6b preparation was used as spermicide to test its effect on motility and immobilization of human spermatozoa. Two semen samples were collected from healthy volunteers in sterile wide-mouth polypropylene containers with a screw cap by self-masturbation on the day of experimentation. Within 1 h of collection, samples were dispensed by mixing a drop of diluted spermicide with a drop of semen and examined under microscope. An approximation is obtained to the highest bacteriocin concentration that does immobilize spermatozoa in 2 min, and the series of dilutions to be used in the test were made in a range below this point. Total sperm count was calculated using a compound microscope (Olympus, 100x) after dilution (1 : 50) of the semen in normal saline. The sperm suspensions were made in small glass tubes, one tube being required for each concentration of the spermicide. The suspensions were made by adding 0.5 mL of semen to each tube of saline or spermicide solution. Samples were placed in an incubator at 37°C for 15 to 30 min to reach that temperature. The percent sperm motility was determined by the progressive (forward) and nonprogressive (vibrating and zig-zag) movement of sperm observed in a compound microscope. The sperm count was calculated using Neubauer haemocytometer from a count of 100–200 sperms using randomly selected (100x) [14].

2.6. Treatment of Spermatozoa with Fermenticin HV6b. Standard protocol of Sutyak et al. [13] was used to determine the effect of purified bacteriocin fermenticin HV6b on the motility of human spermatozoa with little modifications, to measure the effect of fermenticin HV6b on sperm mobility and aggregation after 30 sec exposure time to different concentrations of bacteriocin ranging from 50 to 200 μg/mL of diluted semen sample. The motilities of human spermatozoa cells from random high magnification fields (100x) of the sample were determined in duplicate using atomic force microscope. Results were evaluated according to WHO grade system, and motilities of sperms were divided into four different grades [25]. Grade a sperms have progressive motility. These are the strongest and swim fast in a straight line. Grade b sperms exhibit forward movement but tend to travel in a curved or crooked motion. Grade c sperms show nonprogressive motility because they do not move forward despite the fact that they move their tails, and Grade d sperms have immotile sperms that fail to move at all.

2.7. Tissue Models for Testing Anticancerous Activity of Fermenticin HV6b. Tissue model, namely, HepG2 a hepatocarcinoma cell line, was procured from NCCS, Pune, India. A perpetual cervical cell lines (Hela ATCC CCL2), a breast

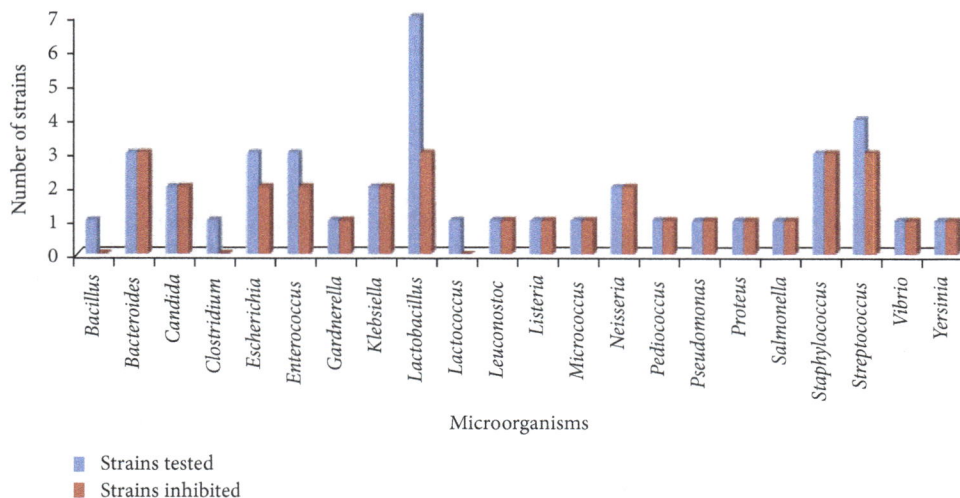

FIGURE 1: Antimicrobial spectrum of fermenticin HV6b.

carcinoma cell line (MCF7 ATCC-HTB-22) of *Homo sapiens*, a spleen lymphoblast cell line (Sp2/0-Ag14 ATCC-CRL-1581) of *Mus musculus,* and kidney embryonal cell line (HEK-293 CRL-1573) of *Homo sapiens* were gifted by Dr. Sanjog Jain, Niper, Mohali. Tissues were seeded in culture flasks containing DMEM and RPMI-1640 medium with 10% fetal bovine serum and 100 μg/mL penicillin and streptomycin and cultured in a humidified 5% CO_2 incubator at 37°C. After reaching 80% confluence, cells were passaged and cultured. Spent culture medium was discarded. The cell layer was rinsed with 0.25% (w/v) trypsin 0.53 mM EDTA solution to remove all traces of serum which may contain trypsin inhibitor. 6.0 to 8.0 mL of growth medium was added, and cells were aspirated by gently pipetting. Tissues were exposed to different concentrations of fermenticin HV6b ranging from 20 μg/mL to 500 μg/mL for 4, 24, and 48 hours. For exposure time over 24 h, the tissues were fed with fresh assay media. After the required exposure time, MTT assay was used to determine overall cell viability. Cell counts of tissue models were checked using haemocytometer. IC-50 (half maximal inhibitory concentration) value for fermenticin HV6b against each cell line is calculated which indicates how much bacteriocin is needed to inhibit the biological system (cell lines) by half [26].

2.8. MTT Viability Assay. The MTT assay was carried out according to the protocol given by Kumar et al. [27]. The viability of the cells exposed to bacteriocin was measured as a direct proportion of the breakdown of yellow compound tetrazolium to dark blue water insoluble formazan. Only the metabolically active cells can show this reaction which can be solubilized with DMSO and then quantified. The absorbance of formazon directly correlates with the number of viable cancerous cells. T sperms have progressive movement which ishe liquid in the plate wells was combined with the liquid from the tissue. Mixture is then assayed spectrophotometrically at 540 nm using 96-well plate ELISA reader to determine level of tetrazolium degradation.

2.9. DNA Fragmentation. DNA fragmentation analysis reveals the ability of fermenticin HV6b to induce apoptosis in cancer cells. It was carried out as per methodology of Kumar et al. [27] where cells (1×10^5) were treated with 1.0 mg/mL fermenticin HV6b for 48 h and then lysed with 250 μL lysis buffer. After incubation at 37°C for 90 min, 200 μg/mL proteinase K and lithium chloride (0.2% w/v) were added and incubated again for 60 min at 50°C. After incubation was over, suspension was centrifuged at 13,000 rpm for 3 min; aqueous phase was transferred to fresh tube containing deproteinizing mixture of phenol, chloroform, and isoamyl alcohol (25 : 24 : 1) and again centrifuged at 3,000 rpm for 3 min. DNA was precipitated from the aqueous phase with 3 volumes of chilled ethanol containing 0.3 M sodium acetate at 4°C. Samples were subjected to electrophoresis in 1% w/v agarose gel using TAE buffer at 50 V and visualised on a UV transilluminator.

2.10. Statistical Analysis. Where appropriate, data are expressed as mean values and standard deviations. Student's *t*-test was used for single comparisons. A probability value of $P < 0.05$ was used as the criterion for statistical significance.

3. Results

3.1. Antimicrobial Spectrum of Fermenticin HV6b. Fermenticin HV6b is capable of inhibiting a wide spectrum of human pathogens including *B. fragilis, B. ovatus, B. vulgatus, C. albicans, C. sporogenes, E. coli, E. faecalis, G. vaginalis, K. pneumoniae, L. mesenteroides, L. monocytogenes, M. flavus, N. gonorrhoeae, N. mucosa, P. aeruginosa, P. mirabilis, Staphylococci, Streptococci, S. typhi,* and *V. cholerae.* It did not show any activity against *B. subtilis, C. perfringens, E. faecalis, L. casei, L. leichmannii, L. plantarum, L. pentosus, L. lactis* subsp. *cremoris, S. agalactiae,* and *Y. enterocolitica.* However, fermenticin HV6b has been reported to exhibit very little growth inhibition of healthy microflora associated with gut and urinary tract as evidenced by a less degree of inhibition in case of *lactobacilli* and *Lactococcus* (Figure 1). Preliminary

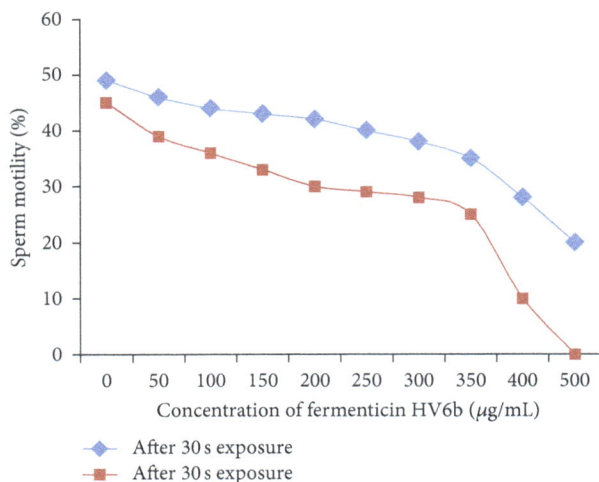

FIGURE 2: Retardation of sperm motility is a concentration-dependent phenomenon.

experiments performed using trypsin and protease-digested, neutralized pure bacteriocin samples led to the idea that inhibitory principle is proteinaceous in nature instead of growth inhibition simply due to acid produced by *L. fermentum* HV6b.

3.2. Inhibition of Sperm Motility.

Fermenticin was shown to significantly ($P < 0.05$) reduce motility of the human spermatozoa in a concentration-dependent manner (Figure 2). Normal untreated spermatozoa have progressive movement which is characteristic of "Grade a" category according to WHO. But upon exposure to higher concentrations of the fermenticin HV6b, coiling, clumping, and agglutination of sperms were observed that falls in Grade d. The effect of bacteriocin on the forward progression of human sperms and their aggregation was observed to be a dose-dependent interaction. Sperms show a steady decrease in forward progression (up to 50%) on exposure to increasing concentration of bacteriocin, with all progression halted at the concentration of 200 μg/mL for fermenticin HV6b as compared to control samples. Tails of sperm cells became curved or coiled as result of bacteriocin treatment, indicating their damage beyond a simple restriction of movement (Figure 3). Coiling of the sperm tails is considered to be an abnormality and may indicate damage to the plasma membrane [13, 14]. In a suspension of spermatozoa treated with sublethal concentration of bacteriocin, the speed was reduced, and if such reduction was compared with a control suspension, it was found to be proportional to the concentration of fermenticin in the test suspension. The degree of reduction in speed would be a convenient method of assessing the relative spermicidal power of purified bacteriocins (Table 1). The results established fermenticin HV6b as a general spermicidal agent.

3.3. Anticancerous Activity of Fermenticin HV6b.

Cytotoxicity of fermenticin HV6b was assessed on several cancerous cell lines with different morphologies and physiology

(Table 2). Figure 4 shows cell viability after incubation for 24 h in a medium containing fermenticin HV6b. MCF-7 and HEK-293 presented a slight sensitivity to this bacteriocin, whereas HeLa, Sp2/O, and Hep G2 cell lines were sensitive at different degrees to the fermenticin HV6b toxic effect. After 24–48 hours of exposure to the fermenticin HV6b, the epithelial tissue models retained only a low level of viability. Total cell viability drops below 50% at any point due to toxicity of the antimicrobial factor. The cell viability of MCF-7 cell was 46% whereas that of HeLa cells was 25%. Sp2/0 cells, HEK-293 cells, and HepG2 cells showed 30%, 38%, and 20% cell viability after treatment with 1 μg/mL fermenticin, respectively. Results showed that fermenticin HV6b is effective against cancerous cell lines, and, therefore, it is suggested to be used for clinical purposes. IC-50 values have been calculated for cytotoxic assay which indicates that 0.9 μg/mL of fermenticin HV6b was required to inhibit the growth of MCF 7 by 50%, whereas IC-50 value of bacteriocin to inhibit HEP G2 was found to be as less as 0.1 μg/mL. Figure 4 shows the modifications in the different cell lines with the treatment of fermenticin HV6b.

3.4. DNA Fragmentation.

Apoptosis in tissues was examined by DNA fragmentation assay. Cells were treated with fermenticin HV6b at 1.0 mg/mL for 48 h. Fermenticin HV6b was reported to increase DNA fragmentation (Figure 5) in cancerous cells. These results provided evidence that fermenticin HV6b induces cell-cycle arrest and apoptosis in cells.

4. Discussion

Human lactobacilli are used as probiotics to restore and maintain a healthy urogenital tract as an alternative to conformist chemotherapy. Present study therefore aimed at investigating *in vitro* efficacy of fermenticin HV6b in inhibiting growth of BV associated and other human pathogens and exploring spermicidal and anticancerous activities so that efficacy of *L. fermentum* HV6b could be established as a general human probiotic agent. Lactobacilli have been recommended as GRAS biotherapeutic agents for cure as well as prevention of human gastrointestinal and vaginal diseases. Colonization of the infected tissue by health promoting LAB particularly prevents infection by synthesizing a variety of antagonistic factors such as bacteriocins, diacetyl, H_2O_2, and fungicidal agents, by competing for available nutrients and mannose sugar, and by interfering pathogen attachment to cell surface receptors. An acidic pH of vagina alone is not sufficient to inhibit vaginal pathogens and to prevent bacterial vaginosis [14]. Thus, bacteriocin based therapeutics are urgently desired to cure such diseases and to overcome problems associated with antibiotic therapy such as diarrhea, poor compliance, and recurrence of vaginal infections. There is increasing body of evidence that indicates potential of GRAS lactic acid bacteria in maintaining and restoring gut homeostatis [28]. Use of living probiotic culture may have prophylactic applications, but use of purified bacteriocins appears to be more attractive for eradicating an established infection [29].

TABLE 1: Dose-dependent inhibition of sperm motility by fermenticin HV6b.

S. no.	Concentration of bacteriocin (μg/mL)	Mobile spermatozoa after 30 sec of exposure	Mobile spermatozoa after 60 sec of exposure	Motility grade after 1 min of exposure	Motility grade after 10 min of exposure
1	0	49 ± 1	45 ± 1	Grade a	Grade a
2	50	46 ± 1	39 ± 2	Grade a	Grade b
3	100	44 ± 2	37 ± 1	Grade a	Grade b
4	150	43 ± 1	33 ± 2	Grade b	Grade b
5	200	42 ± 1	30 ± 1	Grade b	Grade c
7	250	40 ± 3	29 ± 3	Grade c	Grade c
8	300	38 ± 1	28 ± 2	Grade c	Grade d
9	350	35 ± 2	25 ± 2	Grade d	Grade d
10	400	28 ± 2	10 ± 1	Grade d	Grade d
11	500	20 ± 3	0 ± 1	Grade d	Grade d

Each data is mean ± standard deviation; P value < 0.05; F crit (4.4138) < F value (14.5832).

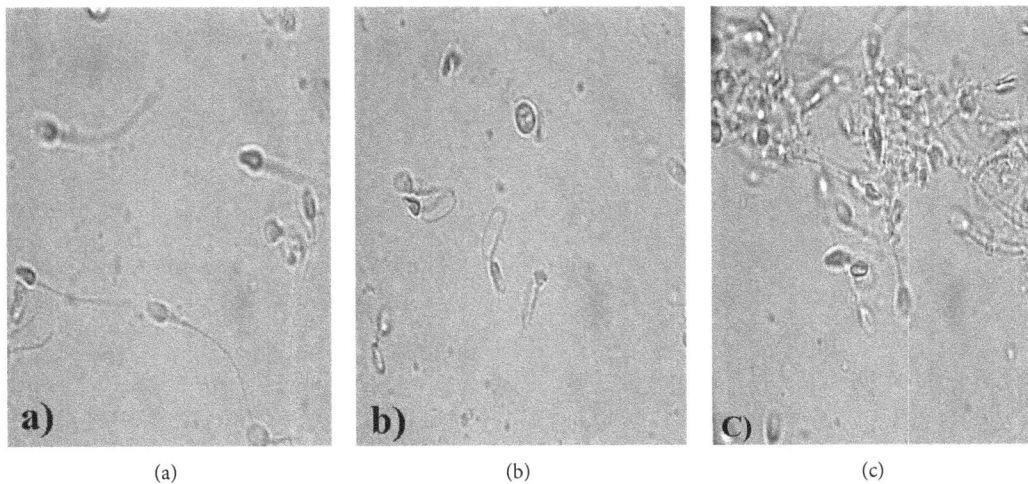

FIGURE 3: Images showing effect of fermenticin HV6b on human spermatozoa: (a) normal spermatozoa; (b) coiled sperm tails; (c) sperm aggregation and immobilization.

Several investigators have isolated and partially purified bacteriocin from different species of lactobacilli [30]. Most of these studies were carried on nonhuman strains which were predominantly isolated from food. In a study, inhibition of urinary tract infections (UTI) pathogens such as *E. faecalis*, *E. faecium*, and *N. gonorrhoeae* was reported by bacteriocin vaginal *Lactobacillus salivarius* strain [21]. Similarly, another study reported killing of a wide range of Gram-positive and Gram-negative pathogenic bacteria by bacteriocin HV219 of *L. lactis* origin which itself was isolated from human vagina [30].

Herein, we report that fermenticin HV6b produced by *L. fermentum* HV6b could target vaginal pathogens while leaving the healthy vaginal microflora intact as evidenced in the present study by performing growth inhibition assays on both human normal (*L. brevis* MTCC1750, *L. bulgaricus* NCDC253, *L. casei* NCIM2651, *L. helveticus* NCIM2126, *L. leichmannii* NCIM2027, *L. pentosus* NCIM2669, *L. plantarum* NCIM2912, *Lactococcus lactis* subsp. *cremoris* MTCC1484,

and *Pediococcus acidilactici* LB42) as well as pathogenic gut flora including *B. fragilis*, *B. ovatus*, *B. vulgatus*, *C. albicans*, *C. sporogenes*, *E. coli*, *E. faecalis*, *L. monocytogenes*, *M. flavus*, *N. gonorrhoeae*, *N. mucosa*, *P. aeruginosa*, *P. mirabilis*, *Staphylococci*, *Streptococci*, *S. typhi*, and *V. cholera*. It is of interest to the food and pharmaceutical industries both as it exhibits such as broad inhibitory spectrum against food-borne pathogens, spoilage organism, and human opportunistic pathogens of gut and urinary tract. Microorganisms such as *S. pyogenes* causing superficial skin infections to life-threatening systemic diseases and *L. monocytogenes* that causes spectic abortion, newborn, and adult septicemia, listeriosis, meningitis, and meningoencephalitis in immune-deficient persons [31] are also susceptible to fermenticin. In accordance with earlier reports, we herein report the utility of fermenticin to control human diseases as the best alternative to antibiotic therapy as it could be safely incorporated into personal care applications aimed at treatment of bacterial vaginosis [32, 33].

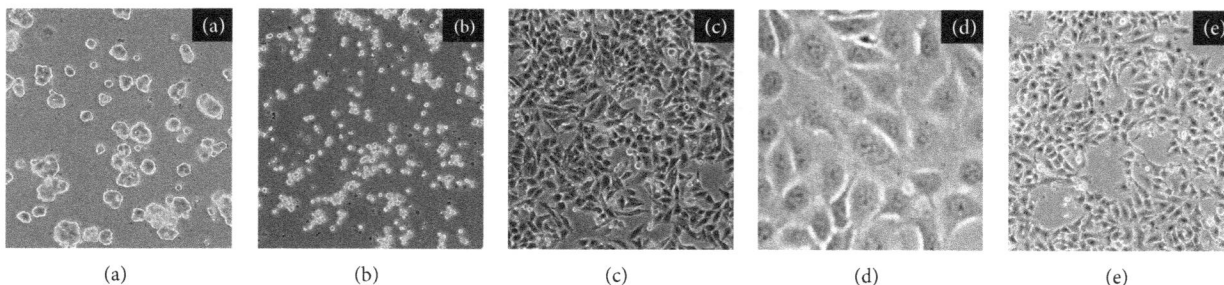

(a)   (b)   (c)   (d)   (e)

FIGURE 4: Morphological properties of cancerous cell lines used in the study: (a) MCF-7; (b) Sp2/0; (c) Hep G2; (d) HEK-293; (e) HeLa.

TABLE 2: Cytotoxic effect of fermenticin on cancerous cells.

Cancer cell lines tested	% cell viability at bacteriocin concentration			IC-50 value (μg/mL)
	0.1 μg/mL	1 μg/mL	10 μg/mL	
MCF-7	88 ± 1.9	46 ± 2.1	10 ± 1.7	0.9 ± 0.016
Sp2/0	70 ± 1.4	30 ± 1.7	9 ± 1.5	0.5 ± 0.012
Hep-G2	50 ± 1.8	20 ± 1.5	10 ± 1.9	0.1 ± 0.014
HEK-293	78 ± 2.4	38 ± 2.8	8 ± 2.6	0.6 ± 0.015
HeLa	65 ± 1.3	25 ± 1.2	6 ± 1.6	0.4 ± 0.013

Each data is mean ± standard deviation; P value < 0.05; F crit (3.2388) < F value (62.3008).

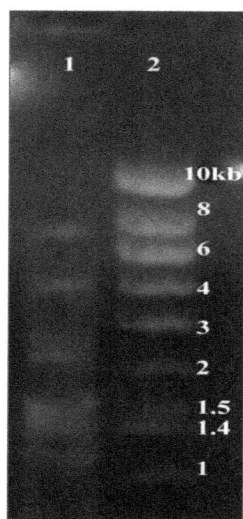

FIGURE 5: Analysis of genomic DNA on 1% agarose gel. Lane 1: fragmented genomic DNA of bacteriocin treated cells. Lane 2: 0.05 to 10 Kb DNA ruler (Novagen).

Results of present study also establishes fermenticin as a general spermicidal agent. Previous studies on subtilosin [13], nisin [34], and pediocin CP2 [14] have already reported them as potent spermicidal agents. Currently, used anticancer drugs have been shown to induce apoptosis in susceptible cells. Apoptosis is an important process of many pathological conditions. A series of studies have provided convincing evidence, suggesting that bacteriocin induces apoptosis of vascular endothelial cells. Principle of apoptosis was described by Vogt [35] which shows it as a programmed death of cells, which may even occur in multicellular organisms. Various biochemical changes such as loss of cell membrane asymmetry and attachment, cell shrinkage, nuclear fragmentation, chromatin condensation, and chromosomal DNA fragmentation take place during apoptosis. DNA fragmentation occurs at an end stage of apoptosis, which includes activation of calcium and magnesium dependent nucleases that degrade genomic DNA. The results presented here indicate cytotoxic effect of fermenticin HV6b on various cancerous cell lines. The cytotoxic effect on cancerous cells from human origin was also reported earlier [27, 36]. The uniqueness of the bacteriocins lies in their interaction with the cell surface without penetrating the target cells, yet affecting cell division and DNA synthesis [37]. Bacteriocins are highly specific in their membrane interaction which is related to the unique receptors found in different bacterial species or types [38].

5. Conclusion

Class IIa bacteriocins have ability to target a relatively wide range of pathogenic bacteria. This is an important attribute of GRAS bacteria that could be exploited to replace antibiotic therapy for treating bacterial vaginosis gut infections and peptic ulcers. Adhesive capacity and colonization of the gastrointestinal tract of humans and animals by probiotic lactobacilli including L. fermentum have been extensively investigated [39–41]. Organism can be delivered at the site of action, that is, gut or vagina, in the lyophilized capsular form where it can multiply and establish itself. Organism has a property to synthesize bacteriocin which can further help the bacteria to eradicate pathogenic organism from the inhabited area through competitive exclusion. Several in vitro studies have established GRAS bacteriocins as potent spermicidal and anticancerous agents. Preliminary experiments with fermenticin HV6b have shown its potency for formulating personal care products. Continued study is however desired for having complete insight into mechanism of killing sensitive bacterial species. Its cytoxicity has been proved for cancerous cell lines and which is attributed through the induction of programmed cell death or apoptosis. In future, this information could be integrated and exploited to fully explore the suitability of fermenticin HV6b as *in vivo* therapeutics against BV and various forms of cancers.

References

[1] P. Madhivanan, K. Krupp, J. Hardin, C. Karat, J. D. Klausner, and A. L. Reingold, "Simple and inexpensive point-of-care tests improve diagnosis of vaginal infections in resource constrained settings," *Tropical Medicine and International Health*, vol. 14, no. 6, pp. 703–708, 2009.

[2] D. A. Eschenbach, "History and review of bacterial vaginosis," *American Journal of Obstetrics and Gynecology*, vol. 169, no. 2, pp. 441–445, 1993.

[3] P. B. Mead, "Epidemiology of bacterial vaginosis," *American Journal of Obstetrics and Gynecology*, vol. 169, no. 2, pp. 446–449, 1993.

[4] Z. D. Ling, Q. Chang, J. W. Lipton, C. W. Tong, T. M. Landers, and P. M. Carvey, "Combined toxicity of prenatal bacterial endotoxin exposure and postnatal 6-hydroxydopamine in the adult rat midbrain," *Neuroscience*, vol. 124, no. 3, pp. 619–628, 2004.

[5] P. E. Hay, D. J. Morgan, C. A. Ison et al., "A longitudinal study of bacterial vaginosis during pregnancy," *British Journal of Obstetrics and Gynaecology*, vol. 101, no. 12, pp. 1048–1053, 1994.

[6] H. Moi, R. Erkkola, F. Jerve et al., "Should male consorts of women with bacterial vaginosis be treated?" *Genitourinary Medicine*, vol. 65, no. 4, pp. 263–268, 1989.

[7] J. G. Lossick, "Treatment of sexually transmitted vaginosis/vaginitis," *Reviews of Infectious Diseases*, vol. 12, supplement 6, pp. S665–S681, 1990.

[8] L. A. Simons, S. G. Amansec, and P. Conway, "Effect of *Lactobacillus fermentum* on serum lipids in subjects with elevated serum cholesterol," *Nutrition, Metabolism and Cardiovascular Diseases*, vol. 16, no. 8, pp. 531–535, 2006.

[9] E. C. Lauritano, M. Gabrielli, E. Scarpellini et al., "Antibiotic therapy in small intestinal bacterial overgrowth: rifaximin versus metronidazole," *European Review for Medical and Pharmacological Sciences*, vol. 13, no. 2, pp. 111–116, 2009.

[10] C. Barbés and S. Boris, "Potential role of *lactobacilli* as prophylactic agents against genital pathogens," *AIDS Patient Care and STDs*, vol. 13, no. 12, pp. 747–751, 1999.

[11] G. Famularo, M. Pieluigi, R. Coccia, P. Mastroiacovo, and C. De Simone, "Microecology, bacterial vaginosis and probiotics: perspectives for bacteriotherapy," *Medical Hypotheses*, vol. 56, no. 4, pp. 421–430, 2001.

[12] L. Maggi, P. Mastromarino, S. Macchia et al., "Technological and biological evaluation of tablets containing different strains of *lactobacilli* for vaginal administration," *European Journal of Pharmaceutics and Biopharmaceutics*, vol. 50, no. 3, pp. 389–395, 2000.

[13] K. E. Sutyak, R. A. Anderson, S. E. Dover et al., "Spermicidal activity of the safe natural antimicrobial peptide subtilosin,"

[14] B. Kumar, P. P. Balgir, B. Kaur, B. Mittu, and N. Garg, "Antimicrobial and spermicidal activity of native and recombinant pediocin CP2: a comparative evaluation," *Archives of Clinical and Microbiology*, vol. 3, no. 3, 2012.

[15] P. Blackburn and B. P. Goldstein, "Applied microbiology," Inc. International Patent Application WO 97/10801, 1995.

[16] T. H. Howell, J. P. Fiorellini, P. Blackburn, S. J. Projan, J. De La Harpe, and R. C. Williams, "The effect of a mouthrinse based on nisin, a bacteriocin, on developing plaque and gingivitis in beagle dogs," *Journal of Clinical Periodontology*, vol. 20, no. 5, pp. 335–339, 1993.

[17] P. McConville, "SmithKline Beecham," Plc. International Patent Application WO 97: 06772, 1995.

[18] R. M. Peek, G. G. Miller, K. T. Tham et al., "Heightened inflammatory response and cytokine expression *in vivo* to *cag*A+ *Helicobacter pylori* strains," *Laboratory Investigation*, vol. 73, no. 6, pp. 760–770, 1995.

[19] C. van Kraaij, W. M. de Vos, R. J. Siezen, and O. P. Kuipers, "Lantibiotics: biosynthesis, mode of action and applications," *Natural Product Reports*, vol. 16, no. 5, pp. 575–587, 1999.

[20] H. Farkas-Himsley, J. Freedman, S. E. Read, S. Asad, and M. Kardish, "Bacterial proteins cytotoxic to HIV-1-infected cells," *AIDS*, vol. 5, no. 7, pp. 905–907, 1991.

[21] B. Kaur, P. P. Balgir, B. Mittu, A. Chauhan, B. Kumar, and N. Garg, "Isolation and In vitro characterization of anti-*Gardnerella vaginalis* bacteriocin producing *Lactobacillus fermentum* HV6b isolated from human vaginal ecosystem," *Internatinal Journal of Fundamental and Applied Sciences*, vol. 1, no. 3, p. 41, 2012.

[22] R. Yang, M. C. Johnson, and B. Ray, "Novel method to extract large amounts of bacteriocins from lactic acid bacteria," *Applied and Environmental Microbiology*, vol. 58, no. 10, pp. 3355–3359, 1992.

[23] L. M. Cintas, J. M. Rodriguez, M. F. Fernandez et al., "Isolation and characterization of pediocin L50, a new bacteriocin from *Pediococcus acidilactici* with a broad inhibitory spectrum," *Applied and Environmental Microbiology*, vol. 61, no. 7, pp. 2643–2648, 1995.

[24] M. J. Pucci, E. R. Vedamuthu, B. S. Kunka, and P. A. Vandenbergh, "Inhibition of *Listeria monocytogenes* by using bacteriocin PA-1 produced by *Pediococcus acidilactici* PAC 1.0," *Applied and Environmental Microbiology*, vol. 54, no. 10, pp. 2349–2353, 1988.

[25] R. D. Wilson, P. M. Fricke, M. L. Leibfried-Rutledge, J. J. Rutledge, C. M. S. Penfield, and K. A. Weigel, "In vitro production of bovine embryos using sex-sorted sperm," *Theriogenology*, vol. 65, no. 6, pp. 1007–1015, 2006.

[26] T. Katla, K. Naterstad, M. Vancanneyt, J. Swings, and L. Axelsson, "Differences in susceptibility of *Listeria monocytogenes* strains to sakacin P, sakacin A, pediocin PA-1, and nisin," *Applied and Environmental Microbiology*, vol. 69, no. 8, pp. 4431–4437, 2003.

[27] B. Kumar, P. P. Balgir, B. Kaur, B. Mittu, and A. Chauhan, "In Vitro cytotoxicity of native and rec-pediocin CP2 against cancer cell lines: a comparative study," *Pharmaceutical Analytical Acta*, vol. 3, p. 183, 2012.

[28] M. Thirabunyanon, "Biotherapy for and protection against gastrointestinal pathogenic infections via action of probiotic bacteria," *Maejo International Journal of Science and Technology*, vol. 5, no. 1, pp. 108–128, 2011.

Infectious Diseases in Obstetrics and Gynecology, vol. 2008, Article ID 540758, 2008.

[29] C. T. Lohans and J. C. Vederas, "Development of class IIa bacteriocins as therapeutic agents," *International Journal of Microbiology*, vol. 2012, Article ID 386410, 13 pages, 2012.

[30] S. D. Todorov, M. Botes, S. T. Danova, and L. M. T. Dicks, "Probiotic properties of *Lactococcus lactis* ssp. *lactis* HV219, isolated from human vaginal secretions," *Journal of Applied Microbiology*, vol. 103, no. 3, pp. 629–639, 2007.

[31] E. J. Ridgway and J. M. Brown, "*Listeria monocytogenes* meningitis in the acquired immune deficiency syndrome-limitations of conventional typing methods in tracing a foodborne source," *Journal of Infection*, vol. 19, no. 2, pp. 167–171, 1989.

[32] S. Boris and C. Barbés, "Role played by *lactobacilli* in controlling the population of vaginal pathogens," *Microbes and Infection*, vol. 2, no. 5, pp. 543–546, 2000.

[33] R. D. Joerger, "Alternatives to antibiotics: bacteriocins, antimicrobial peptides and bacteriophages," *Poultry Science*, vol. 82, no. 4, pp. 640–647, 2003.

[34] K. V. R. Reddy, C. Aranha, S. M. Gupta, and R. D. Yedery, "Evaluation of antimicrobial peptide nisin as a safe vaginal contraceptive agent in rabbits: *in vitro* and *in vivo* studies," *Reproduction*, vol. 128, no. 1, pp. 117–126, 2004.

[35] C. Vogt, *Untersuchungen Uber Die Entwicklungsgeschichte Der Geburtshelferkroete (Alytes Obstetricians)*, Jent & Gassman, Solothurn, Switzerland, 1842.

[36] H. Farkas-Himsley and R. Cheung, "Bacterial proteinaceous products (Bacteriocins) as cytotoxic agents of neoplasia," *Cancer Research*, vol. 36, no. 10, pp. 3561–3567, 1976.

[37] A. Jayawardene and H. Farkas-Himsley, "Vibriocin: a Bacteriocin from Vibrio comma II. Interaction with sensitive bacteria," *MicroBios*, vol. 4, p. 325, 1969.

[38] M. Nomura, "Colicins and related bacteriocins," *Annual Review of Microbiology*, vol. 21, pp. 257–284, 1967.

[39] L. Plant and P. Conway, "Association of *Lactobacillus* spp. with Peyer's patches in mice," *Clinical and Diagnostic Laboratory Immunology*, vol. 8, no. 2, pp. 320–324, 2001.

[40] L. J. Plant and P. L. Conway, "Adjuvant properties and colonization potential of adhering and non-adhering *Lactobacillus* spp. following oral administration to mice," *FEMS Immunology and Medical Microbiology*, vol. 34, no. 2, pp. 105–111, 2002.

[41] S. Kang and P. L. Conway, "Characteristics of the adhesion of PCC® *Lactobacillus fermentum* VRI 003 to Peyer's patches," *FEMS Microbiology Letters*, vol. 261, no. 1, pp. 19–24, 2006.

14

The Promising Future of Chia, *Salvia hispanica* L.

Norlaily Mohd Ali,[1] Swee Keong Yeap,[2] Wan Yong Ho,[1] Boon Kee Beh,[3] Sheau Wei Tan,[2] and Soon Guan Tan[1]

[1] Department of Cell and Molecular Biology, Faculty of Biotechnology and Biomolecular Sciences, University Putra Malaysia, Serdang, 43300 Selangor, Malaysia
[2] Institute of Bioscience, University Putra Malaysia, Serdang, 43300 Selangor, Malaysia
[3] Department of Bioprocess Technology, Faculty of Biotechnology and Biomolecular Sciences, University Putra Malaysia, Serdang, 43300 Selangor, Malaysia

Correspondence should be addressed to Sheau Wei Tan, tansheau@putra.upm.edu.my

Academic Editor: Kazim Husain

With increasing public health awareness worldwide, demand for functional food with multiple health benefits has also increased. The use of medicinal food from folk medicine to prevent diseases such as diabetes, obesity, and cardiovascular problems is now gaining momentum among the public. Seed from *Salvia hispanica* L. or more commonly known as chia is a traditional food in central and southern America. Currently, it is widely consumed for various health benefits especially in maintaining healthy serum lipid level. This effect is contributed by the presence of phenolic acid and omega 3/6 oil in the chia seed. Although the presence of active ingredients in chia seed warrants its health benefits, however, the safety and efficacy of this medicinal food or natural product need to be validated by scientific research. *In vivo* and clinical studies on the safety and efficacy of chia seed are still limited. This paper covers the up-to-date research on the identified active ingredients, methods for oil extraction, and *in vivo* and human trials on the health benefit of chia seed, and its current market potential.

1. Introduction

Salvia hispanica L. (Figure 1), a biannually cultivated plant, is categorized under the mint family (*Labiatae*), superdivision of *Spermatophyta*, and kingdom of *Plantae*. Prominently grown for its seeds, *Salvia hispanica* also produces white or purple flowers. The seed (Figure 2) contains from 25% to 40% oil with 60% of it comprising (omega) ω-3 alpha-linolenic acid and 20% of (omega) ω-6 linoleic acid. Both essential fatty acids are required by the human body for good health, and they cannot be artificially synthesized. Chia can grow up to 1 m tall and has opposite arranged leaves. Chia flowers are small flower (3-4 mm) with small corollas and fused flower parts that contribute to a high self-pollination rate. The seed color varies from black, grey, and black spotted to white, and the shape is oval with size ranging from 1 to 2 mm [1–4]. Wild and domesticated chia differs little. Currently, only *Salvia hispanica* but not other species of the genus *Salvia* can be grown domestically. To prevent the misidentification of *Salvia hispanica* and other species of *Salvia*, clear understanding of the morphological and genotypical differences among them had been proposed as solutions [4, 5]. Locally known for its medicinal uses, *Salvia hispanica* L. acquired the common name chia from the indigenous South American people of the pre-Columbian and Aztec eras [6]. Owing to the fact that it can grow in arid environments, it has been highly recommended as an alternative crop for the field crop industry [7].

Chia seed is composed of protein (15–25%), fats (30–33%), carbohydrates (26–41%), high dietary fiber (18–30%), ash (4-5%), minerals, vitamins, and dry matter (90–93%). It also contains a high amount of antioxidants [8]. Heavy metal analysis showed that chia seed contains them at safe levels, not exceeding the maximum metal levels for food safety, and the seed is also free from mycotoxins [1]. Another key feature of chia seed is that it does not contain gluten [9]. Recent studies on chia seeds have focused on phytochemicals and their extractions from the seed. Only very little studies

FIGURE 1: *Salvia hispanica* L. plant with purple flower and large leaves [10].

have focused on *in vivo* or clinical bioactivities and the safety aspects of chia seed. The aim of this paper is to critically evaluate the health benefits, phytochemical contents, methods of oil extraction, and the current market potential of chia seed as a health food supplement.

2. Phytochemicals in Chia Seed

Various active ingredients including essential fatty acids and phenolic compounds have been identified in chia seed. These active compounds which contribute to the health benefits of chia seeds are summarised in Table 1.

There are many factors that may cause variations in the concentrations of the active compounds in chia seed. One of them is the cultivation area of the plant itself. Differences in the environment, climate changes, availabilities of nutrient, year of cultivation, or soil conditions play crucial roles to the variations [17, 22]. For example, the protein content tends to decrease as the temperature increased [23]. Furthermore, an inverse relationship between altitude and the content of saturated fatty acids (SFAs) had been observed whereby, at low elevation, an increase in fatty acid saturation was noted in areas where the temperature was high [7, 24]. In Argentina, Ayerza [25] demonstrated that temperature largely contributed to the type of fatty acid found in the oil. They found that, during seed development from April to May, an increase in the temperature of the environment brought about a decrease in the polyunsaturated fatty acid (PUFA) content.

Another factor that may contribute to differences in the chemical compositions of chia seed is the developmental stage of the plant. It was shown that the (α-linolenic acid) ALA content decreased by 23% from the early stage to the matured stage of the seed. This concurrently resulted in the increase of linolenic acid (LA) and lignin content [7].

3. Health Benefits of Chia Seed-Animal Studies

Several crops have been commercially recognized as being good sources of oil for dietary use including flaxseed,

rapeseed, sunflower seed, soybean seed, maize, evening primrose, and chia seed. A comparative study using flaxseed, rapeseed, and chia seed as chicken feed had been conducted. Eggs from hens fed with chia had the highest ω-3 ALA content as compared to hens fed with flaxseed or rapeseed [26]. Due to the easier availability and lower price of flaxseed over chia, an attempt to replace chia with flaxseed in laying hen's feed was carried out. The incorporation of flaxseed in the diet resulted in a slight decrease of the ω-3 ALA content of egg yolk [27]. However, the high antinutritional content of flaxseed affected the poultry meat quality.

Besides the utilization of chia in poultry animal targeted for human consumption, it had also been used for animal nutrition by itself. Other than chia seed oil, studies had been done using other grain oil such as flaxseed in broiler feed which also resulted in an increase of fatty acid quality level in broiler's meat [16].

Ayerza and Coates [28] and Fernandez et al. [29] conducted studies concerning the effects of chia seed feeding on rat plasma. Their findings indicated that serum triglycerides (TG) and low-density lipoprotein (LDL) were significantly decreased whereas high-density lipoprotein (HDL) and ω-3 PUFA levels were increased. They also noted that no adverse effects were observed on the rat's thymus and IgE serum level. Furthermore, chia seed feeding was tested in pigs and rabbits, which resulted in an increase of PUFA in meat fats as well as aroma and flavor [30–32]. These are desirable characteristics of human food. In summary, the incorporation of chia seed into animal feed results in an increase of ALA and a decrease of cholesterol levels in meat and eggs. Hence, it is a good substitute source of PUFA to fish and other seed oils. Moreover, atypical organoleptic characteristics such as flavor and smell from marine sources were not found in chia [33]. This showed the superiority of chia seed against other nutritional sources.

4. Health Benefits of Chia Seed-Human Clinical Trials

Correlation between high SFA and low PUFA intake with diseases such as cardiovascular diseases, diabetes, and metabolic syndrome were widely reported [20, 34]. Besides, the additive effect of ALA and n-3 long chain PUFA was observed to exhibit cardioprotective effects in women [35], which led to consequent human clinical studies of chia on disease risk factors. To date, four clinical trials have been carried out, and the details are summarised in Table 2. Among these trials, only that of Nieman et al. [19] showed no health benefits from chia seed. This difference could be due to the treatment durations employed and also the actual biochemical components of the dietary chia seed used in the various studies. Nevertheless, later studies [18, 20, 21] demonstrated well the benefits of chia on human health. However, studies of chia's intake in human diet which take into consideration factors such as lifestyle and genetic variations are still limited. Hence, studies which target these factors should be done in the future.

TABLE 1: Active compounds identified in *Salvia hispanica* L. and their chemical structure.

Active compounds in *Salvia hispanica* L.	Chemical structure	Activities
Omega 3 alpha-linolenic acid; ω-3 ALA (18:3) (*PUFA fatty acids*)		Essential fatty acid Benefits: (1) lowering TG and cholesterol levels, which in turn results in low blood pressure and heart-related diseases [11]; (2) anti-inflammatory activity [12, 13]; (3) cardioprotective and hepatoprotective activities by redistributing lipid away from visceral fat and liver [14, 15]; (4) antidiabetic action; (5) protection against arthritis, autoimmune disease, and cancer [13].
Omega 6 linoleic acid; ω-6 LA (18:2) (*PUFA fatty acids*)		Essential fatty acid This FA has inflammatory, hypertensive, and thrombotic activities. Eicosanoid produced from LA has been associated with CVD and cancer [16]. It works inversely with ALA; thus a balanced ratio of ALA to LA is important in maintaining good health [15, 17].

Table 1: Continued.

Active compounds in *Salvia hispanica* L.	Chemical structure	Activities
Myricetin (flavonols and phenolic acids)		
Quercetin (flavonols and phenolic acids)		Antioxidant, anti-inflammatory, anticancer, and antithrombotic activities [12].
Kaempferol (flavonols and phenolic acids)		

TABLE 1: Continued.

Active compounds in *Salvia hispanica* L.	Chemical structure	Activities
Caffeic acid (flavonols and phenolic acids)		

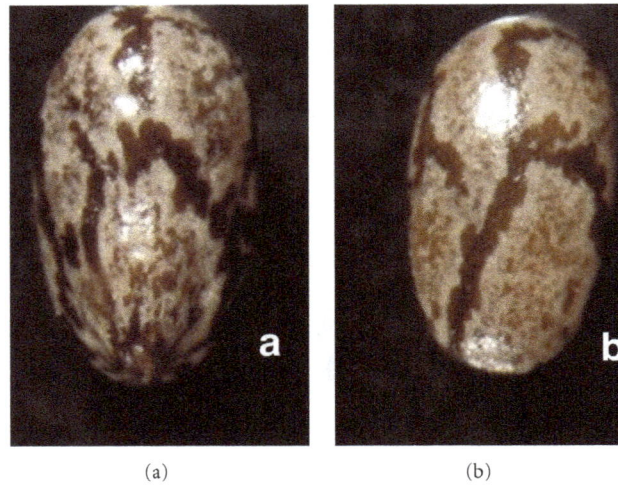

FIGURE 2: *Salvia hispanica* L. seeds with brown stripes color [10]. They can also be found having white and dark seed coat colors [8].

TABLE 2: Human clinical trials of chia seed.

Duration	Mode of trial	Formulation	Results	Ref.
7 weeks	10 postmenopausal women	25 g chia seed/day	Polyunsaturated fatty acid content particularly ALA, and eicosapentaenoic acid (EPA) was elevated after supplementation with milled chia. The result was in agreement with previous studies conducted in hens, rats, and rabbits.	[18]
12 weeks	Single blinded with 76 subjects (placebo 37; chia seed 39)	25 g chia seed in 250 mL water twice/day	Although Nieman et al. have hypothesized that the high dietary fiber and ALA content in chia can promote human weight loss and reduce disease risk factors related to heart disease and obesity, no significant results on weight loss and disease risk factors even though the plasma level of ALA increased.	[19]
2 months	Randomized trial, with control diet (500 kcal for 2 weeks), 67 metabolic syndrome subjects (placebo 35; beverage 32)	Beverage of 235 kcal that contains soy protein, nopal, chia seed, and oat	Body weight loss and reduction of triglyceride and blood glucose levels.	[20]
120 minutes	Randomized, double-blind trial on 11 healthy subjects	50 g white bread containing either 0, 7, 15, or 24 g of chia seed	Reduced postprandial glycemia.	[21]

TABLE 3: Extraction of oil from chia seed.

Methods of extraction	Details
Seed compression	(i) Cold pressing technique and storage at low temperature (4°C) in dark [36]. (ii) Komet screw press at 25–30°C using electrical resistance heating. (1) Pro: better preservation of antioxidant contents (quercetin and myricetin) than solvent extraction [37]. (2) Con: only partial recovery of oil yield [38].
Solvent	(i) Soxhlet method using n-Hexane (less preferable than other methods). (1) Pro: it favors the functional characteristics of the oil such as water holding and absorption capacity, organic molecule absorption, and emulsifying stability. (2) Con: it causes slight loss of antioxidant content [37] and health and environment safety issues of using hexane [38].
Supercritical fluids	(i) Use of carbon dioxide at optimum pressure $P = 408$ and 80°C (more preferable method). (1) Pros: better purity and higher ALA/LA content of the final products [39, 40]. The oil yield can be increased with pressure enhancement, but high temperature will slightly affect it [40].

TABLE 4: Commercial usage of chia seed.

Chia seed usage	Products	Remarks
Animal feed	Chicken	(1) Increased ω-3 alpha-linoleic acid and ω-6 linoleic acid of egg and yolk [26]. (2) Increased ω-3 alpha-linoleic acid and decreased palmitic fatty acid of meat. (3) Taste, sensory evaluation, and production of eggs and broilers were not affected [34, 45–47].
	Pigs and rabbits	Increase of PUFA in meat fats as well as improved aroma, flavor, and digestibility of meat [2, 30, 46–49].
Food formulation	Composite flour (15–20% of chia with corn flour)	Increased total dietary fiber and a decrease in Glycemic Index [41].
	Ingredient for cookies, cereal bars, chips, desserts, breads, jellies, and emulsions	Improved water holding, absorption capacity, and emulsifying stability [1, 37, 50].
Health supplement	Chia seed oil	Topical application for skin diseases such as pruritus and xerotic especially in diabetic and renal dysfunction patients [12].
	Carbohydrate-loading drinks	Enhanced athletes' sports endurance by more than 90 minutes but not athletes' performance [51].
	Supplement for postmenopausal women	Enhanced the levels of ALA and eicosapentaenoic acid (EPA) [18].

5. Extraction of Chia Seed Oil

Chia seed is mainly valued for its oil. Thus, many oil extraction methods had been utilized. Differences in the extraction methods caused variations in the oil yield, quality of fatty acids, fatty acid contents, total dietary fibers, and also antioxidant content. Table 3 summarises the current methods used in the extraction of chia seed oil.

6. Market Potential and Commercial Application of Chia Seed

Functional foods have gained tremendous attention worldwide over the past few years due to the wave of healthy lifestyle changes. One of the reasons for the interest to shift to a healthier lifestyle is the increasing number of people suffering from cardiovascular diseases (CVDs), high blood pressure, obesity, diabetes, and other related diseases. These conditions are commonly due to inactive lifestyle and poor diet where the food consumed daily contains high amounts of saturated fatty acids (SFAs). There are numerous studies which reported on the correlation between high SFA, particularly palmitic acid, and low PUFA intakes with CVD [34]. Traditionally, the now so-called functional foods have been consumed based on their availabilities as daily staple foods. At present, many studies have been done to increase their functionality as high nutrient food supplements. The benefits of functional foods primarily come from the presence of active ingredients and bioactivities of compounds originally present in the plant being still present in the food products after they have been processed to make them suitable for human consumption.

Recently, chia has regained its popularity by becoming one of the main oil sources that contains high levels of PUFA. Chia, which used to be the major food crop of the indigenous peoples of Mexico and Guatemala, is now widely cultivated and commercialized for its (omega) ω-3 alpha-linolenic acid (ALA) content and antioxidant properties. Today, its cultivation is not only limited to the Americas but is also extended to other areas such as Australia and Southeast Asia [6].

At present, chia seed is used as a healthy oil supplement for humans and animals. Table 4 presented a summary of the current commercial usages of chia seed. Human consumption of chia in diet is mainly from the extracted oil through its incorporation into cooking oil, confections, or supplements. In 2000, the US Dietary Guidelines recommend that chia seed can be used as a primary food not exceeding 48 g/day. Chia is commonly consumed as salad from chia sprout, in beverages, cereals, and salad dressing from the seed, or it is eaten raw [41, 42]. The European Commission approved the use of chia seed in bread products with a limit of not more than 5%. Other than bread, the food industry of various countries around the world including US, Canada, Chile, Australia, New Zealand, and Mexico has widely used chia seeds or its oil for different applications such as breakfast cereals, bars, cookie snacks, fruit juices, cake, and yoghurt [43, 44].

Despite of its well-known antioxidant activities and healthy fatty acid profile, consumers are not very aware of chia's benefits until recently. Chia seed production is a major contributor to the Argentine economy being responsible for 24% of its agricultural industry. In 2008, Argentina contributed approximately 4% of the world grain production [52]. Although chia seed has been commercialized for a long time in Argentina, however, due to the comparatively small-scale production there, problems in its availability and sustainability as an edible oil source in the global market exist. The current planting and production of chia seed oil are yet to fully meet the world market demand [17, 53, 54].

7. Summary

Based on the current research findings, chia seed is a good choice of healthy oil to maintain a balanced serum lipid profile. However, unlike vitamin E and coenzyme Q_{10}, *in vivo* clinical bioactivity and safety evaluation of chia seeds are still limited. Furthermore, details on the mechanisms of chia seed's hypolipidemic effects need to be studied and compared with those of the isolated omega 3 and omega 6 fatty acids.

References

[1] J. L. Bresson, A. Flynn, M. Heinonen et al., "Opinion on the safety of "Chia seeds (*Salvia hispanica L.*) and ground whole Chia seeds" as a food ingredient," *The European Food Safety Authority Journal*, vol. 996, pp. 1–26, 2009.

[2] P. G. Peiretti and G. Meineri, "Effects on growth performance, carcass characteristics, and the fat and meat fatty acid profile of rabbits fed diets with chia (*Salvia hispanica L.*) seed supplements," *Meat Science*, vol. 80, no. 4, pp. 1116–1121, 2008.

[3] E. Reyes-Caudillo, A. Tecante, and M. A. Valdivia-López, "Dietary fibre content and antioxidant activity of phenolic compounds present in Mexican chia (*Salvia hispanica L.*) seeds," *Food Chemistry*, vol. 107, no. 2, pp. 656–663, 2008.

[4] J. P. Cahill and M. C. Provance, "Genetics of qualitative traits in domesticated chia (*Salvia hispanica L.*)," *Journal of Heredity*, vol. 93, no. 1, pp. 52–55, 2002.

[5] A. Reales, D. Rivera, J. A. Palazón, and C. Obón, "Numerical taxonomy study of *Salvia sect. Salvia* (Labiatae)," *Botanical Journal of the Linnean Society*, vol. 145, no. 3, pp. 353–371, 2004.

[6] W. Jamboonsri, T. D. Phillips, R. L. Geneve, J. P. Cahill, and D. F. Hildebrand, "Extending the range of an ancient crop, *Salvia hispanica L.*—a new ω3 source," *Genetic Resources and Crop Evolution*, vol. 59, no. 2, pp. 171–178, 2012.

[7] P. G. Peiretti and F. Gai, "Fatty acid and nutritive quality of chia (*Salvia hispanica L.*) seeds and plant during growth," *Animal Feed Science and Technology*, vol. 148, no. 2–4, pp. 267–275, 2009.

[8] V. Y. Ixtaina, S. M. Nolasco, and M. C. Tomás, "Physical properties of chia (*Salvia hispanica L.*) seeds," *Industrial Crops and Products*, vol. 28, no. 3, pp. 286–293, 2008.

[9] M. Bueno, O. di Sapio, M. Barolo, H. Busilacchi, M. Quiroga, and C. Severin, "Quality tests of *Salvia hispanica L.* (Lamiaceae) fruits marketed in the city of Rosario (Santa Fe province, Argentina)," *Boletín Latinoamericano y del Caribe de Plantas Medicinales y Aromáticas*, vol. 9, no. 3, pp. 221–227, 2010.

[10] O. di Sapio, M. Bueno, H. Busilacchi, M. Quiroga, and C. Severin, "Morphoanatomical characterization of *Salvia hispanica L.* (LAMIACEAE) leaf, stem, fruit and seed," *Boletín Latinoamericano Y Del Caribe De Plantas Medicinales Y Aromáticas*, vol. 11, no. 3, pp. 249–2268, 2012.

[11] B. Heuer, Z. Yaniv, and I. Ravina, "Effect of late salinization of chia (*Salvia hispanica*), stock (*Matthiola tricuspidata*) and evening primrose (*Oenothera biennis*) on their oil content and quality," *Industrial Crops and Products*, vol. 15, no. 2, pp. 163–167, 2002.

[12] S. K. Jeong, H. J. Park, B. D. Park, and I. H. Kim, "Effectiveness of topical chia seed oil on pruritus of end-stage renal disease (ESRD) patients and healthy volunteers," *Annals of Dermatology*, vol. 22, no. 2, pp. 143–148, 2010.

[13] A. P. Simopoulos, "Omega-3 fatty acids in inflammation and autoimmune diseases," *Journal of the American College of Nutrition*, vol. 21, no. 6, pp. 495–505, 2002.

[14] H. Poudyal, S. K. Panchal, J. Waanders, L. Ward, and L. Brown, "Lipid redistribution by α-linolenic acid-rich chia seed inhibits stearoyl-CoA desaturase-1 and induces cardiac and hepatic protection in diet-induced obese rats," *Journal of Nutritional Biochemistry*, vol. 23, no. 2, pp. 153–162, 2012.

[15] A. P. Simopoulos, "The omega-6/omega-3 fatty acid ratio, genetic variation, and cardiovascular disease," *Asia Pacific Journal of Clinical Nutrition*, vol. 17, no. 1, pp. 131–134, 2008.

[16] M. Betti, T. I. Perez, M. J. Zuidhof, and R. A. Renema, "Omega-3-enriched broiler meat: 3. Fatty acid distribution between triacylglycerol and phospholipid classes," *Poultry Science*, vol. 88, no. 8, pp. 1740–1754, 2009.

[17] V. Dubois, S. Breton, M. Linder, J. Fanni, and M. Parmentier, "Fatty acid profiles of 80 vegetable oils with regard to their nutritional potential," *European Journal of Lipid Science and Technology*, vol. 109, no. 7, pp. 710–732, 2007.

[18] F. Jin, D. C. Nieman, W. Sha et al., "Supplementation of milled chia seeds increases plasma ALA and EPA in postmenopausal women," *Plant Foods For Human Nutrition*, vol. 67, pp. 105–110, 2010.

[19] D. C. Nieman, E. J. Cayea, M. D. Austin, D. A. Henson, S. R. McAnulty, and F. Jin, "Chia seed does not promote weight loss or alter disease risk factors in overweight adults," *Nutrition Research*, vol. 29, no. 6, pp. 414–418, 2009.

[20] G. C. Martha, R. T. Armando, A. A. Carlos et al., "A dietary pattern including Nopal, Chia seed, soy protein, and oat reduces serum triglycerides and glucose intolerance in patients with metabolic syndrome," *Journal of Nutrition*, vol. 142, no. 1, pp. 64–69, 2012.

[21] V. Vuksan, A. L. Jenkins, A. G. Dias et al., "Reduction in postprandial glucose excursion and prolongation of satiety: possible explanation of the long-term effects of whole grain Salba (*Salvia hispanica L.*)," *European Journal of Clinical Nutrition*, vol. 64, no. 4, pp. 436–438, 2010.

[22] R. Ayerza and W. Coates, "Influence of environment on growing period and yield, protein, oil and α-linolenic content of three chia (*Salvia hispanica L.*) selections," *Industrial Crops and Products*, vol. 30, no. 2, pp. 321–324, 2009.

[23] R. Ayerza h and W. Coates, "Protein content, oil content and fatty acid profiles as potential criteria to determine the origin of commercially grown chia (*Salvia hispanica L.*)," *Industrial Crops and Products*, vol. 34, no. 2, pp. 1366–1371, 2011.

[24] R. Ayerza, "Effects of seed color and growing locations on fatty acid content and composition of two chia (*Salvia hispanica L.*) genotypes," *Journal of the American Oil Chemists' Society*, vol. 87, no. 10, pp. 1161–1165, 2010.

[25] R. Ayerza (h) R., "Oil content and fatty acid composition of chia (*Salvia hispanica L.*) from five northwestern locations in Argentina," *Journal of the American Oil Chemists' Society*, vol. 72, no. 9, pp. 1079–1081, 1995.

[26] A. Antruejo, J. O. Azcona, P. T. Garcia et al., "Omega-3 enriched egg production: the effect of a-linolenic x-3 fatty acid sources on laying hen performance and yolk lipid content and fatty acid composition," *British Poultry Science*, vol. 52, no. 6, pp. 750–760, 2011.

[27] R. Ayerza and W. Coates, "Omega-3 enriched eggs: the influence of dietary α-linolenic fatty acid source on egg production and composition," *Canadian Journal of Animal Science*, vol. 81, no. 3, pp. 355–362, 2000.

[28] R. Ayerza and W. Coates, "Effect of dietary α-linolenic fatty acid derived from chia when fed as ground seed, whole seed and oil on lipid content and fatty acid composition of rat plasma," *Annals of Nutrition and Metabolism*, vol. 51, no. 1, pp. 27–34, 2007.

[29] I. Fernandez, S. M. Vidueiros, R. Ayerza, W. Coates, and A. Pallaro, "Impact of chia (*Salvia hispanica L.*) on the immune system: preliminary study," *Proceedings of the Nutrition Society*, vol. 67, article E12, 2008.

[30] W. Coates and R. Ayerza, "Chia (*Salvia hispanica L.*) seed as an n-3 fatty acid source for finishing pigs: effects on fatty acid composition and fat stability of the meat and internal fat, growth performance, and meat sensory characteristics," *Journal of Animal Science*, vol. 87, no. 11, pp. 3798–3804, 2009.

[31] G. Masoero, G. Sala, G. Meineri, P. Cornale, S. Tassone, and P. G. Peiretti, "Nir spectroscopy and electronic nose evaluation on live rabbits and on the meat of rabbits fed increasing levels of Chia (*Salvia hispanica L.*) seeds," *Journal of Animal and Veterinary Advances*, vol. 7, no. 11, pp. 1394–1399, 2008.

[32] A. Dalle Zotte and Z. Szendro, "The role of rabbit meat as functional food," *Meat Science*, vol. 88, no. 3, pp. 319–331, 2011.

[33] R. Ayerza, "Chia as a new source of ω-3 fatty acids: advantage over other raw materials to produce ω-3 enriched eggs," in *Proceedings of the Symposium on Omega-3 Fatty Acids, Evolution and Human Health*, Washington, DC, USA, September 2002.

[34] R. Ayerza, W. Coates, and M. Lauria, "Chia seed (*Salvia hispanica L.*) as an ω-3 fatty acid source for broilers: influence on fatty acid composition, cholesterol and fat content of white and dark meats, growth performance, and sensory characteristics," *Poultry Science*, vol. 81, no. 6, pp. 826–837, 2002.

[35] M. S. Vedtofte, M. U. Jakobsen, L. Lauritzen et al., "Dietary alpha linolenic acid, linoleic acid and n-3 long-chain PUFA and risk of ischemic heart disease," *The American Journal of Clinical Nutrition.*, vol. 94, pp. 1097–1103, 2011.

[36] V. Y. Ixtaina, S. M. Nolasco, and M. C. Tomàs, "Oxidative Stability of Chia (*Salvia hispanica L.*) Seed Oil: effect of Antioxidants and Storage Conditions," *Journal of the American Oil Chemists' Society*, vol. 89, pp. 1077–1090, 2012.

[37] M. I. Capitani, V. Spotorno, S. M. Nolasco, and M. C. Tomás, "Physicochemical and functional characterization of by-products from chia (*Salvia hispanica L.*) seeds of Argentina," *LWT—Food Science and Technology*, vol. 45, no. 1, pp. 94–102, 2012.

[38] V. Y. Ixtaina, F. Mattea, D. A. Cardarelli, M. A. Mattea, S. M. Nolasco, and M. C. Tomás, "Supercritical carbon dioxide extraction and characterization of Argentinean chia seed oil," *Journal of the American Oil Chemists' Society*, vol. 88, no. 2, pp. 289–298, 2011.

[39] J. A. R. Uribe, J. I. N. Perez, H. C. Kauil, G. R. Rubio, and C. G. Alcocer, "Extraction of oil from chia seeds with supercritical CO_2," *Journal of Supercritical Fluids*, vol. 56, no. 2, pp. 174–178, 2011.

[40] V. Y. Ixtaina, A. Vega, S. M. Nolasco et al., "Supercritical carbon dioxide extraction of oil from Mexican chia seed (*Salvia hispanica L.*): characterization and process optimization," *Journal of Supercritical Fluids*, vol. 55, no. 1, pp. 192–199, 2010.

[41] R. Rendón-Villalobos, A. Ortiz-Sanchez, J. Solorza-Feria, and C. A. Trujillo-Hernandez, "Formulation, physicochemical, nutritional and sensorial evaluation of corn tortillas supplemented with chia seed (*Salvia hispanica L.*)," *Czech Journal of Food Sciences*, vol. 30, no. 2, pp. 118–125, 2012.

[42] W. F. Baughman and G. S. Jamieson, "Chia seed oil," *Oil & Fat Industries*, vol. 6, no. 9, pp. 15–17, 1929.

[43] The Chia Company, "Request for scientific evaluation of substantial equivalence application for the approval of Chia seeds (*Salvia hispanica*L.) from the Chia Company for use in bread," Food Law Consultants, 2010, http://www.food.gov.uk/multimedia/pdfs/thechiacompany.pdf.

[44] R. Borneo, A. Aguirre, and A. E. León, "Chia (*Salvia hispanica L*) gel can be used as egg or oil replacer in cake formulations," *Journal of the American Dietetic Association*, vol. 110, no. 6, pp. 946–949, 2010.

[45] R. Ayerza and W. Coates, "An ω-3 fatty acid enriched chia diet: influence on egg fatty acid composition, cholesterol and oil content," *Canadian Journal of Animal Science*, vol. 79, no. 1, pp. 53–58, 1999.

[46] R. Ayerza and W. Coates, "Dietary levels of chia: influence on yolk cholesterol, lipid content and fatty acid composition for two strains of hens," *Poultry Science*, vol. 79, no. 5, pp. 724–739, 2000.

[47] R. Ayerza and W. Coates, "Dietary levels of chia: influence on hen weight, egg production and sensory quality, for two strains of hens," *British Poultry Science*, vol. 43, no. 2, pp. 283–290, 2002.

[48] G. Meineri and P. G. Peiretti, "Apparent digestibility of mixed feed with increasing levels of chia (*Salvia hispanica L.*) seeds in rabbit diets," *Italian Journal of Animal Science*, vol. 6, no. 1, pp. 778–780, 2007.

[49] G. Meineri, P. Cornale, S. Tassone, and P. G. Peiretti, "Effects of chia (*Salvia hispanica L.*) seed supplementation on rabbit meat quality, oxidative stability and sensory traits," *Italian Journal of Animal Science*, vol. 9, no. 10, pp. 45–49, 2009.

[50] B. L. Olivos-Lugo, M. Á. Valdivia-López, and A. Tecante, "Thermal and physicochemical properties and nutritional value of the protein fraction of mexican chia seed (*Salvia hispanica L.*)," *Food Science and Technology International*, vol. 16, no. 1, pp. 89–96, 2010.

[51] T. G. Illian, J. C. Casey, and P. A. Bishop, "Omega 3 chia seed loading as a means of carbohydrate loading," *Journal of Strength and Conditioning Research*, vol. 25, no. 1, pp. 61–65, 2011.

[52] D. Lema, "Growth and productivity in Argentine agriculture," in *Conference on Causes and Consequences of Global Agricultural Productivity Growth*, Washington, DC, USA, May 2010, http://www.farmfoundation.org/news/articlefiles/1725-Lema.pdf.

[53] R. Ayerza and W. Coates, "Seed yield, oil content and fatty acid composition of three botanical sources of ω-3 fatty acid planted in the Yungas ecosystem of tropical Argentina," *Tropical Science*, vol. 47, no. 4, pp. 183–187, 2007.

[54] W. Coates and R. Ayerza, "Production potential of chia in northwestern Argentina," *Industrial Crops and Products*, vol. 5, no. 3, pp. 229–233, 1996.

15

Rapamycin Conditioning of Dendritic Cells Differentiated from Human ES Cells Promotes a Tolerogenic Phenotype

Kathryn M. Silk,[1] **Alison J. Leishman,**[1] **Kevin P. Nishimoto,**[2]
Anita Reddy,[2] **and Paul J. Fairchild**[1]

[1] *Sir William Dunn School of Pathology, University of Oxford, South Parks Road, Oxford, OX1 3RE, UK*
[2] *Translational Research and Immunology, Geron Corporation, 230 Constitution Drive, Menlo Park, CA 94025, USA*

Correspondence should be addressed to Paul J. Fairchild, paul.fairchild@path.ox.ac.uk

Academic Editor: Ken-ichi Isobe

While human embryonic stem cells (hESCs) may one day facilitate the treatment of degenerative diseases requiring cell replacement therapy, the success of regenerative medicine is predicated on overcoming the rejection of replacement tissues. Given the role played by dendritic cells (DCs) in the establishment of immunological tolerance, we have proposed that DC, rendered tolerogenic during their differentiation from hESC, might predispose recipients to accept replacement tissues. As a first step towards this goal, we demonstrate that DC differentiated from H1 hESCs (H1-DCs) are particularly responsive to the immunosuppressive agent rapamycin compared to monocyte-derived DC (moDC). While rapamycin had only modest impact on the phenotype and function of moDC, H1-DC failed to upregulate CD40 upon maturation and displayed reduced immunostimulatory capacity. Furthermore, coculture of naïve allogeneic T cells with rapamycin-treated H1-DC promoted an increased appearance of CD25^{hi} Foxp3$^+$ regulatory T cells, compared to moDC. Our findings suggest that conditioning of hESC-derived DC with rapamycin favours a tolerogenic phenotype.

1. Introduction

Human embryonic stem cells (hESCs) derived under conditions compliant with their downstream clinical application, serve as a renewable source of cell types that may one day enable the replacement of tissues whose function has become compromised by chronic or degenerative disease [1]. Nevertheless, the routine implementation of cell replacement therapy (CRT) requires strategies to address the immunological barriers encountered by the use of hESC of allogeneic origin [2]. While conventional immunosuppression offers a potential solution to the immunogenicity of hESC-derived tissues, the risks inherent in its protracted use make the induction of transplantation tolerance an attractive alternative.

Dendritic cells (DCs) play a critical role in determining the outcome of antigen presentation to naive T cells, either promoting their activation and subsequent immunity, or favouring the induction of tolerance [3]. The delivery of foreign antigen to DC in the steady state by conjugation to monoclonal antibodies (mAbs) specific for the surface receptor CD205, was, for instance, found to render recipient mice specifically tolerant to the antigen upon subsequent immunization [4]. Such findings have been extended to a transplantation setting by demonstrating how administration of immature donor DC to mice across a minor histocompatibility barrier is sufficient to secure the indefinite survival of donor skin grafts. In this model, the resulting tolerance could be attributed to the polarisation of responding T cells towards a regulatory phenotype, characterised by upregulation of the transcription factor Foxp3 [5]. Such findings, together with early success at inducing tolerance in healthy human volunteers by the administration of immature antigen-pulsed monocyte-derived DC (moDC) [6], augur well for the future use of DC as a conditioning regime in the context of CRT. Indeed, the recent description of protocols for the differentiation of DC from hESC under conditions substantially free of animal products paves the way for such an approach: given that this source of DC would share with the replacement tissue the very alloantigens to which tolerance must be established, their administration in advance of CRT might be anticipated to condition the recipient to accept the transplanted tissue, providing the DC have first been rendered stably tolerogenic [7]. Accordingly, Senju et al. generated DC expressing the inhibitory receptor programmed death ligand 1 (PD-L1) by genetic modification of the parent hESC line [8], a similar approach in the mouse having successfully yielded DC capable of preventing the onset of experimental autoimmune encephalomyelitis by induction of tolerance to myelin antigens [9]. While such a strategy is clearly promising, the administration of genetically modified cells to patients poses additional regulatory barriers, suggesting that exposure of DC to pharmacological agents, known to promote a tolerogenic phenotype, may prove to be a more pragmatic approach [10].

The macrocyclic triene antibiotic, rapamycin, displays potent immunosuppressive properties that are routinely employed to facilitate whole-organ transplantation. In addition to its systemic use, however, rapamycin has been shown to render DC profoundly protolerogenic through inhibition of mammalian target of rapamycin (mTOR) signalling pathways. In the mouse, rapamycin-treated DC display profoundly suppressed allostimulatory capacity in vitro and enhanced propensity for the induction of Foxp3+ regulatory T (Treg) cells [11]. Furthermore, exposure to rapamycin, unlike other immunosuppressive agents, leads to the upregulation of CCR7 by both mouse and human DC and a commensurate increase in responsiveness to CCL19, compatible with their trafficking in vivo to regional lymph nodes [12, 13]. Furthermore, the administration of rapamycin-treated recipient DC pulsed with donor alloantigens has secured the indefinite survival of tissue allografts in various animal models [14–16], the resulting tolerance having been demonstrated to rely on the expansion of antigen-specific Treg cells [17]. Nevertheless, despite its compelling credentials, rapamycin has been reported to exert quite distinct effects on human DC, depending on the source and subset involved [18]. We have, therefore, investigated the compatibility of protocols for the differentiation of DC from the H1 hESC line (H1-DC) with the use of rapamycin. Here we report that H1-DCs are peculiarly sensitive to the immunomodulatory effects of rapamycin, compared with conventional moDC, as evidenced by the specific loss of immunogenicity and enhanced capacity to polarise responding T cells towards a regulatory phenotype. Our findings provide an important first step towards the use of DC differentiated from hESC in the establishment of tolerance to replacement tissues, providing a proof of concept for their future application in regenerative medicine.

2. Materials and Methods

2.1. *Isolation of Primary Cells.* Monocytes and naïve T cells were isolated from peripheral blood mononuclear cells (PBMCs) of buffy coats (NHS Blood Transfusion Service) or from blood provided by volunteers under informed consent using CD14-coated beads or naïve CD4+ T cell selection kit (Miltenyi Biotec). Cell populations were positively selected

or depleted from PBMC using AutoMACS separation according to the manufacturer's instructions.

2.2. *Culture of hESC.* H1 ESCs were cultured in X-VIVO-10 medium (without gentamycin or phenol red, Lonza) supplemented with nonessential amino acids (PAA Laboratories GmbH), 2 mM L-glutamine (PAA Laboratories GmbH), 50 μM 2-mercaptoethanol (Sigma), 0.5 ng/mL recombinant human transforming growth factor β (TGF-β, R&D Systems), and 80 ng/mL recombinant human basic fibroblast growth factor (bFGF, R&D Systems) on 6-well plates, previously coated with Matrigel (phenol red-free, growth factor reduced, BD Biosciences) diluted 1:30 using ice-cold knockout Dulbecco's Modified Eagle's Medium (KO-DMEM, Invitrogen). Supplemented X-VIVO-10 medium was replaced daily except the day following passaging.

Human ESCs were routinely passaged as cell clusters of about 0.5 mm diameter every 4–6 days. For passaging, colonies were incubated in filter-sterilised warm collagenase IV (Invitrogen) until detachment of the stromal cells. Stromal cells were removed by washing with Dulbecco's Phosphate-Buffered Saline (DPBS) and hESC were scraped off into supplemented X-VIVO-10 Medium for 1:5 passaging. All cell cultures were incubated in a humidified incubator at 37°C and 5% CO_2.

2.3. *Differentiation of hESC.* H1 hESCs were plated at 3×10^6 per well of 6-well ultralow attachment (ULA) plates (Costar) in a total volume of 4 mL of X-VIVO-15 medium (Lonza), supplemented with 1 mM sodium pyruvate, nonessential amino acids, 2 mM L-glutamine (all PAA Laboratories GmbH) and 5 μM 2-mercaptoethanol (Sigma). The following growth factors were added: 50 ng/mL recombinant human bone morphogenetic protein-4 (BMP-4, R&D Systems), 50 ng/mL recombinant human vascular endothelial growth factor (VEGF, R&D Systems), 20 ng/mL recombinant human stem cell factor (SCF, R&D Systems), and 50 ng/mL recombinant human granulocyte macrophage-colony stimulating factor (GM-CSF, R&D Systems). After 2-3 days, the medium was topped up with 2 mL of fresh supplemented X-VIVO-15 medium to produce a total volume of 6 mL. Subsequent feeding was performed every 2-3 days by replacing 2-3 mL of old medium with new supplemented X-VIVO-15 medium from which every 5 days a growth factor was removed starting with BMP-4 at day 5, followed by VEGF at day 10 and SCF at day 15 of differentiation [19]. Once macrophage-like cells were observed, 25 ng/mL of IL-4 (Peprotech) was added, which was increased stepwise to 100 ng/mL.

On days 30–35, monocytes were harvested by gentle pipetting, leaving adherent macrophages in the culture dish. The cell suspension was passed through a 70 μm cell strainer (BD Falcon) to remove cellular debris, washed with DPBS and plated at $1-1.5 \times 10^6$ monocytes per well of a 6-well Cellbind plate (Corning) in X-VIVO-15 supplemented with 50 ng/mL GM-CSF and 100 ng/mL IL-4.

2.4. *Derivation of DC from Human Monocytes.* Monocytes were cultured in RPMI 1640 (Invitrogen) supplemented with

2 mM L-glutamine (PAA laboratories GmbH), 50 U/mL penicillin (PAA laboratories GmbH), 50 μg/mL streptomycin (PAA laboratories GmbH), 10% heat-inactivated and filter-sterilised fetal bovine serum (FBS), 50 ng/mL GM-CSF, and 100 ng/mL IL-4 on 6-well Cellbind plates for 6–8 days.

2.5. DC Maturation and Rapamycin Treatment. Two days after monocytes were plated, monocyte-derived and hESC-derived immature DC were treated with 10 ng/mL and 5–7 ng/mL of rapamycin (Sigma), respectively. On day 5, DCs were matured for 48 hr using a maturation cocktail consisting of 50 ng/mL of GM-CSF (R&D Systems), 100 ng/mL IL-4 (R&D Systems), 20 ng/mL IFNγ (R&D Systems), 50 ng/mL TNFα (R&D Systems), 10 ng/mL of IL-1β (R&D Systems), and 1 μg/mL PGE$_2$ (Sigma). On day 6-7, DCs were harvested by gentle pipetting, passed through a 70 μm cell strainer, centrifuged, and resuspended prior to their use in experiments.

2.6. Allogeneic Mixed Leukocyte Reaction (MLR). DCs were incubated in 10 μg/mL mitomycin C (Sigma) in supplemented RPMI 1640 at 37°C for 30 minutes. Cells were washed, resuspended in supplemented RPMI 1640, and plated in triplicate to give either 2.5×10^3 cells, 5×10^3 cells, or 1×10^4 cells in a total volume of 100 μl per well using 96-well round-bottom plates (Corning). Naïve CD4$^+$ T cells were plated at 5×10^4 cells per well to yield a stimulator to responder ratio of 1 : 5, 1 : 10, and 1 : 20 and a total volume of 200 μl/well. Wells containing T cells and mitomycin C-treated DC alone were included as controls for background proliferation of either cell type. Cells were incubated for 5 days at 37°C, after which T cells were pulsed with 0.5 μCi of [^{3}H]-thymidine per well for 18 hr before harvesting.

2.7. DC-T-Cell Cocultures. DC (2×10^5) and 1×10^6 T cells were cocultured in supplemented RPMI 1640 using 24-well Cellbind plates (Corning). After 7 days of coculture, cells were harvested and stained for CD4, CD25, and Foxp3 and analysed by flow cytometry as described below.

2.8. Flow Cytometry. Cells were incubated for 15 min in blocking solution (5% normal rat serum, 0.5% bovine serum albumin, and 0.1% NaN$_3$ in DPBS) on ice. Cells were washed with DPBS containing 1% FBS and 0.1% NaN$_3$ and resuspended in this solution together with one or several of the following fluorescently labelled antibodies: SSEA-4 (clone: MC-813-70, R&D Systems), eZFluor anti-human CD4-FITC and either CD25-APC or CD25-AF488 Cocktail (eBioscience), CD83 (HB15e, AbD Serotec), CD86 (BU63, AbD Serotec), CD40 (LOB7/6, AbD Serotec), PD-L1 (AbD Serotec), CD127 (40131, R&D Systems), CTLA-4 (BNI3, BD Pharmingen), MHC II HLA-DR/DQ/DP (WR18, AbD Serotec), CD80 (MEM-233, AbD Serotec), CD45 (15.2, AbD Serotec), CD14 (MEM18, AbD Serotec), CD11c (BU15, AbD Serotec), and CD13 (AbD Serotec). Cells were incubated at 4°C in the dark for 30–60 minutes. For the last 10 minutes, 250 ng/mL 7-AAD was added. Cells were washed, fixed in 2% formaldehyde, and analysed by flow cytometry.

Intracellular staining was performed according to the manufacturer's instructions using permeabilisation and fixation buffers (eBioscience) and antibodies specific for Oct-4 (240408, R&DSystems) or Foxp3 (eBioscience).

3. Results

3.1. Differentiation and Characterisation of DC from the H1 hESC Line. In order to investigate whether protocols we have established previously for the differentiation of DC from hESC might be compatible with the use of rapamycin, we made use of the well-characterised H1 hESC line. In keeping with its downstream clinical application, H1 was maintained in serum-free medium devoid of animal products and feeder cells, as described previously [20, 21]. Under these conditions, H1 formed compact colonies with clearly defined boarders (Figure 1(a)), the individual cells displaying a high nucleus : cytoplasm ratio and prominent heterochromatin. Flow cytometric analysis revealed expression of the transcription factor Oct-4 and stage-specific embryonic antigen 4 (SSEA-4), both of which are known to strongly correlate with pluripotency (Figure 1(b)).

The differentiation of H1 was directed along the DC lineage in ultralow attachment plates by exposure to a cocktail of growth factors consisting of BMP-4, VEGF, SCF, and GM-CSF, as described previously [19]. The initiation of hematopoiesis was apparent by day 20 of culture, as evidenced by the appearance of CD45$^+$ cells, although the lack of expression of CD13, CD14, and CD11c suggested that commitment to the myeloid lineage had yet to occur (Figure 2(a)). In contrast, by day 27 of culture, a small proportion of cells, residing within a population expressing intermediate levels of CD45, had upregulated these markers, consistent with their progressive commitment to the myeloid lineage (Figure 2(b)). Indeed, from day 28 of culture onwards, cells with the characteristic morphology of human DC could be identified within cultures, either as clusters with prominent veils of cytoplasm or individual cells with long dendrites (Figure 2(c)).

By day 33 of culture, up to 21% of cells had adopted a CD45^{hi} phenotype, the majority of which were CD11c$^+$ (Figure 2(d)). Whereas these cells predominantly expressed MHC class I and CD86, CD83, and MHC class II expression were low, consistent with the phenotype of immature DC. Culture of H1-DC for 2 days in a cocktail of cytokines consisting of GM-CSF, IL-4, IFNγ, TNFα, IL-1β, and PGE$_2$ induced their maturation, as evidenced by the upregulation of CD83, similar to moDC (Figure 2(e)). Although, as previously described, MHC class II was not upregulated by H1-DC to the same extent as their monocyte-derived counterparts [19], surface expression of the costimulatory molecules CD40, CD80, and CD86 was consistent with our previous reports of the ability of this novel source of DC to stimulate proliferative responses among naïve allogeneic T cells [19].

3.2. Rapamycin Reduces the Immunogenicity of H1-DC. We next investigated whether the exposure of H1-DC to rapamycin could promote the acquisition of a protolerogenic phenotype, similar to that described for other populations

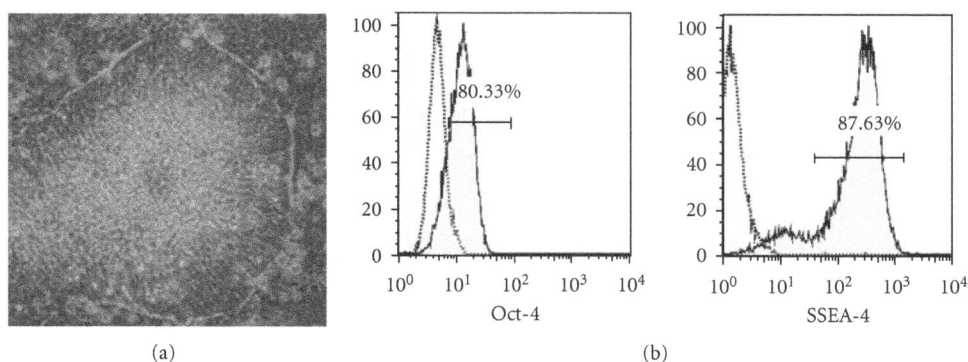

FIGURE 1: Maintenance of the H1 hESC line. (a) Colony of H1 hESC showing the morphology typical of pluripotent stem cells, including prominent boarders (×20 magnification). (b) Expression by H1 hESC of the transcription factor Oct-4 and the surface marker SSEA-4, both of which correlate with pluripotency. Dead cells were removed from flow cytometric analysis using 7-AAD staining. Open histograms represent appropriate isotype controls.

of mouse and human DC [11–13]. Accordingly, we cultured H1-DC with rapamycin for 3 days prior to inducing their maturation with proinflammatory cytokines and assessed their surface phenotype and immunostimulatory capacity in the allogeneic MLR. Whereas the addition of 10 ng/mL of rapamycin to moDC had only a modest impact on their viability, H1-DC proved especially sensitive to its toxicity, undergoing significant levels of apoptosis at concentrations greater than 7 ng/mL, as described in other studies [22]. Nevertheless, careful titration of the compound revealed that exposure of H1-DC to concentrations between 5 and 7 ng/mL exerted immunomodulatory effects without compromising their viability. Interestingly, conditioning of H1-DC with rapamycin did not appear to inhibit their maturation since they upregulated CD83 and CD86 and maintained surface expression of MHC class II and the inhibitory receptor PD-L1 (Figure 3(a)), strongly implicated in the polarisation of naïve T cells towards a Treg phenotype [23]. Significantly, however, H1-DC consistently failed to up-regulate CD40 following exposure to rapamycin, even though higher concentrations of the pharmacological agent had little impact on CD40 expression by moDC (Figure 3(b)). Consistent with their reduced levels of CD40 expression, the immunostimulatory capacity of rapamycin-treated H1-DC was significantly reduced in cocultures with naïve allogeneic T cells (Figure 3(c)). In contrast, 10 ng/mL of rapamycin exerted only modest inhibitory effects on the capacity of moDC to stimulate proliferative responses among naïve allogeneic T cells (Figure 3(c)).

3.3. Rapamycin-Treated H1-DC Polarise Naïve T Cells towards a Regulatory Phenotype. Given the reduced immunostimulatory capacity of rapamycin-treated H1-DC and their acquisition of a CD40^{lo} PD-L1$^+$ phenotype, we next investigated whether their coculture with naïve CD4$^+$ T cells might favour the induction of Treg cells, defined as CD4$^+$CD25^{hi} cells with persistent expression of Foxp3. Although at the outset, T cells enriched for CD4$^+$ cells were predominantly Foxp3$^-$ (Figure 4(a)), coculture with immature H1-DC for 7 days, resulted in up to 8.5% of CD4$^+$CD25^{hi} cells retaining Foxp3

expression by the end of the culture period (Figure 4(b)). When the H1-DC had been matured prior to coculture with naïve allogeneic T cells, the proportion of cells committed to the Treg cell lineage increased marginally to 12.5%. However, the use of H1-DC, which had been induced to mature following exposure to rapamycin, consistently resulted in a significant increase in the induction of Treg cells which represented approximately 26.5% of CD4$^+$CD25^{hi} cells, similar results being obtained in four independent experiments. By contrast, rapamycin conditioning of moDC exerted only a marginal effect on the ability of the cells to polarise responding T cells towards a regulatory phenotype (Figure 4(b)).

Given that the identification of *bona fide* human Treg cells is confounded by the universal upregulation of CD25 by activated T cells and their transient expression of Foxp3, irrespective of final lineage commitment, we investigated whether CD25^{hi} Foxp3$^+$ cells appearing in such cultures displayed other known phenotypic features of Treg cells. Cells coexpressing CD25 and Foxp3 were found to express CTLA-4 (Figure 5(a)), while lacking expression of the α subunit of the IL-7R, CD127 (Figure 5(b)), such a phenotype being strongly suggestive of a regulatory function [24, 25].

4. Discussion

The development of robust protocols for the differentiation of DC from hESC lines, derived under cGMP conditions, offers a potentially unlimited source of cells with little variability between batches, which may be subjected to rigorous quality control. The potent immunostimulatory capacity of DC differentiated in this way has suggested that they will find a likely application in the presentation of tumour associated antigens to the T-cell repertoire, thereby overcoming many of the limitations inherent in the use of moDC for cancer immunotherapy [19]. Nevertheless, given the accumulation of evidence in favour of an additional role played by DC in the establishment and ongoing maintenance of immunological tolerance [3], the availability of DC differentiated from hESC suggests they may enjoy a broader remit. We have, for instance, proposed that hESC-derived DC might be exploited

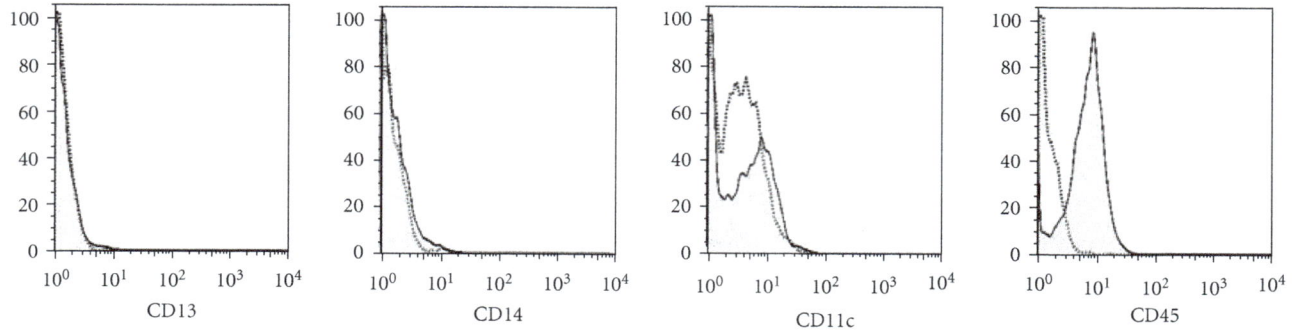

(a)

Gated on CD45^{int}

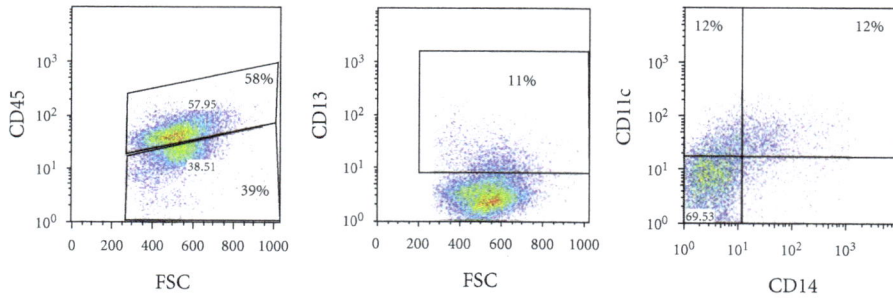

(b)

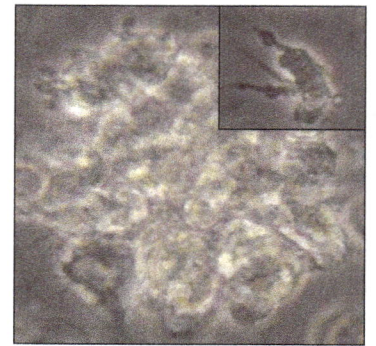

(c)

Gated on CD45^{hi}

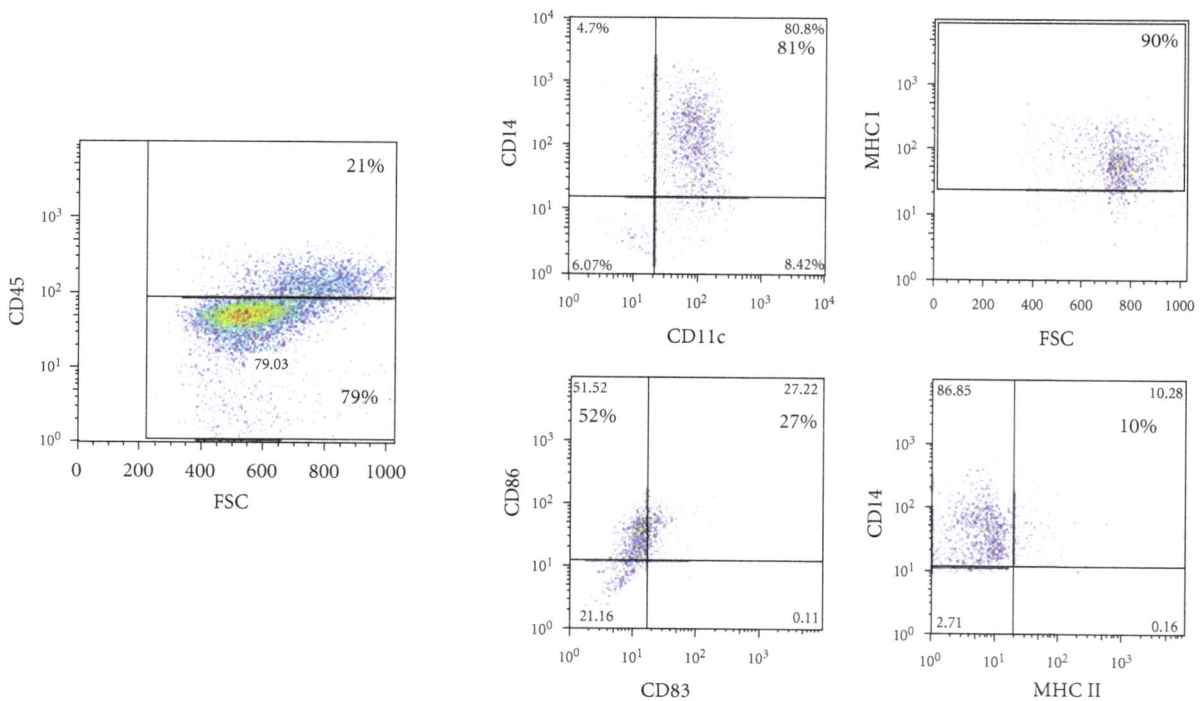

(d)

FIGURE 2: Continued.

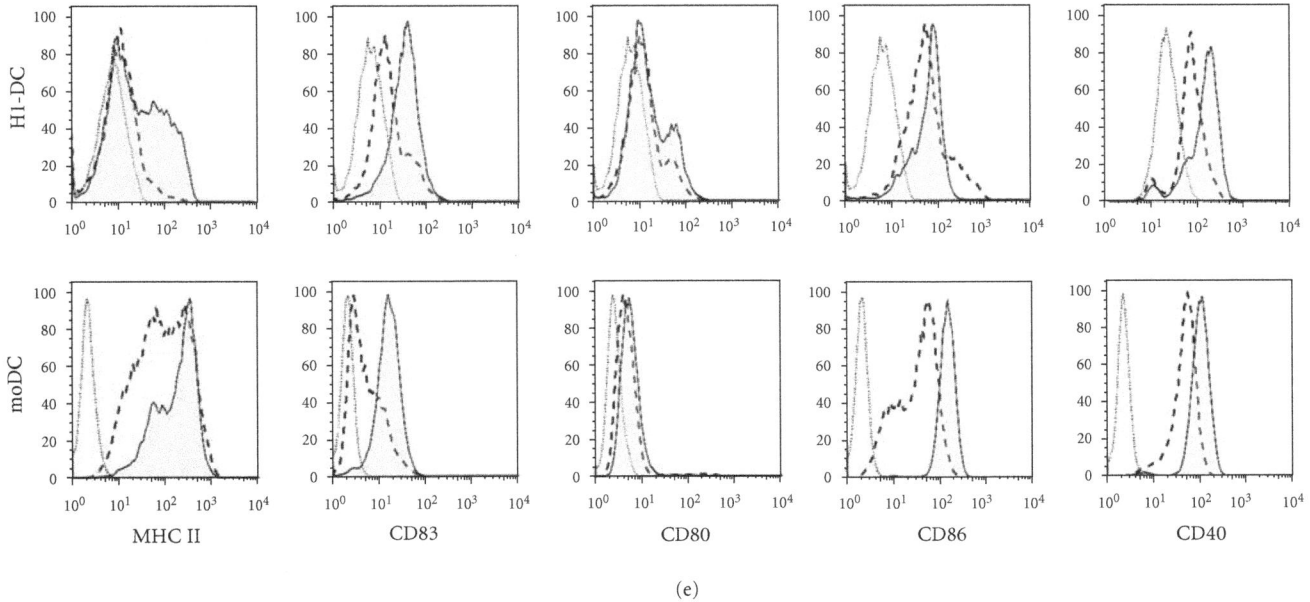

(e)

FIGURE 2: Time course of DC differentiation from H1 hESC. Cells were harvested from cultures at various time points and analysed by flow cytometry for the onset of hematopoiesis and the appearance of DC. (a) Cells harvested at day 20 of culture showing expressing of CD45 but lack of myeloid commitment, as evidenced by staining for CD13, CD14, and CD11c. Open histograms show levels of background staining using isotype-matched control antibodies. (b) Appearance of CD45^{int} cells at day 27 of culture, accompanied by the upregulation of myeloid-specific markers. (c) Photomicrograph, taken at day 28 of culture, showing the morphology of DC, including veils of cytoplasm and long dendrites (inset) ($\times 40$ magnification). (d) Cells harvested at day 33 of culture, showing the appearance of a CD45^{hi} population containing predominantly DC progenitors expressing CD14, CD11c, CD86 and MHC class I. (e) Phenotype of immature and mature H1-DCs compared with human moDC. DCs were cultured either in medium alone or medium supplemented with the maturation cocktail and stained for MHC class II, the maturation marker CD83 and classical costimulatory molecules. Dead cells were excluded from the analysis using 7-AAD. Dashed histograms show the phenotype of immature DCs while the filled histograms represent mature DCs. Open histograms depict background staining using isotype-matched controls.

to induce tolerance to the alloantigens they express, thereby conditioning recipients to accept replacement tissues differentiated from the same parent cell line [7]. This prospect is, however, contingent on the development of clinically compliant strategies to ensure the stable tolerogenicity of DC generated in this way. While the introduction of transgenes, such as PD-L1, at the ESC stage might confer on the resulting DC an immunomodulatory function [8], the additional regulatory hurdles encountered by the administration to patients of genetically modified cells, has fuelled attempts to identify approved pharmacological agents that coerce DC to adopt a protolerogenic phenotype [10].

Rapamycin is one such agent routinely exploited for its immunosuppressive properties in the treatment of allograft rejection but which has been shown to exert a profound effect on the function of individual components of the immune system, including DC. Indeed, treatment of DC with rapamycin in vitro has been demonstrated to arrest them in an immature or semimature state rendering them tolerogenic [11–13]. Accordingly, in various preclinical transplantation models, administration of rapamycin-treated recipient DC, pulsed with a source of donor alloantigens, secured the long-term survival of organ allografts [14–16]. If such a conditioning regime could be applied to DC differentiated from hESC, it may prove feasible to establish operational tolerance to the

alloantigens they endogenously express, in advance of CRT. As a first step towards this goal, we have demonstrated the sensitivity of H1-DC to rapamycin which significantly reduces their immunostimulatory properties in the allogeneic MLR (Figure 3(b)), an in vitro correlate of the direct pathway of alloantigen presentation. Furthermore, rapamycin substantially augments their ability to polarise responding CD4$^+$ T cells towards a regulatory phenotype (Figure 4(b)), as determined by their sustained expression of Foxp3 and adoption of a CTLA4$^+$CD127$^-$ phenotype. Furthermore, our preliminary results indicate that, while maturation of H1-DC induces secretion of high levels of the inflammatory cytokine IL-6 [19], prior exposure to rapamycin significantly reduces IL-6 production, possibly guiding responding T cells away from Th1/Th17 commitment towards a Treg phenotype. These results strongly suggest, therefore, that rapamycin may have the desirable properties of preventing activation of alloreactive T cells through both the direct and indirect pathways of alloantigen presentation, the induction of Treg cells potentially modulating responsiveness to indirectly presented alloantigens that have been reprocessed by endogenous recipient DC. In contrast to our findings with H1-DC, rapamycin-treatment of moDC had only a modest impact on their immunostimulatory capacity and little effect on their surface phenotype. Although our results are contrary to

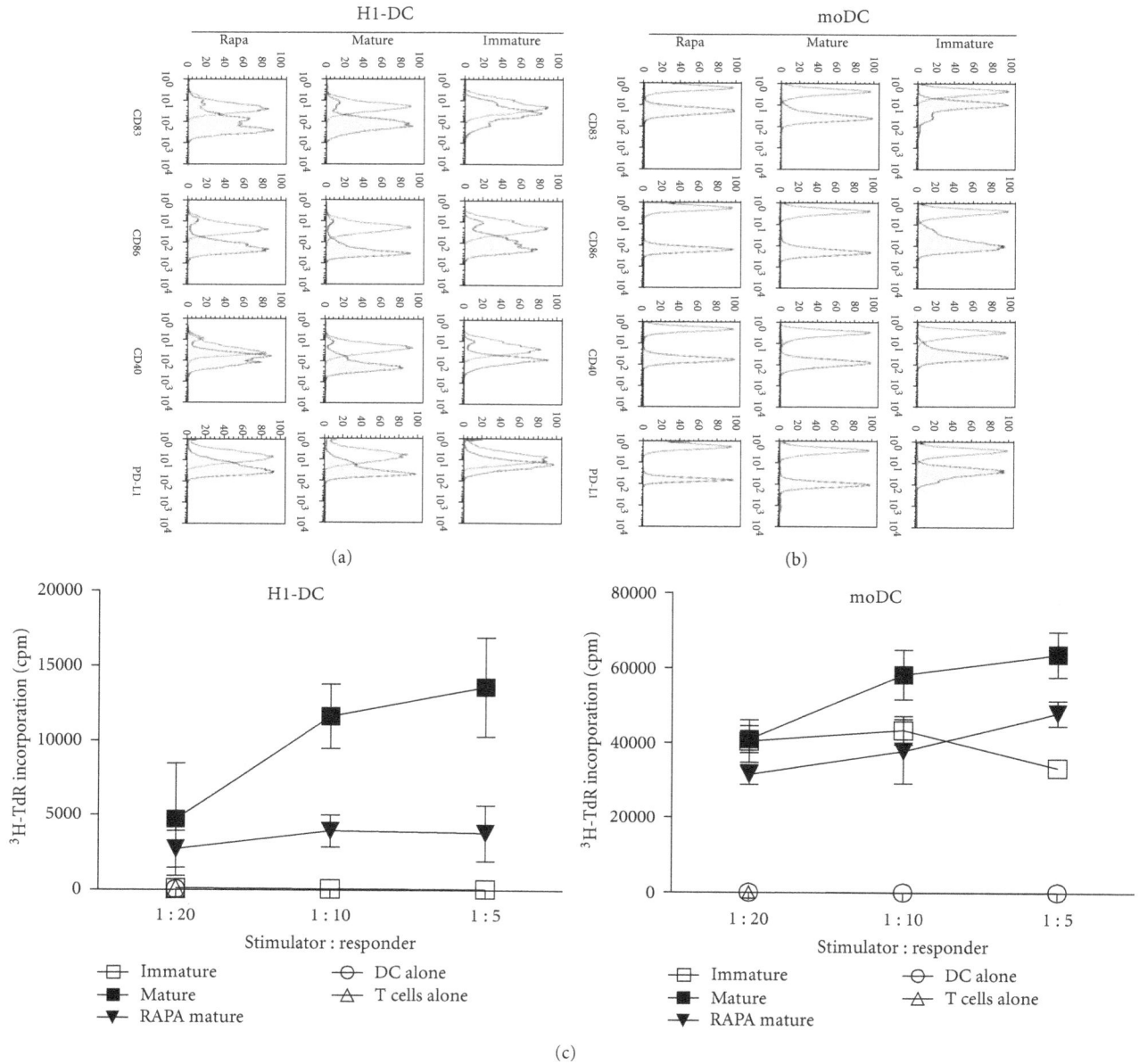

FIGURE 3: Effect of rapamycin (Rapa) on the phenotype and function of H1-DC. DCs were either untreated, matured in response to the maturation cocktail or treated with Rapa for 3 days prior to maturation. (a) H1-DC stained for the expression of the maturation marker CD83, the costimulatory molecules CD86 and CD40, as well as the inhibitory receptor PD-L1. Dead cells were excluded from analysis using 7-AAD. Open histograms represent the level of background staining using appropriate isotype-matched controls. Data from one of 3 independent experiments are shown. (b) Phenotypic analysis of control populations of moDC treated and stained in parallel with rapamycin. (c) Effect of rapamycin on the allostimulatory capacity of DC in the allogeneic MLR. DCs were mitotically-inactivated using mitomycin C and plated in triplicate at a top dose of 10^4 cells per well of a 96-well round-bottomed plate; naïve CD4$^+$ T cells were plated at 5×10^4 cells/well. Cells were incubated for 5 days before pulsing with ^{3}H-thymidine overnight. Graphs show the mean of triplicate cultures ±S.D. Data are shown from one experiment, representative of 3 independent experiments.

some other reports [12, 18], many studies have typically used higher concentrations of rapamycin and regimes for the maturation of moDC involving exposure to bacterial products, such as lipopolysaccharide, which target different intracellular signalling pathways from those solicited upon culture with the cocktail of proinflammatory cytokines used in these studies.

Despite the profound effect that rapamycin exerts on the functional potential of H1-DC, phenotypic analysis of cells treated with the compound was largely unremarkable, with the exception that upregulation of the costimulatory molecule CD40 upon maturation was prevented by prior exposure to the compound. The significance of these findings may lie in the growing appreciation of the role played by CD40 as

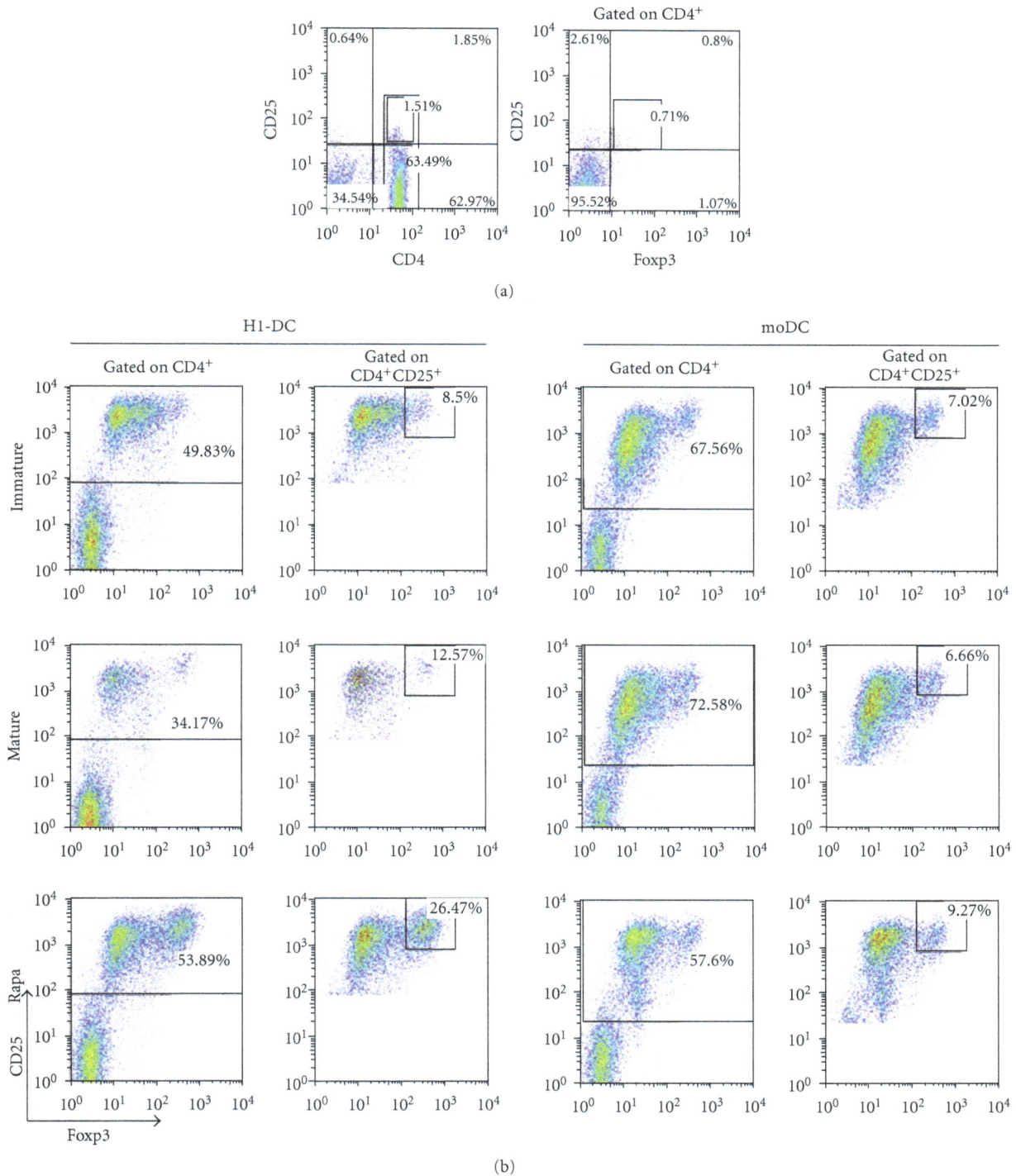

(a)

(b)

FIGURE 4: Enhanced capacity of rapamycin-treated H1-DC to promote Treg induction. (a) The starting population of naive CD4$^+$ T cells was analysed by flow cytometry for the expression of CD25 and Foxp3, markers associated with commitment of T cells to the regulatory T cell lineage. (b) Rapamycin enhances the capacity of H1-DC to induce Treg cells compared to moDC. DCs were either untreated, matured with the maturation cocktail or treated with rapamycin for 3 days prior to maturation. DCs were harvested, washed, and plated at 2×10^5 per well with 10^6 naive CD4$^+$ T cells per well of a 24-well plate to yield a ratio of DC : T cells of 1 : 5. On day 7, cocultures were stained for CD4, CD25, and Foxp3 and analysed. Dead cells were excluded from the analysis using 7-AAD staining. Data from one experiment representative of 4 independent experiments are shown.

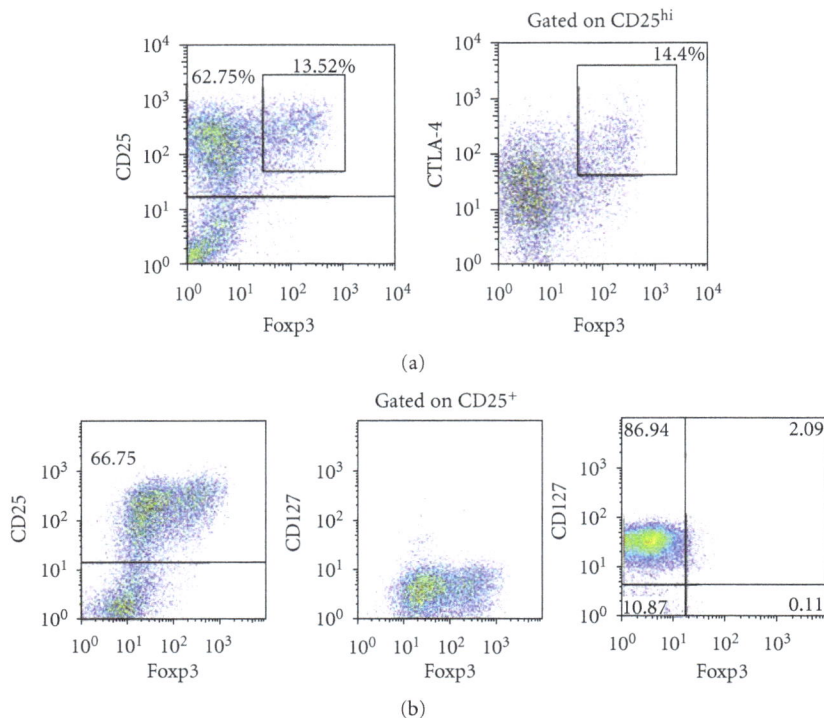

Figure 5: Phenotypic characterisation of putative CD25^{hi} Foxp3$^+$ Treg cells from cocultures of DC and naïve T cells. The CD25^{hi} Foxp3$^+$ population expresses CTLA-4 (a) but lacks expression of CD127 in comparison with control Foxp3$^-$ T cells (b), consistent with the reported phenotype of *bona fide* Treg cells.

the fulcrum on which the balance between tolerance and immunity has been shown to pivot. For instance, the administration to mice of CD40$^{-/-}$ DC laden with foreign antigen was shown to induce profound antigen-specific tolerance upon subsequent immunization, results which are consistent with the induction of a repertoire of Treg cells [26]. Furthermore, in mice receiving foreign antigen chemically conjugated to CD205-specific mAb as a way of delivering antigen to DC in the steady state, the induction of tolerance could be abrogated in favour of systemic immunity by the concomitant administration of agonistic antibodies specific for CD40 [4].

It is the central role played by CD40 in a transplantation setting that underlies the success of strategies for intervening in allograft rejection based on the blockade of CD40-CD154 interactions [27]. Although unanticipated complications associated with the use of mAb specific for CD154 have hindered the application of such a strategy to the clinic; the long-term acceptance of allografts in mice was found to bear the distinctive features of regulation, including linked suppression and infectious tolerance [28, 29]. A conditioning regime that limits the delivery of CD40 signalling by donor DC might, therefore, be anticipated to predispose recipients towards tolerance based on the generation of a repertoire of alloantigen-specific Treg cells. Indeed, the level of expression of CD40 has been shown to be critical in determining the outcome of antigen recognition in a model of *Leishmania donovani* infection, high levels of expression inducing effector T cells, low levels favouring polarisation towards a Treg phenotype [30]. Furthermore, blockade of the

CD40-CD154 axis in combination with rapamycin was shown to achieve tolerance even across a xenogeneic barrier [31]. Although the mechanisms involved were not specifically elucidated in this study, it may be significant in our own experiments that PD-L1 expression by H1-DC was unaffected by rapamycin treatment, being upregulated in response to maturation stimuli, irrespective of prior exposure to the compound. Given the essential role described for PD-L1 in polarisation of naïve T cells towards a Treg cell phenotype [23], it is tempting to speculate that it is by altering the critical balance between costimulatory and inhibitory signals delivered by H1-DC, that rapamycin treatment strongly favours a tolerogenic profile.

5. Conclusions

We have demonstrated previously that hESC may be differentiated into populations of immunogenic DC whose properties may be exploited in regimes of cancer immunotherapy. Here, we extend this paradigm by showing that our protocols are fully compliant with the use of rapamycin which favours a protolerogenic phenotype of the resulting DC. Our results pave the way for the future use of rapamycin-conditioned hESC-derived DC in regimes for the induction of tolerance, as a prelude to CRT.

Conflict of Interests

A. Reddy and K. P. Nishimoto declare a potential financial conflict of interests as employees of Geron Corporation.

Acknowledgments

The authors are grateful to Tim Davies, Naoki Ichiryu, and Simon Hackett for helpful discussions and to Lara Whitmore for administrative assistance. A. J. Leishman is the recipient of an MRC capacity building studentship, awarded to the Oxford Stem Cell Institute. This work was performed under a sponsored research agreement between the University of Oxford and Geron Corporation, with additional support from the Medical Research Council UK, in the form of Grant G0802538 awarded to P. J. Fairchild.

References

[1] I. Klimanskaya, N. Rosenthal, and R. Lanza, "Derive and conquer: sourcing and differentiating stem cells for therapeutic applications," *Nature Reviews Drug Discovery*, vol. 7, no. 2, pp. 131–142, 2008.

[2] P. J. Fairchild, N. J. Robertson, S. L. Minger, and H. Waldmann, "Embryonic stem cells: protecting pluripotency from alloreactivity," *Current Opinion in Immunology*, vol. 19, no. 5, pp. 596–602, 2007.

[3] R. M. Steinman and J. Banchereau, "Taking dendritic cells into medicine," *Nature*, vol. 449, no. 7161, pp. 419–426, 2007.

[4] D. Hawiger, K. Inaba, Y. Dorsett et al., "Dendritic cells induce peripheral T cell unresponsiveness under steady state conditions in vivo," *Journal of Experimental Medicine*, vol. 194, no. 6, pp. 769–779, 2001.

[5] S. F. Yates, A. M. Paterson, K. F. Nolan et al., "Induction of regulatory T cells and dominant tolerance by dendritic cells incapable of full activation," *Journal of Immunology*, vol. 179, no. 2, pp. 967–976, 2007.

[6] M. V. Dhodapkar, R. M. Steinman, J. Krasovsky, C. Munz, and N. Bhardwaj, "Antigen-specific inhibition of effector T cell function in humans after injection of immature dendritic cells," *Journal of Experimental Medicine*, vol. 193, no. 2, pp. 233–238, 2001.

[7] P. J. Fairchild, S. Cartland, K. F. Nolan, and H. Waldmann, "Embryonic stem cells and the challenge of transplantation tolerance," *Trends in Immunology*, vol. 25, no. 9, pp. 465–470, 2004.

[8] S. Senju, H. Suemori, H. Zembutsu et al., "Genetically manipulated human embryonic stem cell-derived dendritic cells with immune regulatory function," *Stem Cells*, vol. 25, no. 11, pp. 2720–2729, 2007.

[9] S. Hirata, S. Senju, H. Matsuyoshi, D. Fukuma, Y. Uemura, and Y. Nishimura, "Prevention of experimental autoimmune encephalomyelitis by transfer of embryonic stem cell-derived dendritic cells expressing myelin oligodendrocyte glycoprotein peptide along with TRAIL or programmed death-1 ligand 1," *Journal of Immunology*, vol. 174, no. 4, pp. 1888–1897, 2005.

[10] A. J. Leishman, K. M. Silk, and P. J. Fairchild, "Pharmacological manipulation of dendritic cells in the pursuit of transplantation tolerance," *Current Opinion in Organ Transplantation*, vol. 16, no. 4, pp. 372–378, 2011.

[11] H. R. Turnquist, G. Raimondi, A. F. Zahorchak, R. T. Fischer, Z. Wang, and A. W. Thomson, "Rapamycin-conditioned dendritic cells are poor stimulators of allogeneic CD4$^+$ T cells, but enrich for antigen-specific Foxp3$^+$ T regulatory cells and promote organ transplant tolerance," *Journal of Immunology*, vol. 178, no. 11, pp. 7018–7031, 2007.

[12] V. Sordi, G. Bianchi, C. Buracchi et al., "Differential effects of immunosuppressive drugs on chemokine receptor CCR7 in human monocyte-derived dendritic cells: selective upregulation by rapamycin," *Transplantation*, vol. 82, no. 6, pp. 826–834, 2006.

[13] W. Reichardt, C. Dürr, D. Von Elverfeldt et al., "Impact of mammalian target of rapamycin inhibition on lymphoid homing and tolerogenic function of nanoparticle-labeled dendritic cells following allogeneic hematopoietic cell transplantation," *Journal of Immunology*, vol. 181, no. 7, pp. 4770–4779, 2008.

[14] J. M. Sacks, Y. R. Kuo, A. Taieb et al., "Prolongation of composite tissue allograft survival by immature recipient dendritic cells pulsed with donor antigen and transient low-dose immunosuppression," *Plastic and Reconstructive Surgery*, vol. 121, no. 1, pp. 37–49, 2008.

[15] R. Ikeguchi, J. M. Sacks, J. V. Unadkat et al., "Long-term survival of limb allografts induced by pharmacologically conditioned, donor alloantigen-pulsed dendritic cells without maintenance immunosuppression," *Transplantation*, vol. 85, no. 2, pp. 237–246, 2008.

[16] T. Taner, H. Hackstein, Z. Wang, A. E. Morelli, and A. W. Thomson, "Rapamycin-treated, alloantigen-pulsed host dendritic cells induce Ag-specific T cell regulation and prolong graft survival," *American Journal of Transplantation*, vol. 5, no. 2, pp. 228–236, 2005.

[17] E. K. Horibe, J. Sacks, J. Unadkat et al., "Rapamycin-conditioned, alloantigen-pulsed dendritic cells promote indefinite survival of vascularized skin allografts in association with T regulatory cell expansion," *Transplant Immunology*, vol. 18, no. 4, pp. 307–318, 2008.

[18] M. Haidinger, M. Poglitsch, R. Geyeregger et al., "A versatile role of mammalian target of rapamycin in human dendritic cell function and differentiation," *Journal of Immunology*, vol. 185, no. 7, pp. 3919–3931, 2010.

[19] S. Y. Tseng, K. P. Nishimoto, K. M. Silk et al., "Generation of immunogenic dendritic cells from human embryonic stem cells without serum and feeder cells," *Regenerative Medicine*, vol. 4, no. 4, pp. 513–526, 2009.

[20] C. Xu, M. S. Inokuma, J. Denham et al., "Feeder-free growth of undifferentiated human embryonic stem cells," *Nature Biotechnology*, vol. 19, no. 10, pp. 971–974, 2001.

[21] Y. Li, S. Powell, E. Brunette, J. Lebkowski, and R. Mandalam, "Expansion of human embryonic stem cells in defined serum-free medium devoid of animal-derived products," *Biotechnology and Bioengineering*, vol. 91, no. 6, pp. 688–698, 2005.

[22] A. M. Woltman, J. W. De Fijter, S. W. A. Kamerling et al., "Rapamycin induces apoptosis in monocyte- and CD34-derived dendritic cells but not in monocytes and macrophages," *Blood*, vol. 98, no. 1, pp. 174–180, 2001.

[23] L. Wang, K. Pino-Lagos, V. C. De Vries, I. Guleria, M. H. Sayegh, and R. J. Noelle, "Programmed death 1 ligand signaling regulates the generation of adaptive Foxp3$^+$CD4$^+$ regulatory T cells," *Proceedings of the National Academy of Sciences of the United States of America*, vol. 105, no. 27, pp. 9331–9336, 2008.

[24] N. Seddiki, B. Santner-Nanan, J. Martinson et al., "Expression of interleukin (IL)-2 and IL-7 receptors discriminates between human regulatory and activated T cells," *Journal of Experimental Medicine*, vol. 203, no. 7, pp. 1693–1700, 2006.

[25] W. Liu, A. L. Putnam, Z. Xu-yu et al., "CD127 expression inversely correlates with FoxP3 and suppressive function of human CD4$^+$ T reg cells," *Journal of Experimental Medicine*, vol. 203, no. 7, pp. 1701–1711, 2006.

[26] K. Hochweller and S. M. Anderton, "Systemic administration of antigen-loaded CD40-deficient dendritic cells mimics

soluble antigen administration," *European Journal of Immu-nology*, vol. 34, no. 4, pp. 990–998, 2004.

[27] C. P. Larsen, E. T. Elwood, D. Z. Alexander et al., "Long-term acceptance of skin and cardiac allografts after blocking CD40 and CD28 pathways," *Nature*, vol. 381, no. 6581, pp. 434–438, 1996.

[28] K. Honey, S. P. Cobbold, and H. Waldmann, "CD40 ligand blockade induces CD4+ T cell tolerance and linked suppression," *Journal of Immunology*, vol. 163, no. 9, pp. 4805–4810, 1999.

[29] L. Graca, K. Honey, E. Adams, S. P. Cobbold, and H. Waldmann, "Cutting edge: anti-CD154 therapeutic antibodies induce infectious transplantation tolerance," *Journal of Immunology*, vol. 165, no. 9, pp. 4783–4786, 2000.

[30] S. Martin, R. Agarwal, G. Murugaiyan, and B. Saha, "CD40 expression levels modulate regulatory T cells in Leishmania donovani infection," *Journal of Immunology*, vol. 185, no. 1, pp. 551–559, 2010.

[31] Y. D. Muller, G. Mai, P. Morel et al., "Anti-CD154 mAB and rapamycin induce T regulatory cell mediated tolerance in rat-to-mouse islet transplantation," *PLoS One*, vol. 5, no. 4, Article ID e10352, 2010.

16

Roles of Organic Acid Anion Secretion in Aluminium Tolerance of Higher Plants

Lin-Tong Yang,[1,2] **Yi-Ping Qi,**[3] **Huan-Xin Jiang,**[2,4] **and Li-Song Chen**[1,2,5]

[1] *Department of Agricultural Resources and Environmental Sciences, College of Resources and Environmental Sciences, Fujian Agriculture and Forestry University, Fuzhou 350002, China*
[2] *Institute of Horticultural Plant Physiology, Biochemistry, and Molecular Biology, Fujian Agriculture and Forestry University, Fuzhou 350002, China*
[3] *Institute of Materia Medica, Fujian Academy of Medical Sciences, Fuzhou 350001, China*
[4] *Department of Life Sciences, College of Life Sciences, Fujian Agriculture and Forestry University, Fuzhou 350002, China*
[5] *Department of Horticulture, College of Horticulture, Fujian Agriculture and Forestry University, Fuzhou 350002, China*

Correspondence should be addressed to Li-Song Chen; lisongchen2002@hotmail.com

Academic Editor: Margarita Rodríguez Kessler

Approximately 30% of the world's total land area and over 50% of the world's potential arable lands are acidic. Furthermore, the acidity of the soils is gradually increasing as a result of the environmental problems including some farming practices and acid rain. At mildly acidic or neutral soils, aluminium (Al) occurs primarily as insoluble deposits and is essentially biologically inactive. However, in many acidic soils throughout the tropics and subtropics, Al toxicity is a major factor limiting crop productivity. The Al-induced secretion of organic acid (OA) anions, mainly citrate, oxalate, and malate, from roots is the best documented mechanism of Al tolerance in higher plants. Increasing evidence shows that the Al-induced secretion of OA anions may be related to the following several factors, including (a) anion channels or transporters, (b) internal concentrations of OA anions in plant tissues, (d) temperature, (e) root plasma membrane (PM) H$^+$-ATPase, (f) magnesium (Mg), and (e) phosphorus (P). Genetically modified plants and cells with higher Al tolerance by overexpressing genes for the secretion and the biosynthesis of OA anions have been obtained. In addition, some aspects needed to be further studied are also discussed.

1. Introduction

Approximately 30% of the world's total land area and over 50% of the world's potential arable lands are acidic [1]. Furthermore, the acidity of the soils is gradually increasing as a result of the environmental problems including some farming practices and acid rain. Soil pH decreased significantly from the 1980s to the 2000s in the major Chinese crop-production areas [2]. Aluminium (Al) is the most abundant metal and the third most abundant element in the earth's crust after oxygen (O) and silicon (Si), comprising approximately 7% of its mass [3]. At mildly acidic or neutral soils, it occurs primarily as insoluble deposits and is essentially biologically inactive. In acidic solutions (pH < 5.0), Al becomes soluble and available to plants in the Al^{3+} and Al(OH)$^{2+}$ forms [4]. Micromolar concentration of Al^{3+} can rapidly inhibit root growth. The subsequent impairments on water and nutrient uptake lead to poor growth and productivity [5]. Therefore, in many acidic soils throughout the tropics and subtropics, Al toxicity is a major factor limiting crop productivity [6]. Many plants have evolved different mechanisms for detoxifying Al externally, including secretion of Al-chelating substances (e.g., organic acid (OA) anions, phosphate (Pi), and phenolic compounds) from the roots, increased pH in the rhizosphere, modified cell wall, redistribution of Al, and efflux of Al [6–9]. Increasing evidence shows that the Al-induced secretion of OA anions from roots is a major mechanism leading to Al tolerance in higher plants [6, 9–15]. In this paper, we review the roles of the Al-induced secretion of OA anions from roots in Al tolerance of higher plants.

2. Aluminium-Induced Secretion of Organic Acid Anions from Roots

The Al-induced secretion of OA anions, mainly citrate, oxalate, and malate, from roots is the best documented mechanism of Al tolerance in higher plants. Plants differ in the species of OA anions secreted, secretion patterns, temperature sensitivity, response to inhibitors, and dose response to Al (see [9, 13], Table 1). Since the first report on the Al-induced secretion of malate from wheat (*Triticum aestivum*) roots [16], increasing evidence shows that many Al-tolerant species or cultivars are able to secrete high levels of citrate, malate, and/or oxalate from roots when exposed to Al, including barley (*Hordeum vulgare*) [17], maize (*Zea mays*) [18], buckwheat (*Fagopyrum esculentum*) [19, 20], rye (*Secale cereale*) [21], soybean (*Glycine max*) [22, 23], *Citrus junos* [24], sorghum (*Sorghum bicolor*) [25], triticale(× *Triticosecale* Wittmark) [26], *Polygonum* spp. [27], *Paraserianthes falcataria* [28], *Lespedeza bicolor* [29], *Citrus grandis*, and *Citrus sinensis* [30, 31]. All these OA anions (citrate, oxalate, and malate) secreted from plant roots can form stable, nontoxic complexes with Al in the rhizosphere, thereby preventing the binding of Al to cellular components, resulting in detoxification of Al [9, 12]. Of the three OA anions, citrate has the highest chelating activity for Al followed by oxalate and malate [9]. The Al-induced secretion of OA anions is localized to the root apex, which is in agreement with the targeting site for Al toxicity [20, 32, 33] and their secretion is highly specific to Al, neither phosphorus (P) deficiency nor other polyvalent cations result in the secretion of OA anions [20, 30, 34–38]. Based on the timing of secretion, two patterns of Al-induced OA anion secretion have been proposed [10, 11]. In Pattern I plants, no discernible delay is observed between the addition of Al and the onset of OA anion secretion such as buckwheat [39], tobacco (*Nicotiana tabacum*) [40], and wheat [37]. In this case, Al may simply activate a transporter in the plasma membrane (PM) to initiate OA anion secretion, and the induction of genes is not required [9, 11]. In Pattern II plants, OA anion secretion is delayed for several hours after exposure to Al such as in rye [22], *Cassia tora* [39], *C. junos* [24], soybean [41], *L. bicolor* [29], and triticale [26]. In this case, Al may induce the expression of genes and the synthesis of proteins involved in OA metabolism or in the transport of OA anions [12]. Yang et al. investigated the effects of a protein-synthesis inhibitor (cycloheximide, CHM) on the Al-induced secretion of OA anions from the roots of buckwheat, a typical Pattern I plant, and *C. tora*, a typical Pattern II plants, suggesting that both *de novo* synthesis and activation of an anion channel are needed for the Al-activated secretion of citrate in *C. tora*, but in buckwheat the PM protein responsible for oxalate secretion preexisted [39]. Although the Al-induced secretion of OA anions has been well documented, there is a lack of correlation between OA anion secretion and Al tolerance in some plant species. For example, the Al-induced secretion of OA anions (citrate and oxalate) cannot account for the genotypic differences in Al tolerance in maize, soybean, and buckwheat cultivars [42–44]. Wenzl et al. observed that the secretion of OA anions from Al-treated signalgrass

(*Brachiaria decumbens*) apices was three- to 30-times smaller than that from Al-treated apices of buckwheat, maize, and wheat (all much more sensitive to Al than signalgrass) [45]. Ishikawa et al. investigated the amount of malate and citrate in Al media of seven plant species (Al tolerance order: *Brachiaria brizantha*, rice (*Oryza sativa*), and tea (*Camellia sinensis*) > maize > pea (*Pisum sativum*) and *C. tora* > barley) and of two cultivars with differential Al tolerance each in five plant species (rice, maize, wheat, pea, and sorghum). They did not observe any correlation of Al tolerance among some plant species or between two cultivars in some plant species with the amount of citrate and malate in Al media [46]. Yang et al. showed that eight oxalate accumulator cultivars from four species including *Amaranthus* spp., buckwheat, spinach (*Spinacia oleracea*), and tomato (*Lycopersicon esculentum*) secreted oxalate rapidly under Al stress, but oxalate secretion was not related to their Al tolerance [47]. Therefore, it is reasonable to assume that some plant species may contain other (stronger) mechanisms, which mask the effect of OA anions and/or that the Al-induced secretion of OA anions is too low to be an effective mechanism [44, 48–50]. In this section, we will discuss several aspects that have been implicated in the regulation of the Al-induced OA anion secretion.

2.1. Anion Channels or Transporters. From the experiments with anion channel and carrier inhibitors, the Al-activated secretion of OA anions is mediated through anion channels and/or carriers [9, 20, 37, 69]. As early as 1995, Ryan et al. observed that inhibitors of anion channels inhibited the Al-activated secretion of malate from wheat roots, providing evidence that Al might activate malate secretion *via* a channel in the PM in the apical cells of Al-tolerant wheat cells [37]. Increasing evidence shows that the influence of anion channel inhibitors on the Al-activated secretion of OA anions depends on the species of OA anions secreted, plant species, inhibitor concentration, and species (see [13, 30, 37, 62], Table 1). Li et al. observed that two citrate carrier inhibitors (pyridoxal $5'$-phosphate (PP) and phenylisothiocyanate (PITC)) effectively inhibited citrate secretion, meaning that the Al-activated citrate from rye roots is mediated by citrate carrier [21]. Yang et al. [36] and Li et al. [63] showed that the Al-activated secretion of citrate from rice bean (*Vigna umbellata*) and *Stylosanthes* spp. roots was inhibited by both anion channel and carrier inhibitors, indicating the possible involvement of both the citrate carrier and anion channel in the Al-activated citrate secretion. Although the use of inhibitors can be indicative of the type of transport protein involved in OA anion secretion, they do not provide definitive evidence because most inhibitors will eventually affect transport processes that can happen nonspecifically depending on the concentration and period of application. The use of patch clamp technique, which directly measures the transport activity, provides a much stronger evidence that anion channels are involved in the secretion of OA anions from roots under Al stress [69–72]. To date, two families of membrane transporters, the Al-activated malate transporter (ALMT) and the multidrug

TABLE 1: Characteristics of the aluminum- (Al-) induced secretion of organic acid (OA) anions from roots of different plant species.

Plant species	OA anions secreted	Secretion pattern	Dose response	Temperature sensitivity	Effective inhibitors	References
Acacia mangium	Citrate	NA	NA	NA	NA	[28]
Acacia auriculiformis	Oxalate, citrate	NA	P	NA	NA	[51]
Arabidopsis thaliana	Citrate, malate	NA	NA	NA	NA	[52, 53]
Barley (Hordeum vulgare)	Citrate	I	A	P	NIF, A9C	[17]
Buckwheat (Fagopyrum esculentum)	Oxalate	I	P	NA	PG	[19, 20, 39, 54]
Cassia tora	Citrate	II	P	NA	CHM	[34, 39, 46]
Citrus grandis and Citrus sinensis	Citrate, malate	I	P	P	CHM (malate), DIDS(1), A9C	[30, 31]
Citrus junos	Citrate	II	P	NA	NA	[24]
Deschampsia flexuosa	Malate	NA	NA	NA	NA	[55]
Eucalyptus camaldulensis	Oxalate, citrate	NA	P	NA	NA	[51]
Galium saxatile	Citrate	NA	P	NA	NA	[55]
Lespedeza bicolor	Citrate, malate	II	P (malate), A (citrate)	NA	A9C, CHM	[29]
Leucaena leucocephala	Citrate	NA	NA	NA	NA	[28]
Maize (Zea mays)	Citrate, malate	NA	P	NA	NIF, DIDS(2)	[18, 56]
	Oxalate	NA	A	NA	NA	[8]
Melaleuca cajuputi	Oxalate, citrate	NA	P	NA	NA	[51]
Melaleuca leucadendra	Oxalate, citrate	NA	P	NA	NA	[51]
Oat (Avena sativa)	Citrate, malate	NA	NA	NA	NA	[19]
Oryza glaberrima	Citrate	NA	NA	NA	NA	[46]
Paraserianthes falcataria	Citrate	NA	NA	NA	NA	[28]
Pea (Pisum sativum)	Citrate	NA	NA	NA	NA	[46]
Polygonum aviculare and Polygonum lapathifolium	Oxalate	I	NA	NA	PG	[35]
Poplar (Populus tremula)	Oxalate, citrate	NA	P	NA	NA	[57]
Radish (Raphanus sativus)	Citrate, malate	NA	NA	NA	NA	[19]
Rape (Brassica napus)	Citrate, malate	NA	NA	NA	PG	[19, 58, 59]
Rice (Oryza sativa)	Citrate	NA	NA	NA	NA	[46]
Rice bean (Viga umbellata)	Citrate	II	NA	NA	A9C, NIF, MA, PITC, CHM	[36]
Rye (Secale cereale)	Citrate, malate	II	P	P	PP, PITC	[21]
Rye	Citrate, malate	I (malate), II (citrate)	P	P (citrate), A (malate)	NA	[60]
Rumex acetosella	Oxalate	NA	P	NA	NA	[55]
Snapbena (Phaseolus vulgaris)	Citrate	NA	NA	NA	NA	[61]
Soybean (Glycine max)	Citrate, malate	II	P	NA	NA	[22, 23, 41]
Soybean	Citrate	II	P	NA	A9C, CHM, MA	[62]
Sorghum (Sorghum bicolor)	Citrate	NA	NA	NA	NA	[53]

TABLE 1: Continued.

Plant species	OA anions secreted	Secretion pattern	Dose response	Temperature sensitivity	Effective inhibitors	References
Spinach (*Spinacia oleracea*)	Oxalate	I	P	NA	NA	[35]
Stylosanthes spp.	Citrate	II	P	NA	A9C, PITC, PG, NIF, DIDS(1), CHM	[63]
Sunflower (*Helianthus annuus*)	Citrate, malate	NA	NA	NA	NA	[64]
Taro (*Colocasia esculenta*)	Oxalate	NA	P	NA	NA	[65]
Tobacco	Citrate	I	P	NA	NA	[40]
Tomato (*Lycopersicon esculentum*)	Oxalate	I	A	NA	PG	[66]
Triticale (×*Triticosecale* Wittmack)	Citrate, malate	II	P	NA	NA	[26]
Veronica officinalis	Citrate	NA	P	NA	NA	[55]
Viscaria vulgaris	Oxalate	NA	P	NA	NA	[55]
Wheat (*Triticum aestivum*)	Malate	I	P	A	NIF, DPC, EA, A9C, NPPB, IAA-94	[21, 37, 67, 68]

A: absent; A9C: anthracene-9-carboxylic acid; CHM: cycloheximide; DIDS(1): 4,4′-diisothiocyanatostilbene -2,2′-disulfonic acid; DIDS(2): 4,4′-dinitrostilbene-2,2′-disulfonic acid; DPC: diphenylamine-2-carboxylic acid; EA: ethacrynic acid; IAA-94, (6,7-dichloro-2-cyclopentyl-2, 3-dihydro-2-methyl-oxo-1H-inden-5-yloxy) acetic acid; NA: not applicable; MA: mersalyl acid; NIF: niflumic acid; NPPB: 5-nitro-2-(3-phenylpropylamino)-benzoic acid; P: present; PG: phenylglyoxal; PITC: phenylisothiocyanate; PP: pyridoxal 5′-phosphate; anion channel inhibitors: A9C, NIF, PG; Citrate carrier inhibitors: MA: PITC, PP; protein synthesis inhibitor: CHM. Two patterns of Al-induced OA anion secretion can be identified on the basis of the timing of secretion. In Pattern I plants, no discernible delay is observed between the addition of Al and the onset of OA anion secretion. In Pattern II plants, OA anion secretion is delayed for several hours after exposure to Al.

and toxin compounds extrusion (MATE) families, have been implicated in the secretion of OA anions from plant roots in response to Al. In 2004, Sasaki et al. first isolated the Al-activated OA anion secretion transporter from wheat (i.e., *Al-activated malate transporter 1, TaALMT1*) [73]. Electrophysiological studies show that TaALMT1 functions as a ligand-activated and voltage-dependent anion channel to facilitate malate secretion across the PM of root cells [67, 74, 75]. Following the cloning of the first Al-activated OA anion secretion transporter, *TaALMT1* homologs have been cloned from rape (*Brassica napus; BnALMT1* and *BnALMT2*) [76], *Arabidopsis thaliana* (*AtMALMT1*) [77], and rye (*ScALMT1*) [78]. Osawa and Matsumoto proposed that protein phosphorylation was associated with the Al-activated malate secretion from wheat root apex and that the OA anion-specific channel was possibly a terminal target that responded to Al signal mediated by phosphorylation [79]. Kobayashi et al. observed that the activation of AtALMT1 by Al was inhibited by staurosporine (kinase inhibitor) and calyculin A (phosphatase inhibitor), and that K252a (serine/threonine protein kinase inhibitor) inhibited the Al-dependent malate secretion without reducing gene expression [38]. Ligaba et al. provided evidence indicating that TaALMT1 activity was regulated by protein kinase C-mediated phosphorylation. They observed that TaALMT1 activity was disrupted when the serine residue at position 384 was replaced with an alanine, and concluded that the serine residue needed to be phosphorylated before TaALMT1 was activated by Al. These results suggest that the activation of ALMT1 by Al may involve reversible protein phosphorylation [80]. However, not all ALMT1-type transporters mediate Al-activated OA responses. For example, *ZmALMT1* isolated from maize was suggested to play a role in anion homeostasis and mineral nutrition, and the activity of this protein was independent of extracellular Al [81]. *ZmALMT2* isolated from maize is a root anion transporter that mediates constitutive root malate secretion and may play a role in mineral nutrient acquisition and transport but not Al tolerance [82]. Three ALMT-type transporters isolated from *A. thaliana* were expressed in leaf mesophyll (*AtALMT9*) or guard cells (*AtALMT6* and *AtALMT12*), implicating a primary role in malate homeostasis and guard cell function [83–85]. Recently, Gruber et al. demonstrated that HvALMT1 from barley likely contributed to the homeostasis of OA anions in the cytosol of guard cells and root cells by transporting them out of the cell or into cytosolic vesicles [86, 87].

In 2007, genes encoding citrate transporters that are members of a different transporter family, the MATE family, were isolated from barley (*HvAACT1*, also designated as *HvMATE1*) [88, 89] and sorghum (*SbMATE*) [53]. The MATE family of transporter proteins is a large and diverse group present widely in bacteria, fungi, plants, and mammals. Evidence shows that MATE proteins function as H^+ or Na^+ coupled antiporters for numerous substances such as flavonoid, anthocyanins, norfloxacin, ethidium bromide, berberine, acriflavine, nicotine, citrate, and Cd^{2+} [88, 90–92]. Plant MATEs can transport substrates other than citrate, which may also play a role in Al tolerance [88, 90–94].

Recently, MATE homologs involved in Al-activated citrate secretion were isolated from *A. thaliana* (*AtMATE*) [95], rye (*ScFRDL2*) [92], maize (*ZmMATE1*) [96], rice (*OsFRDL4*) [97], and rice bean (*VuMATE*) [98]. All these citrate transporters exhibit varying degrees of constitutive expression (i.e., in the absence of Al) except for *VuMATE*, and their expressions are upregulated by Al treatment except for *HvMATE*. Evidence shows that *OsFRDL4* isolated from rice and *AtMATE* isolated from *A. thaliana* are regulated by a C2H2-type zinc finger transcription factor ART1 and STOP1, respectively [95, 97].

In buckwheat, evidence shows that ABA is involved in the secretion of oxalate [99]. ABA activates the anion channel in stomatal guard cells and may play a similar role in the roots [9]. However, no oxalate transporter has been isolated from plants so far.

2.2. Internal Concentrations of Organic Acid Anions in Plant Tissues. The effects of Al on OA metabolism have been investigated in some plant species. In an Al tolerant maize single cross, exposure to increasing level of Al led to a strong (over 3-fold) increase in root tip citrate concentration and a significant activation of citrate secretion, which saturated at a rate close to 0.5 nmol citrate h^{-1} $root^{-1}$ occurring at $80\,\mu M$ Al^{3+} activity, with the half-maximal rate of citrate secretion occurring at about $20\,\mu M$ Al^{3+} activity [72]. Ligaba et al. demonstrated that Al-treated rape roots had increased *in vitro* activities of citrate synthase (CS, EC 4.1.3.7), malate dehydrogenase (MDH, EC 1.1.1.37) and phospho*enol*pyruvate carboxylase (PEPC, EC 4.1.1.31), and concentrations of citrate and malate, together with decreased respiration rate, concluding that the Al-induced accumulation and subsequent secretion of citrate and malate were associated with both increased biosynthesis and reduced catabolism [59]. In an Al tolerant tree species, *P. facataria*, the Al-induced increases in both secretion and accumulation of citrate were accompanied by increased mitochondrial CS (mCS) activity and enhanced *mCS* expression, indicating that the increased amount of citrate is produced in response to Al [28]. Aluminium treatments resulted in an increase in CS activity and a decrease in aconitase (ACO, EC 4.2.1.3) activity in the root tips of *C. tora*, accompanied by an increase in citrate concentration. However, the activities of NADP-isocitrate dehydrogenase (NADP-IDH, EC 1.1.1.42), MDH, and PEPC were unaffected by Al. It was suggested that Al-regulation of both CS and ACO activities might be responsible for the Al-induced increase in both secretion and accumulation of citrate [100]. Yang et al. reported that CS activity in soybean root apex was increased by 16% when exposed to Al, but the activities of PEPC and NADP-IDH and the concentration of citrate were unaffected. They suggested that the Al-induced increase in CS activity resulted in the increased secretion of citrate [41]. The unchanged concentration of citrate in the Al-treated roots likely reflected the balance between citrate synthesis and secretion in the root apex. In rye, the activity of CS in the root tip increased by 30% when exposed to Al, and the Al-induced increase in the synthesis of citrate appeared to be responsible for the enhanced secretion of citrate from roots [21]. In an Al-tolerant soybean cultivar, mitochondrial MDH and CS activities increased and ACO activity decreased with the increasing of Al concentration and duration of Al treatment. The Al-induced citrate secretion was inhibited by the CS inhibitor suramin and enhanced by the ACO inhibitor fluorocitric acid. Transcript level of the mitochondrial CS increased in soybean roots in response to Al, whereas the expression of ACO showed no significant difference. These results indicate that Al triggers OA metabolic responses in mitochondria of soybean roots, which support the sustained secretion of citrate [101]. Based on the above results, it is reasonable to believe that altered OA metabolism is involved in the Al-induced secretion of OA anions. However, it is not immediately obvious that simply increasing internal OA will lead to increased secretion, because in any case some transport processes must somewhat be involved in the Al-induced secretion of OA anions. For example, Gaume et al. showed that the Al-tolerant maize cultivar had higher root concentrations and higher root secretion of citrate, malate, and succinate compared with the Al-sensitive one. Increased PEPC activity in root apices after Al exposure partially explained the differences of OA anion concentrations in the roots. However, the increased secretion was not proportional to the OA anion concentrations in the roots. The concentrations of citrate, malate, and succinate in the roots of both cultivars increased by a factor of 2 to 4, whereas the secretion of these anions increased by 2 to 20. They suggested that the secretion of OA anions might be mediated by transporters that are either activated or induced by Al [102]. In a study with two lines of triticale differing in the Al-induced secretion of malate and citrate and in Al tolerance, the concentrations of citrate (root apices and mature root segments) and malate (mature segments only) in roots increased in response to Al, but similar changes were observed in the two lines. The Al-induced changes in *in vitro* activities of CS, PEPC, NAD-MDH, and NADP-IDH were similar in the sensitive and resistant lines in both root apices and mature root segments. These results suggest that the Al-induced secretion of malate and citrate from triticale roots is not regulated by their internal levels in the roots or by the capacity of root tissues to synthesize them [103]. Yang et al. reported that the root concentration of oxalate was poorly related with its secretion among some oxalate accumulators such as *Amaranth* spp., buckwheat, spinach, and tomato [47]. Recently, we observed that Al decreased or did not affect the concentrations of malate and citrate in roots of two citrus species having different tolerance to Al, indicating that the Al-induced secretion of citrate and malate is poorly related to their internal levels in roots [30, 31, 104–106].

Transgenic plants and cells have provided additional evidence that OA metabolism can contribute to the Al-induced secretion of OA anions and Al tolerance [6, 107]. Modulation of OA metabolism enhanced Al tolerance and secretion of citrate and/or malate in transgenic tobacco and papaya (*Carica papaya*) plants overexpressing a *Pseudomonas aeruginosa* CS [108], rape plants [58], and carrot (*Daucus carota*) cells [109] overexpressing a mitochondrial CS (*mCS*) from *A. thaliana*, *Nicotiana benthamiana* plants overexpressing an *mCS* from *C. junos* [24], tobacco plants overexpressing a cytosolic *MDH*

from *A. thaliana*, an *MDH* from *Escherichia coli* [110] and a *CS* from rice mitochondria [111], alfalfa (*Medicago sativa*) plants overexpressing a alfalfa nodule-enhanced form of *MDH* (*neMDH*) [112], and tobacco plants overexpressing *pyruvate phosphate dikinase* (*PPDK*, EC 2.7.9.1) gene from *Mesembryanthemum crystallinum* [113]. However, Delhaize et al. showed that the expression of a *P. aeruginosa* CS in tobacco did not result in enhanced citrate accumulation or secretion, despite generating transgenic tobacco lines that expressed the CS protein at up to a 100-fold greater level than the previously described CSb lines [40, 108]. They concluded that the activity of the *P. aeruginosa* CS in transgenic tobacco is either sensitive to environmental conditions or that the improvements in Al tolerance and P nutrition observed previously are due to some other variable [40].

In an Al-tolerant soybean cultivar, the Al-induced root secretion of citrate increased steadily when exposed to continuous light, and only low citrate secretion was observed under 24 h in the continuous dark. The rate of Al-induced citrate secretion decreased at 6 h after the shoots were excised. The rate of citrate secretion by shoot-excised roots was 3-times lower than that of their respective controls (Al treatment in plants with shoots) during the 6–9 h after 50 μM treatment and 6-times lower during the 9–12 h. These results indicate that the shoots play a role in the Al-induced citrate secretion through providing the carbon source and/or energy for citrate synthesis in the roots [41]. Neumann and Römheld reported that P deficiency strongly increased the concentrations of carboxylic acids in chickpea (*Cicer arietinum*) and white lupin (*Lupinus albus*) roots, but only had small effects on the accumulation of carboxylates in shoots, and suggested that the ability to accumulate carboxylic acids in roots depended on the partitioning of carboxylic acids or related precursors between roots and shoots [114]. Quantification of soybean root enzymes involved in OA metabolism displayed only a 16% increase in CS activity 6 h after Al treatment with no differences in other enzymes; hence citrate may be transported from the shoots to the roots [41].

2.3. Temperature. Yang et al. observed that the Al-induced secretion of citrate and malate by the roots of *C. grandis* and *C. sinensis* seedlings was inhibited by low temperature, indicating that an energy dependent process may be involved in the Al-induced secretion of OA anions [30]. A similar result has been obtained in Al-tolerant barley [17]. However, the Al-induced secretion of citrate in rye (Pattern II) was decreased by low temperature, but the Al-induced secretion of malate in wheat (Pattern I) was unaffected by low temperature [21]. Recently, Li et al. reported that, in rye, the Al-induced secretion of malate belonged to Pattern I and was not inhibited, while the Al-induced secretion of citrate belonged to Pattern II and was affected by low temperature [60]. Further research is needed to elucidate the mechanism.

2.4. Root Plasma Membrane H^+-ATPase. Since PM H^+-ATPase plays a critical role in energizing and regulating an array of secondary transporters [115, 116], the modulation of PM H$^+$-ATPase activity may be involved in the Al-induced secretion of OA anions. In two soybean cultivars, the Al-induced activity of root PM H$^+$-ATPase paralleled the secretion of citrate. The Al-induced increase in PM H$^+$-ATPase activity was caused by a transcriptional and translational regulation. Both activity and expression of root PM H$^+$-ATPase were higher in the Al-tolerant than in the Al-sensitive cultivar. Aluminum activated the threonine-oriented phosphorylation of PM H$^+$-ATPase in a dose- and time-dependent manner. The relationship between the Al-induced secretion of citrate and the activity of PM H$^+$-ATPase was further demonstrated by an analysis of PM H$^+$-ATPase transgenic *A. thaliana*. When grown on Murashige and Skoog medium containing 30 μM Al, transgenic plants of *A. thaliana* overpressing PM H$^+$-ATPase secreted more citrate compared with wild-type *A. thaliana* [117]. Ahn et al. showed that after 4 h *in vivo* treatment with 2.6 μM Al, PM H$^+$-ATPase activity and H$^+$-transport rate were decreased and ζ potential was depolarized in PM vesicles from root tips of Al-sensitive wheat cultivar (ES8) but not of Al-tolerant ET8. They concluded that the Al-induced secretion of malate from wheat roots was accompanied by changes in PM surface potential and activation of H$^+$-ATPase [118]. However, the Al-induced changes of root PM H$^+$-ATPase activity were not associated with oxalate secretion in two tomato cultivars differing in the ability to secrete oxalate under Al stress [66]. Other studies showed Al inhibited root PM H$^+$-ATPase activity in barley [119], squash (*Cucurbita pepo*) [120], and rice bean [121].

2.5. Magnesium. Magnesium (Mg) can ameliorate Al toxicity, but the mechanism by which Al alleviates it remains obscure [122, 123]. Long-term secretion of OA anions requires continuous biosynthesis of OAs inside the root cells. In this regard, cytoplasmic Mg^{2+} is pivotal for the activation of many enzymes (e.g., CS, PEPC, IDH, malic enzyme (ME, EC 1.1.1.40), and MDH) involved in OA biosynthesis and degradation [122]. In soybean, micromolar concentration of Mg in the treatment solution alleviated Al toxicity by enhancing citrate biosynthesis and secretion by roots. Increased production and secretion by soybean roots in response to Mg might promote both external and internal detoxification by formation of Al-citrate complexes [123]. In rice bean, Mg could stimulate the Al-induced secretion of citrate from roots thus alleviating the inhibition of root growth by Al. The stimulation of citrate secretion by Mg might result from the restoration of root PM H$^+$-ATPase activity by Mg [121].

2.6. Phosphorus. Phosphorus deficiency is another major factor limiting plant growth in acidic soils [6]. Evidence has shown that Al toxicity can be alleviated by P supply in some plants, including *C. grandis* [30, 31], sorghum [124], maize [102], and *L. bicolor* [125]. There are several authors investigating the effects of P on the Al-induced secretion of OA anions from roots, but the results are somewhat different. Yang et al. showed that the Al-induced secretion of citrate and malate by excised roots from Al-treated *C. grandis* and

C. sinensis seedlings decreased with increasing P supply, whereas P supply increased or had no effect on the concentrations of both citrate and malate in Al-treated roots [30, 31]. The decreased secretion of OA anions due to P application can be due to the amelioration of Al toxicity by P rather than due to decreased root accumulation of OA anions. In two maize cultivars, the Al-induced increases in root activity of PEPC, root concentrations, and secretion of OA anions were decreased in plants pretreated with higher P concentrations during the 21 days prior to Al treatment [102]. In two cowpea genotypes of contrasting Al tolerance, Al enhanced malate secretion from root apices of both genotypes. Phosphorus deficiency increased the Al-induced secretion of malate by roots only in the Al-tolerant genotype IT89KD-391 [126]. In an Al-tolerant leguminous shrub, *L. bicolor*, the Al-induced secretion of citrate and malate under P sufficiency was less than that under P deficiency [125]. The above results indicate that the enhancement of Al tolerance by P is not associated with an increased secretion of OA anions from roots. However, P-sufficient rape plants displayed more pronounced Al-induced accumulation and secretion of citrate and malate in roots than P-deficient plants. Interestingly, the degree of inhibition of Al-induced root elongation was more or less the same in both P-sufficient and P-deficient plants. It was suggested that the severity of Al toxicity in P-deficient plants was masked by the stimulating effect of P deficiency on root elongation [59]. Using four soybean genotypes differing in P efficiency, Liao et al. investigated the effects of Al and P interactions on OA anion secretion by roots grown in homogeneous and heterogeneous nutrient solutions. In the homogenous solution experiments, P enhanced Al tolerance in four soybean genotypes, but greatly decreased the Al-induced citrate and malate secretion by roots. The two P-efficient genotypes displayed more Al tolerance than the two P-inefficient genotypes under high-P condition, but no significant genotypic difference was found in the secretion of OA anions under both low- and high-P conditions. The secretion of OA anions in a homogenous solution may not reflect the ability of soybean plants to detoxify exogenous Al. At the early stages of the heterogeneous nutrient solution experiment, P greatly increased the rates of the Al-activated citrate and malate secretion from the taproot tips of the four genotypes and the Al tolerance for the two P-efficient genotypes, and the two P-efficient genotypes secreted more malate from the taproot apices under high-P condition. They concluded that, at the early stage of heterogeneous nutrient solution experiment, P might increase the Al-activated secretion of OA anions, thus enhancing Al tolerance [23]. In two soybean cultivars and one rye cultivar, P deficiency did not increase the Al-induced secretion of citrate by roots [60, 127]. In soybean, short-term P deficiency (4 days) followed by Al treatment led to 50% increase in the Al-induced citrate secretion, while longer-term (10 days) P deficiency followed by Al treatment reduced the Al-induced citrate secretion to trace amounts [128]. However, in another study with soybean, Yang et al. showed that application of Al induced a greater citrate secretion rate in the Al-tolerant cultivar than in the Al-sensitive cultivar independently of the P status of the plants [22]. Dong et al. showed that long-term (14 days)

P deficiency followed by Al stress (7 h) had no effect on the Al-induced secretion of citrate from soybean roots [127]. This disagreement was attributed to the differences in the plant materials and experimental methods used [128]. Thus, it appears that the influence of P on the Al-induced secretion of OA anions depends on the time of exposure to Al, growth conditions, and plant species or cultivar.

2.7. Other Factors. Yang et al. showed that sodium nitroprusside (SNP, a nitric oxide (NO) donor) increased the Al-induced secretion of malate and citrate by excised roots from Al-treated *C. grandis* seedlings and that the stimulatory effects of SNP on the Al-induced secretion of malate and citrate might be involved in the SNP-induced amelioration of Al toxicity [105]. There are several papers reporting that NO regulates K^+ and Ca^{2+} channels in plants [129, 130]. The stimulation of OA anion secretion by SNP might result from its direct effect on anion channels, because SNP did not enhance the accumulation of malate and citrate in the roots [105]. Chen et al. demonstrated that H_2S played an ameliorative role in protecting soybean plants against Al toxicity by increasing citrate secretion and citrate transporter gene expression and enhancing the expression of PM H^+-ATPase [131].

3. Genetic Engineering Technology for the Secretion and Biosynthesis of Organic Acid Anions

A common agricultural practice for acidic soils is to apply lime to raise soil pH. However, the option is not economically feasible for poor farmers, nor is it an effective strategy for alleviating subsoil acidity [132]. A complementary approach to liming practice is to tailor plants to suit acidic soils by identifying and/or developing plants with improved tolerance to Al in acidic soils. The best documented mechanism for plant Al tolerance is the Al-induced secretion of OA anions from roots [6, 9], although it is not the only tolerance mechanism [43, 77]. The production of transgenic plants with an enhanced ability to secrete OA anions appears to be an appealing strategy to produce Al tolerant plants. The two main approaches for increasing OA anion secretion are to increase OA synthesis and to increase OA anion transport across the PM [107]. Table 2 summarizes the attempts to obtain transgenic plants or cells with higher Al tolerance by overexpressing genes involved in the biosynthesis and the secretion of OA anions.

The early attempts to enhance Al tolerance focused on OA synthesis because the genes encoding transporters for OA anions were not cloned at the time. The first influential report of such an approach was from De La Fuente et al. [108] who overexpressed a *P. aeruginosa CS* in tobacco and papaya. Overproduction of citrate was shown to result in Al tolerance in transgenic tobacco and papaya through increasing citrate secretion. Delhaize et al. [40] could not repeat the original study of De La Fuente et al. [108] in the same and other tobacco lines expressing the *CS* from *P. aeruginosa* at higher levels. The authors also observed that *CS*-expressing alfalfa

did not show an improved Al tolerance. They concluded that the expression of the *CS* in plants is unlikely to be a robust and easily reproducible strategy for enhancing the Al tolerance of crop and pasture species. Recently, Barone et al. assessed the Al tolerance of 15 transgenic alfalfa overexpressing the *CS* from *P. aeruginosa* by *in vitro* root growth, hydroponics, or soil assay. They deemed that *CS* overexpression could be a useful tool to enhance Al tolerance, but the type of assay used was critical to properly evaluate the transgenic phenotype [133]. Overexpression of mitochondrial *CS* also resulted in increased citrate secretion and enhanced Al tolerance in *A. thaliana* [134], rape [58], *N. benthamiana* [24], and tobacco [111]. In tobacco, overexpression of cytosolic *MDH* genes from *A. thaliana* and from *E. coli* led to increased malate secretion and enhanced Al tolerance [110]. Zhang et al. reported that overexpression of a *C. junos MDH* in tobacco conferred Al tolerance [135]. In alfalfa, overexpression of an alfalfa nodule-enhanced form of *MDH* (*neMDH*) resulted in increased synthesis and secretion of OA anions and enhanced Al tolerance. However, the transgenic alfalfa overexpressing an alfalfa *PEPC* did not show increased accumulation and secretion of OA anions [112]. Trejo-Téllez et al. showed that overexpression of an *M. crystallinum PPDK* in tobacco roots increased the exudation of OA anions, with a concomitant decrease in plant Al accumulation [113]. It should be noted that OA metabolism may not be a limiting factor for Al-induced secretion of OA anions in some plant species because only a small portion of internal OAs is secreted in response to Al [9]. In addition, the Al-induced secretion of OA anions must somewhat be associated with some transport processes [9, 107]. Therefore, the effect that modifying internal OA concentrations has on the level of plant Al tolerance may be small or not observed in some plant species.

Transgenic plants overexpressing genes encoding transporters for OA anions have been widely studied since the first major gene (*TaALMT1*) was cloned from wheat [73]. *TaALMT1* expression in rice, cultured tobacco cells, barley, and *A. thaliana* led to increased Al-activated malate secretion and enhanced Al tolerance for all except for rice [73, 136, 137]. Overexpression of *TaALMT1* in wheat conferred greater Al-activated malate secretion from the roots and improved Al tolerance, which was kept in the T1 and T2 generations [138]. This is the first report of a major food crop being stably transformed for greater Al tolerance. Homologs of *TaALMT1* cloned from *A. thaliana*, rape, barley, and rye [76–78, 87] could also be utilized to increase plant Al tolerance. For example, expression of *BnALMT1* and *BnALMT2* in tobacco cultured cells [76], *HvALMT1* in barley [87], and *AtALMT1* in *A. thaliana* [107] resulted in increased malate secretion and enhanced Al tolerance. Increases in Al tolerance have been achieved by overexpressing citrate transporters. Expression of *SbMATE1* [53], *FRD3* [94], and *ZmMATE1* [96] in *A. thaliana*, *HvAACT1* in tobacco [88], and *VuMATE* in tomato [98] enhanced the secretion of citrate and Al tolerance.

Arabidopsis thaliana has one gene encoding for a type I H^+-pyrophosphatase (AVP1, *Arabidopsis* Vacuolar Pyrophosphatase 1) and another gene encoding for a type II H^+-pyrophosphatase (AVP2) [139, 142]. Yang et al. reported

TABLE 2: Transgenic plants or cells with higher aluminum- (Al-) tolerance overexpressing genes for the biosynthesis and the secretion of organic acid (OA) anions.

Genes	Origins	Transgenic plants or cells	Increased secretions of OA anions	Al tolerance	References
		Papaya (Carica papaya)	NA	+	[108]
		Tobacco (Nicotiana tabacum)	Citrate	+	[108]
	Pseudomonas aeruginosa	Tobacco	No	No	[40]
		Alfalfa (Medicago sativa)	No	No	[40]
Citrate synthase (CS)		Alfalfa	NA	+	[133]
		Rape (Brassica napus)	Citrate	+	[58]
	Arabidopsis thaliana	Carrot (Daucus carota) cell	Citrate	+	[109]
	Citrus junos	Nicotiana benthamiana	Citrate	+	[24]
	Carrot	A. thaliana	Citrate	+	[134]
	Rice (Oryza sativa)	Tobacco	Citrate	+	[111]
	C. junos	Tobacco	NA	+	[135]
Malate dehydrogenase (MDH)	A. thaliana, Escherichia coli	Tobacco	Malate	+	[110]
Nodule-enhanced form of MDH (neMDH)	Alfalfa	Alfalfa	Citrate, oxalate, malate, succinate, acetate	+	[112]
Phosphoenolpyruvate carboxylase (PEPC)	Alfalfa	Alfalfa	No	No	[112]
Pyruvate phosphate dikinase (PPDK)	Mesembryanthemum crystallinum	Tobacco	Citrate, malate	+	[113]
		Tobacco cells	Malate	+	[73]
		Rice	Malate	No	[73]
TaALMT1	Wheat (Triticum aestivum)	Barley	Malate	+	[136, 137]
		Wheat	Malate	+	[138]
		A. thaliana	Malate	+	[107]
AtALMT1	A. thaliana	A. thaliana	Malate	+	[107]
BnALMT1 and BnALMT1	Rape	Tobacco cells	Malate	+	[76]
HvALMT1	Barley	Barley	Malate	+	[87]
HvAACT1	Barley	Tobacco	Citrate	+	[88]
SbMATE	Sorghum (Sorghum bicolor)	A. thaliana	Citrate	+	[53]
FRD3	A. thaliana	A. thaliana	Citrate	+	[94]
ZmMATE1	Maize	A. thaliana	Citrate	+	[96]
VuMATE	Vigna umbellata	Tomato	Citrate	+	[98]
Plasma membrane H^+-ATPase	A. thaliana	A. thaliana	Citrate	NA	[117]
Type 1 H^+-pyrophosphatase (AVP1)	A. thaliana	A. thaliana, tomato, and rice	Malate	+	[139]

NA: not applicable; No: no change in secretion of OA anions or Al tolerance.

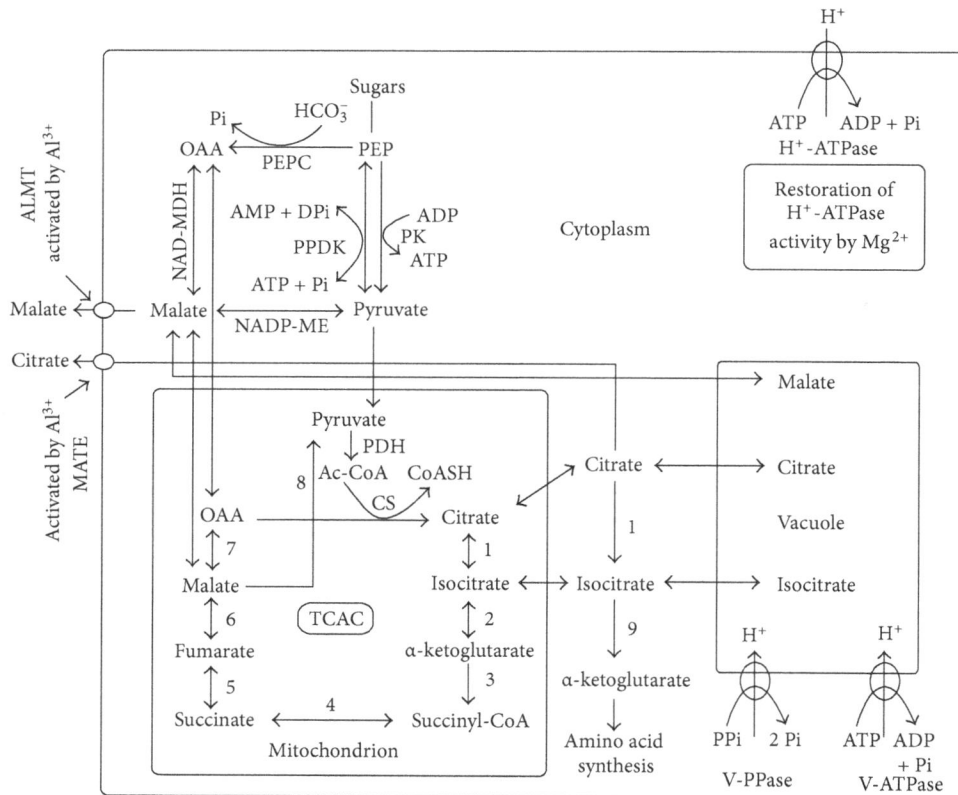

FIGURE 1: A diagram showing the reactions and processes involved in the accumulation and secretion of organic acid (OA) anions in aluminium- (Al-) treated plants. Ac-CoA: acetyl-CoA: ALMT, Al-activated malate transporter; CS: citrate synthase; DPi: diphosphate; MATE, multidrug and toxin compounds extrusion; NAD-MDH: NAD-malate dehydrogenase; NADP-ME: NADP-malic enzyme; OAA: oxaloacetate; PDH: pyruvate dehydrogenase; PEP: phospho*enol*pyruvate; PEPC: PEP carboxylase; Pi: phosphate; PK: pyruvate kinase; PPDK: pyruvate Pi dikinase; PPi: pyrophosphate; TCAC: tricarboxylic acid cycle; V-ATPase: tonoplast adenosine triphosphatase; V-PPiase, tonoplast pyrophosphatase; 1, aconitase (ACO); 2, NAD-isocitrate dehydrogenase (NAD-IDH); 3, α-ketoglutarate dehydrogenase; 4, succinate thiokinase; 5, succinate dehydrogenase; 6, fumarase; 7, NAD-MDH; 8, NAD-malic enzyme (NAD-ME); 9, NADP-IDH (redrawn from Delhaize et al. [14], Anoop et al. [58], Bose et al. [122], Lin et al. [140], and Mariano et al. [141]).

that overexpression of *AVP1* in *A. thaliana*, tomato, and rice exhibited greater tolerance to Al and higher level of citrate and malate secretion compared with controls and that increased AVP1-dependent H$^+$ extrusion appeared to be charge balanced by the enhancement of K uptake and the release of OA from roots [139].

4. Concluding Remark

Aluminium toxicity is one of the most deleterious factors for plant growth in acidic soils, which comprise approximately 30% of the world's total land area and over 50% of the world's potential arable lands [1]. In recent years, there has been significant progress in our understanding of the physiological and molecular mechanisms of Al tolerance in higher plants. In particular, the Al-induced secretion of OA anions from roots has been widely studied by many researchers in different plants because it is a major mechanism leading to Al tolerance in higher plants, but the mechanisms which lead to the accumulation and secretion of OA anions are not fully understood (Figure 1). Modulation of

OA metabolism and activation of anion channels have been suggested to be involved in the Al-activated secretion of OA anions and transgenic plants or cells overexpressing genes for biosynthesis and secretion of OA anions have displayed increased secretion of OA anions and enhanced Al tolerance (Table 2). It is clear that both ALMT1 proteins from wheat, rye, *A. thaliana*, and rape and MATE proteins from barley and sorghum require Al to activate their function [107]. However, the mechanisms of this activation remain unclear, although evidence shows that the induction of *ALMT1* expression by Al may involve reversible phosphorylation [79, 80, 107] and that *OsFRDL4* and *AtMATE* are regulated by a C2H2-type zinc finger transcription factor ART1 and STOP1, respectively [95, 97]. Although anion channels may play an important role in the Al-induced secretion of oxalate in some plants [47], no oxalate transporter has been isolated from plants so far. Organic acid anions secreted from the roots of P-deficient plants have been shown to mainly result from increased PEPC activity in the shoots [143], but relatively few studies have investigated the roles of shoots in the Al-induced secretion of OA anions. Although some transgenic plants overexpressing *PEPC* or *CS* did not show enhanced accumulation and secretion of OA anions [40, 112], genetically modified plants with higher Al tolerance by overexpressing genes for the secretion and the biosynthesis of OA anions may still be a potentially rewarding area of research in the future. Recent work showed that the Al-activated malate and citrate transporters from the MATE and ALMT families functioned independently to confer Al tolerance of *A. thaliana* [95] and that overexpression of *TaALMT1* in barley, which is very sensitive to Al and does not possess an Al-activated secretion of malate, had an Al-activated secretion of malate with properties similar to those of Al-tolerant wheat and enhanced Al tolerance [136, 137]. Therefore, an additive or a synergistic effect on Al tolerance may be achieved by overexpressing two or more anion transporters regulating the secretion of different OA anions at the same time. If the secretion of OA anions is limited by their supply, genes for OA anion synthesis can be cotransformed with genes for OA anion transport, which may produce transgenic plants with higher levels of Al tolerance. Finally, many transgenic plants with significantly increased Al tolerance will be produced through the collaboration between plant breeders and plant physiologists.

Acknowledgments

This work was financially supported by the earmarked fund for China Agriculture Research System and the National Natural Science Foundation of China (no. 30771487).

References

[1] H. R. von Uexküll E and Mutert, "Global extent, development and economic impact of acid soils," in *Plant-Soil Interactions at Low PH: Principles and Management*, R. A. Date, N. J. Grundon, and G. E. Raymet, Eds., pp. 5–19, Kluwer Academic, Dordrecht, The Netherlands, 1995.

[2] J. H. Guo, X. J. Liu, Y. Zhang et al., "Significant acidification in major chinese croplands," *Science*, vol. 327, no. 5968, pp. 1008–1010, 2010.

[3] C. D. Foy, R. L. Chaney, and M. C. White, "The physiology of metal toxicity in plants," *Annual Review of Plant Physiology*, vol. 29, pp. 511–566, 1978.

[4] T. B. Kinraide, "Identity of the rhizotoxic aluminium species," *Plant and Soil*, vol. 134, no. 1, pp. 167–178, 1991.

[5] L. V. Kochian, "Cellular mechanisms of aluminum toxicity and resistance in plants," *Annual Review of Plant Physiology and Plant Molecular Biology*, vol. 46, pp. 237–260, 1995.

[6] L. V. Kochian, O. A. Hoekenga, and M. A. Piñeros, "How do crop plants tolerate acid soils? Mechanisms of aluminum tolerance and phosphorous efficiency," *Annual Review of Plant Biology*, vol. 55, pp. 459–493, 2004.

[7] D. M. Pellet, L. A. Papernik, and L. V. Kochian, "Multiple aluminum-resistance mechanisms in wheat roles of root apical phosphate and malate exudation," *Plant Physiology*, vol. 112, no. 2, pp. 591–597, 1996.

[8] P. S. Kidd, M. Llugany, C. Poschenrieder, B. Gunsé, and J. Barceló, "The role of root exudates in aluminium resistance and silicon-induced amelioration of aluminium toxicity in three varieties of maize (*Zea mays* L.)," *Journal of Experimental Botany*, vol. 52, no. 359, pp. 1339–1352, 2001.

[9] J. F. Ma, "Syndrome of aluminum toxicity and diversity of aluminum resistance in higher plants," *International Review of Cytology*, vol. 264, pp. 225–252, 2007.

[10] J. F. Ma, "Role of organic acids in detoxification of aluminum in higher plants," *Plant and Cell Physiology*, vol. 41, no. 4, pp. 383–390, 2000.

[11] J. F. Ma, "Physiological mechanisms of Al resistance in higher plants," *Soil Science and Plant Nutrition*, vol. 51, no. 5, pp. 609–612, 2005.

[12] J. F. Ma, P. R. Ryan, and E. Delhaize, "Aluminium tolerance in plants and the complexing role of organic acids," *Trends in Plant Science*, vol. 6, no. 6, pp. 273–278, 2001.

[13] J. F. Ma and J. Furukawa, "Recent progress in the research of external Al detoxification in higher plants: a minireview," *Journal of Inorganic Biochemistry*, vol. 97, no. 1, pp. 46–51, 2003.

[14] E. Delhaize, J. F. Ma, and P. R. Ryan, "Transcriptional regulation of aluminium tolerance genes," *Trends in Plant Science*, vol. 17, no. 6, pp. 341–348, 2012.

[15] C. Inostroza-Blancheteau, Z. Rengel, M. Alberdi et al., "Molecular and physiological strategies to increase aluminum resistance in plants," *Molecular Biology Reports*, vol. 39, no. 3, pp. 2069–2079, 2012.

[16] T. Kitagawa, T. Morishita, Y. Tachibana, H. Namai, and Y. Ohta, "Differential aluminum resistance of wheat varieties and organic acid secretion," *Japanese Journal of Soil Science and Plant Nutrition*, vol. 57, no. 4, pp. 352–358, 1986.

[17] Z. Zhao, J. F. Ma, K. Sato, and K. Takeda, "Differential Al resistance and citrate secretion in barley (*Hordeum vulgare* L.)," *Planta*, vol. 217, no. 5, pp. 794–800, 2003.

[18] D. M. Pellet, D. L. Grunes, and L. V. Kochian, "Organic acid exudation as an aluminum-tolerance mechanism in maize (*Zea mays* L.)," *Planta*, vol. 196, no. 4, pp. 788–795, 1995.

[19] S. J. Zheng, J. F. Ma, and H. Matsumoto, "Continuous secretion of organic acids is related to aluminium resistance during relatively long-term exposure to aluminium stress," *Physiologia Plantarum*, vol. 103, no. 2, pp. 209–214, 1998.

[20] S. J. Zheng, J. F. Ma, and H. Matsumoto, "High aluminum resistance in buckwheat: I. Al-induced specific secretion of oxalic acid from root tips," *Plant Physiology*, vol. 117, no. 3, pp. 745–751, 1998.

[21] X. F. Li, J. F. Ma, and H. Matsumoto, "Pattern of aluminum-induced secretion of organic acids differs between rye and wheat," *Plant Physiology*, vol. 123, no. 4, pp. 1537–1544, 2000.

[22] Z. M. Yang, M. Sivaguru, W. J. Horst, and H. Matsumoto, "Aluminium tolerance is achieved by exudation of citric acid from roots of soybean (*Glycine max*)," *Physiologia Plantarum*, vol. 110, no. 1, pp. 72–77, 2000.

[23] H. Liao, H. Wan, J. Shaff, X. Wang, X. Yan, and L. V. Kochian, "Phosphorus and aluminum interactions in soybean in relation to aluminum tolerance. Exudation of specific organic acids from different regions of the intact root system," *Plant Physiology*, vol. 141, no. 2, pp. 674–684, 2006.

[24] W. Deng, K. Luo, Z. Li, Y. Yang, N. Hu, and Y. Wu, "Overexpression of *c* mitochondrial citrate synthase gene in *Nicotiana benthamiana* confers aluminum tolerance," *Planta*, vol. 230, no. 2, pp. 355–365, 2009.

[25] L. V. Kochian, M. A. Piñeros, and O. A. Hoekenga, "The physiology, genetics and molecular biology of plant aluminum resistance and toxicity," *Plant and Soil*, vol. 274, no. 1-2, pp. 175–195, 2005.

[26] J. F. Ma, S. Taketa, and Z. M. Yang, "Aluminum tolerance genes on the short arm of chromosome 3R are linked to organic acid release in triticale," *Plant Physiology*, vol. 122, no. 3, pp. 687–694, 2000.

[27] J. F. You, Y. F. He, J. L. Yang, and S. J. Zheng, "A comparison of aluminum resistance among *Polygonum* species originating on strongly acidic and neutral soils," *Plant and Soil*, vol. 276, no. 1-2, pp. 143–151, 2005.

[28] H. Osawa and K. Kojima, "Citrate-release-mediated aluminum resistance is coupled to the inducible expression of mitochondrial citrate synthase gene in *Paraserianthes falcataria*," *Tree Physiology*, vol. 26, no. 5, pp. 565–574, 2006.

[29] X. Y. Dong, R. F. Shen, R. F. Chen, Z. L. Zhu, and J. F. Ma, "Secretion of malate and citrate from roots is related to high Al-resistance in *Lespedeza bicolor*," *Plant and Soil*, vol. 306, no. 1-2, pp. 139–147, 2008.

[30] L. T. Yang, H. X. Jiang, N. Tang, and L. S. Chen, "Mechanisms of aluminum-tolerance in two species of citrus: secretion of organic acid anions and immobilization of aluminum by phosphorus in roots," *Plant Science*, vol. 180, no. 3, pp. 521–530, 2011.

[31] L. T. Yang, H. X. Jiang, Y. P. Qi, and L. S. Chen, "Differential expression of genes involved in alternative glycolytic pathways, phosphorus scavenging and recycling in response to aluminum and phosphorus interactions in citrus roots," *Molecular Biology Reports*, vol. 39, no. 5, pp. 6353–6366, 2012.

[32] P. R. Ryan, J. M. Ditomaso, and L. V. Kochian, "Aluminium toxicity in roots: an investigation of spatial sensitivity and the role of the root cap," *Journal of Experimental Botany*, vol. 44, no. 2, pp. 437–446, 1993.

[33] M. Sivaguru and W. J. Horst, "The distal part of the transition zone is the most aluminum-sensitive apical root zone of maize," *Plant Physiology*, vol. 116, no. 1, pp. 155–163, 1998.

[34] J. F. Ma, S. J. Zheng, and H. Matsumoto, "Specific secretion of citric acid induced by Al stress in *Cassia tora* L.," *Plant and Cell Physiology*, vol. 38, no. 9, pp. 1019–1025, 1997.

[35] J. L. Yang, S. J. Zheng, Y. F. He, and H. Matsumoto, "Aluminium resistance requires resistance to acid stress: a case study with spinach that exudes oxalate rapidly when exposed to Al stress," *Journal of Experimental Botany*, vol. 56, no. 414, pp. 1197–1203, 2005.

[36] J. L. Yang, L. Zhang, Y. Y. Li, J. F. You, P. Wu, and S. J. Zheng, "Citrate transporters play a critical role in aluminium-stimulated citrate efflux in rice bean (*Vigna umbellata*) roots," *Annals of Botany*, vol. 97, no. 4, pp. 579–584, 2006.

[37] P. R. Ryan, E. Delhaize, and P. J. Randall, "Characterisation of Al-stimulated efflux of malate from the apices of Al-tolerant wheat roots," *Planta*, vol. 196, no. 1, pp. 103–110, 1995.

[38] Y. Kobayashi, O. A. Hoekenga, H. Itoh et al., "Characterization of *AtALMT1* expression in aluminum-inducible malate release and its role for rhizotoxic stress tolerance in *Arabidopsis*," *Plant Physiology*, vol. 145, no. 3, pp. 843–852, 2007.

[39] J. L. Yang, S. J. Zheng, Y. F. He, J. F. You, L. Zhang, and X. H. Yu, "Comparative studies on the effect of a protein-synthesis inhibitor on aluminium-induced secretion of organic acids from Fagopyrum esculentum Moench and *Cassia tora* L. roots," *Plant, Cell and Environment*, vol. 29, no. 2, pp. 240–246, 2006.

[40] E. Delhaize, D. M. Hebb, and P. R. Ryan, "Expression of a *Pseudomonas aeruginosa* citrate synthase gene in tobacco is not associated with either enhanced citrate accumulation or efflux," *Plant Physiology*, vol. 125, no. 4, pp. 2059–2067, 2001.

[41] Z. M. Yang, H. Nian, M. Sivaguru, S. Tanakamaru, and H. Matsumoto, "Characterization of aluminium-induced citrate secretion in aluminium-tolerant soybean (*Glycine max*) plants," *Physiologia Plantarum*, vol. 113, no. 1, pp. 64–71, 2001.

[42] H. Nian, Z. Yang, H. Huang, X. Yan, and H. Matsumoto, "Citrate secretion induced by aluminum stress may not be a key mechanism responsible for differential aluminum tolerance of some soybean genotypes," *Journal of Plant Nutrition*, vol. 27, no. 11, pp. 2047–2066, 2004.

[43] M. A. Piñeros, J. E. Shaff, H. S. Manslank, V. M. Carvalho Alves, and L. V. Kochian, "Aluminum resistance in maize cannot be solely explained by root organic acid exudation. A comparative physiological study," *Plant Physiology*, vol. 137, no. 1, pp. 231–241, 2005.

[44] S. J. Zheng, J. L. Yang, Y. F. Yun et al., "Immobilization of aluminum with phosphorus in roots is associated with high aluminum resistance in buckwheat," *Plant Physiology*, vol. 138, no. 1, pp. 297–303, 2005.

[45] P. Wenzl, G. M. Patiño, A. L. Chaves, J. E. Mayer, and I. M. Rao, "The high level of aluminum resistance in signalgrass is not associated with known mechanisms of external aluminum detoxification in root apices," *Plant Physiology*, vol. 125, no. 3, pp. 1473–1484, 2001.

[46] S. Ishikawa, T. Wagatsuma, R. Sasaki, and P. Ofei-Manu, "Comparison of the amount of citric and malic acids in Al media of seven plant species and two cultivars each in five plant species," *Soil Science and Plant Nutrition*, vol. 46, no. 3, pp. 751–758, 2000.

[47] J. L. Yang, L. Zhang, and S. J. Zheng, "Aluminum-activated oxalate secretion does not associate with internal content among some oxalate accumulators," *Journal of Integrative Plant Biology*, vol. 50, no. 9, pp. 1103–1107, 2008.

[48] A. N. Famoso, R. T. Clark, J. E. Shaff, E. Craft, S. R. McCouch, and L. V. Kochian, "Development of a novel aluminum tolerance phenotyping platform used for comparisons of cereal aluminum tolerance and investigations into rice aluminum tolerance mechanisms," *Plant Physiology*, vol. 153, no. 4, pp. 1678–1691, 2010.

[49] J. L. Yang, X. F. Zhu, C. Zheng, Y. J. Zhang, and S. J. Zheng, "Genotypic differences in Al resistance and the role of cell-wall pectin in Al exclusion from the root apex in *Fagopyrum tataricum*," *Annals of Botany*, vol. 107, no. 3, pp. 371–378, 2011.

[50] J. L. Yang, Y. Y. Li, Y. J. Zhang et al., "Cell wall polysaccharides are specifically involved in the exclusion of aluminum from the rice root apex," *Plant Physiology*, vol. 146, no. 2, pp. 602–611, 2008.

[51] N. T. Nguyen, K. Nakabayashi, J. Thompson, and K. Fujita, "Role of exudation of organic acids and phosphate in aluminum tolerance of four tropical woody species," *Tree Physiology*, vol. 23, no. 15, pp. 1041–1050, 2003.

[52] O. A. Hoekenga, T. J. Vision, J. E. Shaff et al., "Identification and characterization of aluminum tolerance loci in *Arabidopsis* (Landsberg erecta x Columbia) by quantitative trait locus mapping. A physiologically simple but genetically complex trait," *Plant Physiology*, vol. 132, no. 2, pp. 936–948, 2003.

[53] J. V. Magalhaes, J. Liu, C. T. Guimarães et al., "A gene in the multidrug and toxic compound extrusion (MATE) family confers aluminum tolerance in sorghum," *Nature Genetics*, vol. 39, no. 9, pp. 1156–1161, 2007.

[54] J. F. Ma, S. J. Zheng, H. Matsumoto, and S. Hiradate, "Detoxifying aluminium with buckwheat," *Nature*, vol. 390, no. 6660, pp. 569–570, 1997.

[55] M. Schöttelndreier, M. M. Norddahl, L. Ström, and U. Falkengren-Grerup, "Organic acid exudation by wild herbs in response to elevated Al concentrations," *Annals of Botany*, vol. 87, no. 6, pp. 769–775, 2001.

[56] M. Kollmeier, P. Dietrich, C. S. Bauer, W. J. Horst, and R. Hedrich, "Aluminum activates a citrate-permeable anion channel in the aluminum-sensitive zone of the maize root apex. A comparison between an aluminum-sensitive and an aluminum-resistant cultivar," *Plant Physiology*, vol. 126, no. 1, pp. 397–410, 2001.

[57] R. Qin, Y. Hirano, and I. Brunner, "Exudation of organic acid anions from poplar roots after exposure to Al, Cu and Zn," *Tree Physiology*, vol. 27, no. 2, pp. 313–320, 2007.

[58] V. M. Anoop, U. Basu, M. T. McCammon, L. McAlister-Henn, and G. J. Taylor, "Modulation of citrate metabolism alters aluminum tolerance in yeast and transgenic canola overexpressing a mitochondrial citrate synthase," *Plant Physiology*, vol. 132, no. 4, pp. 2205–2217, 2003.

[59] A. Ligaba, H. Shen, K. Shibata, Y. Yamamoto, S. Tanakamaru, and H. Matsumoto, "The role of phosphorus in aluminium-induced citrate and malate exudation from rape (*Brassica napus*)," *Physiologia Plantarum*, vol. 120, no. 4, pp. 575–584, 2004.

[60] Y. Y. Li, Y. X. Yu, X. L. Tang, G. Z. Ling, M. H. Gu, and Q. C. Huang, "Studies on the mechanisms responsible for rye cultivar differences in Al resistance," *Plant Nutrition and Fertilizer Science*, vol. 14, no. 6, pp. 1070–1075, 2008.

[61] S. C. Miyasaka, J. George Buta, R. K. Howell, and C. D. Foy, "Mechanism of aluminum tolerance in snapbeans: root exudation of citric acid," *Plant Physiology*, vol. 96, no. 3, pp. 737–743, 1991.

[62] J. F. You, N. N. Hou, M. Y. Xu, H. M. Zhang, and Z. M. Yang, "Citrate transporters play an important role in regulating aluminum-induced citrate secretion in *Glycine max*," *Biologia Plantarum*, vol. 54, no. 4, pp. 766–768, 2010.

[63] X. F. Li, F. H. Zuo, G. Z. Ling et al., "Secretion of citrate from roots in response to aluminum and low phosphorus stresses in *Stylosanthes*," *Plant and Soil*, vol. 325, no. 1, pp. 219–229, 2009.

[64] N. E. Saber, A. M. Abdel-Moneim, and S. Y. Barakat, "Role of organic acids in sunflower tolerance to heavy metals," *Biologia Plantarum*, vol. 42, no. 1, pp. 65–73, 1999.

[65] Z. Ma and S. C. Miyasaka, "Oxalate exudation by Taro in response to Al," *Plant Physiology*, vol. 118, no. 3, pp. 861–865, 1998.

[66] J. L. Yang, X. F. Zhu, Y. X. Peng, C. Zheng, F. Ming, and S. J. Zheng, "Aluminum regulates oxalate secretion and plasma membrane H$^+$-ATPase activity independently in tomato roots," *Planta*, vol. 234, no. 2, pp. 281–291, 2011.

[67] W. H. Zhang, P. R. Ryan, and S. D. Tyerman, "Malate-permeable channels and cation channels activated by aluminum in the apical cells of wheat roots," *Plant Physiology*, vol. 125, no. 3, pp. 1459–1472, 2001.

[68] E. Delhaize, P. R. Ryan, and R. J. Randall, "Aluminum tolerance in wheat (*Triticum aestivum* L.). II. Aluminum-stimulated excretion of malic acid from root apices," *Plant Physiology*, vol. 103, no. 3, pp. 695–702, 1993.

[69] P. R. Ryan, M. Skerrett, G. P. Findlay, E. Delhaize, and S. D. Tyerman, "Aluminum activates an anion channel in the apical cells of wheat roots," *Proceedings of the National Academy of Sciences of the United States of America*, vol. 94, no. 12, pp. 6547–6552, 1997.

[70] L. A. Papernik and L. V. Kochian, "Possible involvement of AL-Induced electrical signals in Al tolerance in wheat," *Plant Physiology*, vol. 115, no. 2, pp. 657–667, 1997.

[71] M. A. Piñeros and L. V. Kochian, "A patch-clamp study on the physiology of aluminum toxicity and aluminum tolerance in maize. Identification and characterization of Al^{3+}-induced anion channels," *Plant Physiology*, vol. 125, no. 1, pp. 292–305, 2001.

[72] M. A. Piñeros, J. V. Magalhaes, V. M. Carvalho Alves, and L. V. Kochian, "The physiology and biophysics of an aluminum tolerance mechanism based on root citrate exudation in maize," *Plant Physiology*, vol. 129, no. 3, pp. 1194–1206, 2002.

[73] T. Sasaki, Y. Yamamoto, B. Ezaki et al., "A wheat gene encoding an aluminum-activated malate transporter," *Plant Journal*, vol. 37, no. 5, pp. 645–653, 2004.

[74] M. Yamaguchi, T. Sasaki, M. Sivaguru et al., "Evidence for the plasma membrane localization of Al-activated malate transporter (ALMT1)," *Plant and Cell Physiology*, vol. 46, no. 5, pp. 812–816, 2005.

[75] M. A. Piñeros, G. M. A. Cançado, and L. V. Kochian, "Novel properties of the wheat aluminum tolerance organic acid transporter (TaALMT1) revealed by electrophysiological characterization in *Xenopus oocytes*: functional and structural implications," *Plant Physiology*, vol. 147, no. 4, pp. 2131–2146, 2008.

[76] A. Ligaba, M. Katsuhara, P. R. Ryan, M. Shibasaka, and H. Matsumoto, "The *BnALMT1* and *BnALMT2* genes from rape encode aluminum-activated malate transporters that enhance the aluminum resistance of plant cells," *Plant Physiology*, vol. 142, no. 3, pp. 1294–1303, 2006.

[77] O. A. Hoekenga, L. G. Maron, M. A. Piñeros et al., "*AtALMT1*, which encodes a malate transporter, is identified as one of several genes critical for aluminum tolerance in Arabidopsis," *Proceedings of the National Academy of Sciences of the United States of America*, vol. 103, no. 25, pp. 9738–9743, 2006.

[78] G. Fontecha, J. Silva-Navas, C. Benito et al., "Candidate gene identification of an aluminum-activated organic acid transporter gene at the Alt4 locus for aluminum tolerance in rye (*Secale cereale* L.)," *Theoretical and Applied Genetics*, vol. 114, no. 2, pp. 249–260, 2007.

[79] H. Osawa and H. Matsumoto, "Possible involvement of protein phosphorylation in aluminum-responsive malate efflux from wheat root apex," *Plant Physiology*, vol. 126, no. 1, pp. 411–420, 2001.

[80] A. Ligaba, L. Kochian, and M. Piñeros, "Phosphorylation at S384 regulates the activity of the TaALMT1 malate transporter that underlies aluminum resistance in wheat," *Plant Journal*, vol. 60, no. 3, pp. 411–423, 2009.

[81] M. A. Piñeros, G. M. A. Cançado, L. G. Maron, S. M. Lyi, M. Menossi, and L. V. Kochian, "Not all ALMT1-type transporters mediate aluminum-activated organic acid responses: the case of ZmALMT1, an anion-selective transporter," *Plant Journal*, vol. 53, no. 2, pp. 352–367, 2008.

[82] A. Ligaba, L. Maron, J. Shaff, L. V. Kochian, and M. Piñeros, "Maize ZmALMT2 is a root anion transporter that mediates constitutive root malate efflux," *Plant Cell & Environment*, vol. 35, no. 7, pp. 1185–1200, 2012.

[83] P. Kovermann, S. Meyer, S. Hörtensteiner et al., "The *Arabidopsis* vacuolar malate channel is a member of the ALMT family," *Plant Journal*, vol. 52, no. 6, pp. 1169–1180, 2007.

[84] S. Meyer, P. Mumm, D. Imes et al., "AtALMT12 represents an R-type anion channel required for stomatal movement in

Arabidopsis guard cells," *Plant Journal*, vol. 63, no. 6, pp. 1054–1062, 2010.

[85] S. Meyer, J. Scholz-Starke, A. De Angeli et al., "Malate transport by the vacuolar AtALMT6 channel in guard cells is subject to multiple regulation," *Plant Journal*, vol. 67, no. 2, pp. 247–257, 2011.

[86] B. D. Gruber, P. R. Ryan, A. E. Richardson et al., "HvALMT1 from barley is involved in the transport of organic anions," *Journal of Experimental Botany*, vol. 61, no. 5, pp. 1455–1467, 2010.

[87] B. D. Gruber, E. Delhaize, A. E. Richardson et al., "Characterisation of *HvALMT1* function in transgenic barley plants," *Functional Plant Biology*, vol. 38, no. 2, pp. 163–175, 2011.

[88] J. Furukawa, N. Yamaji, H. Wang et al., "An aluminum-activated citrate transporter in barley," *Plant and Cell Physiology*, vol. 48, no. 8, pp. 1081–1091, 2007.

[89] J. Wang, H. Raman, M. Zhou et al., "High-resolution mapping of the Alp locus and identification of a candidate gene HvMATE controlling aluminium tolerance in barley (*Hordeum vulgare* L.)," *Theoretical and Applied Genetics*, vol. 115, no. 2, pp. 265–276, 2007.

[90] H. Omote, M. Hiasa, T. Matsumoto, M. Otsuka, and Y. Moriyama, "The MATE proteins as fundamental transporters of metabolic and xenobiotic organic cations," *Trends in Pharmacological Sciences*, vol. 27, no. 11, pp. 587–593, 2006.

[91] M. Morita, N. Shitan, K. Sawada et al., "Vacuolar transport of nicotine is mediated by a multidrug and toxic compound extrusion (MATE) transporter in *Nicotiana tabacum*," *Proceedings of the National Academy of Sciences of the United States of America*, vol. 106, no. 7, pp. 2447–2452, 2009.

[92] K. Yokosho, N. Yamaji, and J. F. Ma, "Isolation and characterisation of two MATE genes in rye," *Functional Plant Biology*, vol. 37, no. 4, pp. 296–303, 2010.

[93] J. V. Magalhaes, "How a microbial drug transporter became essential for crop cultivation on acid soils: aluminium tolerance conferred by the multidrug and toxic compound extrusion (MATE) family," *Annals of Botany*, vol. 106, no. 1, pp. 199–203, 2010.

[94] T. P. Durrett, W. Gassmann, and E. E. Rogers, "The FRD3-mediated efflux of citrate into the root vasculature is necessary for efficient iron translocation," *Plant Physiology*, vol. 144, no. 1, pp. 197–205, 2007.

[95] J. Liu, J. V. Magalhaes, J. Shaff, and L. V. Kochian, "Aluminum-activated citrate and malate transporters from the MATE and ALMT families function independently to confer *Arabidopsis* aluminum tolerance," *Plant Journal*, vol. 57, no. 3, pp. 389–399, 2009.

[96] L. G. Maron, M. A. Piñeros, C. T. Guimarães et al., "Two functionally distinct members of the MATE (multi-drug and toxic compound extrusion) family of transporters potentially underlie two major aluminum tolerance QTLs in maize," *Plant Journal*, vol. 61, no. 5, pp. 728–740, 2010.

[97] K. Yokosho, N. Yamaji, and J. F. Ma, "An Al-inducible *MATE* gene is involved in external detoxification of Al in rice," *The Plant Journal*, vol. 68, no. 6, pp. 1061–1069, 2011.

[98] X. Y. Yang, J. L. Yang, Y. A. Zhou et al., "A de novo synthesis citrate transporter, *Vigna umbellate* multidrug and toxic compound extrusion, implicates in Al-activated citrate efflux in rice bean (*Vigna umbellata*) root apex," *Plant Cell & Environment*, vol. 34, no. 12, pp. 2138–2148, 2011.

[99] J. F. Ma, W. Zhang, and Z. Zhao, "Regulatory mechanism of Al-induced secretion of organic acids anions-involvement of ABA in the Al-induced secretion of oxalate in buckwheat," in *Plant Nutrition-Food Security and Sustainability of Agro-Ecosystems Through Basis and Applied Research*, W. J. Hortst, H. Flessa, B. Sattelmacher et al., Eds., pp. 486–487, Kluwer Academic, Dordrecht, The Netherland, 2001.

[100] Z. M. Yang, H. Yang, J. Wang, and Y. S. Wang, "Aluminum regulation of citrate metabolism for Al-induced citrate efflux in the roots of *Cassia tora* L.," *Plant Science*, vol. 166, no. 6, pp. 1589–1594, 2004.

[101] M. Xu, J. You, N. Hou, H. Zhang, G. Chen, and Z. Yang, "Mitochondrial enzymes and citrate transporter contribute to the aluminium-induced citrate secretion from soybean (*Glycine max*) roots," *Functional Plant Biology*, vol. 37, no. 4, pp. 285–295, 2010.

[102] A. Gaume, F. Mächler, and E. Frossard, "Aluminium resistance in two cultivars of *Zea mays* L.: root exudation of organic acids and influence of phosphorus nutrition," *Plant and Soil*, vol. 234, no. 1, pp. 73–81, 2001.

[103] J. E. Hayes and J. F. Ma, "Al-induced efflux of organic acid anions is poorly associated with internal organic acid metabolism in triticale roots," *Journal of Experimental Botany*, vol. 54, no. 388, pp. 1753–1759, 2003.

[104] L. S. Chen, N. Tang, H. X. Jiang, L. T. Yang, Q. Li, and B. R. Smith, "Changes in organic acid metabolism differ between roots and leaves of *Citrus grandis* in response to phosphorus and aluminum interactions," *Journal of Plant Physiology*, vol. 166, no. 18, pp. 2023–2034, 2009.

[105] L. T. Yang, Y. P. Qi, L. S. Chen et al., "Nitric oxide protects sour pummelo (*Citrus grandis*) seedlings against aluminum-induced inhibition of growth and photosynthesis," *Environmental and Experimental Botany*, vol. 83, pp. 1–13, 2012.

[106] L. T. Yang, L. S. Chen, H. Y. Peng, P. Guo, P. Wang, and C. L. Ma, "Organic acid metabolism in *Citrus grandis* leaves and roots is differently affected by nitric oxide and aluminum interactions," *Scientia Horticulturae*, vol. 133, no. 1, pp. 40–46, 2012.

[107] P. R. Ryan, S. D. Tyerman, T. Sasaki et al., "The identification of aluminium-resistance genes provides opportunities for enhancing crop production on acid soils," *Journal of Experimental Botany*, vol. 62, no. 1, pp. 9–20, 2011.

[108] J. M. De La Fuente, V. Ramírez-Rodríguez, J. L. Cabrera-Ponce, and L. Herrera-Estrella, "Aluminum tolerance in transgenic plants by alteration of citrate synthesis," *Science*, vol. 276, no. 5318, pp. 1566–1568, 1997.

[109] H. Koyama, E. Takita, A. Kawamura, T. Hara, and D. Shibata, "Over expression of mitochondrial citrate synthase gene improves the growth of carrot cells in Al-phosphate medium," *Plant and Cell Physiology*, vol. 40, no. 5, pp. 482–488, 1999.

[110] Q. F. Wang, Y. Zhao, Q. Yi, K. Z. Li, Y. X. Yu, and L. M. Chen, "Overexpression of malate dehydrogenase in transgenic tobacco leaves: enhanced malate synthesis and augmented Al-resistance," *Acta Physiologiae Plantarum*, vol. 32, no. 6, pp. 1209–1220, 2010.

[111] Y. Han, W. Zhang, B. Zhang, S. Zhang, W. Wang, and F. Ming, "One novel mitochondrial citrate synthase from *Oryza sativa* L. can enhance aluminum tolerance in transgenic tobacco," *Molecular Biotechnology*, vol. 42, no. 3, pp. 299–305, 2009.

[112] M. Tesfaye, S. J. Temple, D. L. Allan, C. P. Vance, and D. A. Samac, "Overexpression of malate dehydrogenase in transgenic alfalfa enhances organic acid synthesis and confers tolerance to aluminum," *Plant Physiology*, vol. 127, no. 4, pp. 1836–1844, 2001.

[113] L. I. Trejo-Téllez, R. Stenzel, F. C. Gómez-Merino, and J. M. Schmitt, "Transgenic tobacco plants overexpressing pyruvate phosphate dikinase increase exudation of organic acids and decrease accumulation of aluminum in the roots," *Plant and Soil*, vol. 326, no. 1, pp. 187–198, 2010.

[114] G. Neumann and V. Römheld, "Root excretion of carboxylic acids and protons in phosphorus-deficient plants," *Plant and Soil*, vol. 211, no. 1, pp. 121–130, 1999.

[115] M. Arango, F. Gévaudant, M. Oufattole, and M. Boutry, "The plasma membrane proton pump ATPase: the significance of gene subfamilies," *Planta*, vol. 216, no. 3, pp. 355–365, 2003.

[116] T. E. Sondergaard, A. Schulz, and M. G. Palmgren, "Energization of transport processes in plants. Roles of the plasma membrane H$^+$-ATPase," *Plant Physiology*, vol. 136, no. 1, pp. 2475–2482, 2004.

[117] H. Shen, L. F. He, T. Sasaki et al., "Citrate secretion coupled with the modulation of soybean root tip under aluminum stress. Up-regulation of transcription, translation, and threonine-oriented phosphorylation of plasma membrane H$^+$-ATPase," *Plant Physiology*, vol. 138, no. 1, pp. 287–296, 2005.

[118] S. J. Ahn, Z. Rengel, and H. Matsumoto, "Aluminum-induced plasma membrane surface potential and H$^+$-ATPase activity in near-isogenic wheat lines differing in tolerance to aluminum," *New Phytologist*, vol. 162, no. 1, pp. 71–79, 2004.

[119] H. Matsumoto, "Inhibition of proton transport activity of microsomal membrane vesicles of barley roots by aluminum," *Soil Science and Plant Nutrition*, vol. 34, no. 4, pp. 499–506, 1988.

[120] S. J. Ahn, M. Sivaguru, H. Osawa, G. C. Chung, and H. Matsumoto, "Aluminum inhibits the H$^+$-ATPase activity by permanently altering the plasma membrane surface potentials in squash roots," *Plant Physiology*, vol. 126, no. 4, pp. 1381–1390, 2001.

[121] J. L. Yang, J. F. You, Y. Y. Li, P. Wu, and S. J. Zheng, "Magnesium enhances aluminum-induced citrate secretion in rice bean roots (*Vigna umbellata*) by restoring plasma membrane H$^+$-ATPase activity," *Plant and Cell Physiology*, vol. 48, no. 1, pp. 66–73, 2007.

[122] J. Bose, O. Babourina, and Z. Rengel, "Role of magnesium in alleviation of aluminium toxicity in plants," *Journal of Experimental Botany*, vol. 62, no. 7, pp. 2251–2264, 2011.

[123] I. R. Silva, T. J. Smyth, D. W. Israel, C. D. Raper, and T. W. Rufty, "Magnesium ameliorates aluminum rhizotoxicity in soybean by increasing citric acid production and exudation by roots," *Plant and Cell Physiology*, vol. 42, no. 5, pp. 546–554, 2001.

[124] K. Tan and W. G. Keltjens, "Interaction between aluminium and phosphorus in sorghum plants—I. Studies with the aluminium sensitive sorghum genotype TAM428," *Plant and Soil*, vol. 124, no. 1, pp. 15–23, 1990.

[125] Q. B. Sun, R. F. Shen, X. Q. Zhao, R. F. Chen, and X. Y. Dong, "Phosphorus enhances Al resistance in Al-resistant *Lespedeza bicolor* but not in Al-sensitive *L. cuneata* under relatively high Al stress," *Annals of Botany*, vol. 102, no. 5, pp. 795–804, 2008.

[126] M. Jemo, R. C. Abaidoo, C. Nolte, and W. J. Horst, "Aluminum resistance of cowpea as affected by phosphorus-deficiency stress," *Journal of Plant Physiology*, vol. 164, no. 4, pp. 442–451, 2007.

[127] D. Dong, X. Peng, and X. Yan, "Organic acid exudation induced by phosphorus deficiency and/or aluminium toxicity in two contrasting soybean genotypes," *Physiologia Plantarum*, vol. 122, no. 2, pp. 190–199, 2004.

[128] H. Nian, S. J. Ahn, Z. M. Yang, and H. Matsumoto, "Effect of phosphorus deficiency on aluminium-induced citrate exudation in soybean (*Glycine max*)," *Physiologia Plantarum*, vol. 117, no. 2, pp. 229–236, 2003.

[129] V. Casolo, E. Petrussa, J. Krajňáková, F. Macrì, and A. Vianello, "Involvement of the mitochondrial K$^+$ATP channel in H$_2$O$_2$ or NO-induced programmed death of soybean suspension cell cultures," *Journal of Experimental Botany*, vol. 56, no. 413, pp. 997–1006, 2005.

[130] S. Sokolovski, A. Hills, R. Gay, C. Garcia-Mata, L. Lamattina, and M. R. Blatt, "Protein phosphorylation is a prerequisite for intracellular Ca^{2+} release and ion channel control by nitric oxide and abscisic acid in guard cells," *Plant Journal*, vol. 43, no. 4, pp. 520–529, 2005.

[131] J. Chen, W. H. Wang, F. H. Wu et al., "Hydrogen sulfide alleviates aluminum toxicity in barley seedlings," *Plant and Soil*. In press.

[132] I. M. Rao, R. S. Zeigler, R. Vera, and S. Sarkarung, "Selection and breeding for acid-soil tolerance in crops," *BioScience*, vol. 43, no. 7, pp. 454–465, 1993.

[133] P. Barone, D. Rosellini, P. LaFayette, J. Bouton, F. Veronesi, and W. Parrott, "Bacterial *citrate synthase* expression and soil aluminum tolerance in transgenic alfalfa," *Plant Cell Reports*, vol. 27, no. 5, pp. 893–901, 2008.

[134] H. Koyama, A. Kawamura, T. Kihara, T. Hara, E. Takita, and D. Shibata, "Overexpression of mitochondrial citrate synthase in *Arabidopsis thaliana* improved growth on a phosphorus-limited soil," *Plant and Cell Physiology*, vol. 41, no. 9, pp. 1030–1037, 2000.

[135] M. Zhang, X. Y. Luo, W. Q. Bai et al., "Characterization of malate dehydrogenase gene from *Citrus junos* and its transgenic tobacco's tolerance to aluminium toxicity," *Acta Horticulturae Sinica*, vol. 35, no. 12, pp. 1751–1758, 2008.

[136] E. Delhaize, P. R. Ryan, D. M. Hebb, Y. Yamamoto, T. Sasaki, and H. Matsumoto, "Engineering high-level aluminum tolerance in barley with the *ALMT1* gene," *Proceedings of the National Academy of Sciences of the United States of America*, vol. 101, no. 42, pp. 15249–15254, 2004.

[137] E. Delhaize, P. Taylor, P. J. Hocking, R. J. Simpson, P. R. Ryan, and A. E. Richardson, "Transgenic barley (*Hordeum vulgare* L.) expressing the wheat aluminium resistance gene (*TaALMT1*) shows enhanced phosphorus nutrition and grain production when grown on an acid soil," *Plant Biotechnology Journal*, vol. 7, no. 5, pp. 391–400, 2009.

[138] J. F. Pereira, G. Zhou, E. Delhaize, T. Richardson, M. Zhou, and P. R. Ryan, "Engineering greater aluminium resistance in wheat by over-expressing *TaALMT1*," *Annals of Botany*, vol. 106, no. 1, pp. 205–214, 2010.

[139] H. Yang, J. Knapp, P. Koirala et al., "Enhanced phosphorus nutrition in monocots and dicots over-expressing a phosphorus-responsive type I H$^+$-pyrophosphatase," *Plant Biotechnology Journal*, vol. 5, no. 6, pp. 735–745, 2007.

[140] Z. H. Lin, L. S. Chen, R. B. Chen et al., "Root release and metabolism of organic acids in tea plants in response to phosphorus supply," *Journal of Plant Physiology*, vol. 168, no. 7, pp. 644–652, 2011.

[141] E. D. Mariano, R. A. Jorge, W. G. Keltjens, and M. Menossi, "Metabolism and root exudation of organic acid anions under aluminium stress," *Brazilian Journal of Plant Physiology*, vol. 17, no. 1, pp. 157–172, 2005.

[142] Y. M. Drozdowicz, J. C. Kissinger, and P. A. Rea, "AVP2, a sequence-divergent, K$^+$-insensitive H$^+$-translocating inorganic

pyrophosphatase from Arabidopsis," *Plant Physiology*, vol. 123, no. 1, pp. 353–362, 2000.

[143] E. Hoffland, R. van den Boogaard, J. Nelemans, and G. Findenegg, "Biosynthesis and root exudation of citric and malic acids in phosphate-starved rape plants," *New Phytologist*, vol. 122, no. 4, pp. 675–680, 1992.

17

Production and Characterization of Keratinolytic Protease from New Wool-Degrading *Bacillus* Species Isolated from Egyptian Ecosystem

Mohamed A. Hassan,[1] **Bakry M. Haroun,**[2] **Amro A. Amara,**[1] **and Ehab A. Serour**[1,3]

[1] *Protein Research Department, Genetic Engineering and Biotechnology Research Institute,*
   City of Scientific Research and Technological Applications, New Borg Al-Arab, P.O. Box. 21934, Alexandria, Egypt
[2] *Botany and Microbiology Department, Faculty of Science (Boys), Al-Azhar University, Cairo, Egypt*
[3] *King Abdulaziz City for Science and Technology, Riyadh, Saudi Arabia*

Correspondence should be addressed to Mohamed A. Hassan; m adelmicro@yahoo.com

Academic Editor: Periasamy Anbu

Novel keratin-degrading bacteria were isolated from sand soil samples collected from Minia Governorate, Egypt. In this study, the isolates were identified as *Bacillus amyloliquefaciens* MA20 and *Bacillus subtilis* MA21 based on morphological and biochemical characteristics as well as 16S rRNA gene sequencing. *B. amyloliquefaciens* MA20 and *B. subtilis* MA21 produced alkaline keratinolytic serine protease when cultivated in mineral medium containing 1% of wool straight off sheep as sole carbon and nitrogen source. The two strains were observed to degrade wool completely to powder at pH 7 and 37°C within 5 days. Under these conditions the maximum activity of proteases produced by *B. amyloliquefaciens* MA20 and *B. subtilis* MA21 was 922 and 814 U/ml, respectively. The proteases exhibited optimum temperature and pH at 60°C and 9, respectively. However, the keratinolytic proteases were stable in broad range of temperature and pH values towards casein Hammerstein. Furthermore the protease inhibitor studies indicated that the produced proteases belong to serine protease because of their sensitivity to PMSF while they were inhibited partially in presence of EDTA. The two proteases are stable in most of the used organic solvents and enhanced by metals suggesting their potential use in biotechnological applications such as wool industry.

1. Introduction

Keratins are classified as fibrous proteins called scleroproteins that occur abundantly in epithelial cells. These proteins are insoluble in water, weak acid and alkali, and organic solvents and are insensitive to the attack of common proteolytic enzymes such as trypsin or pepsin [1]. The animal remains rich in α-keratin such as animal skin, hair, claws, horns, and wools.

The important property of these proteins is the presence of high cystine content that differentiates keratins from other structural proteins such as collagen and elastin. Both a high cystine content as well as a high content of glycine, proline, serine, and acidic amino acids and a low content of lysine, histidine, and methionine (or their lack) as well as the absence of tryptophan are also characteristic of keratins [2, 3]. Numerous disulfide cystine bonds present in keratin to bind peptide chains and packed as α-helices as in hair and wool or in β-sheet arrangements as in case of feathers. The disulfide linkage and the tight secondary structure of keratins make them difficult to be hydrolysed by common proteolytic enzymes [4].

The major problem of α-keratin hydrolysis is the presence a high numbers of disulfide bonds that make it insoluble in nature and resistant to proteases hydrolysis [5]. Keratinolytic protease enzymes are spread in nature and elaborated by different groups of microorganisms that can be isolated from polluted area with keratin wastes [6]. A vast variety of Gram-positive bacteria including *Bacillus, Lysobacter, Nesternokia,*

Kocuria, and *Microbacterium* as well as a few strains of Gram-negative bacteria such as *Xanthomonas*, *Stenotrophomonas*, and *Chryseobacterium* are confined as keratin degraders [7–10]. Most of keratin degrading bacteria belong to the genus of *Bacillus* [11].

The keratinous substrate such as feather and wool can be degraded in basal medium by microorganisms which are capable of utilize keratin as sole carbon and nitrogen source [12]. Keratinolytic proteases have broad substrate specificity where they have the ability to hydrolyze soluble protein such as casein, gelatin, and bovine serum albumin. Additionally, they can hydrolyze the insoluble protein including feather, silk, and wool [13]. Keratinolytic proteases mostly belong to serine or metalloproteases showing sequence similarity with subtilisin group of proteases [14, 15]. In recent years, more demands to keratinolytic proteases are increasing due to their multitude in industrial applications such as the feed, fertilizer, detergent, and textile industries. The present study describes the isolation and identification of new *Bacillus amyloliquefaciens* MA20 and *B. subtilis* MA21 strains from Egyptian ecosystem that grow well on wool as sole carbon source. Moreover the two strains are able to degrade wool and their enzymes can be used to improve the wool quality. This paper includes full characterization of the keratinolytic protease which explains that the best environmental conditions can be used to improve the wool quality in industry.

2. Materials and Methods

2.1. Sample Collection. Different types of samples were collected from different Governorates of Egyptian ecosystem included Alexandria, Behera, Qaliubia, Mania, and Asiut. These fresh samples were varied such as soil, sand soil, humus, waste wool, and rhizosphere under olive trees. These samples were collected in sterile falcon tubes and transported to the microbiological lab in (City of Scientific Research and Technological Applications).

2.2. Strains Isolation. The bacterial strains were isolated by suspending 1 g of soil samples in a 10 mL sterile 0.85% (w/v) saline solution and then treated for 20 min at 80°C. This will enable the isolation of the spore-forming bacteria. Luria-Bertani (LB) agar medium with 1% (w/v) skim milk was used for their cultivation by spreading 0.1 mL of each 10^{-5} and 10^{-6} dilutions. The plates then were incubated for 24 hours at different temperatures [4, 16]. The colonies which give clear zones formed by hydrolysis of skim milk were picked. Pure bacterial isolates were obtained by reculturing individual colonies several times on fresh LB agar medium to produce single colony from each.

2.3. Strains Selection. Twelve selected strains isolated according to the diameter of clear zone were cultured on medium containing 0.5 g NaCl, 0.3 g K_2HPO_4, 0.4 g KH_2PO_4, and 10 g wool per liter; pH 7 and incubated for 5 days at 37°C. The wool was used as sole carbon and nitrogen source for detecting potent strains that have the ability to degrade the wool completely. Three strains degraded the wool and the

supernatants of their culture were assayed on plate containing 1% gelatin powder which is soluble in phosphate buffer pH 7. After determining the existence of the activity (by the clear zone of the supernatants), *Bacillus* sp. MA20 and *Bacillus* sp. MA21 were selected and preserved for further investigation.

2.4. Bacterial Identification. While the phenotypic characteristics and isolation method of the two selected isolates indicate that they are related to *Bacillus* group but further identification was conducted.

The strains identification are included the spore morphology, Gram stain, and motility. The morphological and physiological characteristics of the bacterial isolates were compared with the data from Bergey's Manual of Determinative Bacteriology [17].

2.5. Scanning of Bacillus sp. MA20 and Bacillus sp. MA21 by Scanning Electron Microscope. The bacterial smear was prepared by centrifuging the bacterial cultures at 12,000 rpm for 20 min. The pellets were washed 2 times by saline solution. The pellet was suspended in sterilized distilled H_2O. The bacterial film was prepared and fixed on glass slides till complete drying. The smear was coated with gold using sputter coater. The golden coated sample was scanned at 20 KV acceleration voltages at room temperature.

2.6. Genetic Identification and Differentiation

2.6.1. DNA Extraction. The genomic DNA of *Bacillus* strains was isolated using modification method from Sambrook et al. [18].

2.6.2. Identification by 16S Ribosomal RNA (rRNA)

PCR Amplification according to Sambrook et al. [18]. The PCR amplification reactions were performed in a total volume of 50 μL. Each reaction mixture contained the following solutions: 2 μL of DNA (40 ng), 1 μL of 10 pmol forwarded 16S-rRNA primer (5′-AAATGGAGGAAGGTGGGGAT-3′); 1 μL of 10 pmol reverse 16S rRNA primer (5′-AGGAGGTGATCCAACCGCA-3′); 0.8 μL of 12.5 mM (dNTP's); 5 μL of PCR buffer included $MgCl_2$, and 0.2 μL Taq polymerase (1 Unit) and water-free DNAse and RNAse were added up to 50 μL. The PCR apparatus was programmed as follows: 3 min denaturation at 95°C, followed by 35 cycles that consisted of 1 min at 95°C, 1 min at 58°C, and 1 min at 72°C, and a final extension was 10 min at 72°C. The products of the PCR amplification were analyzed by agarose gel electrophoresis (2%).

2.6.3. PCR Cleanup and 16S rRNA Sequencing. The PCR products were cleaned up for DNA sequencing following the method described by Sambrook et al. [18]. Automated DNA sequencing based on enzymatic chain terminator technique, developed by Sanger et al. [19], was carried out using 3130X DNA Sequencer (Genetic Analyzer, Applied Biosystems, Hitachi, Japan).

Production and Characterization of Keratinolytic Protease from New Wool-Degrading Bacillus Species
Isolated from Egyptian Ecosystem

159

2.6.4. Phylogenetic Analysis.

Similarity analysis of the nucleotides was performed by BLAST searches against sequences available in GenBank. For phylogenetic tree construction, multiple sequences were obtained from GenBank and the alignments were performed using MEGA 5 software version 5.1 [20].

2.7. Keratinolytic Protease Production.

Bacillus sp. MA20 and Bacillus sp. MA21 were first inoculated in liquid LB medium to produce large amount of cells. After 18 hrs, the colony forming units (CFU)/mL culture were $3 * 10^6$, and 2% volume of the liquid medium was transferred to the production medium using 250 mL flask containing 100 mL of the production medium. The production medium containing (w/v) NaCl, 0.5 g/L; K_2HPO_4, 0.3 g/L; KH_2PO_4, 0.4 g/L; wool, 10 g/L; and the pH was adjusted 7.0–7.2 using 2N of NaOH and HCl [16]. The cultivated media were incubated at 37°C and 200 rpm for 5 days. Culture supernatants were obtained by centrifugation at 12,000 rpm and 4°C for 30 min. The different supernatants which contain the crude enzymes were used in assay and analysis of enzymes.

2.8. Enzymes Assay

2.8.1. Detection of the Proteolytic Activity on Plates.

The crude enzyme of Bacillus sp. MA20 and Bacillus sp. MA21 was screened for their proteolytic activity using agar well diffusion plate method described by Amara et al., after modification [21]. One gram of gelatin powder was suspended in 100 mL phosphate buffer pH 7 and autoclaved. After sterilization, the soluble component was added to sterile water containing agar (18 gm agar/L). The suspension then stirred gently and distributed in Petri dishes (25 mL/plate). After complete solidification of the agar on plates, wells were punched out of the agar, by using a clean sterile cork borer. The base of each hole was sealed with a drop of melted sterile water agar (15 g agar per liter H_2O) using sterile Pasteur pipette. Fifty μL of each bacterial supernatant was added to each well and preincubated at 4°C for 2 hrs and then overnight incubated at different temperatures.

2.8.2. Visualization of the Enzyme Clear Zone.

Coomassie blue (0.25%, w/v) in methanol-acetic acid-water 5 : 1 : 4 (v/v/v) was used in plates staining to visualize the gelatin hydrolysis where 10 mL was added to each plate and incubated in room temperature for 15 min followed by removing the staining solution from the plates surfaces and washing gently by distilled water. Then the plates were destained using destining solution (66 mL methanol, 20 mL acetic acid, and 114 mL H_2O_{bidest}) for a suitable time [22]. Also extracellular protease detection was determined according to Vermelho et al., [23] with modification. Staining was performed with 0.1% amido black in methanol-acetic acid-water 30 : 10 : 60 (v/v/v) for 5–20 min (according to the quality of amido black stain) at room temperature. Regions of enzyme activity were detected as clear areas, indicating that hydrolysis of the substrates has been occurred.

2.8.3. Enzyme Assay Spectrophotometrically

Preparation of Casein in Different Buffers with Different pH. Casein Hammarstein was weighted in a quantity of 0.325 g and dissolved in 50 mL of different buffers at different pH. The mixture was dissolved by heating gently to 80–90°C without boiling in water bath according to Amara and Serour [24] or using hot plate with stirring for more accurate because casein is highly stick with the walls of the container. The following buffers were used:

(i) sodium phosphate, pH 6-7-8 (0.1 M)

(ii) glycine/NaOH pH 9-10 (0.1 M)

(iii) sodium phosphate dibasic/NaOH pH 11 (0.05 M)

(iv) KCl/NaOH pH 12 (0.2 M).

2.8.4. Optimum Temperature of the Enzymes.

Ten μL of each supernatant which contains the crude enzyme was added to 490 μL of the casein Hammarsten soluble in buffer pH 7. The enzymes-substrate mixture was incubated at different temperatures of 20, 30, 40, 50, 60, 70, and 80°C using water bath for 15 min. After the incubation period the enzyme reaction stopped by adding 500 μL of 10% trichloroacetic acid (TCA). The mixture was allowed to stand in ice for 10 min and then centrifuged at 13,000 rpm for 10 min. The absorbance of each sample was determined spectrophotometrically at 280 nm and their tyrosine content derived from the tyrosine standard curve which was carried out according to Amara and Serour [24] and the enzyme activity was determined as Unit/mL.

2.8.5. Optimum pH of the Enzymes.

Ten μL of each supernatant which contains the crude enzyme was added to the 490 μL of the casein Hammarsten soluble in different pH values. The enzymes-substrate mixture was incubated at 60°C which acts as optimum temperature in a water bath for 15 min. At the end of incubation period the enzyme reaction stopped by adding 500 μL of 10% trichloroacetic acid (TCA). The mixture was allowed to stand in ice for 10 min and then centrifuged at 13,000 rpm for 10 min. The absorbance of each sample was determined spectrophotometrically at 280 nm and their tyrosine content derived from the tyrosine standard curve and the enzyme activity determined as Unit/mL.

2.8.6. Confirmation of Optimum pH and Temperature.

The previous reaction was performed using casein Hammarsten soluble in Glycine/NaOH pH 9. The enzyme substrate mixture was incubated at various temperature of 20, 30, 40, 50, 60, 70 and 80°C in water bath for 15 min. The enzyme activity was calculated as mentioned previously.

2.8.7. The Optimized Enzymes Reaction.

The optimized reaction is a mixing between 10 uL of crude enzyme and 490 μL of the casein Hammarsten soluble in Glycine/NaOH pH 9. The mixture was incubated at 60°C in water bath for 15 min. The reaction was stopped with 500 μL of 10% TCA. The mixture was allowed to stand in ice for 10 min and then was centrifuged at 13,000 rpm for 10 min. The enzyme activity was determined as mentioned above using tyrosine standard curve.

2.8.8. Thermal Stability.
The thermostability was carried out by preincubating the crude enzyme solution at a temperature range of 4°C to 80°C for 0.0 to 24 hrs. The residual activity was measured with standard enzyme reaction. The control is enzyme reacted at zero time (consider as 100%) [16].

2.8.9. pH Stability.
The pH stability was determined by preincubating the enzyme solution in buffers with different pH values (3–12) at room temperature from 0.0 to 48 hrs. The residual activity was measured with standard enzyme reaction. The control is enzyme reacted at zero time and considered as 100% [16]. The following buffers were used:

(i) citrate buffer pH 3-4-5-6 (0.1 M)

(ii) sodium phosphate pH 7-8 (0.1 M)

(iii) glycine/NaOH pH 9-10 (0.1 M)

(iv) sodium phosphate dibasic/NaOH pH 11 (0.05 M)

(v) KCl/NaOH pH 12 (0.2 M).

2.8.10. Effect of Inhibitors.
The inhibitors were added to supernatants which contain the produced enzyme and were incubated for 30 min at 30°C before being tested for proteolytic activity. Protease inhibitors phenylmethanesulphonyl fluoride (PMSF), ethylenediaminetetraacetic acid (EDTA), β-mercaptoethanol, and the detergent sodium dodecyl sulfate (SDS) were used. The inhibitors stocks were prepared in distilled water except that PMSF was prepared by using isopropanol. The final concentrations of PMSF, EDTA, and β-Mercaptoethanol are 5 mM and 1 mM while SDS is 0.5% and 0.1% (w/v). The control was enzyme mixed with distilled water instead of inhibitors. Control activity was considered to be 100% [25].

2.8.11. Effect of Metal Ions.
The effect of metal ions on protease activity was investigated using two concentrations of 5 mM and 10 mM (final concentration). The metal stock solutions were prepared in distilled water and diluted to the appropriate concentrations. The enzyme solution was mixed with the different metal solutions and incubated for 30 min at 30°C before assay. A control was also included where the enzyme was mixed with distilled water instead of metal solution. Control activity was considered to be 100% [25]. $ZnCl_2$, $MgCl_2$, $CuSO_4$, Urea, $HgCl_2$, $CaCl_2$, $BaCl_2$, Guanidin HCl, and $MnCl_2$ were used.

2.8.12. Effect of Solvents.
The effect of the different solvents (Methanol, Ethanol, DMSO, Isopropanol, Tween 20, and Triton X100) on protease activity was investigated using a concentration of 1% and 0.5% (final concentration) [25].

2.9. First-Dimension Protein Electrophoresis

2.9.1. Sodium Dodecyl Sulfate Polyacrylamid Gel Electrophoresis (SDS-PAGE).
Characterization of proteins and evaluation of the protein enrichment process SDS-PAGE was performed in a discontinuous SDS-PAGE vertical slab gel electrophoresis apparatus as described by Laemmli [26]. Discontinuous SDS-PAGE consisted of a stacking gel (5%, w/v, pH 6.8) and a separating gel (12%, w/v, pH 8.8). The separating gel was prepared in a 1 mm slab gel (10 × 10 cm).

2.10. Gel Staining Using Silver Stain.
The gels were stained using silver nitrate staining methods as described by Blum et al. [27].

2.11. Zymogram for the Detection of Protease Activity Using SDS-PAGE.
As above in the case of protease zymography using SDS PAGE and the separating gel concentration was 12%. The samples were mixed with 5X sample buffer without β-mercaptoethanol and without boiling.

After running the gel at 80 V the SDS-PAGE gel was stripped off from the gel plate and soaked in 1% triton X100 for 2 hrs to change the solution every 1 hour or in 2.5% triton X100 for 30 min for removing SDS and renaturating the enzymes.

The gel was washed three times with tape water and soaked in 1% gelatin powder solubilized in Glycine/NaOH buffer pH 9 for 90 min at 60°C. The gel was washed with buffer for 10 min before staining.

Then the gel was stained with Coomassie blue stain for 2 hrs. After that the gel was destained till the active bands appear.

2.12. Zymogram for Detecting Protease Activity Using Native PAGE.
The gel was carried out using all components of the SDS-PAGE and the same conditions but without SDS as well as the buffers described above. The samples were prepared by mixing with 5X sample buffer for native PAGE and without boiling. The gel was washed with tap water and soaked into 1% gelatin powder as mentioned previously. Then the gel was stained with Coomassie blue stain for 2 hrs. After that, the gel was destained till the active bands appear.

2.13. Two-Dimension Polyacrylamide Gel Electrophoresis (2D-PAGE).
First-dimension isoelectric focusing (IEF) was performed using Ettan IPGphor3 and the second-dimension SDS-PAGE was applied using vertical electrophoresis of Ettan DALTtwelve System. The experiment was carried out using the operation instruction of manufacturer (GE healthcare company). The gels were stained with Coomassie stain where Coomassie blue R-250 (0.02 g) in methanol-glacial acetic acid-distilled water (10 mL, 5 mL, and 100 mL resp.) was used.

3. Results and Discussion

3.1. Strain Isolation.
Microbial keratinolytic protease has been described for various biotechnological applications in food, detergent, textiles, and leather industries, and yet the

Production and Characterization of Keratinolytic Protease from New Wool-Degrading Bacillus Species Isolated from Egyptian Ecosystem

161

growing demand for these enzymes necessitates the screening for novel keratinolytic microorganisms with potential applications [28, 29]. Keratinolytic protease has been described for several species of *Bacillus* [30, 31] due to the broad distribution of keratinase among these genera, and this study focused on keratinolytic protease production from them.

A total of 48 pure cultures of spore-forming bacteria were isolated and purified which obtained from different samples collected from Governorates of Egypt. All isolates were screened using selective method for *Bacillus* isolation. The proteolysis activities of all the isolates were detected using the plate test method containing LB agar medium with 1% (w/v) skim milk. Among the isolates analyzed, 12 isolates exhibited proteolytic activity in which they had a halo diameter of fivefold longer than the colony diameter. All isolates have the proteolytic activity but do not have the ability to degrade wool, for that the twelve selected isolates were grown using a medium which contain (w/v) NaCl, 0.5 g/L; K_2HPO_4, 0.3 g/L; KH_2PO_4, 0.4 g/L; wool, 10 g/L; and the pH was adjusted at 7.0–7.2 using 2N NaOH and HCl. The flasks were incubated for 5 days until the wool has been completely degraded by some isolates. Three *Bacillus* isolates show the ability to degrade the wool and gained the names *Bacillus* sp. MA20, *Bacillus* sp. MA21, and *Bacillus* sp. MA10. In the last screening step, the obtained supernatants from the cultivation of three above *Bacillus* strains culture media were assayed using agar well diffusion in Petri dishes including gelatin powder plate which suspended in phosphate buffer pH 7 as in Figure 1. *Bacillus* sp. MA20 and *Bacillus* sp. MA21 were selected for studying the keratinolytic protease enzyme based on the diameter of clear zone (Figure 1). By refering to their isolation source, the two selected strains were found to be isolated from Minia Governorates. This method is simple but proves to be efficient for determining the best enzyme-producing isolates. Out of the three strains, the two which were given the best result have been subjected to further identification.

3.2. Strains Identification

3.2.1. Identification by Morphological and Biochemical Tests.
The morphological and physiological characteristics of the isolated strains were compared with the data from Bergey's Manual of Determinative Bacteriology [17] who revealed that *Bacillus* sp. MA20 and *Bacillus* sp. MA21 are matching with those of *B. subtilis* group. In previous articles on taxonomy, species included in the *B. subtilis* group are the following: *B. velezensis, B. atrophaeus, B. mojavensis, B. malacitensis, B. axarquiensis, B. nematocida, B. vallismortis, B. subtilis,* and *B. amyloliquefaciens* [32–34].

The two strains are aerobic, motile, and Gram-positive rods. The smears of *Bacillus* sp. MA20 and *Bacillus* sp. MA21 were scanned by scanning electron microscope which indicates the bacterial size of the two *Bacillus* strains which was measured by slime view program software.

The results of biochemical tests of the *Bacillus* sp. MA20 and *Bacillus* sp. MA21 indicated that they are related to

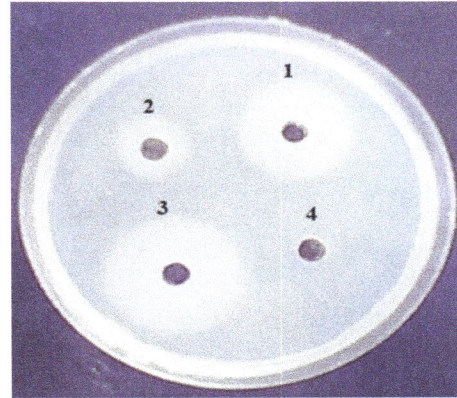

FIGURE 1: Proteolytic activity of the supernatants obtained from three *Bacillus* sp. on gelatin suspended in phosphate buffer pH 7. The well number (1) is supernatant from *Bacillus* sp. MA21, number (2) is from *Bacillus* sp. MA10, number (3) is from *Bacillus* sp. MA20, and (4) is control free of enzyme.

B. subtilis, and *B. amyloliquefaciens* which summarized in Table 1.

3.3. Identification Based on Genetic Materials (DNA)

3.3.1. DNA Extraction.
The isolated DNA was analyzed by gel electrophoresis and the quality of the DNA for each sample has been identified for further investigation.

3.3.2. Identification by 16S Ribosomal RNA (rRNA).
The amplified 16S rRNA gene from the DNA of *Bacillus* sp. MA20 and *Bacillus* sp. MA21 was determined using 2% agarose gel. The size of the amplified fragments was determined by using size standard (Gene ruler 50 bp–1031 bp DNA ladder). The PCR products were visualized under UV light and photographed using gel documentation system. Approximately 380 bp of 16S rRNA gene was amplified.

The PCR products were purified and sequenced using 16S forward primer. The sequences of *Bacillus* sp. MA20 and *Bacillus* sp. MA21 were deposited in national center for biotechnology information (NCBI GenBank) under the Accession numbers (HQ115599.1–HQ115600.1), respectively. The basic local alignment search tool (BLAST) algorithm was used to retrieve for homologous sequences in GenBank.

The *Bacillus* sp. MA20 revealed 98% identity to *Bacillus amyloliquefaciens* and *Bacillus subtilis* while *Bacillus* sp. MA21 revealed 97% identity to *Bacillus subtilis* and *Bacillus amyloliquefaciens*. Based on the morphological, biochemical, and molecular characteristics, the *Bacillus* sp. MA20 and *Bacillus* sp. MA21 were designated as *B. amyloliquefaciens* MA20 and *B. subtilis* MA21, respectively.

A phylogenetic tree based on the comparison of 16S rRNA sequences of reference strains was constructed. The phylogenetic analysis was performed with 341 bp sequences using the software MEGA 5 [20], using the neighbour-joining method and based on Jukes-Cantor distances. The branching

TABLE 1: Morphological and biochemical properties of *Bacillus* sp. MA20 and *Bacillus* sp. MA21.

Characteristics	*Bacillus* sp. MA20	*Bacillus* sp. MA21
Morphological		
Shape	Rods	Rods
Gram stain	G+ve	G+ve
Motility	Motile	Motile
spore formation	+ve	+ve
Growth		
Growth temperature	15°C–50°C	15°C–60°C
Growth pH	5–8	5–8
Biochemical tests		
Oxidase	+ve	+ve
Catalase	+ve	+ve
Voges-Proskauer	+ve	+ve
Indol production	−ve	−ve
Nitrate reduced to nitrite	+ve	+ve
Hydrolysis of		
Casein	+ve	+ve
Gelatin	+ve	+ve
Wool	+ve	+ve
Starch	+ve	+ve
Acid from		
Glucose	+ve	+ve
Arabinose	+ve	+ve
Xylose	+ve	+ve
Mannitol	+ve	+ve
Gas from glucose	−ve	−ve
Utilization of		
Citrate	+ve	+ve
Propionate	−ve	−ve

pattern was checked by 500 bootstrap replicates (Figures 2 and 3).

3.4. Media Screening for Keratinolytic Protease Production.
Keratinolytic proteases are largely produced in a basal medium with keratinous substrates, and most of the organisms could utilize keratin sources such as feather and wool as the sole source of carbon and nitrogen [35, 36]. *B. amyloliquefaciens* MA20 and *B. subtilis* MA21 were tested on eight nutrient media. The selected media is medium (1) which is containing (w/v) NaCl, 0.5 g/L; K_2HPO_4, 0.3 g/L; KH_2PO_4, 0.4 g/L; wool, 10 g/L, and the pH was adjusted 7.0–7.2 using 2N of NaOH and HCl.

Korniłłowicz-Kowalska, (1997) and Amara and Serour (2008) reported that the mass loss of the keratin substrate is the most reliable indicator of microbial keratinolytic abilities. *B. amyloliquefaciens* MA20 and B. *subtilis* MA21 have the ability to degrade wool completely after incubation for 5 days and remain it as powders in the bottom of flasks [24, 37].

3.5. Enzyme Production and Characterization.
The keratinolytic protease enzyme was produced using production medium mentioned above for each of *B. amyloliquefaciens* MA20 and *B. subtilis* MA21, and the crude enzymes were characterized.

3.6. Detection of the Proteolytic Activity on Plates.
The crude enzymes of *B. amyloliquefaciens* MA20 and *B. subtilis* MA21 were assayed by agar well diffusion methods on plates containing gelatin soluble in glycine/NaOH buffer of pH 9. The Coomassie blue and amido black bind to gelatin and whole plate giving the colour of dye except the hydrolysis areas which appear as transparent without dye (Figure 4).

3.7. Characterization of Keratinolytic Protease Enzyme

3.7.1. Influence of pH and Temperature.
The effect of temperature and pH on enzymes activity and stability was determined. The optimum temperature and pH for protease activity of enzymes produced by *B. amyloliquefaciens* MA20 and *B. subtilis* MA21 was found to be 60°C, 9.0, respectively.

The enzymes activity was investigated at pH 7 and different temperatures (Figure 5). The proteolytic activities were determined at optimized temperature 60°C and different pH (Figure 6). For confirmation from the optimized reaction, the enzymes activity was assayed at pH 9 with different temperatures (Figure 7).

Results revealed that the keratinolytic protease from *B. amyloliquefaciens* MA20 and *B. subtilis* MA21 is similar to those produced using bacteria, actinomycetes, and fungi and has a pH optimum in a neutral-to-alkaline range [38, 39]. The optimal temperature for activity was also found in the usual range for keratinolytic protease (30–80°C).

The maximum activity of protease enzyme produced by *B. amyloliquefaciens* MA20 was 922 U/mL at pH 9 and 60°C while the maximum activity of protease enzyme produced by *B. subtilis* MA21 was determined as 814 U/mL at the same conditions. The alkaline pH of the keratinolytic protease enzyme from *B. amyloliquefaciens* MA20 and *B. subtilis* MA21 suggests a positive biotechnological potential.

The enzymes were stable at temperatures between 4 and 70°C. The enzyme produced by *B. amyloliquefaciens* MA20 is more thermostable than enzyme produced by *B. subtilis* MA21 (Figures 8 and 9). The results of thermal stability indicated that the keratinolytic protease from *B. amyloliquefaciens* MA20 is more thermostable than the enzyme produced by *B. subtilis* MA21.

The thermal stability studies give the enzyme from *B. amyloliquefaciens* MA20 the advantage for using in industrial applications which are the main objective of this study.

The crude enzyme from *B. amyloliquefaciens* MA20 is described as stable over a broad pH range of 4.0–12.0, but the best stable pH is 9 and 10 (Figure 10). Figure 11 indicates that the enzyme from *B. subtilis* MA21 was stable in pH range (5.0–12.0) with high stability at pH 9. The stability of keratinolytic proteases produced by *B. amyloliquefaciens* MA20 and *B. subtilis* MA21 has been suggested to offer great advantages for

Production and Characterization of Keratinolytic Protease from New Wool-Degrading Bacillus Species
Isolated from Egyptian Ecosystem

163

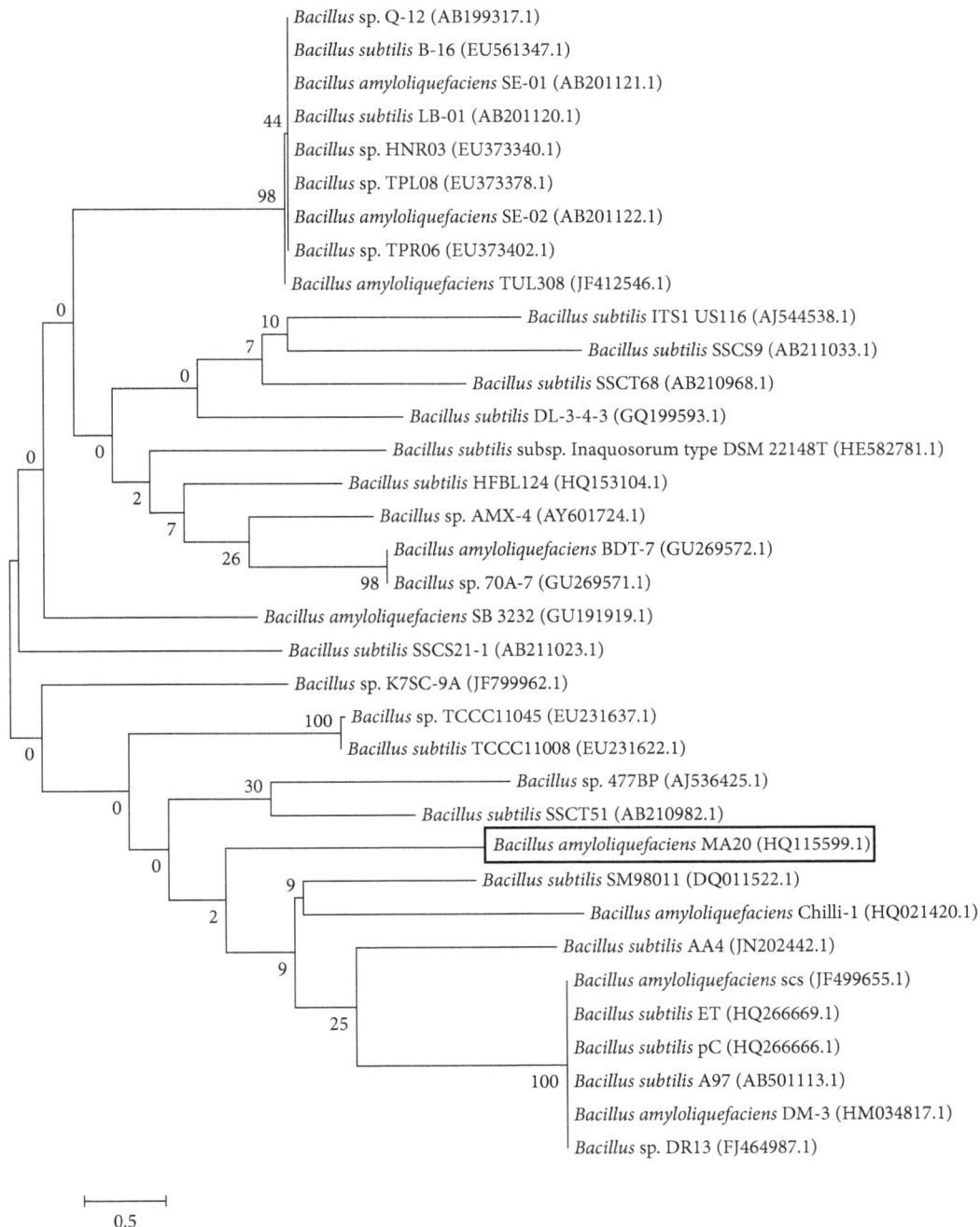

FIGURE 2: Phylogenetic position of *Bacillus amyloliquefaciens* MA20 within the genus *Bacillus*. The branching pattern was generated by neighbor-joining tree method. The Genbank accession numbers of the 16S rRNA nucleotide sequences are indicated in brackets. The number of each branch indicates the bootstrap values. The bar indicates a Jukes-Cantor distance of 0.5.

industrial purposes such as wastewater treatment and leather tanning [40].

3.8. Influence of Protease Inhibitors, Solvents, and Metal Ions on Enzymes Activity.
Mostly keratinolytic proteases belong to the subtilisin family of serine proteases with cysteine

proteases, which have higher activity on casein [41]. The keratinolytic proteases produced by *Bacillus* sp. are often serine-proteases, such as the enzymes produced by *B. licheniformis* [42], *B. pseudofirmus* [43], and *B. subtilis* [44, 45]. Protease activity of enzymes prepared from *B. amyloliquefaciens* MA20 and *B. subtilis* MA21 was completely inhibited by serine protease inhibitor (PMSF). The result indicated the presence

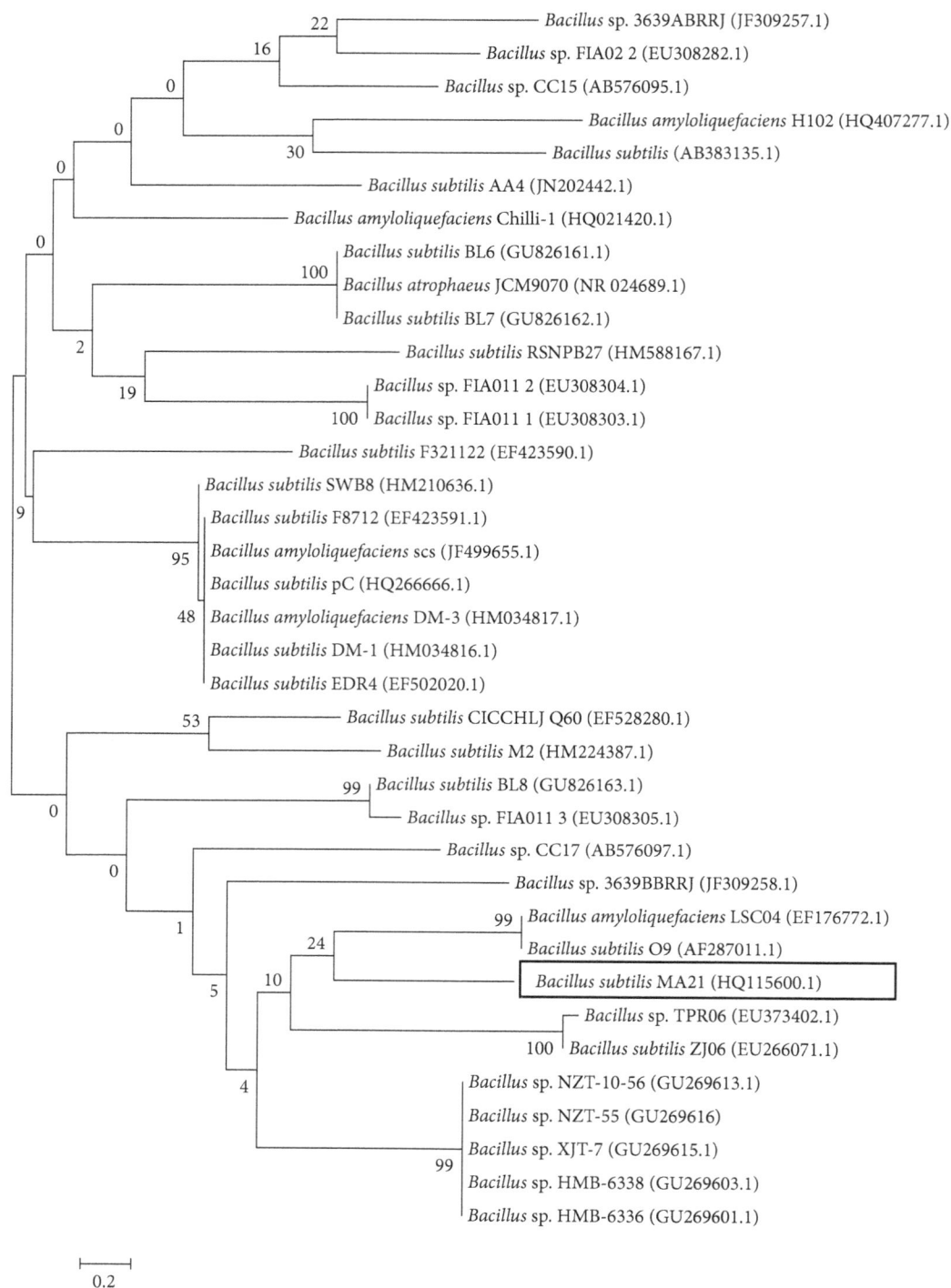

FIGURE 3: Phylogenetic position of *Bacillus subtilis* MA21 within the genus *Bacillus*. The branching pattern was generated by neighbor-joining tree method. The Genbank accession numbers of the 16S rRNA nucleotide sequences are indicated in brackets. The number of each branch indicates the bootstrap values. The bar indicates a Jukes-Cantor distance of 0.2.

of the serine group in the enzyme active site. The enzymes activity was partially inhibited by EDTA (Tables 2 and 3). This suggests that the keratinolytic protease from the *Bacillus* strain belongs to keratinolytic serine protease family.

The stability of keratinolytic proteases in presence of SDS acts as a positive advantage of enzymes feature because it indicated the possibility of using them in different industrial purposes as detergent industry, leather industry, and wool

Production and Characterization of Keratinolytic Protease from New Wool-Degrading Bacillus Species
Isolated from Egyptian Ecosystem

165

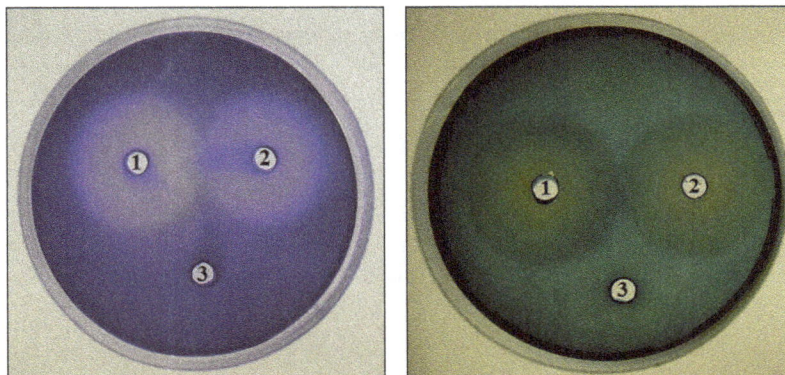

FIGURE 4: Proteolytic activity of crude enzymes produced by *B. amyloliquefaciens* MA20 and *B. subtilis* MA21 using gelatin as substrate and both of Coomassie blue and amido black stains. Blue plate stained with Coomassie blue while green plate stained with Amido black. The well (1) refers to enzymes produced by *B. amyloliquefaciens* MA20, (2) refers to enzymes produced by *B. subtilis* MA21, and (3) is inactive enzyme.

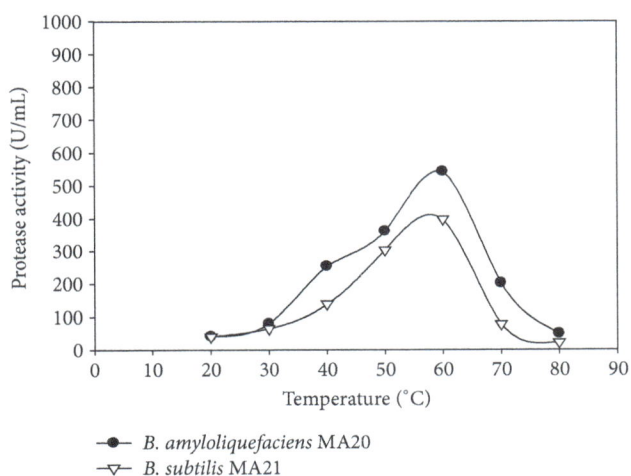

FIGURE 5: Effect of temperatures at pH 7 on protease activity of crude enzymes produced by *B. amyloliquefaciens* MA20 and *B. subtilis* MA21.

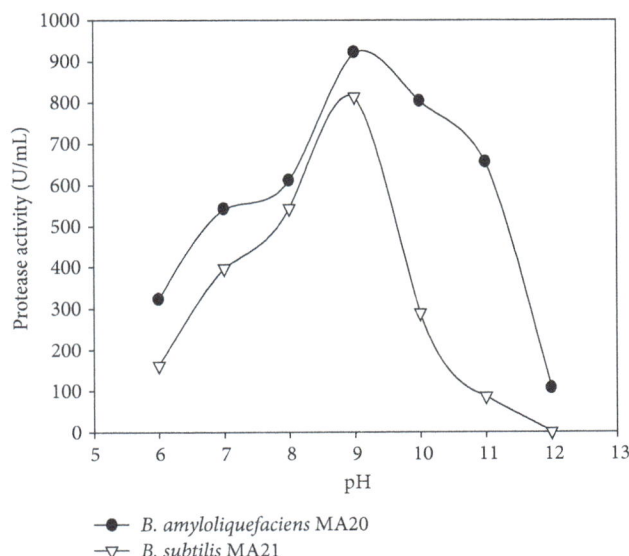

FIGURE 6: Effect of pH at 60°C on protease activity of crude enzymes produced by *B. amyloliquefaciens* MA20 and *B. subtilis* MA21.

improvement. The stability toward SDS is important because a few authors reported that SDS-stable enzymes are also not generally available except for a few strains such as *Bacillus clausii* I-52 [46] and *Bacillus* sp. RGR-14 [47]. The effects of various solvents and metal ions on enzyme activity were examined in order to find which ions are stimulators and which are inhibitors of the catalytic process.

The metal ions were used with two final concentrations (5.0 mM and 10 mM) while the used solvents with final concentrations (1% and 0.5%). The effects of metal ions and solvents on enzyme activities are summarized in Tables 2 and 3.

3.9. First-Dimension Protein Electrophoresis.

The production of extracellular keratinolytic proteases by *B. amyloliquefaciens* MA20 and *B. subtilis* MA21 was evaluated by SDS-PAGE and zymogram analysis. Giongo et al. found multiple bands after zymogram of keratinolytic protease produced by three strains of *Bacillus sp*. P6, P7, and P11 using feather degrading medium [25]. Growth of the two strains in presence 10 g/L of wool resulted in the production of multiple proteases, as observed by a zymogram on gelatin (Figures 12 and 13). Multiple clear zones were observed and detected using (PAGE Ruler Prestained Protein Ladder) which indicated that the proteolytic activity was not due to a single protein. The native-PAGE zymogram of enzymes from the 2 strains was performed and compared with protein ladder (Figures 14 and 15). The zymogram using native page was carried out for detecting the bands of protein where the purification performs using native protein. Multiple bands could be detected on native-PAGE zymogrm.

3.10. Two-Dimension Protein Electrophoresis.

The two-dimensional polyacrylamide gel electrophoresis (2D-PAGE)

TABLE 2: Effect of protease inhibitors, metal ions, and solvents on proteolytic activity of *B. amyloliquefaciens* MA20.

Substance	Final concentration	Relative activity (%)	Final concentration	Relative activity (%)
Control	0	100	0	100
β-mercaptoethanol	5 mM	18	1 mM	98
PMSF	5 mM	0	1 mM	0
EDTA	5 mM	35	1 mM	97
SDS	0.5%	54	0.1%	122
$ZnCl_2$	10 mM	153	5 mM	156
$MgCl_2$	10 mM	119	5 mM	95
$CuSO_4$	10 mM	266	5 mM	204
Urea	10 mM	114	5 mM	139
$HgCl_2$	10 mM	100.4	5 mM	107
$CaCl_2$	10 mM	104	5 mM	107
$BaCl_2$	10 mM	115	5 mM	116
Guanidin HCl	10 mM	103	5 mM	132
$MnCl_2$	10 mM	71	5 mM	162
Methanol	1%	90	0.5%	100
Ethanol	1%	103	0.5%	100
DMSO	1%	90	0.5%	103
Isopropanol	1%	98	0.5%	108
Tween 20	1%	86	0.5%	94
Triton X100	1%	61	0.5%	65

TABLE 3: Effect of protease inhibitors, metal ions, and solvents on proteolytic activity of *B. subtilis* MA21.

Substance	Final concentration	Relative activity (%)	Final concentration	Relative activity (%)
Control	0	**100**	0	100
β-mercaptoethanol	5 mM	48	1 mM	70
PMSF	5 mM	0	1 mM	0
EDTA	5 mM	13	1 mM	96
SDS	0.5%	25	0.1%	99
$ZnCl_2$	10 mM	**92**	5 mM	93
$MgCl_2$	10 mM	**22**	5 mM	98
$CuSO_4$	10 mM	**23**	5 mM	73
Urea	10 mM	**24**	5 mM	52
$HgCl_2$	10 mM	**27**	5 mM	37
$CaCl_2$	10 mM	**65**	5 mM	142
$BaCl_2$	10 mM	**29**	5 mM	69
Guanidin HCl	10 mM	**42**	5 mM	49
$MnCl_2$	10 mM	**64**	5 mM	67
Methanol	1%	92	0.5%	99
Ethanol	1%	92	0.5%	97
DMSO	1%	78	0.5%	96
Isopropanol	1%	98	0.5%	101
Tween 20	1%	78	0.5%	86
Triton X100	1%	57	0.5%	133

TABLE 4: Two-dimension protein gel electrophoresis report of crude enzymes produced by *B. amyloliquefaciens* MA20 and *B. subtilis* MA21.

Gels	Spots	Minimum gray	Maximum gray	Columns	Rows	Pixel width	Pixel height
B. amyloliquefaciens MA20	237	23	209	2512	1510	353	353
B. subtilis MA21	291	0	255	2336	1452	353	353

Production and Characterization of Keratinolytic Protease from New Wool-Degrading Bacillus Species
Isolated from Egyptian Ecosystem

167

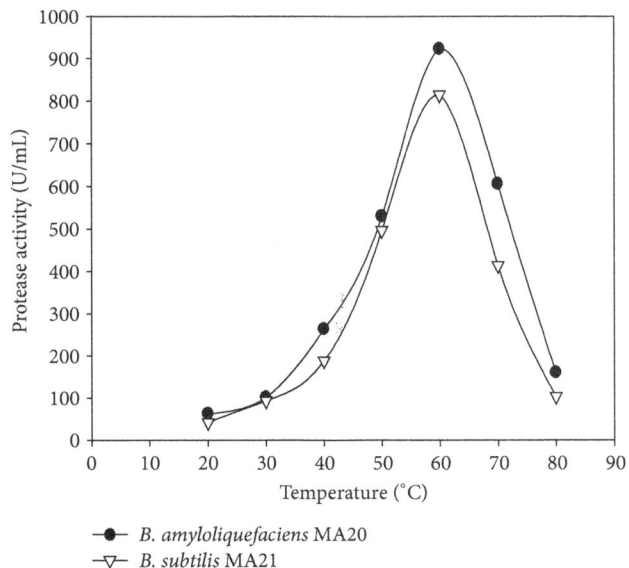

FIGURE 7: Effect of temperatures at pH 9 on protease activity of crude enzymes produced by *B. amyloliquefaciens* MA20 and *B. subtilis* MA21.

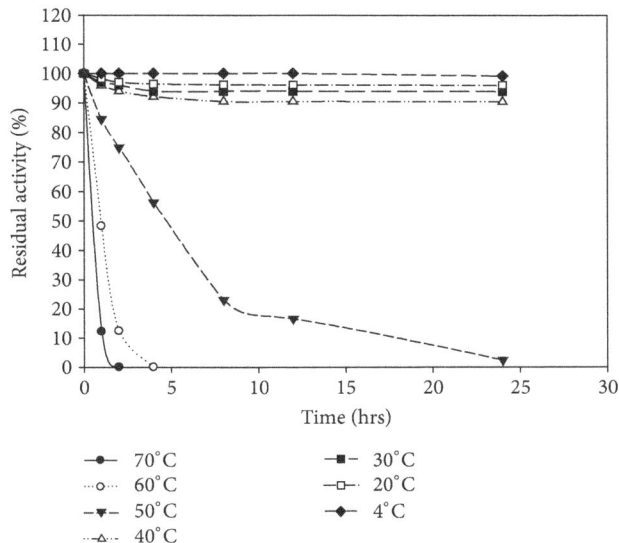

FIGURE 9: Thermal stability of crude enzyme produced by *B. subtilis* MA21.

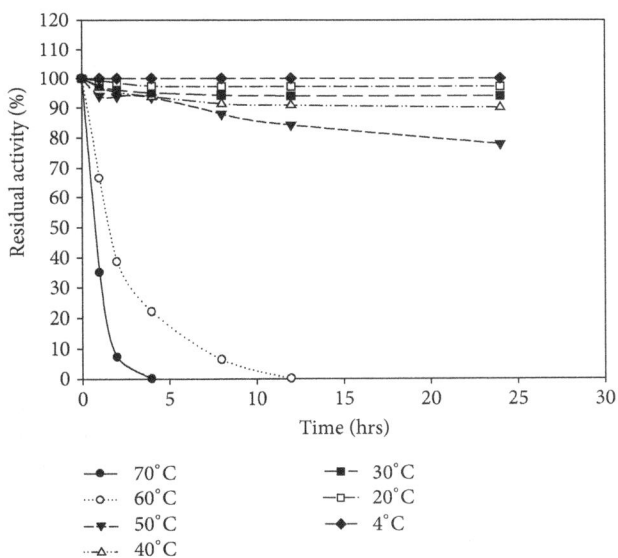

FIGURE 8: Thermal stability of crude enzyme produced by *B. amyloliquefaciens* MA20.

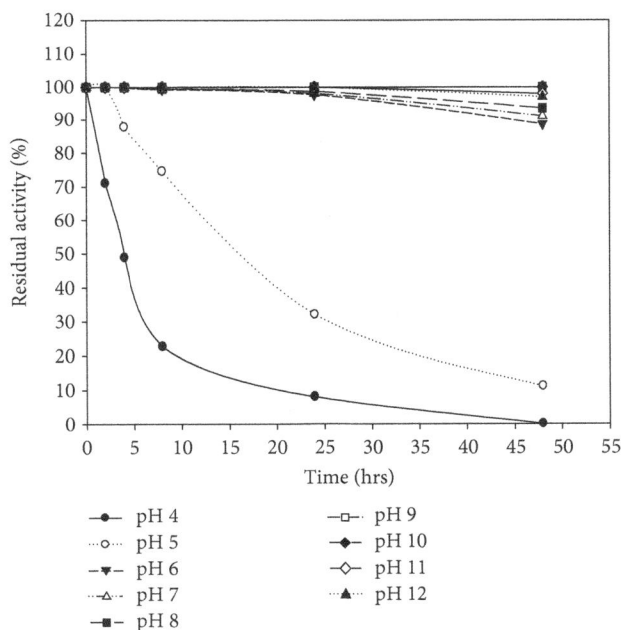

FIGURE 10: pH stability of crude enzyme produced by *B. amyloliquefaciens* MA20.

is an advanced technique that depends on protein separation firstly according to the pH and secondly based on the molecular weight. This technique was carried out to differentiate between the crude keratinolytic proteases obtained from *B. amyloliquefaciens* MA20 and *B. subtilis* MA21. After the separation based on the pH, the strips were applied on gel for separating according to molecular weight. The gels were stained with Coomassie blue for detecting the protein spots.

In order to classify the keratinolytic protease enzyme, a systemic comparison of 2D maps of proteases was conducted

with image master 2D Platinum 6 software. The two gels were photographed by high quality scanner image (Figure 16).

The results were analyzed using image master 2D Platinum 6 software. The spots were detected before matching between the two gels and the report was obtained (Table 4). Every spot refers to the presence of one protein. The gel which was loaded with keratinolytic protease from *B. amyloliquefaciens* MA20 had 237 spots while the gel applied with *B. subtilis* MA21 had 291 spots.

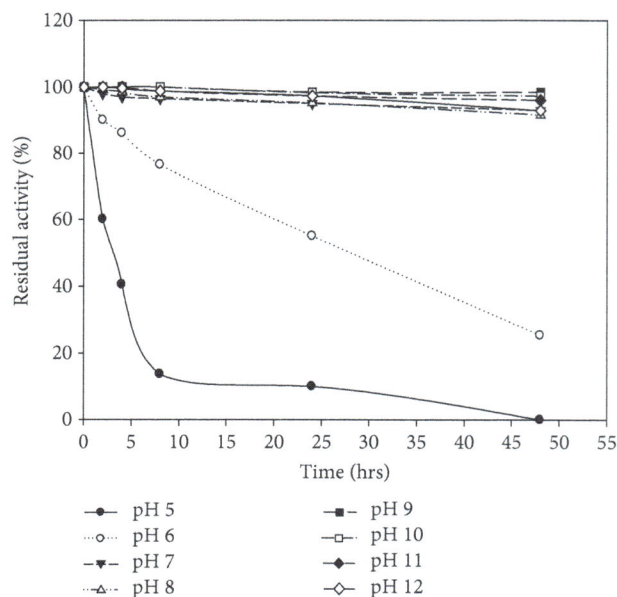

FIGURE 11: pH stability of crude enzyme produced by *B. subtilis* MA21.

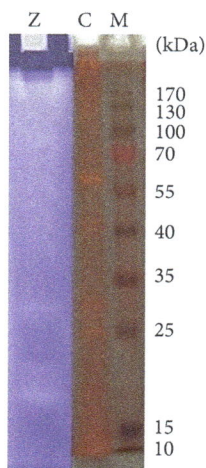

FIGURE 12: SDS-PAGE and zymogram analysis of keratinolytic protease enzyme from *B. amyloliquefaciens* MA20. The lane (M) is prestained protein ladder, lane (C) is protein pattern of crude enzyme, and lane (Z) is zymogram of enzyme.

The 2 gels were matched and consider gel resulted from crude enzymes of *B. amyloliquefaciens* MA20 as reference and the final report that summarizes the difference between 2 gels was obtained (Table 5). The 2 gels were matched and the percent of matching between them was 13.25%.

4. Conclusion

This paper described in details different methods that lead to the production of keratinolytic protease from two *Bacillus* sp. strains. Different methods and assays ranging from simple to advanced were carried out to prove that the enzymes have

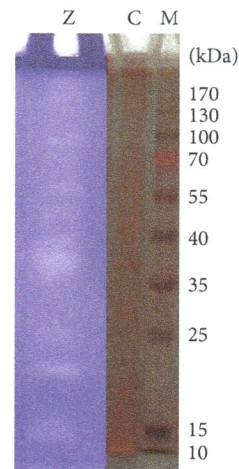

FIGURE 13: SDS-PAGE and zymogram analysis of keratinolytic protease enzyme from *B. subtilis* MA21. The lane (M) is prestained protein ladder, lane (C) is protein pattern of crude enzyme, and lane (Z) is zymogram of enzyme.

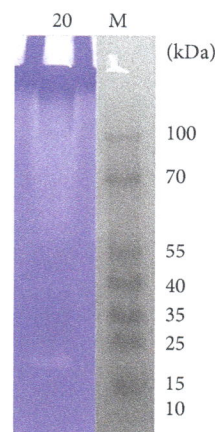

FIGURE 14: Native-PAGE zymogram analysis of keratinolytic protease enzyme from *B. amyloliquefaciens* MA20. The lane (M) is prestained protein ladder, and lane (20) is zymogram of enzyme.

TABLE 5: Match statistics report of crude enzymes produced by *B. amyloliquefaciens* MA20 and *B. subtilis* MA21 after two-dimension protein gel electrophoresis.

Gel name	Gel name	Number of matches	Percent matches
B. amyloliquefaciens MA20 (reference)	*B. subtilis* MA21	35	13.2576

the ability to degrade wool. The two used strains have been selected based on that they belong to *Bacillus* sp. and show the best keratinolytic activities. The study included molecular and bioinformatic tools to identify the two *Bacillus* sp. strains. 16S rRNA, phylogenetic tree, Blast search for nucleotide similarity, scanning electron microscope, SDS-PAGE, and 2D-PAGE have been used to differentiate the both strains

Production and Characterization of Keratinolytic Protease from New Wool-Degrading Bacillus Species
Isolated from Egyptian Ecosystem

169

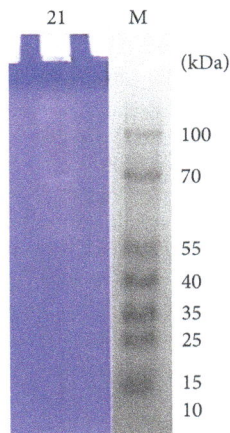

FIGURE 15: Native-PAGE zymogram analysis of keratinolytic protease enzyme from *B. subtilis* MA21. The lane (M) is prestained protein ladder, and lane (21) is zymogram of enzyme.

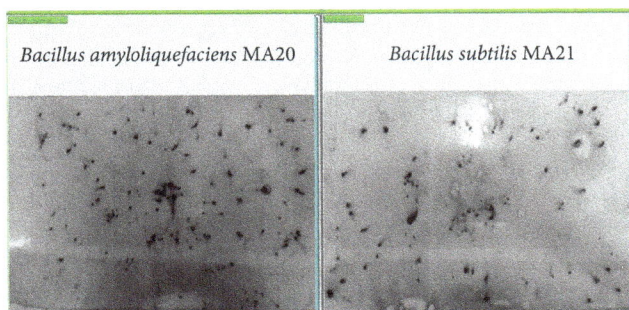

FIGURE 16: Two-dimension gel of enzymes from *B. amyloliquefaciens* MA20 and *B. subtilis* MA21.

and their produced enzymes. The study succeeded in characterization of *Bacillus* sp. strains which were named *Bacillus amyloliquefaciens* MA20 and *Bacillus subtilis* MA21. The enzyme activities either on agar diffusion plates or through the enzyme activity bioassays under different experimental conditions proved that the both strains are able to produce different keratinolytic protease enzymes. *B. amyloliquefaciens* MA20 and *B. subtilis* MA21 could be used in large-scale production of keratinolytic protease enzymes where these enzymes were stable at temperature range between 4 and 70°C, and over a wide range of pH values (4–12), as well as, stable against organic solvents and detergents. The characters of keratinolytic enzymes produced by *B. amyloliquefaciens* MA20 and *B. subtilis* MA21 could play an important role especially in industrial applications; therefore this research acts as preliminary studies for applying the keratinolytic proteases in wool quality improvement.

References

[1] T. Korniłłowicz-Kowalska and J. Bohacz, "Biodegradation of keratin waste: theory and practical aspects," *Waste Management*, vol. 31, pp. 1689–1701, 2011.

[2] R. D. B. Fraser and D. A. D. Parry, "Macrofibril assembly in trichocyte (hard α-) keratins," *Journal of Structural Biology*, vol. 142, no. 2, pp. 319–325, 2003.

[3] L. N. Jones, "Hair structure anatomy and comparative anatomy," *Clinics in Dermatology*, vol. 19, no. 2, pp. 95–103, 2001.

[4] P. Pillai and G. Archana, "Hide depilation and feather disintegration studies with keratinolytic serine protease from a novel *Bacillus subtilis* isolate," *Applied Microbiology and Biotechnology*, vol. 78, no. 4, pp. 643–650, 2008.

[5] V. Filipello Marchisio, "Karatinophilic fungi: their role in nature and degradation of keratinic substrates," in *Biology of Dermatophytes and Other Keratinophilic Fungi*, R. K. S. Kushawaha and J. Guarro, Eds., vol. 17, pp. 86–92, Revista Iberoamericana de Micología, 2000.

[6] R. Gupta and P. Ramnani, "Microbial keratinases and their prospective applications:an overview," *Applied Microbiology and Biotechnology*, vol. 70, pp. 21–33, 2006.

[7] S. Sangali and A. Brandelli, "Feather keratin hydrolysis by a *Vibrio sp.* strain kr2," *Journal of Applied Microbiology*, vol. 89, no. 5, pp. 735–743, 2000.

[8] C. H. De Toni, M. F. Richter, J. R. Chagas, J. A. P. Henriques, and C. Termignoni, "Purification and characterization of an alkaline serine endopeptidase from a feather-degrading *Xanthomonas maltophilia* strain," *Canadian Journal of Microbiology*, vol. 48, no. 4, pp. 342–348, 2002.

[9] F. S. Lucas, O. Broennimann, I. Febbraro, and P. Heeb, "High diversity among feather-degrading bacteria from a dry meadow soil," *Microbial Ecology*, vol. 45, no. 3, pp. 282–290, 2003.

[10] S. Yamamura, Y. Morita, Q. Hasan et al., "Characterization of a new keratin-degrading bacterium isolated from deer fur," *Journal of Bioscience and Bioengineering*, vol. 93, no. 6, pp. 595–600, 2002.

[11] E. H. Brutt and J. M. Ichida, "Keratinase produced by *Bacillus licheniformis*," US Patent 5,877,000, 1999.

[12] A. Gousterova, D. Braikova, I. Goshev et al., "Degradation of keratin and collagen containinq wastes by newly isolated thermoactinomycetes or by alkaline hydrolysis," *Letters in Applied Microbiology*, vol. 40, no. 5, pp. 335–340, 2005.

[13] K. L. Evans, J. Crowder, and E. S. Miller, "Subtilisins of *Bacillus spp.* hydrolyze keratin and allow growth on feathers," *Canadian Journal of Microbiology*, vol. 46, no. 11, pp. 1004–1011, 2000.

[14] I. N. S. Dozie, C. N. Okeke, and N. C. Unaeze, "A thermostable, alkaline-active, keratinolytic proteinase from *Chrysosporium keratinophilum*," *World Journal of Microbiology and Biotechnology*, vol. 10, no. 5, pp. 563–567, 1994.

[15] A. Riffel, F. Lucas, P. Heeb, and A. Brandelli, "Characterization of a new keratinolytic bacterium that completely degrades native feather keratin," *Archives of Microbiology*, vol. 179, no. 4, pp. 258–265, 2003.

[16] B. Zhang, D. Jiang, W. Zhou, H. Hao, and T. Niu, "Isolation and characterization of a new *Bacillus sp.* 50-3 with highly alkaline keratinase activity from Calotes versicolor faeces," *World Journal of Microbiology and Biotechnology*, vol. 25, no. 4, pp. 583–590, 2009.

[17] J. G. Holt, N. R. Krieg, P. H. A. Sneath, J. T. Staley, and S. T. Williams, *Bergey's Manual of Determinative Bacteriology*, Williams & Wilkins, Baltimore, Md, USA, 9th edition, 1994.

[18] J. Sambrook, E. F. Fritsch, and T. Maniatis, *Molecular Cloning: A Laboratory Manual*, Cold Spring Harbor Laboratory, Cold Spring Harbor, New York, NY, USA, 2nd edition, 1989.

[19] F. Sanger, S. Nicklen, and A. R. Coulson, "DNA sequencing with chain-terminating inhibitors," *Proceedings of the National Academy of Sciences of the United States of America*, vol. 74, no. 12, pp. 5463–5467, 1977.

[20] K. Tamura, D. Peterson, N. Peterson, G. Stecher, M. Nei, and S. Kumar, "MEGA5: molecular evolutionary genetics analysis using maximum likelihood, evolutionary distance, and maximum parsimony methods," *Molecular Biology and Evolution*, vol. 28, pp. 2731–2739, 2011.

[21] A. A. Amara, R. S. Slem, and S. A. M. Shabeb, "The possibility to use crude proteases and lipases as biodetergent," *Global Journal of Biotechnology & Biochemistry*, vol. 4, no. 2, pp. 104–114, 2009.

[22] W. Mao, R. Pan, and D. Freedman, "High production of alkaline protease by *Bacillus licheniformis* in a fed-batch fermentation using a synthetic medium," *Journal of Industrial Microbiology*, vol. 11, no. 1, pp. 1–6, 1992.

[23] A. B. Vermelho, M. N. L. Meirelles, A. Lopes, S. D. G. Petinate, A. A. Chaia, and M. H. Branquinha, "Detection of extracellular proteases from microorganisms on agar plates," *Memorias do Instituto Oswaldo Cruz*, vol. 91, no. 6, pp. 755–760, 1996.

[24] A. A. Amara and A. E. Serour, "Wool quality improvement using thermophilic crude proteolytic microbial enzymes," *American-Eurasian Journal of Agricultural & Environmental Sciences*, vol. 3, no. 4, pp. 554–560, 2008.

[25] J. L. Giongo, F. S. Lucas, F. Casarin, P. Heeb, and A. Brandelli, "Keratinolytic proteases of *Bacillus* species isolated from the Amazon basin showing remarkable de-hairing activity," *World Journal of Microbiology and Biotechnology*, vol. 23, no. 3, pp. 375–382, 2007.

[26] U. K. Laemmli, "Cleavage of structural proteins during the assembly of the head of bacteriophage T4," *Nature*, vol. 227, no. 5259, pp. 680–685, 1970.

[27] H. Blum, H. Beier, and J. H. Gross, "Improved silver staining of plant proteins, RNA and DNA in polyacrylamide gels," *Electrophoresis*, vol. 8, pp. 93–99, 1987.

[28] R. Gupta, Q. Beg, and P. Lorenz, "Bacterial alkaline proteases: molecular approaches and industrial applications," *Applied Microbiology and Biotechnology*, vol. 59, no. 1, pp. 15–32, 2002.

[29] J. Cortez, P. L. R. Bonner, and M. Griffin, "Transglutaminase treatment of wool fabrics leads to resistance to detergent damage," *Journal of Biotechnology*, vol. 117, no. 1, pp. 379–386, 2005.

[30] J. M. Kim, W. J. Lim, and H. J. Suh, "Feather-degrading Bacillus species from poultry waste," *Process Biochemistry*, vol. 37, no. 3, pp. 287–291, 2001.

[31] M. Rozs, L. Manczinger, C. Vágvölgyi, and F. Kevei, "Secretion of a trypsin-like thiol protease by a new keratinolytic strain of *Bacillus licheniformis*," *FEMS Microbiology Letters*, vol. 205, no. 2, pp. 221–224, 2001.

[32] M. S. Roberts, L. K. Nakamura, and F. M. Cohan, "*Bacillus mojavensis* sp. nov., distinguishable from *Bacillus subtilis* by sexual isolation, divergence in DNA sequence, and differences in fatty acid composition," *International Journal of Systematic Bacteriology*, vol. 44, no. 2, pp. 256–264, 1994.

[33] M. S. Roberts, L. K. Nakamura, and F. M. Cohan, "*Bacillus vallismortis* sp. nov., a close relative of *Bacillus subtilis*, isolated from soil in Death Valley, California," *International Journal of Systematic Bacteriology*, vol. 46, no. 2, pp. 470–475, 1996.

[34] L. T. Wang, F. L. Lee, C. J. Tai, and H. Kasai, "Comparison of gyrB gene sequences, 16S rRNA gene sequences and DNA-DNA hybridization in the Bacillus subtilis group," *International Journal of Systematic and Evolutionary Microbiology*, vol. 57, no. 8, pp. 1846–1850, 2007.

[35] I. Szabó, A. Benedek, I. Mihály Szabó, and G. Barabás, "Feather degradation with a thermotolerant Streptomyces graminofaciens strain," *World Journal of Microbiology and Biotechnology*, vol. 16, no. 3, pp. 253–255, 2000.

[36] A. Gushterova, E. Vasileva-Tonkova, E. Dimova, P. Nedkov, and T. Haertlé, "Keratinase production by newly isolated Antarctic actinomycete strains," *World Journal of Microbiology and Biotechnology*, vol. 21, no. 6-7, pp. 831–834, 2005.

[37] T. Korniłłowicz-Kowalska, "Studies on the decomposition of keratin wastes by saprotrophic microfungi. P.I. Criteria for evaluating keratinolytic activity," *Acta Mycologica*, vol. 32, pp. 51–79, 1997.

[38] B. Bockle, B. Galunsky, and R. Muller, "Characterization of a keratinolytic serine proteinase from *Streptomyces pactum* DSM 40530," *Applied and Environmental Microbiology*, vol. 61, no. 10, pp. 3705–3710, 1995.

[39] P. Bressollier, F. Letourneau, M. Urdaci, and B. Verneuil, "Purification and characterization of a keratinolytic serine proteinase from *Streptomyces albidoflavus*," *Applied and Environmental Microbiology*, vol. 65, no. 6, pp. 2570–2576, 1999.

[40] H. Takami, Y. Nogi, and K. Horikoshi, "Reidentification of the keratinase-producing facultatively alkaliphilic *Bacillus sp.* AH-101 as *Bacillus halodurans*," *Extremophiles*, vol. 3, no. 4, pp. 293–296, 1999.

[41] S. Sangali and A. Brandelli, "Isolation and characterization of a novel feather-degrading bacterial strain," *Applied Biochemistry and Biotechnology A*, vol. 87, no. 1, pp. 17–24, 2000.

[42] X. Lin, D. W. Kelemen, E. S. Miller, and J. C. H. Shih, "Nucleotide sequence and expression of kerA, the gene encoding a keratinolytic protease of *Bacillus licheniformis* PWD-1," *Applied and Environmental Microbiology*, vol. 61, no. 4, pp. 1469–1474, 1995.

[43] M. Kojima, M. Kanai, M. Tominaga, S. Kitazume, A. Inoue, and K. Horikoshi, "Isolation and characterization of a feather-degrading enzyme from *Bacillus pseudofirmus* FA30-01," *Extremophiles*, vol. 10, no. 3, pp. 229–235, 2006.

[44] T. I. Zaghloul, "Cloned *Bacillus subtilis* alkaline protease (aprA) gene showing high level of keratinolytic activity," *Applied Biochemistry and Biotechnology A*, vol. 70–72, pp. 199–205, 1998.

[45] H. J. Suh and H. K. Lee, "Characterization of a keratinolytic serine protease from bacillus subtilis KS-1," *Protein Journal*, vol. 20, no. 2, pp. 165–169, 2001.

[46] H. S. Joo, C. G. Kumar, G. C. Park, S. R. Paik, and C. S. Chang, "Oxidant and SDS-stable alkaline protease from *Bacillus clausii* I-52: production and some properties," *Journal of Applied Microbiology*, vol. 95, no. 2, pp. 267–272, 2003.

[47] R. Oberoi, Q. K. Beg, S. Puri, R. K. Saxena, and R. Gupta, "Characterization and wash performance analysis of an SDS-stable alkaline protease from a *Bacillus sp.*," *World Journal of Microbiology and Biotechnology*, vol. 17, no. 5, pp. 493–497, 2001.

18

Electric-Field-Directed Self-Assembly of Active Enzyme-Nanoparticle Structures

Alexander P. Hsiao[1] and Michael J. Heller[1,2]

[1] Department of Bioengineering, University of California San Diego, La Jolla, CA 92093-0412, USA
[2] Department of NanoEngineering, University of California San Diego, La Jolla, CA 92093-0412, USA

Correspondence should be addressed to Michael J. Heller, mheller@ucsd.edu

Academic Editor: Seunghun Hong

A method is presented for the electric-field-directed self-assembly of higher-order structures composed of alternating layers of biotin nanoparticles and streptavidin-/avidin-conjugated enzymes carried out on a microelectrode array device. Enzymes included in the study were glucose oxidase (GOx), horseradish peroxidase (HRP), and alkaline phosphatase (AP); all of which could be used to form a light-emitting microscale glucose sensor. Directed assembly included fabricating multilayer structures with 200 nm or 40 nm GOx-avidin-biotin nanoparticles, with AP-streptavidin-biotin nanoparticles, and with HRP-streptavidin-biotin nanoparticles. Multilayered structures were also fabricated with alternate layering of HRP-streptavidin-biotin nanoparticles and GOx-avidin-biotin nanoparticles. Results showed that enzymatic activity was retained after the assembly process, indicating that substrates could still diffuse into the structures and that the electric-field-based fabrication process itself did not cause any significant loss of enzyme activity. These methods provide a solution to overcome the cumbersome passive layer-by-layer assembly methods to efficiently fabricate higher-order active biological and chemical hybrid structures that can be useful for creating novel biosensors and drug delivery nanostructures, as well as for diagnostic applications.

1. Introduction

With recent advances in the assembly of nanoparticles (NPs) into higher-order structures and components, the ability to incorporate biologically active molecules has become more important [1, 2]. Considerable research efforts are now directed towards the fabrication and integration of biologically active molecules into NP structures that could be used in drug delivery, biological and chemical sensors, and diagnostics. In most cases, these higher-order structures are fabricated with passive layer-by-layer (LBL) techniques to self-assemble the molecules into organized structures through specific interactions including covalent binding, gold-thiol interactions, electrostatic interactions, and protein-ligand binding [3–11]. However, passive processes are concentration dependent and these methods require complex processes and long incubation times in high concentration solutions of molecules. Moreover, in order to direct the assembly onto specific sites, blocking agents or physical patterning such as lithography is necessary [12]. To circumvent these issues, active processes have been developed, including DC electrophoretic deposition and magnetic-field-assisted deposition [13–18]. Also, work has been carried out on the use of AC dielectrophoretic techniques to manipulate NPs [19–21]. The application of electric fields allows for rapid, site-directed concentration of macromolecules, polymers, and NPs to enhance the self-assembly process. Such methods have been employed to produce colloidal aggregates as well as pattern NPs atop electrode surfaces [17, 22–29]. In addition, nonspecific binding and high background, which play a crucial role in the incorporation and detection of biological molecules, can be reduced with electrode patterns which direct the molecules toward the active site where deposition is preferred and away from nonactive regions. More recently, the method of electrophoretic deposition has been applied to biological components. This powerful tool has enabled devices to be made which utilize the electric fields to enhance DNA hybridization, to form protein layers for biosensors, and to pattern cells [30–36]. Recently we have shown the ability to construct higher-order NP structures by electric-field-directed self-assembly through the specific interactions of complementary DNA sequences as well as through protein-molecule interactions (Figures 1(a) and 1(b)) [14, 37, 38]. We now present the ability to integrate active enzymes into these NP structures by directed electrophoretic means (Figure 1(c)), thus providing a new bottom-up fabrication method for patterning and constructing structures from NPs in a rapid and combinatorial fashion atop a microarray.

2. Materials and Methods

2.1. CMOS Microarray Setup. An ACV 400 CMOS electronic microarray (Nanogen, Inc.), shown in Figure 2, which consists of 400 individually controllable 55 μm-diameter platinum electrodes was used for all layer assembly experiments. The microarray chip is overcoated by the manufacturer with a streptavidin-embedded polyacrylamide hydrogel which serves as a permeation layer. The device was computer controlled using ACV400 software. The software allowed each electrode to be configured to independently source 0 to 5 V or 0 to 1 μA per electrode, with each electrode on the array capable of being independently biased.

2.2. Chip Preparation. To prepare the chip surface as depicted in the cross-section in Figure 2, the microarray chip was first washed by pipetting 20 μL deionized water (dH$_2$O, Millipore, 18 MΩ) onto and off the chip a total of 10 times. Subsequently, 20 μL of a 2 μM biotin-dextran (Sigma) solution in dH$_2$O was pipetted onto the chip and allowed to incubate for 30 minutes at room temperature. The chip was then washed again with dH$_2$O, followed by incubation with 20 μL of a 1 mg/mL solution of streptavidin (Sigma) in dH$_2$O for 30 minutes at room temperature. Finally, the chip was washed with 100 mM L-histidine buffer and kept moist prior to use.

2.3. Preparation of NPs and Enzymes. Yellow-green fluorescent biotin-coated NPs, 200 nm and 40 nm in diameter (Molecular Probes, ex505, em515), were diluted to 0.01% (38 pM for 200 nm NPs and 4.7 nM for 40 nm NPs) in 100 mM L-histidine buffer. This suspension was vortexed and sonicated in a water bath for 15 minutes just prior to use to break up any aggregates. Additionally, glucose oxidase-avidin (GOx-avidin, Rockland) was diluted to 30 nM, streptavidin-alkaline phosphatase (streptavidin-AP, Sigma) was diluted to 40 nM, and streptavidin-peroxidase (streptavidin-HRP, Sigma) was diluted to 95 nM in 100 mM L-histidine buffer just prior to use.

2.4. DC Electric-Field-Directed Assembly of Streptavidin/ Avidin Enzymes and Biotin NPs. NP and enzyme addressing conditions are derived from previous work [14]. In brief, 20 μL of the 200 nm biotin NP solution or enzyme solution was pipetted onto the chip. The selected electrodes were biased positive and activated with a constant DC current

of 0.25 μA for 15 seconds to concentrate and assemble the particles or enzymes atop the activated electrodes. The solution was then removed and the chip washed with 20 μL of L-histidine buffer a total of three times. Assembly of the layer structures was achieved by alternating the addressing of biotin NPs with streptavidin/avidin enzymes. Every structure was capped with a final layer of biotin NPs. Different layer structures include layers of biotin NPs and GOx-avidin, layers of biotin NPs and streptavidin-AP, layers of biotin NPs and streptavidin-HRP, and layers of biotin NPs with alternate GOx-avidin and streptavidin-HRP to produce bienzyme structures. Identical conditions were employed to assemble layers of 40 nm biotin NPs with streptavidin-AP.

2.5. Monitoring of Layer Assembly by Fluorescence and ImageJ Calculations. Monitoring of layer growth was done by real-time imaging on an epifluorescent Leica microscope, with a Hamamatsu Orca-ER CCD using a custom LabVIEW interface. Images were acquired throughout the layering process and processed in ImageJ. For analysis, each image had its background subtracted with a rolling ball radius of 50. The image was then inverted and threshold fixed using the IsoData threshold. Manual adjustments were made to include as many electrodes as possible. A corresponding mask was generated to ensure each measured electrode area was identical. Raw integrated density values for each electrode were then acquired by mapping the data in the original image to the generated mask image.

2.6. Verification of Enzyme Activity via X-Ray Film. The verification of enzyme activity was performed on chips composed of alternate layers of 200 nm biotin NPs with either GOx-avidin or streptavidin-AP as the enzyme layers. All 400 electrodes on the array were activated to maximize the total number of fabrication sites for the layer structures. For the structures assembled with GOx-avidin, a reaction solution consisting of 227 mM glucose (Sigma), 8.4 mM luminol (Fluka), and 0.1 mg/mL peroxidase (Sigma) in 0.035 M Tris-HCl (pH 8.4) was prepared. The chips were washed with 100 mM L-histidine buffer and then with 0.1 M Tris-HCl (pH 8.0). Subsequently, 15 μL of the reaction solution was pipetted onto the chip surface. For chips with layer structures assembled using streptavidin-AP, the chips were washed with 100 mM L-histidine buffer and then 15 μL of CDP-star chemiluminescent reagent (Sigma) was dispensed onto the chips.

The chips were then wrapped in plastic wrap to prevent solution loss and placed into a cassette with X-ray film (Denville Scientific) for overnight exposure. The film was developed in a Hope MicroMax developer, scanned, and analyzed using ImageJ. The relative intensity from each chip was normalized to a chip that did not undergo layer assembly which was cleaned, prepared with the appropriate reaction solution, and exposed overnight as well.

2.7. Environmental Scanning Electron Microscopy (ESEM) of the Enzyme-NP Layers. After assembly of the enzyme-NP layers, the chip was washed multiple times with 100 mM

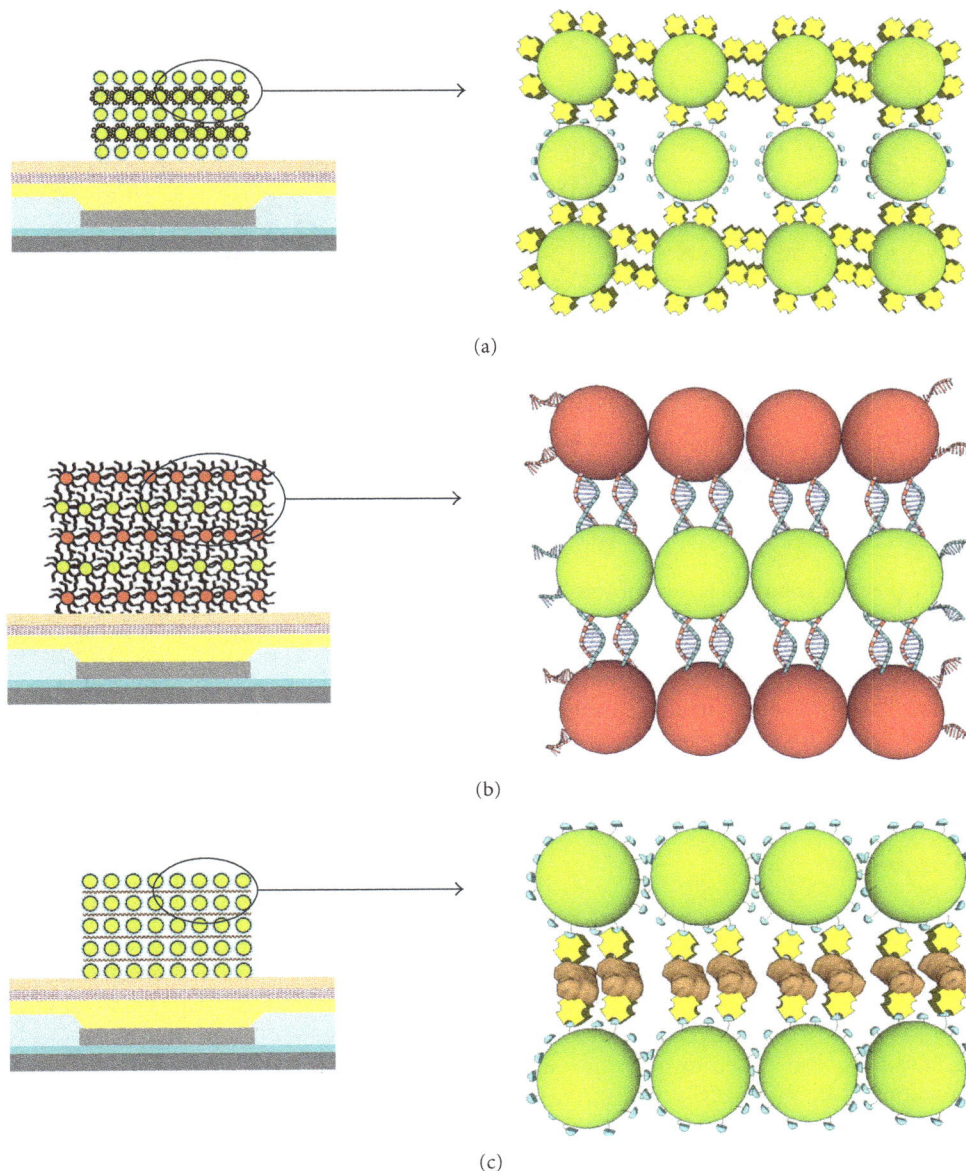

(a)

(b)

(c)

FIGURE 1: Electric-field-directed assembly (layering) of biomolecule NPs by different binding mechanisms: (a) NP layering with alternate biotin (blue)-functionalized NPs and streptavidin (yellow)-functionalized NPs. (b) NP layering by hybridization of complementary DNA sequences. (c) NP layering of biotin-functionalized NPs with streptavidin-functionalized enzymes (brown) (image not to scale).

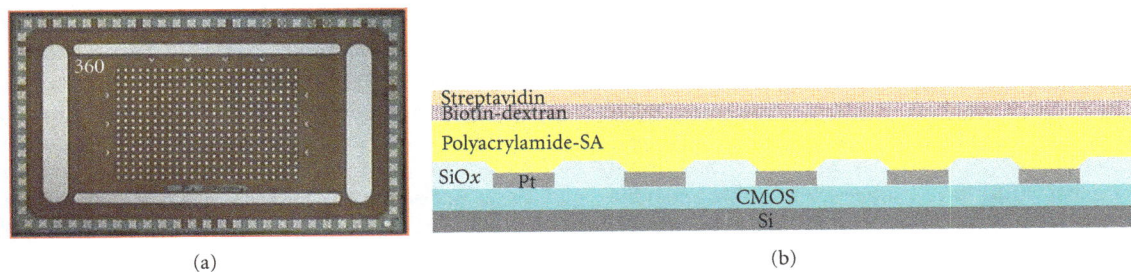

(a)   (b)

FIGURE 2: Images of the 400 site platinum electrode CMOS microelectronic array and a cross-section of the structure. The microarray is 4 mm × 7 mm, and each microelectrode is 55 μm in diameter.

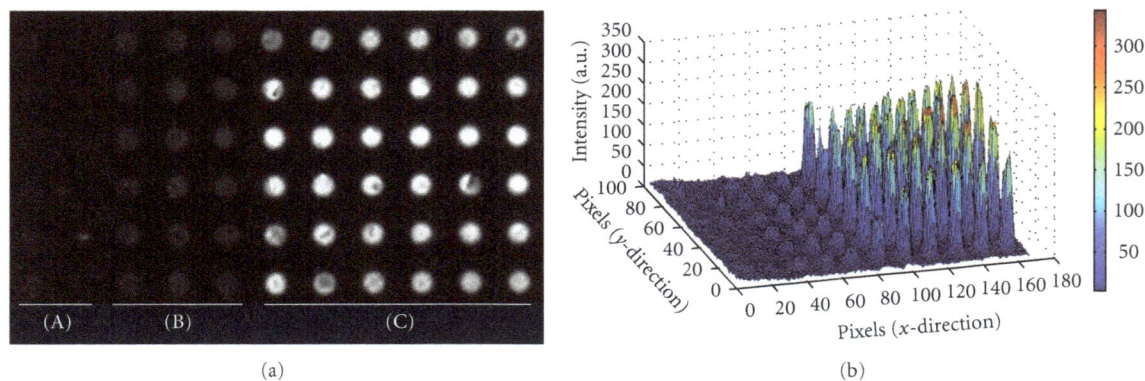

(a)   (b)

FIGURE 3: (a) Fluorescence image of a section of the CMOS microarray after addressing 39 combined layers of biotin NPs and GOx-avidin. (A) No current applied. (B) Current applied ONLY when biotin NPs were addressed. (C) Current applied when BOTH biotin NPs and GOx-avidin were addressed. (b) Corresponding MATLAB plot of the relative fluorescence intensity (z-axis) of each electrode.

L-histidine buffer and then all solution was removed from the surface to allow the chip to dry. Chips were then coated with either 40 nm of gold sputtered via a Denton Discovery 18 sputter system or 40 nm of chromium via Denton IV desktop sputter coater. Fractures were introduced into the structures by careful cutting with a razor blade. Images were then acquired on a Phillips XL30 ESEM using a 10 kV beam in high vacuum mode.

3. Results and Discussion

3.1. Assembly of Enzyme-NP Layers and Verification of Proper Layer Formation. The assembly of NP layers was monitored by epifluorescence imaging; however, because only the biotin NPs are fluorescent, it was first important to verify that the NP-enzyme layers were forming as proposed by alternate layering of enzymes and NPs, as opposed to formation due to nonspecific interactions of the biotin NPs to themselves. This was done by organizing the electrodes into three specific regions, as shown in Figure 3(a). Region A consisted of microelectrodes which were never activated. This section served as a negative control to measure the amount of passive binding to the chip surface that would occur simply due to the presence of NPs and enzyme during alternate addressing steps. Region B consisted of microelectrodes only activated when the biotin NPs were addressed. This region served to measure the amount of non-specific binding of the NPs to themselves. Additionally, it served to show the amount of passive assembly that could occur if no enzyme was actively addressed to these microelectrode sites. Finally, region C consisted of microelectrodes which were activated during all addressing steps of NPs and enzymes. Microelectrodes in this region were expected to have proper formation of enzyme-NP layers. The results in Figures 3(a) and 3(b) indicate that the microelectrodes in region A have a fluorescent signal near that of the background, which is the surface of the chip between the electrodes, thus indicating that a very low number of fluorescent biotin NPs passively bound to the streptavidin surface at these sites. Microelectrodes

in region B, which were only activated when biotin NPs were addressed, have a low level fluorescent signal and the microelectrodes in region C, which were activated when both NPs and GOx-avidin were addressed, have a high level fluorescent intensity indicating that multiple layers of NPs formed in region C. Comparison of fluorescence intensities between the three regions suggests that in order to construct higher-order structures both NPs and enzymes must be addressed to the same site, as in region C. If only biotin NPs are addressed, as in region B, the NPs will not bind to one another and no higher-order structures are formed; therefore, there is only low fluorescence intensity from the first layer of biotin NPs assembled onto the streptavidin chip surface. These results were verified with all three enzyme types and with both 200 nm and 40 nm NPs.

To corroborate with the fluorescence data, Figure 4 shows environmental scanning electron microscopy (ESEM) images of three microelectrode sites; one each for region A, B, and C after addressing 31 total layers of 200 nm biotin NPs and streptavidin-AP as well as 21 layers of 40 nm biotin NPs and streptavidin-AP on separate microarray chips. The microelectrodes from region A show only a small number of passively attached biotin NPs. The electrodes from region B show nearly a complete monolayer of biotin NPs, despite being exposed to 16 total addressing steps of biotin NPs. This demonstrates that there is little non-specific binding of the biotin particles to themselves; so despite the electric field directing additional NPs onto the first layer of NPs, they do not stick and are removed during the wash steps. The electrode from region C shows a high number of NPs assembled atop each other. Thus, active directed concentration of both the streptavidin/avidin enzyme and biotin NPs is necessary to assemble the higher-order structures and the layer assembly process does indeed proceed as designed. Additionally, the lack of particles on region A's microelectrodes further verifies that electric-field-directed assembly is efficient and can overcome the diffusion-limited process of passive LBL assembly. Each assembly step only required 15 seconds with NP and enzyme concentrations

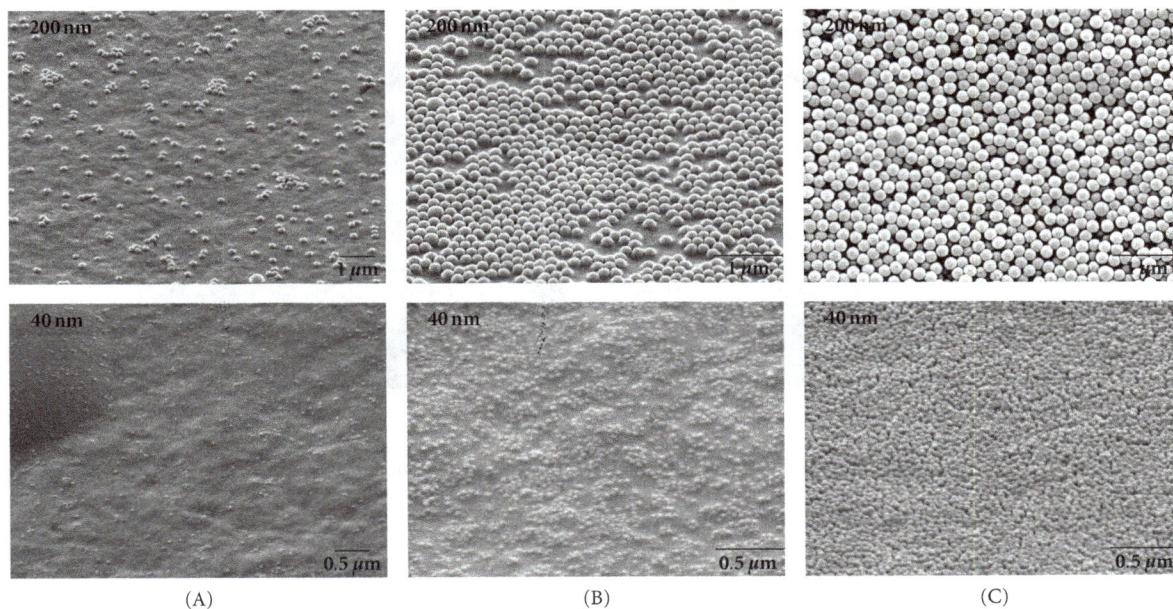

FIGURE 4: ESEM images of a microelectrode in each region (A), (B), and (C) after enzyme-NP assembly. Top row: microelectrodes after assembling 31 total alternating layers of 200 nm biotin NPs and streptavidin-AP. Bottom row: microelectrodes after assembling 21 layers of 40 nm biotin NPs and streptavidin-AP.

in the pM and nM range. At these time scales and NP and enzyme concentrations, no layers could be formed passively on the region A microelectrodes.

These results show that the electric field directed assembly technology is easily scalable to NPs of various sizes. This allows for tuning of the porosity of the final structures which may help control the (enzyme) substrate turnover and reaction kinetics, both of which would play crucial roles in biosensor devices. For drug delivery particles, the porosity will play a paramount role in the drug release profile. Moreover, we believe that integration of various types of NPs with different biomolecules would also be achievable as long as the proper binding elements are in place. Using multiple sized NPs would enable multiple porosities through the structure which may be needed to optimize reaction rates in multienzyme structures. Particles such as quantum dots could be incorporated to enhance detection. Moreover, using other biomolecules such as antibodies or DNA would allow the creation of a wide array of biosensors.

3.2. Monitoring NP Layer Assembly and Quality of Layers.
Real-time layer assembly was monitored by visualizing increasing fluorescence intensity atop the microelectrode sites. Figure 5 shows a plot of the mean integrated density of fluorescence per microelectrode for microelectrodes in regions B and C of a microarray after 9, 19, 29, 39, and 47 total layers of 200 nm biotin NPs with alternate addressing of both GOx-avidin and streptavidin-HRP. From the plot, it is evident that the fluorescence for microelectrodes in region B, microelectrodes activated only when biotin NPs were addressed, maintains roughly the same fluorescence intensity throughout the layering experiments. These results further

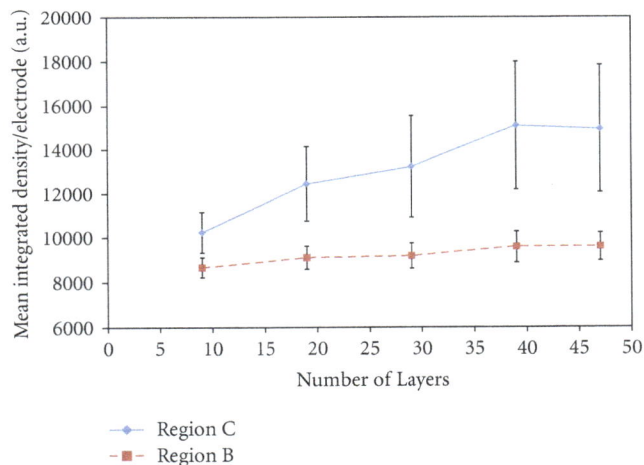

FIGURE 5: Plot showing the calculated mean integrated density per microelectrode for microelectrodes in regions B and C in bienzyme layer structures of 9, 19, 29, 39, and 47 total layers of 200 nm biotin NPs with alternate enzyme layer addressing of GOx-avidin and streptavidin-HRP.

substantiate the results in Figures 3 and 4 that without active electric-field-directed assembly of streptavidin/avidin-conjugated enzymes onto the biotin NPs there is no further layer assembly. Additionally, these results verify that multiple types of enzymes can be incorporated into the same structure as long as they are properly functionalized. In this case, there is a streptavidin-functionalized HRP and an avidin-functionalized GOx, both of which can bind to the biotin on the NPs and facilitate layer formation. The plot shows

(a)

(b)

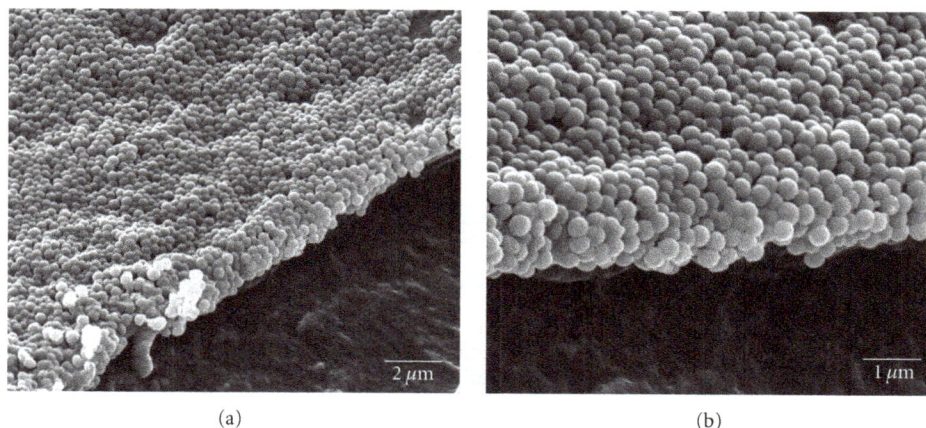

FIGURE 6: ESEM images of 200 nm biotin NPs layered with GOx-avidin at introduced cuts showing the layering of NPs.

a trend of increasing mean fluorescence for microelectrodes in region C as the total number of layers increases. This is what is expected because as the number of layers increases there are more total fluorescent NPs on each microelectrode. The plot in Figure 5, however, does have quite a large amount of variability, which could be attributed to many factors. One factor could be the stoichiometry of the streptavidin conjugation to the enzyme. Streptavidin-HRP was conjugated at a $1:1$ ratio and streptavidin-AP at a $2:1$ ratio according to the manufacturer's specifications. Streptavidin-AP thus has 4 more available biotin binding sites per enzyme molecule. This increased availability of binding sites makes attachment to biotin NPs more robust and can lead to an increased quality of uniformity of NP layers. Thus, to enhance binding, enzymes can be conjugated with a higher ratio of streptavidin/avidin per enzyme. In addition, as the number of layers increases the stresses on the layer structure increase and the structure could shear or break apart more easily during washes. It is sometimes seen that atop a specific microelectrode the fluorescence intensity would suddenly decrease and this effect was believed to be due to layer fracture and particle loss. Again, a higher stoichiometry of streptavidin to protein would increase the binding interactions between layers and help to prevent structure fracture and NP loss. Finally, another factor could be attributed to nonuniformity in the electric field across the microarray chip or even across an individual microelectrode. This would also lead to variations in NP and enzyme assembly.

ESEM images, as seen in Figure 6, obtained at the edge of introduced fractures reveal the layering of the NPs atop the hydrogel layer. From these micrographs, it is evident that the assembled structures have variability in surface topography making it difficult to clearly distinguish one layer from the next. This is mostly attributed to the particle packing orientation as each additional layer of NPs packs onto the layer below. Additionally, this could be due to NP loss during the introduction of a fracture, during the sputtering of the metal overlayer for ESEM imaging, or even during the imaging process itself. In addition, there may be loss during

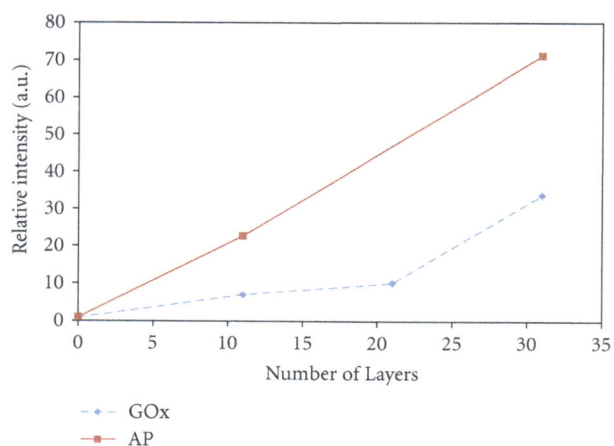

FIGURE 7: Plot of the relative intensity of chemiluminescent signal obtained from chips addressed with 0, 11, 21, and 31 layers of 200 nm biotin NPs and GOx-avidin or 0, 11, and 31 layers with streptavidin-AP.

washes and variations in binding across the electrode during the assembly process.

3.3. Retention of Enzyme Activity.

Retention of enzyme activity after layer assembly was evaluated by incubating the microarray chips with the appropriate chemiluminescent substrate and then exposing the chips to X-ray film. The results of the scanned and analyzed X-ray film detection of the enzyme-NP layers are shown in Figure 7. Data was collected from chips layered with 200 nm biotin NPs and GOx-avidin with 0, 11, 21, and 31 total layers as well as chips layered with 200 nm biotin NPs and streptavidin-AP at 0, 11, and 31 layers. The results show increasing activity detected with increasing numbers of layers. This trend is seen with both types of enzymes, and this indicates that the total enzyme activity can be tuned simply by altering the number of enzyme layers incorporated into each structure. Similar results could not be obtained from

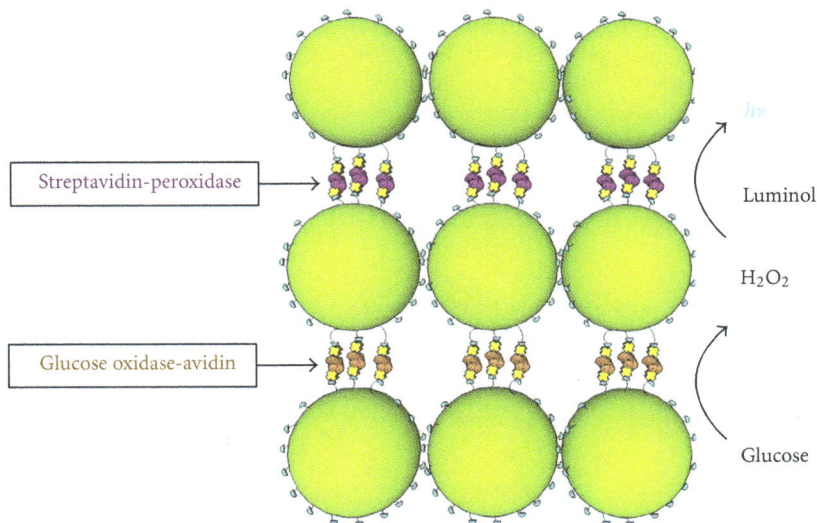

FIGURE 8: Coupling of bi-enzyme NP layers. The incorporation of both streptavidin-HRP and GOx-avidin into the same layer structure may allow for chemical coupling of the layers. The oxidation of glucose by GOx produces hydrogen peroxide which is then a substrate for the chemiluminescent oxidation of luminol, which generates light that can be detected.

bi-enzyme structures, consisting of both GOx-avidin and streptavidin-HRP, as illustrated in Figure 8. This may be due to a number of reasons including poor reagent and substrate quality, poor layer quality, poor structure porosity, insufficient enzyme incorporation into the layers, and a poor detection scheme. A bi-enzyme structure requires optimization due to the coupling of multiple reaction steps. If any one of the reactions is inefficient, then the overall signal may not be detectable. In addition, the products from the first reaction must be able to effectively diffuse to the second set of enzymes; thus, the enzyme layering order may be of importance. Additionally, an important aspect of producing active NP layers is the ability to sensitively detect their activity. The X-ray film used in the detection method verified in proof of principle that the biological activity of the molecules could be retained after assembly. More sensitive methods, including amperometric detection or highly sensitive imaging, beyond the capabilities of the microelectronic array and imaging system we had available, would allow for a better detection scheme to monitor total activity for each fabricated structure. Nonetheless, the presence of a measurable enzyme activity from the single enzyme structures verifies that the application of an electric field is not only efficient for structure assembly but also gentle enough to preserve the functionality of the enzymes.

Altogether the results showing enzyme-nanoparticle layer assembly and enzyme activity retention demonstrate an efficient and effective method of fabricating biological or chemical sensors. Site-specific layer assembly, demonstrated in this study as well as previous studies, means that multiple types of enzyme-nanoparticle structures can be fabricated on each chip in a combinatorial manner [37]. Additionally, various types of enzymes, proteins, or other biomolecules could be used in conjunction with a wide array of particle types as long as they have complementary binding mechanisms, such as the biotin-streptavidin scheme used here. This would allow for production of high-density microarray sensors capable of analyzing hundreds of analytes at a time.

4. Conclusion

We have successfully demonstrated the ability to fabricate higher-order enzyme-NP structures by electric-field-directed self-assembly. Through the application of electric-field-directed assembly, alternating layers of 200 nm or 40 nm biotin NPs and streptavidin/avidin enzymes have been assembled up to 47 layers. These structures included multilayer structures with 200 nm or 40 nm GOx-avidin-biotin NPs, with AP-streptavidin-biotin NPs, and with HRP-streptavidin-biotin NPs. The electrophoretic assembly method atop a microelectronic array allows for site-specific fabrication from low concentration solutions of enzymes and NPs. The concentration effect due to the electrophoretic deposition results in rapid layer assembly with minimal passive non-specific binding on inactive sites across the chip. Moreover, the enzymatic activity of the biological molecules was preserved in the assembled structures. In addition, we have assembled structures consisting of multiple enzyme types, GOx-avidin and streptavidin-HRP, which demonstrates the potential of multilevel reactions or detection schemes, including chemiluminescence and bioluminescence. This method of fabrication now provides an efficient mechanism of creating biologically and chemically active NP structures from individual components much more efficiently than traditional passive layer-by-layer methods. Assembly of these structures in a combinatorial manner to specific sites on the chip, using a wide array of biomolecules (proteins and DNA) and nanoparticles, would allow for fabrication of high-density microarray sensors for high-throughput analysis. The ability to incorporate multiple types of molecules along with the potential of liftoff, which enables the detachment of these structures from the surface, renders them more versatile as dispersible biosensors, diagnostic tools, and drug delivery vehicles.

Acknowledgments

The authors thank Nanogen, Inc. for supplying microelectronic arrays and the Nanochip 400 system, Dr. Dietrich Dehlinger for assistance and training with the system and methods, Juhi Saha for her help in experiments, and the UCSD Nano3 facility and personnel for training and support on the ESEM.

References

[1] S. Guo and S. Dong, "Biomolecule-nanoparticle hybrids for electrochemical biosensors," *TrAC—Trends in Analytical Chemistry*, vol. 28, no. 1, pp. 96–109, 2009.

[2] K. Ariga, Q. Ji, and J. Hill, "Enzyme-encapsulated layer-by-layer assemblies: current status and challenges toward ultimate nanodevices," in *Advances in Polymer Science*, F. Caruso, Ed., vol. 229, pp. 51–87, Springer, Berlin, Germany, 2010.

[3] N. K. Chaki and K. Vijayamohanan, "Self-assembled monolayers as a tunable platform for biosensor applications," *Biosensors and Bioelectronics*, vol. 17, no. 1-2, pp. 1–12, 2002.

[4] Y. Kobayashi and J. I. Anzai, "Preparation and optimization of bienzyme multilayer films using lectin and glyco-enzymes for biosensor applications," *Journal of Electroanalytical Chemistry*, vol. 507, no. 1-2, pp. 250–255, 2001.

[5] B. Limoges, J. M. Savéant, and D. Yazidi, "Avidin-biotin assembling of horseradish peroxidase multi-monomolecular layers on electrodes," *Australian Journal of Chemistry*, vol. 59, no. 4, pp. 257–259, 2006.

[6] Y. Lvov, K. Ariga, I. Ichinose, and T. Kunitake, "Assembly of multicomponent protein films by means of electrostatic layer-by-layer adsorption," *Journal of the American Chemical Society*, vol. 117, no. 22, pp. 6117–6123, 1995.

[7] M. Onda, K. Ariga, and T. Kunitake, "Activity and stability of glucose oxidase in molecular films assembled alternately with polyions," *Journal of Bioscience and Bioengineering*, vol. 87, no. 1, pp. 69–75, 1999.

[8] M. Onda, Y. Lvov, K. Ariga, and T. Kunitake, "Sequential actions of glucose oxidase and peroxidase in molecular films assembled by layer-by-layer alternate adsorption," *Biotechnology and Bioengineering*, vol. 51, no. 2, pp. 163–167, 1996.

[9] K. L. Prime and G. M. Whitesides, "Self-assembled organic monolayers: Model systems for studying adsorption of proteins at surfaces," *Science*, vol. 252, no. 5010, pp. 1164–1167, 1991.

[10] S. V. Rao, K. W. Anderson, and L. G. Bachas, "Controlled layer-by-layer immobilization of horseradish peroxidase," *Biotechnology and Bioengineering*, vol. 65, no. 4, pp. 389–396, 1999.

[11] T. Hoshi, N. Sagae, K. Daikuhara, K. Takahara, and J. I. Anzai, "Multilayer membranes via layer-by-layer deposition of glucose oxidase and Au nanoparticles on a Pt electrode for glucose sensing," *Materials Science and Engineering C*, vol. 27, no. 4, pp. 890–894, 2007.

[12] X. M. Zhao, "Soft lithographic methods for nano-fabrication," *Journal of Materials Chemistry*, vol. 7, no. 7, pp. 1069–1074, 1997.

[13] S. Bharathi and M. Nogami, "A glucose biosensor based on electrodeposited biocomposites of gold nanoparticles and glucose oxidase enzyme," *Analyst*, vol. 126, no. 11, pp. 1919–1922, 2001.

[14] D. A. Dehlinger, B. D. Sullivan, S. Esener, and M. J. Heller, "Electric-field-directed assembly of biomolecular-derivatized nanoparticles into higher-order structures," *Small*, vol. 3, no. 7, pp. 1237–1244, 2007.

[15] S. Dey, K. Mohanta, and A. J. Pal, "Magnetic-field-assisted layer-by-layer electrostatic assembly of ferromagnetic nanoparticles," *Langmuir*, vol. 26, no. 12, pp. 9627–9631, 2010.

[16] M. Shao, X. Xu, J. Han et al., "Magnetic-field-assisted assembly of layered double hydroxide/metal porphyrin ultrathin films and their application for glucose sensors," *Langmuir*, vol. 27, no. 13, pp. 8233–8240, 2011.

[17] M. Trau, D. A. Seville, and I. A. Aksay, "Field-induced layering of colloidal crystals," *Science*, vol. 272, no. 5262, pp. 706–709, 1996.

[18] K. D. Barbee, A. P. Hsiao, M. J. Heller, and X. Huang, "Electric field directed assembly of high-density microbead arrays," *Lab on a Chip*, vol. 9, no. 22, pp. 3268–3274, 2009.

[19] R. Krishnan, D. A. Dehlinger, G. J. Gemmen, R. L. Mifflin, S. C. Esener, and M. J. Heller, "Interaction of nanoparticles at the DEP microelectrode interface under high conductance conditions," *Electrochemistry Communications*, vol. 11, no. 8, pp. 1661–1666, 2009.

[20] R. Krishnan and M. J. Heller, "An AC electrokinetic method for enhanced detection of DNA nanoparticles," *Journal of Biophotonics*, vol. 2, no. 4, pp. 253–261, 2009.

[21] R. Krishnan, B. D. Sullivan, R. L. Mifflin, S. C. Esener, and M. J. Heller, "Alternating current electrokinetic separation and detection of DNA nanoparticles in high-conductance solutions," *Electrophoresis*, vol. 29, no. 9, pp. 1765–1774, 2008.

[22] R. C. Bailey, K. J. Stevenson, and J. T. Hupp, "Assembly of micropatterned colloidal gold thin films via microtransfer molding and electrophoretic deposition," *Advanced Materials*, vol. 12, no. 24, pp. 1930–1934, 2000.

[23] L. Besra and M. Liu, "A review on fundamentals and applications of electrophoretic deposition (EPD)," *Progress in Materials Science*, vol. 52, no. 1, pp. 1–61, 2007.

[24] T. Haruyama and M. Aizawa, "Electron transfer between an electrochemically deposited glucose oxidase/Cu[II] complex and an electrode," *Biosensors and Bioelectronics*, vol. 13, no. 9, pp. 1015–1022, 1998.

[25] A. L. Rogach, N. A. Kotov, D. S. Koktysh, J. W. Ostrander, and G. A. Ragoisha, "Electrophoretic deposition of latex-based 3D colloidal photonic crystals: a technique for rapid production of high-quality opals," *Chemistry of Materials*, vol. 12, no. 9, pp. 2721–2726, 2000.

[26] L. Shi, Y. Lu, J. Sun et al., "Site-selective lateral multilayer assembly of bienzyme with polyelectrolyte on ITO electrode based on electric field-induced directly layer-by-layer deposition," *Biomacromolecules*, vol. 4, no. 5, pp. 1161–1167, 2003.

[27] Y. Solomentsev, M. Böhmer, and J. L. Anderson, "Particle clustering and pattern formation during electrophoretic deposition: a hydrodynamic model," *Langmuir*, vol. 13, no. 23, pp. 6058–6061, 1997.

[28] M. Trau, D. A. Saville, and I. A. Aksay, "Assembly of colloidal crystals at electrode interfaces," *Langmuir*, vol. 13, no. 24, pp. 6375–6381, 1997.

[29] S. R. Yeh, M. Seul, and B. I. Shraiman, "Assembly of ordered colloidal aggregates by electric-field-induced fluid flow," *Nature*, vol. 386, no. 6620, pp. 57–59, 1997.

[30] D. R. Albrecht, V. L. Tsang, R. L. Sah, and S. N. Bhatia, "Photo- and electropatterning of hydrogel-encapsulated living cell arrays," *Lab on a Chip*, vol. 5, no. 1, pp. 111–118, 2005.

[31] J. Cheng, E. L. Sheldon, A. Uribe et al., "Preparation and hybridization analysis of DNA/RNA from E. coli on microfabricated bioelectronic chips," *Nature Biotechnology*, vol. 16, no. 6, pp. 541–546, 1998.

[32] C. F. Edman, D. E. Raymond, D. J. Wu et al., "Electric field directed nucleic acid hybridization on microchips," *Nucleic Acids Research*, vol. 25, no. 24, pp. 4907–4914, 1997.

[33] C. Gurtner, E. Tu, N. Jamshidi et al., "Microelectronic array devices and techniques for electric field enhanced DNA hybridization in low-conductance buffers," *Electrophoresis*, vol. 23, no. 10, pp. 1543–1550, 2002.

[34] A. Kueng, C. Kranz, and B. Mizaikoff, "Amperometric ATP biosensor based on polymer entrapped enzymes," *Biosensors and Bioelectronics*, vol. 19, no. 10, pp. 1301–1307, 2004.

[35] R. G. Sosnowski, E. Tu, W. F. Butler, J. P. O'Connell, and M. J. Heller, "Rapid determination of single base mismatch mutations in DNA hybrids by direct electric field control," *Proceedings of the National Academy of Sciences of the United States of America*, vol. 94, no. 4, pp. 1119–1123, 1997.

[36] M. J. Heller, D. A. Dehlinger, and B. D. Sullivan, "Parallel assisted assembly of multilayer DNA and protein nanoparticle structures using a CMOS electronic array," in *International Symposium on DNA-Based Nanoscale Integration*, vol. 859 of *AIP Conference Proceedings*, pp. 73–81, May 2006.

[37] D. Dehlinger, B. Sullivan, S. Esener, D. Hodko, P. Swanson, and M. J. Heller, "Automated combinatorial process for nanofabrication of structures using bioderivatized nanoparticles," *Journal of the Association for Laboratory Automation*, vol. 12, no. 5, pp. 267–276, 2007.

[38] D. A. Dehlinger, B. D. Sullivan, S. Esener, and M. J. Heller, "Directed hybridization of DNA derivatized nanoparticles into higher order structures," *Nano Letters*, vol. 8, no. 11, pp. 4053–4060, 2008.

19

Enhancement of Phosphate Absorption by Garden Plants by Genetic Engineering: A New Tool for Phytoremediation

Keisuke Matsui,[1] **Junichi Togami,**[2] **John G. Mason,**[3]
Stephen F. Chandler,[4] **and Yoshikazu Tanaka**[1]

[1] Research Institute, Suntory Global Innovation Center Limited, 1-1-1 Wakayama-dai, Shimamoto-cho, Mishima-gun,
Osaka 618-8503, Japan
[2] Safety Science Institute, Quality Assurance Division, Suntory Business Expert Limited, 57 Imaikami-cho, Nakahara-ku,
Kanagawa Kawasaki 211-0067, Japan
[3] Biosciences Research Division, Department of Environment & Primary Industries, AgriBio, Centre for AgriBioscience,
5 Ring Road, La Trobe University, Bundoora, VIC 3083, Australia
[4] School of Applied Sciences, RMIT University, P.O. Box 71, Bundoora, VIC 3083, Australia

Correspondence should be addressed to Keisuke Matsui; keisuke_matsui@suntory.co.jp

Academic Editor: Ana Moldes

Although phosphorus is an essential factor for proper plant growth in natural environments, an excess of phosphate in water sources causes serious pollution. In this paper we describe transgenic plants which hyperaccumulate inorganic phosphate (Pi) and which may be used to reduce environmental water pollution by phytoremediation. AtPHR1, a transcription factor for a key regulator of the Pi starvation response in *Arabidopsis thaliana*, was overexpressed in the ornamental garden plants *Torenia, Petunia, and Verbena*. The transgenic plants showed hyperaccumulation of Pi in leaves and accelerated Pi absorption rates from hydroponic solutions. Large-scale hydroponic experiments indicated that the enhanced ability to absorb Pi in transgenic torenia (AtPHR1) was comparable to water hyacinth a plant that though is used for phytoremediation causes overgrowth problems.

1. Introduction

Water pollution has become a serious problem around the world. Contamination by toxic substances such as endocrine disruptors and heavy metals and excessive inflows of phosphorus, nitrogen and other elements all contribute to water pollution. Eutrophication is one of the major problems associated with water pollution and is caused by inflow of excess amounts of nutrients (especially phosphorus and nitrogen) [1]. The sources of excessive amounts of phosphorus and nitrogen are agricultural run-off, sewage, industrial effluents, and natural erosion from soil and rocks. Eutrophication is due to rapid growth of phytoplankton causing algal blooms or "red tides," the result of which are serious environmental problems such as bad odor and fish death as a result of oxygen depletion and accumulation of toxic cyanotoxins [2].

Phosphorus can be removed by physical, chemical, and biological methods [3–6]. Physical and chemical methods (e.g., electrolytic, crystallization, filtration, and aggregation/separation methods) are superior in terms of removal efficiency and throughput capacity. However, these methods require complicated equipment and large quantities of chemicals, resulting in high cost and environmental burdens. A biological method, the anaerobic-anoxic-oxic method (A2O), is one of the advanced activated sludge methods and has been widely examined in sewage plants. However this method is also very expensive [7], and presently, there are no practically useable technologies to remove inorganic ions such as phosphorus and nitrogen during sewage treatment using activated sludge methods. Thus, though various types of water purification systems have been developed for water and sewage plants [8], these technologies are often difficult to apply directly to aquatic environments due to cost and the need for special equipment. Eutrophication therefore remains a problem.

Concurrently with improving sewage treatment technology, a low-cost and highly efficient method is still needed for sustainable water purification in aquatic environments. A treatment for environmental pollution using plants (phytoremediation) is a possible solution [9, 10]. Since phosphorus is an essential and often limiting nutritive substance for plants, plants actively absorb it from environments through the roots. Phytoremediation of aquatic systems has been attempted using water plants such as water hyacinth and *Phragmites*, as these plants absorb phosphorus relatively efficiently in comparison to terrestrial plants, and they also grow rapidly [11]. However, the high cost of collection and disposal of water plants (especially water hyacinth) presents difficulties in habitat management, and the impact of the plants on preexisting ecosystems hamper their wide application. In addition, the ability of these water plants to eliminate phosphorus in aquatic ecosystems is still inadequate as an even higher efficiency is needed for effective phytoremediation.

Inorganic phosphate (Pi) transporter is a key component in Pi absorption by plant roots. In *Arabidopsis thaliana*, 9 high-affinity transporters are known [12]. One of these, AtPHT1, encodes a cell membrane-located Pi transporter with high affinity for Pi. It has been reported that overexpression of AtPHT1 in cultured cells of *Nicotiana* leads to an acceleration of Pi absorption and an increase in cell growth rate [13]. In contrast, when the same Pi transporter was overexpressed in *Hordeum vulgare*, an increase in absorption of Pi was not observed [14]. These two contradicting reports suggest that merely increasing the number of Pi transporters does not necessarily lead to enhanced Pi absorption.

Several Pi starvation-related genes have been identified in *A. thaliana* mutants [15]. One of the known control factors which function when plants enter a state of Pi starvation is the AtPHR1 gene. AtPHR1 gene encodes a transcription factor which activates the transcription of genes in response to states of Pi starvation [16]. Recently, it is reported that overexpression of AtPHR1 in *A. thaliana* increases the Pi concentration in aerial plant parts [17].

In this study, we introduced the AtPHR1 gene into the garden plants *Torenia*, *Petunia*, and *Verbena*, in order to enhance Pi absorption. Small and large-scale hydroponic trials with transgenic torenia plants expressing the AtPHR1 gene were performed. We demonstrate for the first time that over expression of the AtPHR1 gene results in enhanced Pi absorption rate in different plant species. The AtPHR1 transgenic plants can possibly facilitate effective phytoremediation in polluted aquatic environments.

2. Materials and Methods

2.1. Plant Materials. Plants of *Torenia hybrida* cv. Summer Wave blue, *Petunia hybrida* cv. Surfinia purple mini, and *Verbena hybrida* cv. Temari scarlet (Suntory Flowers, Ltd.) were grown in soil and supplied with full nutrients every week in a green house or a growth chamber in controlled conditions (22–25°C, 12 hours light).

2.2. Constructs for Expression in Plants and Plant Transformation. Molecular biology techniques were employed according to the methods described by Sambrook et al. [18], unless otherwise specified.

The AtPHR1 gene was amplified by PCR using primers PHRf (5′-ATGGAGGCTCGTCCAGTTCAT-3′) and PHRr (5′-TCAATTATCGATTTTGGGACGC-3′) and subcloned into the pCR2.1 vector using a TOPO-TA cloning kit (Life Technologies) according to the manufacturer's instructions. A fragment of the AtPHR1 gene was inserted into binary vector pBinPLUS [19] which contains an enhanced cauliflower mosaic virus 35S promoter [20] and a nopaline synthase (nos) terminator. This plasmid was named pSPB1898.

Transformation with transformation vector pSPB1898 was carried out as described previously for Torenia [21], Petunia [22], and Verbena [23] using *Agrobacterium tumefaciens* strain AGL0 [24].

RNAs were extracted from leaves of the obtained recombinant plants using the RNeasy Plant Mini Kit (Qiagen). Positive strains were selected by RT-PCR.

2.3. Method for Measuring Phosphorus Concentration. Phosphorus concentration was measured according to a modified method of Ames [25]. Leaves were weighed (approximately 100 mg per sample) and inserted into a 2 mL tube for crushing with zirconia beads (4 mm diameter), at −80°C. The frozen sample was taken to room temperature, and 500 μL of 1% (v/v) acetic acid was added to each tube. The mixture was then shaken and crushed for 6 minutes using a TissueLyser (Qiagen). After crushing, the mixture was centrifuged at 15,000 rpm for 5 minutes using a desktop centrifuge to obtain 500 μL of supernatant. This Pi extract was diluted with distilled water (from 10 to 100-fold dilution) to a final concentration of 800 μL. To this solution, 160 μL of measuring buffer (1.25 M sulfuric acid, 30 mM ascorbic acid, 0.405 mg/mL antimony potassium tartrate, and 24 mg/mL ammonium molybdate) was added, and the mixture was stirred well and left for 10 minutes. The absorbance was measured at 880 nm using a spectrophotometer BioSpec-mini (Shimadzu, Japan). The amount of phosphorus in 1 g of leaf was calculated from phosphorus concentration and weight of the sample. For calculations on a dry weight basis, samples were dried at 80°C for about 2 days.

An independent Student's t-test was used to compare differences between host and transgenic plants. All tests were two-sided, and $P < 0.05$ was considered statistically significant. Data are the mean ± SD from at least three different samples.

2.4. Hydroponic Experiment. Wild-type torenia or transgenic torenia was grown on a support made of polystyrene foam with holes to allow the root systems of the plants to grow into the hydroponic solution. Plants were floated on 5 liter of hydroponic solution (0.5 mM KNO_3, 0.2 mM $MgSO_4$, 0.2 mM $Ca(NO_3)_2$, 0.161 mM KPO_4, 5 μM Fe-EDTA, 7 μM H_3BO_3, 1.4 μM $MnCl_2$, 0.05 μM $CuSO_4$, 0.1 mM $ZnSo_4$, 0.02 μM Na_2MoO_4, 1 μM NaCl, and 0.001 μM $CoCl_2$). The initial phosphorus concentration in the hydroponic solution was 5 mg/L. Four plants were used in each support. The Pi concentration in the hydroponic solution was measured each day. Since the fluid volume of the hydroponic solution decreased due to transpiration and evaporation, on every fourth day, deionised water was added to the solution. For

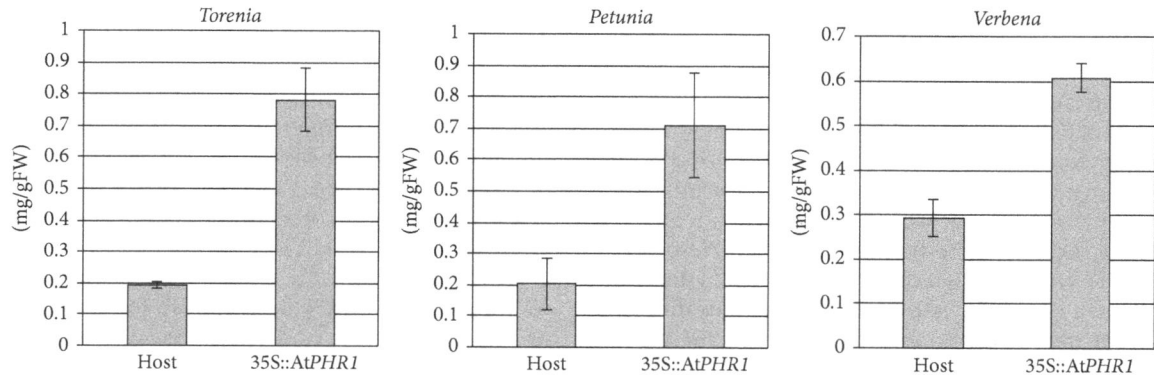

FIGURE 1: Phosphorus measurements of At*PHR1* transgenic plants. Phosphorus concentrations in the leaves of At*PHR1* transgenic plants of potted torenia, petunia, and verbena were measured. The longitudinal axis shows the phosphorus amounts per gram fresh weight (mg/gFW). Significant differences in means between host and transgenic plants were detected for all three species.

large container experiments, the same solution was used, but the volume of hydroponic solution was 400 liter, and 13 plants were used per container. The volume of each container was adjusted with deionised water on a weekly basis.

3. Results and Discussion

3.1. Overexpression of AtPHR1 Enhances Pi Accumulation and Absorption in Transgenic Plants.
It has been shown in *A. thaliana* that over expression of At*PHR1* causes enhanced Pi accumulation in aerial parts [17]. To examine whether At*PHR1* is effective in otherplant species, we transformed torenia, petunia, and verbena with At*PHR1*. These plants were transformed with the plasmid pSPB1898, which contains the At*PHR1* gene under the control of the constitutive 35S promoter. We screened over 30 transgenic plants for each species for the presence of the transgene with RT-PCR and for leaf Pi concentration 4 weeks after potting up from tissue culture. Concentration of phosphorus per fresh leaf weight was then measured for selected lines. In each of the 3 plant species, phosphorus concentration in the leaves of the transgenic plants was 2 to 3-fold higher than that of control host plants (Figure 1).

We examined other Pi starvation-related genes (At*PHT1;1*, At*PHT1;2*, At*IPS1*, and At*PHO1*) from *A. thaliana* by constitutively overexpressing them in transgenic torenia and petunia (data not shown). None of these transgenic plants showed enhanced Pi accumulation. This result is consistent with the observation that over-expression of the Pi transporter did not cause any change to Pi accumulation in *H. vulgare* [14]. Thus, we focused on At*PHR1* in the following experiments.

To confirm that introduction of the At*PHR1* gene accelerates Pi absorption rates, we grew plants of a transgenic torenia line in a hydroponic system. Torenia was chosen as this plant grows luxuriantly and roots tolerate being submerged in water. The torenia plants were grown in 5 liters of hydroponic solution containing 5 mg/L phosphorus for 1 to 2 months in a green house or a growth chamber. The phosphorus concentration of the hydroponic solutions was measured daily. The superior transgenic line expressing At*PHR1*

(35S::At*PHR1*) showed enhanced Pi absorption from the hydroponic solution (Figure 2(a)). Enhanced accumulation of Pi in the transgenic leaves was also confirmed by measurements of leaf phosphorus concentration (Figure 2(b)). The phosphorus concentration of the hydroponic solution in which 35S::At*PHR1* was grown decreased during the two weeks of the experiment. The Pi absorption rate observed for 35S::At*PHR1* was up to 0.091 mgP/day/plant in this experiment compared to 0.056 mgP/day/plant for the host (Figure 2(a)). This result suggests that the enhanced Pi accumulation observed in the potted At*PHR1* transgenic torenia plants is mainly due to enhanced Pi absorption rate.

To see if the decrease of Pi concentration in the hydroponic solution was also reflected in an increase in Pi accumulation in the plant, Pi accumulation in the aerial parts of the plants was measured. Three plants each of the transgenic and the host torenia were hydroponically cultivated in the solution containing 5 mg/L phosphorus for about 2 months. The aerial parts of those plants were collected and dried on the phosphorus concentration measured (Figure 2(b)). The Pi concentration in the transgenic plants was approximately 2.5-fold that of the host.

We weighed aerial and root parts of the tested plants after each hydroponic experiment. Even though slightly less weight was measured in the host, there was no statistically significant difference between the transgenic and host (Figure 2(c)). This suggests that excessively absorbed Pi is not used for plant growth but is accumulated and stored in the aerial part of the plants. As a result, overexpression of At*PHR1* does not retard plant growth. Since the transgenic plants did not show any morphological or reproductive abnormalities, overexpression of the At*PHR1* gene can enhance Pi accumulation with no negative effects on plant growth.

3.2. Limitation of Pi Capacity.
Sections of dead tissues in the leaves were often observed in transgenic torenia during the 4 weeks of the hydroponic experiments (Figures 3(a)–3(c)). We collected the dead sections and compared them to the unaffected areas of the leaves from the same plants. The harvested leaves were dried and then measured for phosphorus concentration. The phosphorus concentration in

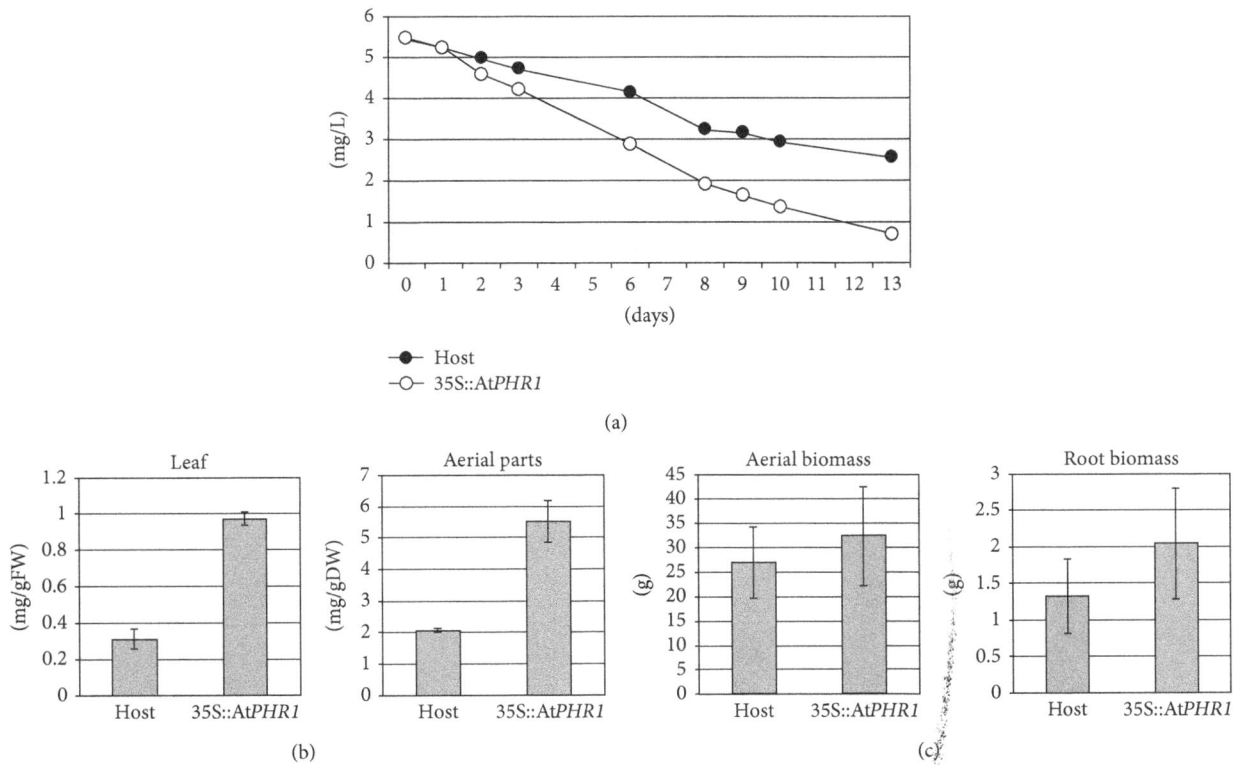

FIGURE 2: Pi accumulation and growth properties of AtPHR1 transgenic torenia. (a) Changes of Pi concentration in hydroponic solutions. The phosphorus concentration in a hydroponic solution in which host (filled circle) and AtPHR1 transgenic torenia (empty circle) were cultured was measured. The longitudinal axis shows the phosphorus concentration (mg/L), and the horizontal axis shows the number of days after exchange of the hydroponic solution. (b) Pi concentration in the leaves and aerial parts of hydroponically-cultivated torenia. The longitudinal axis shows the phosphorus concentration per gram fresh weight of samples (mg/gFW) (left) and the phosphorus concentration per gram dry weight of samples (mg/gDW) (right). There were significant differences in means between host and transgenic plants. (c) Comparison of growth rate. Weights of aerial parts and root parts of the torenia plants were measured at the end of hydroponic experiments. There was no statistically significant difference between transgenic and host.

the dead sections was slightly higher than that of unaffected portions of leaves (Figure 3(d)). Since excess Pi may cause cell toxicity [26], the death may have been the result of exceeding a critical limit of Pi concentration in the torenia leaf cells. It thus appears that the critical limit of Pi accumulation level in AtPHR1 transgenic torenia is approximately 20 mg/gDW. One possible way to overcome the death of leaf tissues due to high Pi accumulation is to convert Pi to a nontoxic form of phosphorus that is phytic acid. Genetic modification could be used to achieve this, resulting in transgenic plants accumulating even more Pi than reported here.

3.3. Large-Scale Hydroponic Experiment.
To access the potential for phytoremediation using the transgenic torenia at a larger scale, we performed longer term hydroponic experiments. Thirteen torenia plants were put each into 400-liter tub and incubated for approximately 2 months (Figure 4). There was no significant difference in average biomass between transgenic and host plants after 65 days incubation (Figure 4 and Table 1). However, approximately 3-fold more Pi accumulation was seen in the transgenic plant when compared to the host. This confirmed that transgenic torenia

TABLE 1: Comparison of phosphate absorption performances. Phosphorus content, total biomass, and absorption rate after 65 days of the hydroponic experiment are indicated. Data are the mean ± SD from 13 plants. Values of water hyacinth were calculated from values listed in [27, 28].

	Phosphorus in leaf (mg/gFW)	Total biomass (g/plant)	Absorption rate (mg/plant/day)
Host	0.18 ± 0.11	396.34 ± 146.06	1.08
35S::AtPHR1	0.69 ± 0.20	382.95 ± 178.85	4.15
Water hyacinth	0.38		1.79

shows the accelerated absorption as well as accumulation of Pi in the leaves when grown on a larger scale. From the daily calculation of Pi accumulation of the transgenic torenia plant, Pi accumulation rates were able to be compared to water hyacinth (Table 1). The AtPHR1 transgenic torenia showed an equivalent efficiency of Pi accumulation to that of water hyacinth [27, 28].

Overexpression of AtPHR1 gene might drive a Pi starvation response in the transgenic plants. As a result, excessive

FIGURE 3: Dead tissue in At*PHR1* transgenic torenia. (a) Host plant at the end of hydroponic experiment. (b) At*PHR1* transgenic plant after 4 weeks of hydroponic experiment. (c) Magnified image of (b). Arrowheads indicate partially dead sections. (d) Phosphorus concentration in unaffected and dead areas from leaves of host and At*PHR1* transgenic plants. The longitudinal axis shows the phosphorus concentration per gram dry weight of sample (mg/gDW).

amounts of Pi accumulated in transgenic leaves. In *A. thaliana*, At*PHR1* gene is not transcriptionally regulated even under Pi starvation condition [17]. Since the key mechanism of the Pi starvation response is still debatable in *Arabidopsis thalinana* [17, 29], it is difficult to postulate why overexpression of At*PHR1* is effective for Pi uptake in other species. We have isolated orthologous Pi starvation-related genes (At*PHR1*, At*IPS1*, At*PHT1;1*, and At*PHO2*) in torenia and examined expression pattern of these genes (data not shown). We could not detect any differences between transgenic torenia and host plants. Overexpression of At*PHR1* may interfere with the proper posttranscriptional modification of the endogenous At*PHR1* counterpart, possibly through competitive inhibition.

Since phosphorus is expected to be exhausted as a natural resource within a hundred year [30], it is necessary to recover phosphorus from the environment, especially in polluted areas. Currently, over 90% of the produced phosphorus in the world is used as fertilizers. Therefore, it is most reasonable to recover phosphorus from fertilized soils and agricultural run-offs. Phytoremediation is a suitable method for such a recycling process, in addition to cleaning up phosphorus from the aquatic environment. One of the critical problems of phytoremediation is the cost of the disposal of the plant [31]. The plant used for phytoremediation was in many cases simply discarded without being used as a source of Pi. Ideally, plants containing high accumulation of Pi can be returned to soils of agricultural land without processing and can be directly

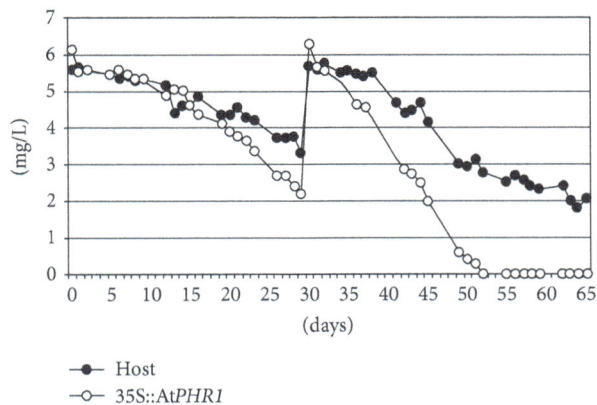

(a)

(b)

(c)

FIGURE 4: Large-scale hydroponic experiments. (a) Changes in Pi concentration in hydroponic solutions. The phosphorus concentration in a hydroponic solution in which host (filled circle) and AtPHR1 transgenic torenia (empty circle) were cultured was measured. The longitudinal axis shows the phosphorus concentration (mg/L), and the horizontal axis shows the number of days. Hydroponic solutions were fully exchanged 30 days after starting the experiment. (b) Large-scale experiment (0 day). (c) Large-scale experiment (65 days).

used as fertilizer. However, at present, absorbing ability of the existing plants used for phytoremediation is not efficient enough to be used as Pi sources for agriculture in this way. In this study, the AtPHR1 transgenic plants accumulated a high level of Pi. Therefore, applications of AtPHR1 transgenic plants for phytoremediation of water could be cost-effective. Moreover, the Pi recycling ability of flowers and ornamental plants for gardening can be increased by means of AtPHR1 gene introduction, and thereby purifying water with plants having both ornamental beauty and high purification ability.

4. Conclusions

In this study, we prove the feasibility of using AtPHR1 as an enhancer of Pi uptake in transgenic plants. By introducing AtPHR1 to garden plants, amounts of Pi accumulation and absorption of Pi were increased to rates approximately 3-fold higher than host plant. There was no significant reduction in biomass or morphology of the transgenic plant expressing AtPHR1. Taken together, these observations indicate that the AtPHR1 gene will be valuable for production of hyperaccumulator plants for the purification of waters polluted with Pi. In addition, an improved appearance of purification sites can be provided by using ornamental plants with many flowers, as shown in Figure 4(c).

Conflict of Interests

The authors have no conflict of interests to declare.

Acknowledgments

The authors thank Mses. Keiko Takeda, Masumi Taniguchi, and Sarah Parsons for producing the transgenic plants and Mses. Chika Shimadzu, Kumi Takemura, Miyuki Ogawa, and Kim Stevenson for their technical assistance. The authors thank Dr. Robert A. Ludwig for providing A. tumefaciens Agl0 and appreciate Mr. Masayasu Yoshikawa for his critical reading of the paper.

References

[1] V. H. Smith, G. D. Tilman, and J. C. Nekola, "Eutrophication: impacts of excess nutrient inputs on freshwater, marine, and terrestrial ecosystems," Environmental Pollution, vol. 100, no. 1–3, pp. 179–196, 1998.

[2] V. H. Smith and D. W. Schindler, "Eutrophication science: where do we go from here?" Trends in Ecology and Evolution, vol. 24, no. 4, pp. 201–207, 2009.

[3] C. Vohlaa, M. Koiva, H. J. Bavor et al., "Filter materials for phosphorus removal from wastewater in treatment wetlands—A review," Ecological Engineering, vol. 37, no. 1, pp. 70–89, 2011.

[4] J. QU, "Research progress of novel adsorption processes in water purification: a review," *Journal of Environmental Sciences*, vol. 20, no. 1, pp. 1–13, 2008.

[5] D. W. de Haas, M. C. Wentzel, and G. A. Ekama, "The use of simultaneous chemical precipitation in modified activated sludge systems exhibiting biological excess phosphate removal part 1: literature review," *Water SA*, vol. 26, no. 4, pp. 439–452, 2000.

[6] F. Y. Wang, V. Rudolph, and Z. H. Zhu, "Sewage Sludge technologies," *Encyclopedia of Ecology*, pp. 3227–3242, 2008.

[7] Y. Peng, X. Wang, W. Wu, J. Li, and J. Fan, "Optimisation of anaerobic/anoxic/oxic process to improve performance and reduce operating costs," *Journal of Chemical Technology and Biotechnology*, vol. 81, no. 8, pp. 1391–1397, 2006.

[8] M. A. Shannon, P. W. Bohn, M. Elimelech, J. G. Georgiadis, B. J. Marĩas, and A. M. Mayes, "Science and technology for water purification in the coming decades," *Nature*, vol. 452, no. 7185, pp. 301–310, 2008.

[9] E. Pilon-Smits, "Phytoremediation," *Annual Review of Plant Biology*, vol. 56, pp. 15–39, 2005.

[10] M. Luqman, T. M. Batt, A. Tanvir et al., "Phytoremediation of polluted water by trees: a review," *African Journal of Agricultural Research*, vol. 8, no. 17, pp. 1591–1595, 2013.

[11] P. Gupta, S. Roy, and A. B. Mahindrakar, "Treatment of water using water hyacinth, water lettuce and vetiver grass—a review," *Resources and Environment*, vol. 2, no. 5, pp. 202–215, 2012.

[12] C. Rausch and M. Bucher, "Molecular mechanisms of phosphate transport in plants," *Planta*, vol. 216, no. 1, pp. 23–37, 2002.

[13] N. Mitsukawa, S. Okumura, Y. Shirano et al., "Overexpression of an *Arabidopsis thaliana* high-affinity phosphate transporter gene in tobacco cultured cells enhances cell growth under phosphate-limited conditions," *Proceedings of the National Academy of Sciences of the United States of America*, vol. 94, no. 13, pp. 7098–7102, 1997.

[14] A. L. Rae, J. M. Jarmey, S. R. Mudge, and F. W. Smith, "Over-expression of a high-affinity phosphate transporter in transgenic barley plants does not enhance phosphate uptake rates," *Functional Plant Biology*, vol. 31, no. 2, pp. 141–148, 2004.

[15] C. A. Ticconi and S. Abel, "Short on phosphate: plant surveillance and countermeasures," *Trends in Plant Science*, vol. 9, no. 11, pp. 548–555, 2004.

[16] V. Rubio, F. Linhares, R. Solano et al., "A conserved MYB transcription factor involved in phosphate starvation signaling both in vascular plants and in unicellular algae," *Genes and Development*, vol. 15, no. 16, pp. 2122–2133, 2001.

[17] L. Nilsson, R. Müller, and T. H. Nielsen, "Increased expression of the MYB-related transcription factor, *PHR1*, leads to enhanced phosphate uptake in *Arabidopsis thaliana*," *Plant, Cell and Environment*, vol. 30, no. 12, pp. 1499–1512, 2007.

[18] J. Sambrook, E. F. Fritsch, and T. Maniatis, *Molecular Cloning: A Laboratory Manual*, Cold Spring Harbor Laboratory Press, 1989.

[19] F. A. van Engelen, J. W. Molthoff, A. J. Conner, J. P. Nap, A. Pereira, and W. J. Stiekema, "pBINPLUS: an improved plant transformation vector based on pBIN19," *Transgenic Research*, vol. 4, no. 4, pp. 288–290, 1995.

[20] I. Mitsuhara, M. Ugaki, H. Hirochika et al., "Efficient promoter cassettes or enhanced expression of foreign genes in dicotyledonous and monocotyledonous plants," *Plant and Cell Physiology*, vol. 37, no. 1, pp. 49–59, 1996.

[21] R. Aida and M. Shibata, "*Agrobacterium*-mediated transformation of torenia (*Torenia fournieri*)," *Breeding Science*, vol. 45, no. 1, pp. 71–74, 1995.

[22] R. B. Horsch, J. E. Fry, N. L. Hoffmann, D. Eichholtz, S. G. Rogers, and R. T. Fraley, "A simple and general method for transferring genes into plants," *Science*, vol. 227, no. 4691, pp. 1229–1231, 1985.

[23] M. Tamura, J. Togami, K. Ishiguro et al., "Regeneration of transformed verbena (*Verbena × hybrida*) by *Agrobacterium tumefaciens*," *Plant Cell Reports*, vol. 21, no. 5, pp. 459–466, 2003.

[24] G. R. Lazo, P. A. Stein, and R. A. Ludwig, "A DNA transformation-competent *Arabidopsis* genomic library in *Agrobacterium*," *Bio/Technology*, vol. 9, no. 10, pp. 963–967, 1991.

[25] B. N. Ames, "Assay of inorganic phosphate, total phosphate and phosphatases," *Methods in Enzymology*, vol. 8, pp. 115–118, 1966.

[26] D. T. Clarkson and C. B. Scattergood, "Growth and phosphate transport in barley and tomato plants during the development of, and recovery from, phosphate-stress," *Journal of Experimental Botany*, vol. 33, no. 5, pp. 865–875, 1982.

[27] W. T. Haller and D. L. Sutton, "Effect of pH and high phosphorus concentrations on growth of water hyacinth," *Hyacinth Control Journal*, pp. 59–61, 1973.

[28] M. Morii, Y. Doyama, and J. Katayama, "On the absorption of nitrogen and phosphorus from water by water hyacinth, *Eichhornia crassipes* (Mart.) Solms," *Bulletin of the Osaka Agricultural Research Center*, vol. 26, pp. 11–15, 1990.

[29] R. Bustos, G. Castrillo, F. Linhares et al., "A central regulatory system largely controls transcriptional activation and repression responses to phosphate starvation in arabidopsis," *PLoS Genetics*, vol. 6, no. 9, Article ID e1001102, 2010.

[30] A. Maggio, J.-P. Malingreau, A.-K. Bock et al., "NPK: Will there be enough plant nutrients to feed a world of 9 billion in 2050?" Publication Office of the European Union, 2012, http://publications.jrc.ec.europa.eu/repository/handle/111111111/25770.

[31] A. Sas-Nowosielska, R. Kucharski, E. Małkowski, M. Pogrzeba, J. M. Kuperberg, and K. Kryński, "Phytoextraction crop disposal—an unsolved problem," *Environmental Pollution*, vol. 128, no. 3, pp. 373–379, 2004.

20

Quantification of Human and Animal Viruses to Differentiate the Origin of the Fecal Contamination Present in Environmental Samples

Sílvia Bofill-Mas, Marta Rusiñol, Xavier Fernandez-Cassi, Anna Carratalà, Ayalkibet Hundesa, and Rosina Girones

Laboratory of Viruses Contaminants of Water and Food, Department of Microbiology, Faculty of Biology, Avenida Diagonal 643, 08028 Barcelona, Catalonia, Spain

Correspondence should be addressed to Sílvia Bofill-Mas; sbofill@ub.edu

Academic Editor: Caroline Rigotto

Many different viruses are excreted by humans and animals and are frequently detected in fecal contaminated waters causing public health concerns. Classical bacterial indicator such as *E. coli* and enterococci could fail to predict the risk for waterborne pathogens such as viruses. Moreover, the presence and levels of bacterial indicators do not always correlate with the presence and concentration of viruses, especially when these indicators are present in low concentrations. Our research group has proposed new viral indicators and methodologies for determining the presence of fecal pollution in environmental samples as well as for tracing the origin of this fecal contamination (microbial source tracking). In this paper, we examine to what extent have these indicators been applied by the scientific community. Recently, quantitative assays for quantification of poultry and ovine viruses have also been described. Overall, quantification by qPCR of human adenoviruses and human polyomavirus JC, porcine adenoviruses, bovine polyomaviruses, chicken/turkey parvoviruses, and ovine polyomaviruses is suggested as a toolbox for the identification of human, porcine, bovine, poultry, and ovine fecal pollution in environmental samples.

1. Fecal Contamination of the Environment

Significant numbers of human microbial pathogens are present in urban sewage and may be considered environmental contaminants. Viruses, along with bacteria and protozoa in the intestine or in urine, are shed and transported through the sewer system. Although most pathogens can be removed by sewage treatment, many are discharged in the effluent and enter receiving waters. Point-source pollution enters the environment at distinct locations, through a direct route of discharge of treated or untreated sewage. Nonpoint sources of contamination are of significant concern with respect to the dissemination of pathogens and their indicators in the water systems. They are generally diffuse and intermittent and may be attributable to the run-off from urban and agricultural areas, leakage from sewers and septic systems, storm water, and sewer overflows [1–3].

Even in highly industrialized countries, viruses that infect humans prevail throughout the environment, causing public health concerns and leading to substantial economic losses. Many orally transmitted viruses produce subclinical infection and symptoms in only a small proportion of the population. However, some viruses may give rise to life-threatening conditions, such as acute hepatitis in adults, as well as severe gastroenteritis in small children and the elderly. The development of disease is related to the infective dose of the viral agent, the age, health, immunological and nutritional status of the infected individual (pregnancy, presence of other infections or diseases), and the availability of health care. Human pathogenic viruses in urban wastewater may potentially include human adenoviruses (HAdVs) and human polyomaviruses (HPyVs), which are detected in all geographical areas and throughout the year, and enteroviruses, noroviruses, rotaviruses, astroviruses, hepatitis A, and hepatitis E viruses, with variable prevalence in different geographical areas and/or periods of the year.

Moreover, with the venue of novel metagenomic techniques, new viruses are being discovered in the recent years that may be present in sewage and potentially contaminate the environment being transmitted to humans [4, 5].

Failures in controlling the quality of water used for drinking, irrigation, aquaculture, food processing, or recreational purposes have been associated to gastroenteritis and other diseases outbreaks in the population [6, 7]. Detailed knowledge about the contamination sources is needed for efficient and cost-effective management strategies to minimize fecal contamination in watersheds and foods, evaluation of the effectiveness of best management practices, and system and risk assessment as part of the water and food safety plans recommended by the World Health Organization [8, 9].

Microbial source tracking (MST) plays a very important role in enabling effective management and remediation strategies. MST includes a group of methodologies that aim to identify, and in some cases quantify, the dominant sources of fecal contamination in the environment and, more specifically, in water resources [10, 11]. Molecular techniques, specifically nucleic acid amplification procedures, provide sensitive, rapid, and quantitative analytical tools for studying specific pathogens, including new emergent strains and indicators. Quantitative PCR (qPCR) is used to evaluate the microbiological quality of water [12] and the efficiency of virus removal in drinking and wastewater treatment plants [13, 14] and as a quantitative MST tool [15].

Between a wide range of MST candidate tools (reviewed in [16–18]), the use of human and animal viruses analyzed by qPCR as fecal indicators and MST tools will be the focus of this review.

2. Indicators of Fecal Contamination

Fecal pollution is a primary health concern in the environment, in water, and in food. The use of index microorganisms (whose presence points to the possible occurrence of a similar pathogenic organism) and indicator microorganisms (whose presence represents a failure affecting the final product) to assess the microbiological quality of waters or food is well established and has been practiced for almost a century [19].

Classic microbiological indicators such as fecal coliforms, *Escherichia coli*, and enterococci are the indicators most commonly analyzed to evaluate the level of fecal contamination. However, whether these bacteria are suitable indicators of the occurrence and concentration of pathogens such as viruses and protozoa cysts has been questioned for the following reasons: (i) indicator bacteria are more sensitive to inactivation through treatment processes and by sunlight than viral or protozoan pathogens; (ii) nonexclusive fecal source; (iii) ability to multiply in some environments; (iv) inability to identify the source of fecal contamination; (v) and low correlation with the presence of pathogens.

Various authors concluded that these indicators could fail to predict the risk for waterborne pathogens including viruses [20, 21]. Moreover, the levels of bacterial indicators do not always correlate with the concentration of viruses, especially when these indicators are present in low concentrations [22, 23].

Those viruses that are transmitted via contaminated food or water are typically stable because they lack the lipid envelopes that render other viruses vulnerable to environmental agents. Moreover since viruses usually respond to a host specific behavior, their detection may provide data for MST.

The fact that rapid methods are required and that, moreover, many pathogens cannot be cultivated in the laboratory has led to the development of new methodologies for the study of pathogens and new proposed indicators of fecal contamination in water and food. These are based on the implementation of molecular techniques that are rapid and sensitive but may pick up both infectious and noninfectious (dead) types. Quantitative PCR assays are being considered by US-EPA as a rapid analytical tool [24]. A review focused on the application of qPCR in the detection of microorganisms in water has been recently published by Botes and coworkers [25].

3. Quantification of Human and Animal Viruses as a Tool-Box for Determining Presence and Origin of Fecal Contamination in Waters

The high stability of viruses in the environment, their host specificity, persistent infections, and high prevalence of some viral infections throughout the year strongly support the use of rapid cost-effective sensitive molecular techniques for the identification and quantification of DNA viruses which can be used as complementary indicators of fecal and urine (hereinafter "fecal") contamination and as MST tools. Detection of excreted DNA viruses may allow the development of cost-effective protocols with more accurate quantification of contaminating sources compared to RNA viruses. This is due to the greater accuracy of qPCR and its/their lower sensitivity to inhibitors, as reverse transcriptase is not used when amplifying DNA viruses.

Our research group has proposed new viral parameters and methodologies for the detection and quantification of human and animal DNA viruses as fecal indicators as well as MST tools. The first viral markers proposed were DNA viruses such as human and animal adenoviruses and polyomaviruses, and the assays developed for their detection were based on qualitative PCR [22, 81–84], and more recently qPCR techniques have been developed for not only detecting but also quantifying these viruses in environmental samples [28–31, 37].

Several research groups are currently using these parameters for analysis of viral contamination in water and as MST tools. One of the objectives of this review is to examine available data, so far, on the application of specific DNA viral indicators proposed many years ago (human adenoviruses, JC polyomavirus, porcine adenovirus, and bovine polyomavirus) and to evaluate its usefulness as quantitative tools for determining the origin of the fecal contamination in different countries.

Quantification of Human and Animal Viruses to Differentiate the Origin of the Fecal Contamination Present in Environmental Samples

189

TABLE 1: Oligonucleotide primers and probes used for the detection and quantification of viral indicators.

Primers and probes	Virus	Position[a]	Reference	Sequence (5′-3′)
ADF	Human adenovirus (HAdV)	18869–18887	[26]	CWTACATGCACATCKCSGG
ADR		18919–18937		CRCGGGCRAAYTGCACCAG
ADP1		18889–18916		FAM-CCGGGCTCAGGTACTCCGAGGCGTCCT-BHQ1
JE3F	JC polyomavirus (JCPyV)	4317–4339	[27]	ATGTTTGCCAGTGATGATGAAAA
JE3R		4251–4277		GGAAAGTCTTTAGGGTCTTCTACCTTT
JE3P		4313–4482		FAM-AGGATCCCAACACTCTACCCCACCTAAAAAGA-BHQ1
QB-F1-1	Bovine polyomavirus (BPyV)	2122–2144	[28]	CTAGATCCTACCCTCAAGGGAAT
QB-R1-1		2177–2198		TTACTTGGATCTGGACACCAAC
QB-P1-2		2149–2174		FAM-GACAAAGATGGTGTGTATCCTGTTGA-BHQ1
Q-PAdV-F	Porcine adenovirus (PAdV)	20701–20718	[29]	AACGGCCGCTACTGCAAG
Q-PAdV-R		20751–20768		CACATCCAGGTGCCGC
Q-PAdV-P		20722–20737		FAM-AGCAGCAGGCTCTTGAGG-BHQ1
qOv_F	Ovine polyomavirus (OPyV)	VP1[b] region	[30]	CAGCTGYAGACATTGTGG
qOv_R				TCCAATCTGGGCATAAGATT
qOv_P				FAM-ATGATTACCAAGCCAGACAGTGGG-BHQ1
Q-PaV-F	Chicken/turkey parvovirus (ChPV/TuPV)	3326–3345	[31]	AGTCCACGAGATTGGCAACA
Q-PaV-R		3388–3407		GCAGGTTAAAGATTTTCACG
Q-PaV-Pr		3356–3378		6FAM-AATTATTCGAGATGGCGCCCACG-BHQ1

[a]The sequence positions are referred to strains J01917.1 (HAdV), NC_001699.1 (JCPyV), D13942 (BPyV), AJ237815 (PAdV), and GU214706 (ChPV/TuPV) from Genbank. [b]VP1: virion protein 1.

Our group recently developed quantitative PCR (qPCR) assays for the quantification of chicken/turkey parvoviruses and ovine polyomaviruses which, together with those previously proposed for human, bovine and porcine fecal contamination, might constitute a tool box for studying the presence and origin of fecal contamination in environmental samples (Table 1).

4. Treatment of Water Samples for Quantification of Viruses

A wide range of concentration methods have been described to recover viruses from water samples. These methods seek to concentrate viruses from large volumes (up to 1000 L) to smaller volumes ranging from 10 mL to 100 μL. Most of the methods used are based on adsorption-elution processes using membranes, filters, or matrixes like glass wool [58, 62, 85]. However, they are two-step methods that can be cumbersome and could hamper the simultaneous processing of a large number of samples. In order to eliminate the bottleneck associated with two-step methods, and when volumes of 1–10 L are analyzed, a one-step concentration is used in our laboratory. The method was initially designed to concentrate viruses from seawater samples [38]. Briefly, the method is based on the addition of a preflocculated skimmed-milk solution to the volume of sample to be concentrated. The pH is then adjusted to 3.5 with HCl 1 N and the sample is then stirred for 8 h to allow the viruses to be adsorbed into the

skimmed-milk flocs at room temperature (RT). Then flocs are recovered by centrifugation at 8,000 ×g for 30 min at 4°C. The supernatants are carefully removed without disturbing the sediment and the pellet is dissolved in phosphate buffer (pH 7.5). Preconditioning of the conductivity of the samples may be needed when applying the method to the concentration of viruses from freshwater samples [86] and a variation of the method has also been reported for sewage samples [87]. The method has proven to be efficient and reproducible, and by applying this method we have been able to concentrate virus from different water matrices [4, 52, 53, 77, 88, 89].

Enzymatic inhibition of the PCR is also a matter to have into consideration when testing environmental samples. Specific qPCR kits designed for working with environmental samples are available commercially. Analyzing neat but also diluted nucleic acids extraction is also recommended as well as introducing controls of inhibition in the assays performed [28].

Although some of these viruses, such as some types of human adenoviruses, may grow in cell culture, other viruses may not and/or cell culture assays take too long to produce rapid results. Some authors use nucleases treatment to destroy free genomes or genomes contained into damaged viral particles before nucleic acid extraction and qPCR in order to quantify only potentially infective viral particles [90–92].

A flowchart summarizing the steps to follow to test an environmental sample for the presence of viral indicators is

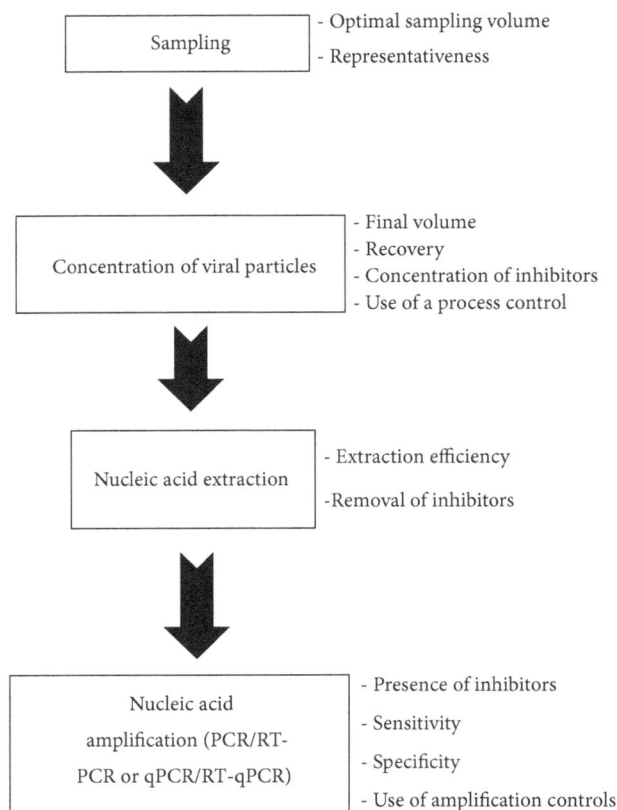

FIGURE 1: Flowchart of the method to detect and quantify viral indicators in the environment by PCR-based methods.

represented in Figure 1. Critical points to which attention should be paid are also summarized in the Figure 1.

5. Quantitative PCR of Human Adenoviruses and Human Polyomavirus JC: A Tool to Determine Human Fecal Pollution in Water Matrices

Some viruses, such as human polyomaviruses (HPyVs) and adenoviruses (HAdVs), infect humans during childhood, thereby establishing, some of them, persistent infections. They are excreted in high quantities in the feces or urine of a high percentage of individuals.

The Adenoviridae family has a double-stranded DNA genome of approximately 35 000 base pairs (bp) surrounded by a 90–100 nm nonenveloped icosahedral shell with fiber-like projections from each vertex. Adenovirus infection may be caused by consumption of contaminated water or food, or by inhalation of aerosols from contaminated waters such as those used for recreational purposes. HAdV comprises 7 species with 52 types, which are responsible for both enteric illnesses and respiratory and eye infections [93].

Quantitative-based qPCR techniques used for the quantification of HAdV have been mainly designed to target the hexon protein and through degeneration of some nucleotides

been able to amplify all HAdV types. In some cases and since HAdV types 40 and 41 are the ones etiologically associated to gastroenteritis as well as to a high prevalence in environmental samples, assays based on the sole detection of this two types have also been developed (Table 2).

Some of the more commonly used qPCR assays have been described by Hernroth et al. [26] with modifications [37] and Heim et al. [35]. We have previously compared both methods obtaining higher quantification in wastewater samples when applying the first one [37].

Table 2 summarizes quantitative HAdV data obtained by testing by qPCR different types of environmental samples.

Polyomaviruses are small and icosahedral viruses, with a circular double-stranded DNA genome of approximately 5000 bp that infect several species of vertebrates. JCPyV is ubiquitously distributed worldwide and antibodies against it are detected in over 80% of humans [94]. Kidney and bone marrow are sites of latent infection with JCPyV, which is excreted in the urine by healthy individuals [95, 96]. The pathogenicity of the virus is commonly associated with progressive multifocal leukoencephalopathy (PML) in immunocompromised states, and it has attracted new attention due to JCPyV reactivation and pathogenesis in some patients of autoimmune diseases under treatment with immunomodulators [97, 98]. JCPyV is ubiquitously distributed and antibodies against JC virus are detected in over 80% of human population worldwide. BKPyV, the other classical human polyomavirus, causes nephropathy in renal transplant recipients and other immunosuppressed individuals. It is also excreted in urine and thus is present in wastewater, although its prevalence is lower than that of JCPyV [82], JCPyV is more frequently excreted than BKPyV. Is for these reasons that the specific polyomaviral marker in use in our laboratory is based on the quantification of JCPyV [37]. The assay developed by McQuaig et al. [63] that targets JC and BK human polyomaviruses (HPyVs) has also been extensively used (Table 3). We have tested both assays in diverse types of environmental samples obtaining equivalent results (data not shown). Results obtained when applying these assays to environmental samples support the applicability of the proposed indicators as molecular markers of the microbiological quality of water and they would fulfill the conditions defined for a human fecal/urine indicator. Harwood et al. [65], in a study using PCR, suggest that human polyomaviruses were the most specific human marker for MST among many other tools analyzed.

Overall, studies show that HAdV has the highest prevalence in environmental samples while JC polyomavirus (or HPyV) qPCR assays have the best specificity. For this reason we propose the analysis of both viruses, HAdV and JCPyV, to determine human fecal pollution of environmental samples (Table 1). It is important to point out that the proposed markers are selected for its stable excretion all over the year in all geographical areas. However, in some cases the numbers of specific pathogens in high excretion periods, such as rotaviruses or noroviruses, may exceed the numbers of HAdV [99].

Quantification of Human and Animal Viruses to Differentiate the Origin of the Fecal Contamination Present in Environmental Samples

191

TABLE 2: HAdV quantification studies in environmental water matrices.

Authors [Reference]	qPCR detection method [Reference]	Matrices analyzed	Main results
He and Jiang, 2005 [32]	He and Jiang, 2005 [32]	Sewage and coastal waters	Mean values in sewage 8.1E + 05 GC/L. Serotypes 1–5, 9, 16, 17, 19, 21, 28, 37, 40, 41
Choi and Jiang, 2005 [33]	He and Jiang, 2005 [32]	River	2–4 logs GC/L, 16% positive samples
Haramoto et al., 2005 [34]	Heim et al., 2003 [35]	River	45% positive samples (29/64)
Albinana-Gimenez et al., 2006 [36]	Hernroth et al., 2002 [26]	River and sewage	River used as a source of water presented 4E + 02 GC/L
Bofill-Mas et al., 2006 [37]	Hernroth et al., 2002 [26]	Sewage, effluent, and biosolids	High HAdV quantities in sewage, effluent, and biosolids. $t90$ and $t99$ of 60.9 and 132.3 days
Calgua et al., 2008 [38]	Hernroth et al., 2002 [26]	Seawater	New skimmed-milk flocculation method to concentrate, mean values of 1.26E + 03 GC/L
Albinana-Gimenez et al., 2009 [39]	Hernroth et al., 2002 [26]	River and drinking-water treatment plants	90% positive for river water, mean values 1E + 01–1E + 04 GC/L
Dong et al., 2010 [40]	Heim et al., 2003 [35], and by Ko et al., 2005 [41]	Sewage, drinking water, and river and recreational waters	Adenovirus detected from all water types. 10/10 positives in sewage (1.87E + 03–4.6E + 06 GC/L), 5/6 positives in recreational waters (1.70E + 01–1.19E + 03 GC/L)
Hamza et al., 2009 [42]	Heim et al., 2003 [35]	River and sewage	97.5% positive river water samples (1.0E + 07–1.7E + 08 GC/L)
Ogorzaly et al., 2009 [43]	Hernroth et al., 2002 [26]	River	100% positive samples (1.0E + 04/l)
Bofill-Mas et al., 2010 [44]	Hernroth et al., 2002 [26]	Seawater	3.2E + 03 GC/L, HAdV41 the most prevalent
Haramoto et al., 2010 [45]	Ko et al., 2005 [41]	River water	HAdV more prevalent (61.1%) than JCPyV (11.1%)
Jurzik et al., 2010 [46]	Heim et al., 2003 [35]	Surface waters	96.3% positive samples (mean 2.9E + 03 GC/L and maximum of 7.3E + 05 GC/L)
Ogorzaly et al., 2010 [47]	Hernroth et al., 2002 [26]	Groundwater	HAdV was the most stable between MS2 and GA phages analyzed in groundwater
Rigotto et al., 2010 [48]	Hernroth et al., 2002 [26]	Seawater, lagoon brackish water, sewage, and drinking water	64.2% positive values (54/84)
Schlindwein et al., 2010 [49]	Hernroth et al., 2002 [26]	Sewage, effluent, and sludge	4.6E + 07–1.2E + 09 GC/L in sludge, 5E + 04–1.3E + 07 GC/L in sewage, and 3.1E + 05–5.4E + 05 GC/L in effluent
Aslan et al., 2011 [50]	Xagoraraki et al., 2007 [51]	Surface waters	2–4 logs GC/L, 36% positives (HAdV 40/41)
Calgua et al., 2011 [52]	Hernroth et al., 2002 [26]	Seawater	Mean values 1–3 logs GC/L
Guerrero-Latorre et al., 2011 [53]	Hernroth et al., 2002 [26]	River and groundwater	Low levels of HAdV in 4/16 groundwater samples
Hamza et al., 2011 [54]	Heim et al., 2003 [35]	River and sewage	3E + 03 GC/L in river and 1.0E + 07–1.7E + 08 GC/L in sewage
Kokkinos et al., 2011 [55]	Hernroth et al., 2002 [26]	Sewage	45.8% positive samples (22/48) in sewage. Main serotypes 8, 40, and 41
Souza et al., 2011 [56]	Hernroth et al., 2002 [26]	Seawater	HAdV as the most prevalent in seawater
Wong and Xagoraraki, 2011 [57]	Heim et al., 2003 [35]	Manure and sewage sludge	Concentrations of E. coli and Enterococcus correlate to HAdV ($P \geq 0.05$) in sludge samples
Wyn-Jones et al., 2011 [58]	Hernroth et al., 2002 [26]	Recreational water	36.4% positive samples, more prevalent than noroviruses (9.4%)
Garcia et al., 2012 [59]	Hernroth et al., 2002 [26]	River (source water)	100% prevalence (1E + 07 GC/L)
Fongaro et al., 2012 [60]	Hernroth et al., 2002 [26]	Lagoon	96% positive samples (46/48)
Rodriguez-Manzano et al., 2012 [13]	Hernroth et al., 2002 [26]	Raw sewage, secondary and terciary effluents	100% positive samples for HAdV in all steps of the treatment. Removal of HAdV within primary and secondary treatments 1.03 log 10 (89%) and UV disinfection process 0.13 log 10 (11%)
Ye et al., 2012 [61]	Heim et al., 2003 [35]	River and drinking water	100% positive samples (24/24). Mean values in river 2.28E + 04 GC/L

TABLE 3: JCPyV (or HPyV) quantification studies in environmental water matrices.

Authors [Reference]	qPCR detection method [Reference]	Matrices analyzed	Main results
Albinana-Gimenez et al., 2006 [36]	Pal et al., 2006 [27]	Sewage and river	100% positive samples in sewage (5/5) and river (9/9). Mean values $2.6E + 06$ and $2.7E + 01$ GC/L, respectively
Bofill-Mas et al., 2006 [37]	Pal et al., 2006 [27]	Sewage, effluent, and sludge	99% positive samples. T99 of 127.3 days
Albinana-Gimenez et al., 2009 [62]	Pal et al., 2006 [27]	River	48% positive samples in river water
Albinana-Gimenez et al., 2009 [39]	Pal et al., 2006 [27]	River and drinking-water treatment plant (DWTP)	48% positive samples (different steps of the DWTP) with mean values $1E + 01$ to $1E + 03$ GC/L
McQuaig et al., 2009 [63]	McQuaig et al., 2009 [63]	Sewage, fresh to marine water, animal waste	Mean values in sewage $3.0E + 07$ GC/L
Hamza et al., 2009 [42]	Biel et al., 2000 [64]	River	Detected (as JC and BK) in 97.5% of the samples
Harwood et al., 2009 [65]	McQuaig et al., 2009 [63]	River, animal feces, and seawater	No detection of HPyV in animal feces No correlation with *Enterococcus*
Ahmed et al., 2009 [66]	McQuaig et al., 2009 [63]	Sewage	100% host specificity
Abdelzaher et al., 2010 [67]	McQuaig et al., 2009 [63]	Seawater	The FIB levels exceeded regulatory guidelines during one event, and this was accompanied by detection of HPyVs and pathogens
Ahmed et al., 2010 [68]	McQuaig et al., 2009 [63]	Sewage and seawater	JC and BK are highly host-specific viruses and high titers are found in sewage
Bofill-Mas et al., 2010 [44]	Pal et al., 2006 [27]	River and sewage	Sewage ranges from $8.3E + 04$ to $8.5E + 06$ GC/L (7/7) River ranges from $4.4E + 03$ to $1.4E + 04$ GC/L (7/7)
Fumian et al., 2010 [69]	Pal et al., 2006 [27]	Sewage and effluent	JCPyV detected in 96% and 43% of raw and treated sewage, respectively
Haramoto et al., 2010 [45]	Pal et al., 2006 [27]	River	JCPyV prevalence 11.1%, BKPyV not detected
Jurzik et al., 2010 [46]	Biel et al., 2000 [64], and modified by Hamza et al., 2009 [42]	River	68.8% were positive for HPyV
Gibson et al., 2011 [70]	McQuaig et al., 2009 [63]	River and drinking water	HPyV were detected in one groundwater, three-surface water, and one drinking-water sample. No correlation with FIB
Hamza et al., 2011 [54]	Biel et al., 2000 [64]	River and sewage	River $5.0E + 01$–$3.8E + 04$ GC/L, sewage $5.7E + 07$–$5.7E + 08$ GC/L
Hellein et al., 2011 [71]	McQuaig et al., 2009 [63]	Seawater, sewage, and animal feces	Presence of HPyV in all sewage samples and in one freshwater sample
Kokkinos et al., 2011 [55]	McQuaig et al., 2009 [63]	Sewage	68.8% positive values (33/48) for JC and BK
Wong and Xagoraraki, 2011 [72]	McQuaig et al., 2009 [63]	Manure sewage and sludge	HPyV concentrations were slightly lower than *Escherichia coli* and *Enterococcus* ($P < 0.05$)
Chase et al., 2012 [73]	McQuaig et al., 2009 [63]	Recreational waters	HPyV detection near septic systems
Fongaro et al., 2012 [60]	McQuaig et al., 2009 [63]	Lagoon	21% positive samples
Gordon et al., 2013 [74]	McQuaig et al., 2009 [63]	Estuarine to marine waters and sewage spills	HPyV demonstrated the ability to detect domestic sewage contamination in water
Rodriguez-Manzano et al., 2012 [13]	Hernroth et al., 2002 [26]	Raw sewage, secondary and tertiary effluent	JCPyV in raw sewage (6/6) with an average concentration of $5.44E + 05$ GC/L. Not detected in the tertiary effluent
McQuaig et al., 2012 [75]	McQuaig et al., 2009 [63]	Seawater	Mean values $5E + 02$ to $3.55E + 05$ GC/L
Staley et al., 2012 [76]	Staley et al., 2012 [76]	Sewage, river	100% and 64% positive samples of sewage and river samples, respectively

Quantification of Human and Animal Viruses to Differentiate the Origin of the Fecal Contamination Present in Environmental Samples

193

TABLE 4: Quantification of PAdV and BPyV in environmental samples.

Authors [Reference]	qPCR detection method [Reference]	Matrices analyzed	Main results
Hundesa et al., 2009 [29]	PAdV, Hundesa et al., 2009 [29]	River, slaughterhouse, and urban sewage	100% positive samples in slaughterhouse sewage (1.56 + 03 GC/L) and 100% in river (8.38 GC/L)
Hundesa et al., 2010 [28]	BPyV, Hundesa et al., 2010 [28]	River, slaughterhouse, and urban sewage	91% positive samples in slaughterhouse sewage (2.95E + 03 GC/L) and 50% in river (3.06E + 02 GC/L)
Bofill-Mas et al., 2011 [77]	BPyV, Hundesa et al., 2010 [28]	Groundwater	1/4 well water positive for BPyV (7.74×10^2 GC/L)
Wolf et al., 2010 [78]	PAdV, Wolf et al., 2010 [78]	River	50% positive river water samples
Wong and Xagoraraki, 2011 [57]	BPyV, Wong and Xagoraraki 2011 [57]	Sewage	100% positive for manure and wastewater, 5.6% positive for feces samples
Viancelli et al., 2012 [79]	PAdV, Hundesa et al., 2009 [29]	Manure	66% of the samples collected in the SMTS and in 78% of the samples collected in the DU system
Viancelli et al., 2013 [80]	PAdV, Hundesa et al., 2009 [29]	Manure	PAdV were more prevalent than other viruses and can possibly be considered as indicators of manure contamination

6. Quantitative PCR of Animal Viruses: Determining Porcine, Bovine, Poultry, or Ovine Pollution Origin in Environmental Samples

Since porcine adenoviruses (PAdVs) and bovine polyomaviruses (BPyVs) were proposed as porcine and bovine fecal indicators [83, 84], several studies have shown that these viruses are widely disseminated in the swine and bovine population, respectively, although they do not produce clinically severe diseases (Table 4).

In 2009 and 2010, quantitative assays for the quantification of these viruses were described to be applied to environmental samples [28, 29].

The results of these studies showed that BPyV and PAdV were quantified in a high percentage of the samples in which their presence was potentially expected, whereas samples used as negative templates were negative. BPyV and PAdV were found to be distributed in slaughterhouse wastewater and sludge, and in river water from farm-contaminated areas, but not in urban wastewater collected in areas without agricultural activities nor in hospital wastewater [28, 84, 100]. These results support the specificity and applicability of the BPyV and PAdV assays for tracing bovine and porcine fecal contamination in environmental samples, respectively. Quantitative data present in the literature on the presence of these viruses in environmental samples are summarized in Table 4.

Recently, the quantification of chicken/turkey parvoviruses (Ch/TuPVs), highly prevalent in healthy chickens and turkey's from different geographical areas [101–103], has been reported as a candidate MST tool for the identification of poultry originated pollution in environmental samples [31]. A quantitative PCR assay targeting the Ch/TuPV VP1/VP2 region was developed (Table 1) and the viruses detected in 73% of pooled chicken stool samples from the different geographical areas tested (Spain, Greece, and Hungary). Also, chicken slaughterhouse raw wastewater samples and raw urban sewage samples downstream of the slaughterhouse tested positive. The specificity of the designed assays was further studied by testing a wide selection of animal samples (feline, canine, porcine, bovine, ovine, duck, and gull) as well as by testing hospital sewage and urban sewage from areas without poultry industry. These results indicate that Ch/TuPVs may be suitable viral indicators of poultry fecal contamination and that these viruses are being disseminated into the environment.

More recently, the quantification of ovine polyomavirus (OPyV), a newly described virus, has been reported as a candidate tool to identify an ovine fecal/urine origin of fecal pollution [30]. Putative OPyV DNA was amplified from ovine urine and faecal samples using a broad-spectrum nested PCR (nPCR) designed by Johne and coworkers [104]. A specific qPCR assay (Table 1) has been developed and applied to faecal and environmental samples, including sheep slurries, slaughterhouse wastewater effluents, urban sewage, and river water samples. Successful quantification of OPyV was achieved in sheep urine samples, sheep slaughterhouse wastewater, and downstream sewage effluents. The assay was specific and was negative in samples of human, bovine, goat, swine, and chicken origin. Ovine faecal pollution was detected in river water samples by applying the designed methods. These results provide a quantitative tool for the analysis of OPyV as a suitable viral indicator of sheep faecal contamination that may be present in the environment.

7. Conclusions

Specific qPCR assays for the quantification of DNA viruses have been proposed as specific and sensitive assays to quantify human, porcine, bovine polyomavirus, poultry, and ovine fecal contamination in environmental samples.

Quantitative data is being accumulated on the presence and concentration of the proposed viral markers in environmental samples in many different countries. Future efforts should be directed towards developing standard procedures and reference materials for a reproducible application of these tools.

Meanwhile, these assays can be used to evaluate the microbiological quality of water and the efficiency of pathogen removal in drinking and wastewater treatment plants and in MST studies.

References

[1] M. J. Brownell, V. J. Harwood, R. C. Kurz, S. M. McQuaig, J. Lukasik, and T. M. Scott, "Confirmation of putative stormwater impact on water quality at a Florida beach by microbial source tracking methods and structure of indicator organism populations," *Water Research*, vol. 41, no. 16, pp. 3747–3757, 2007.

[2] J. R. Stewart, R. J. Gast, R. S. Fujioka et al., "The coastal environment and human health: microbial indicators, pathogens, sentinels and reservoirs," *Environmental Health*, vol. 7, no. 2, article S3, 2008.

[3] B. Bercu, L. C. Van De Werfhorst, J. L. Murray, and P. A. Holden, "Sewage exfiltration as a source of storm drain contamination during dry weather in urban watersheds," *Environmental Science and Technology*, vol. 45, no. 17, pp. 7151–7157, 2011.

[4] P. G. Cantalupo, B. Calgua, G. Zhao et al., "Raw sewage harbors diverse viral populations," *mBio*, vol. 2, no. 5, pp. 180–192, 2011.

[5] T. F. Ng, R. Marine, C. Wang et al., "High variety of known and new RNA and DNA viruses of diverse origins in untreated sewage," *Journal of Virology*, vol. 86, no. 22, pp. 12161–12175, 2012.

[6] R. G. Sinclair, E. L. Jones, and C. P. Gerba, "Viruses in recreational water-borne disease outbreaks: a review," *Journal of Applied Microbiology*, vol. 107, no. 6, pp. 1769–1780, 2009.

[7] K. D. Mena and C. P. Gerba, "Waterborne adenovirus," *Reviews of Environmental Contamination and Toxicology*, vol. 198, pp. 133–167, 2009.

[8] World Health Organization, *Guidelines for Drinking-Water Quality*, WHO, Geneva, Switzerland, 3rd edition, 2004.

[9] USEPA, "Microbial source tracking guide document," Tech. Rep. EPA/600/R-05/064, USA, Environmental Protection Agency, 2005.

[10] K. G. Field, J. A. Cotruvo, A. Dufour et al., *Faecal Source Identification in Waterborne Zoonoses: Identification, Causes and Control*, IWA Publishing, London, UK, 2004.

[11] T. T. Fong and E. K. Lipp, "Enteric viruses of humans and animals in aquatic environments: health risks, detection, and potential water quality assessment tools," *Microbiology and Molecular Biology Reviews*, vol. 69, no. 2, pp. 357–371, 2005.

[12] R. Girones, M. A. Ferrús, J. L. Alonso et al., "Molecular detection of pathogens in water—the pros and cons of molecular techniques," *Water Research*, vol. 44, no. 15, pp. 4325–4339, 2010.

[13] J. Rodriguez-Manzano, J. L. Alonso, M. A. Ferrús et al., "Standard and new faecal indicators and pathogens in sewage treatment plants, microbiological parameters for improving the control of reclaimed water," *Water Science and Technology*, vol. 66, no. 12, pp. 2517–2523, 2012.

[14] D. S. Francy, E. A. Stelzer, R. N. Bushon et al., "Comparative effectiveness of membrane bioreactors, conventional secondary treatment, and chlorine and UV disinfection to remove microorganisms from municipal wastewaters," *Water Research*, vol. 46, no. 13, pp. 4164–4178, 2012.

[15] D. M. Stoeckel and V. J. Harwood, "Performance, design, and analysis in microbial source tracking studies," *Applied and Environmental Microbiology*, vol. 73, no. 8, pp. 2405–2415, 2007.

[16] T. M. Scott, J. B. Rose, T. M. Jenkins, S. R. Farrah, and J. Lukasik, "Microbial source tracking: current methodology and future directions," *Applied and Environmental Microbiology*, vol. 68, no. 12, pp. 5796–5803, 2002.

[17] P. Roslev and A. S. Bukh, "State of the art molecular markers for fecal pollution source tracking in water," *Applied Microbiology and Biotechnology*, vol. 89, no. 5, pp. 1341–1355, 2011.

[18] K. Wong, T. T. Fong, K. Bibby, and M. Molina, "Application of enteric viruses for fecal pollution source tracking in environmental waters," *Environmental International*, vol. 15, no. 45, pp. 151–164, 2012.

[19] G. J. Medema, W. Hoogenboezem, A. J. Van Der Veer, H. A. M. Ketelaars, W. A. M. Hijnen, and P. J. Nobel, "Quantitative risk assessment of Cryptosporidium in surface water treatment," *Water Science and Technology*, vol. 47, no. 3, pp. 241–247, 2003.

[20] C. P. Gerba, S. M. Goyal, R. L. LaBelle, I. Cech, and G. F. Bodgan, "Failure of indicator bacteria to reflect the occurrence of enteroviruses in marine waters," *American Journal of Public Health*, vol. 69, no. 11, pp. 1116–1119, 1979.

[21] E. K. Lipp, S. A. Farrah, and J. B. Rose, "Assessment and impact of microbial fecal pollution and human enteric pathogens in a coastal community," *Marine Pollution Bulletin*, vol. 42, no. 4, pp. 286–293, 2001.

[22] S. Pina, M. Puig, F. Lucena, J. Jofre, and R. Girones, "Viral pollution in the environment and in shellfish: human adenovirus detection by PCR as an index of human viruses," *Applied and Environmental Microbiology*, vol. 64, no. 9, pp. 3376–3382, 1998.

[23] N. Contreras-Coll, F. Lucena, K. Mooijman et al., "Occurrence and levels of indicator bacteriophages in bathing waters throughout Europe," *Water Research*, vol. 36, no. 20, pp. 4963–4974, 2002.

[24] O. C. Shanks, M. Sivaganesan, L. Peed et al., "Interlaboratory comparison of real-time PCR protocols for quantification of general fecal indicator bacteria," *Environmental Science and Technology*, vol. 46, no. 2, pp. 945–953, 2012.

[25] M. Botes, M. de Kwaadsteniet, and T. E. Cloete, "Application of quantitative PCR for the detection of microorganisms in water," *Analytical and Bioanalytical Chemistry*, vol. 405, no. 1, pp. 91–108, 2013.

[26] B. E. Hernroth, A. C. Conden-Hansson, A. S. Rehnstam-Holm, R. Girones, and A. K. Allard, "Environmental factors influencing human viral pathogens and their potential indicator organisms in the blue mussel, *Mytilus edulis*: the first Scandinavian report," *Applied and Environmental Microbiology*, vol. 68, no. 9, pp. 4523–4533, 2002.

[27] A. Pal, L. Sirota, T. Maudru, K. Peden, and A. M. Lewis, "Real-time, quantitative PCR assays for the detection of virus-specific DNA in samples with mixed populations of polyomaviruses," *Journal of Virological Methods*, vol. 135, no. 1, pp. 32–42, 2006.

[28] A. Hundesa, S. Bofill-Mas, C. Maluquer de Motes et al., "Development of a quantitative PCR assay for the quantitation of bovine polyomavirus as a microbial source-tracking tool," *Journal of Virological Methods*, vol. 163, no. 2, pp. 385–389, 2010.

[29] A. Hundesa, C. Maluquer de Motes, N. Albinana-Gimenez et al., "Development of a qPCR assay for the quantification of porcine adenoviruses as an MST tool for swine fecal contamination in the environment," *Journal of Virological Methods*, vol. 158, no. 1-2, pp. 130–135, 2009.

[30] M. Rusiñol, A. Carratalà, A. Hundesa et al., "Description of a novel viral tool to identify and quantify ovine faecal pollution in the environment," *Science of the Total Environment*, vol. 458-460, pp. 355–360, 2013.

Quantification of Human and Animal Viruses to Differentiate the Origin of the Fecal Contamination Present in
Environmental Samples

195

[31] A. Carratalà, M. Rusinol, A. Hundesa et al., "A novel tool for specific detection and quantification of chicken/turkey parvoviruses to trace poultry fecal contamination in the environment," *Applied Environmental Microbiology*, vol. 78, no. 20, pp. 7496–7499, 2012.

[32] J. W. He and S. Jiang, "Quantification of enterococci and human adenoviruses in environmental samples by real-time PCR," *Applied Environmental Microbiology*, vol. 71, no. 5, pp. 2250–2255, 2005.

[33] S. Choi and S. C. Jiang, "Real-time PCR quantification of human adenoviruses in urban rivers indicates genome prevalence but low infectivity," *Applied and Environmental Microbiology*, vol. 71, no. 11, pp. 7426–7433, 2005.

[34] E. Haramoto, H. Katayama, K. Oguma, and S. Ohgaki, "Application of cation-coated filter method to detection of noroviruses, enteroviruses, adenoviruses, and torque teno viruses in the Tamagawa River in Japan," *Applied and Environmental Microbiology*, vol. 71, no. 5, pp. 2403–2411, 2005.

[35] A. Heim, C. Ebnet, G. Harste, and P. Pring-Åkerblom, "Rapid and quantitative detection of human adenovirus DNA by real-time PCR," *Journal of Medical Virology*, vol. 70, no. 2, pp. 228–239, 2003.

[36] N. Albinana-Gimenez, P. Clemente-Casares, S. Bofill-Mas, A. Hundesa, F. Ribas, and R. Girones, "Distribution of human polyomaviruses, adenoviruses, and hepatitis E virus in the environment and in a drinking-water treatment plant," *Environmental Science and Technology*, vol. 40, no. 23, pp. 7416–7422, 2006.

[37] S. Bofill-Mas, N. Albinana-Gimenez, P. Clemente-Casares et al., "Quantification and stability of human adenoviruses and polyomavirus JCPyV in wastewater matrices," *Applied and Environmental Microbiology*, vol. 72, no. 12, pp. 7894–7896, 2006.

[38] B. Calgua, A. Mengewein, A. Grunert et al., "Development and application of a one-step low cost procedure to concentrate viruses from seawater samples," *Journal of Virological Methods*, vol. 153, no. 2, pp. 79–83, 2008.

[39] N. Albinana-Gimenez, M. P. Miagostovich, B. Calgua, J. M. Huguet, L. Matia, and R. Girones, "Analysis of adenoviruses and polyomaviruses quantified by qPCR as indicators of water quality in source and drinking-water treatment plants," *Water Research*, vol. 43, no. 7, pp. 2011–2019, 2009.

[40] Y. Dong, J. Kim, and G. D. Lewis, "Evaluation of methodology for detection of human adenoviruses in wastewater, drinking water, stream water and recreational waters," *Journal of Applied Microbiology*, vol. 108, no. 3, pp. 800–809, 2010.

[41] G. Ko, N. Jothikumar, V. R. Hill, and M. D. Sobsey, "Rapid detection of infectious adenoviruses by mRNA real-time RT-PCR," *Journal of Virological Methods*, vol. 127, no. 2, pp. 148–153, 2005.

[42] I. A. Hamza, L. Jurzik, A. Stang, K. Sure, K. Überla, and M. Wilhelm, "Detection of human viruses in rivers of a densly-populated area in Germany using a virus adsorption elution method optimized for PCR analyses," *Water Research*, vol. 43, no. 10, pp. 2657–2668, 2009.

[43] L. Ogorzaly, A. Tissier, I. Bertrand, A. Maul, and C. Gantzer, "Relationship between F-specific RNA phage genogroups, faecal pollution indicators and human adenoviruses in river water," *Water Research*, vol. 43, no. 5, pp. 1257–1264, 2009.

[44] S. Bofill-Mas, B. Calgua, P. Clemente-Casares et al., "Quantification of human adenoviruses in European recreational waters," *Food and Environmental Virology*, vol. 2, no. 2, pp. 101–109, 2010.

[45] E. Haramoto, M. Kitajima, H. Katayama, and S. Ohgaki, "Real-time PCR detection of adenoviruses, polyomaviruses, and torque teno viruses in river water in Japan," *Water Research*, vol. 44, no. 6, pp. 1747–1752, 2010.

[46] L. Jurzik, I. A. Hamza, W. Puchert, K. Überla, and M. Wilhelm, "Chemical and microbiological parameters as possible indicators for human enteric viruses in surface water," *International Journal of Hygiene and Environmental Health*, vol. 213, no. 3, pp. 210–216, 2010.

[47] L. Ogorzaly, I. Bertrand, M. Paris, A. Maul, and C. Gantzer, "Occurrence, survival, and persistence of human adenoviruses and F-specific RNA phages in raw groundwater," *Applied and Environmental Microbiology*, vol. 76, no. 24, pp. 8019–8025, 2010.

[48] C. Rigotto, M. Victoria, V. Moresco et al., "Assessment of adenovirus, hepatitis A virus and rotavirus presence in environmental samples in Florianopolis, South Brazil," *Journal of Applied Microbiology*, vol. 109, no. 6, pp. 1979–1987, 2010.

[49] A. D. Schlindwein, C. Rigotto, C. M. O. Simões, and C. R. M. Barardi, "Detection of enteric viruses in sewage sludge and treated wastewater effluent," *Water Science and Technology*, vol. 61, no. 2, pp. 537–544, 2010.

[50] A. Aslan, I. Xagoraraki, F. J. Simmons, J. B. Rose, and S. Dorevitch, "Occurrence of adenovirus and other enteric viruses in limited-contact freshwater recreational areas and bathing waters," *Journal of Applied Microbiology*, vol. 111, no. 5, pp. 1250–1261, 2011.

[51] I. Xagoraraki, D. H. W. Kuo, K. Wong, M. Wong, and J. B. Rose, "Occurrence of human adenoviruses at two recreational beaches of the great lakes," *Applied and Environmental Microbiology*, vol. 73, no. 24, pp. 7874–7881, 2007.

[52] B. Calgua, C. R. M. Barardi, S. Bofill-Mas, J. Rodriguez-Manzano, and R. Girones, "Detection and quantitation of infectious human adenoviruses and JC polyomaviruses in water by immunofluorescence assay," *Journal of Virological Methods*, vol. 171, no. 1, pp. 1–7, 2011.

[53] L. Guerrero-Latorre, A. Carratala, J. Rodriguez-Manzano, B. Calgua, A. Hundesa, and R. Girones, "Occurrence of waterborne enteric viruses in two settlements based in Eastern Chad: analysis of hepatitis E virus, hepatitis A virus and human adenovirus in water sources," *Journal of Water Health*, vol. 9, no. 3, pp. 515–524, 2011.

[54] I. A. Hamza, L. Jurzik, K. Überla, and M. Wilhelm, "Methods to detect infectious human enteric viruses in environmental water samples," *International Journal of Hygiene and Environmental Health*, vol. 214, no. 6, pp. 424–436, 2011.

[55] P. A. Kokkinos, P. G. Ziros, G. Mpalasopoulou, A. Galanis, and A. Vantarakis, "Molecular detection of multiple viral targets in untreated urban sewage from Greece," *Virology Journal*, vol. 8, article 195, 2011.

[56] D. S. Souza, A. P. Ramos, F. F. Nunes et al., "Evaluation of tropical water sources and mollusks in southern Brazil using microbiological, biochemical, and chemical parameters," *Ecotoxicology and Environmental Safety*, vol. 76, no. 2, pp. 153–161, 2011.

[57] K. Wong and I. Xagoraraki, "Evaluating the prevalence and genetic diversity of adenovirus and polyomavirus in bovine waste for microbial source tracking," *Applied Microbiology and Biotechnology*, vol. 90, no. 4, pp. 1521–1526, 2011.

[58] A. P. Wyn-Jones, A. Carducci, N. Cook et al., "Surveillance of adenoviruses and noroviruses in European recreational waters," *Water Research*, vol. 45, no. 3, pp. 1025–1038, 2011.

[59] L. A. Garcia, A. Viancelli, C. Rigotto et al., "Surveillance of human and swine adenovirus, human norovirus and swine circovirus in water samples in Santa Catarina, Brazil," *Journal Water of Health*, vol. 10, no. 3, pp. 445–452, 2012.

[60] G. Fongaro, M. A. Nascimento, A. Viancelli, D. Tonetta, M. M. Petrucio, and C. R. Barardi, "Surveillance of human viral contamination and physicochemical profiles in a surface water lagoon," *Water and Science Technology*, vol. 66, no. 12, pp. 2682–2687, 2012.

[61] X. Y. Ye, X. Ming, Y. L. Zhang et al., "Real-time PCR detection of enteric viruses in source water and treated drinking water in Wuhan, China," *Current Microbiology*, vol. 65, no. 3, pp. 244–253, 2012.

[62] N. Albinana-Gimenez, P. Clemente-Casares, B. Calgua, J. M. Huguet, S. Courtois, and R. Girones, "Comparison of methods for concentrating human adenoviruses, polyomavirus JC and noroviruses in source waters and drinking water using quantitative PCR," *Journal of Virological Methods*, vol. 158, no. 1-2, pp. 104–109, 2009.

[63] S. M. McQuaig, T. M. Scott, J. O. Lukasik, J. H. Paul, and V. J. Harwood, "Quantification of human polyomaviruses JC virus and BK Virus by TaqMan quantitative PCR and comparison to other water quality indicators in water and fecal samples," *Applied and Environmental Microbiology*, vol. 75, no. 11, pp. 3379–3388, 2009.

[64] S. S. Biel, T. K. Held, O. Landt et al., "Rapid quantification and differentiation of human polyomavirus DNA in undiluted urine from patients after bone marrow transplantation," *Journal of Clinical Microbiology*, vol. 38, no. 10, pp. 3689–3695, 2000.

[65] V. J. Harwood, M. Brownell, S. Wang et al., "Validation and field testing of library-independent microbial source tracking methods in the Gulf of Mexico," *Water Research*, vol. 43, no. 19, pp. 4812–4819, 2009.

[66] W. Ahmed, A. Goonetilleke, D. Powell, K. Chauhan, and T. Gardner, "Comparison of molecular markers to detect fresh sewage in environmental waters," *Water Research*, vol. 43, no. 19, pp. 4908–4917, 2009.

[67] A. M. Abdelzaher, M. E. Wright, C. Ortega et al., "Presence of pathogens and indicator microbes at a non-point source subtropical recreational marine beach," *Applied and Environmental Microbiology*, vol. 76, no. 3, pp. 724–732, 2010.

[68] W. Ahmed, A. Goonetilleke, and T. Gardner, "Human and bovine adenoviruses for the detection of source-specific fecal pollution in coastal waters in Australia," *Water Research*, vol. 44, no. 16, pp. 4662–4673, 2010.

[69] T. M. Fumian, F. R. Guimarães, B. J. P. Vaz et al., "Molecular detection, quantification and characterization of human polyomavirus JC from waste water in Rio de Janeiro, Brazil," *Journal of Water and Health*, vol. 8, no. 3, pp. 438–445, 2010.

[70] K. E. Gibson, M. C. Opryszko, J. T. Schissler, Y. Guo, and K. J. Schwab, "Evaluation of human enteric viruses in surface water and drinking water resources in southern Ghana," *The American Journal of Tropical Medicine and Hygiene*, vol. 84, no. 1, pp. 20–29, 2011.

[71] K. N. Hellein, C. Battie, E. Tauchman, D. Lund, O. A. Oyarzabal, and J. E. Lepo, "Culturebased indicators of fecal contamination and molecular microbial indicators rarely correlate with Campylobacter spp. in recreational waters," *Journal of Water Health*, vol. 9, no. 4, pp. 695–707, 2011.

[72] K. Wong and I. Xagoraraki, "A perspective on the prevalence of DNA enteric virus genomes in anaerobic-digested biological wastes," *Environmental Monitoring and Assessment*, vol. 184, no. 8, pp. 5009–5016, 2011.

[73] E. Chase, J. Hunting, C. Staley, and V. J. Harwood, "Microbial source tracking to identify human and ruminant sources of faecal pollution in an ephemeral Florida river," *Journal of Applied Microbiology*, vol. 113, no. 6, pp. 1396–1406, 2012.

[74] K. V. Gordon, M. Brownell, S. Y. Wang et al., "Relationship of human-associated microbial source tracking markers with Enterococci in Gulf of Mexico waters," *Water Research*, vol. 47, no. 3, pp. 996–1004, 2013.

[75] S. M. McQuaig, J. Griffith, and V. J. Harwood, "The association of fecal indicator bacteria with human viruses and microbial source tracking markers at coastal beaches impacted by non-point source pollution," *Applied Environmental Microbiology*, vol. 78, no. 18, pp. 6423–6432, 2012.

[76] C. Staley, K. V. Gordon, M. E. Schoen, and V. J. Harwood, "Performance of two quantitative PCR methods for microbial source tracking of human sewage and implications for microbial risk assessment in recreational waters," *Applied Environmental Microbiology*, vol. 78, no. 20, pp. 7317–7326, 2012.

[77] S. Bofill-Mas, A. Hundesa, B. Calgua, M. Rusiñol, C. M. de Motes, and R. Girones, "Costeffective method for microbial source tracking using specific human and animal viruses," *Journal of Visual Experiments*, no. 58, 2011.

[78] S. Wolf, J. Hewitt, and G. E. Greening, "Viral multiplex quantitative PCR assays for tracking sources of fecal contamination," *Applied and Environmental Microbiology*, vol. 76, no. 5, pp. 1388–1394, 2010.

[79] A. Viancelli, L. A. T. Garcia, A. Kunz, R. Steinmetz, P. A. Esteves, and C. R. M. Barardi, "Detection of circoviruses and porcine adenoviruses in water samples collected from swine manure treatment systems," *Research in Veterinary Science*, vol. 93, no. 1, pp. 538–543, 2012.

[80] A. Viancelli, A. Kunz, R. L. R. Steinmetz et al., "Performance of two swine manure treatment systems on chemical composition and on the reduction of pathogens," *Chemosphere*, vol. 90, no. 4, pp. 1539–1544, 2013.

[81] M. Puig, J. Jofre, F. Lucena, A. Allard, G. Wadell, and R. Girones, "Detection of adenoviruses and enteroviruses in polluted waters by nested PCR amplification," *Applied and Environmental Microbiology*, vol. 60, no. 8, pp. 2963–2970, 1994.

[82] S. Bofill-Mas and S. Pina R, "Documenting the epidemiologic patterns of polyomaviruses in human populations by studying their presence in urban sewage," *Applied Environmental Microbiology*, vol. 66, no. 1, pp. 238–245, 2000.

[83] C. M. De Motes, P. Clemente-Casares, A. Hundesa, M. Martín, and R. Girones, "Detection of bovine and porcine adenoviruses for tracing the source of fecal contamination," *Applied and Environmental Microbiology*, vol. 70, no. 3, pp. 1448–1454, 2004.

[84] A. Hundesa, C. Maluquer De Motes, S. Bofill-Mas, N. Albinana-Gimenez, and R. Girones, "Identification of human and animal adenoviruses and polyomaviruses for determination of sources of fecal contamination in the environment," *Applied and Environmental Microbiology*, vol. 72, no. 12, pp. 7886–7893, 2006.

[85] P. Vilaginès, B. Sarrette, G. Husson, and R. Vilaginès, "Glass wool for virus concentration at ambient water pH level," *Water Sience and Technology*, vol. 27, no. 3-4, pp. 299–306, 1993.

[86] B. Calgua, J. Rodriguez-Manzano, A. Hundesa et al., "New Methods from the concentration of viruses from urban sewage using quantitative PCR," *Journal of Virological Methods*, vol. 187, no. 2, pp. 215–221, 2013.

Quantification of Human and Animal Viruses to Differentiate the Origin of the Fecal Contamination Present in Environmental Samples

197

[87] B. Calgua, T. Fumian, M. Rusiñol et al., "Detection and quantification of classic and emerging viruses by skimmed-milk flocculation and PCR in river water from two geographical areas," *Water Research*, vol. 47, no. 8, pp. 2797–2810, 2013.

[88] S. Bofill-Mas, J. Rodriguez-Manzano, B. Calgua, A. Carratala, and R. Girones, "Newly described human polyomaviruses Merkel Cell, KI and WU are present in urban sewage and may represent potential environmental contaminants," *Virology Journal*, vol. 7, article 141, 2010.

[89] B. Calgua, J. Rodriguez-Manzano, A. Hundesa et al., "New methods for the concentration of viruses from urban sewage using quantitative PCR," *Journal of Virological Methods*, vol. 187, no. 2, pp. 215–221, 2013.

[90] S. Nuanualsuwan and D. O. Cliver, "Pretreatment to avoid positive RT-PCR results with inactivated viruses," *Journal of Virological Methods*, vol. 104, no. 2, pp. 217–225, 2002.

[91] B. M. Pecson, L. V. Martin, and T. Kohn, "Quantitative PCR for determining the infectivity of bacteriophage MS2 upon inactivation by heat, UV-B radiation, and singlet oxygen: advantages and limitations of an enzymatic treatment to reduce false-positive results," *Applied and Environmental Microbiology*, vol. 75, no. 17, pp. 5544–5554, 2009.

[92] A. De Abreu-Correa, A. Carratala, C. R. Monte-Baradi, M. Calvo, R. Girones, and S. Bofill-Mas, "Comparative inactivation of Murine Norovirus, Human Adenovirus, and JC Polyomavirus by chlorine in seawater," *Applied and Environmental Microbiology*, vol. 78, no. 18, pp. 6450–6457, 2012.

[93] S. C. Jiang, "Human adenoviruses in water: occurrence and health implications: a critical review," *Environmental Science and Technology*, vol. 40, no. 23, pp. 7132–7140, 2006.

[94] T. Weber, P. E. Klapper, G. M. Cleator et al., "Polymerase chain reaction for detection of JC virus DNA in cerebrospinal fluid: A Quality Control Study," *Journal of Virological Methods*, vol. 69, no. 1-2, pp. 231–237, 1997.

[95] T. Kitamura, Y. Aso, N. Kuniyoshi, K. Hara, and Y. Yogo, "High incidence of urinary JC virus excretion in nonimmunosuppressed older patients," *Journal of Infectious Diseases*, vol. 161, no. 6, pp. 1128–1133, 1990.

[96] I. J. Koralnik, D. Boden, V. X. Mai, C. I. Lord, and N. L. Letvin, "JC virus DNA load in patients with and without progressive multifocal leukoencephalopathy," *Neurology*, vol. 52, no. 2, pp. 253–260, 1999.

[97] J. R. Berger, S. A. Houff, and E. O. Major, "Monoclonal antibodies and progressive multifocal leukoencephalopathy," *mAbs*, vol. 1, no. 6, pp. 583–589, 2009.

[98] T. A. Yousry, E. O. Major, C. Ryschkewitsch et al., "Evaluation of patients treated with natalizumab for progressive multifocal leukoencephalopathy," *The New England Journal of Medicine*, vol. 354, no. 9, pp. 924–933, 2006.

[99] M. P. Miagostovich, F. F. M. Ferreira, F. R. Guimarães et al., "Molecular detection and characterization of gastroenteritis viruses occurring naturally in the stream waters of Manaus, Central Amazônia, Brazil," *Applied and Environmental Microbiology*, vol. 74, no. 2, pp. 375–382, 2008.

[100] S. Bofill-Mas, B. Calgua, J. Rodriguez-Manzano et al., "Cost-effective Applications of human and animal viruses and microbial source-tracking tools in surface waters and groundwater in Faecal Indicators and pathogens," in *Proceedings of the Fédération Internationale des Patrouilles de Ski Conference (FIPs '11)*, Royal Society of Chemistry, London, UK, 2011.

[101] M. Bidin, I. Lojkić, Z. Bidin, M. Tiljar, and D. Majnarić, "Identification and phylogenetic diversity of parvovirus circulating in commercial chicken and turkey flocks in Croatia," *Avian Diseases*, vol. 55, no. 4, pp. 693–696, 2011.

[102] E. A. Palade, J. Kisary, Z. Benyeda et al., "Naturally occurring parvoviral infection in Hungarian broiler flocks," *Avian Pathology*, vol. 40, no. 2, pp. 191–197, 2011.

[103] L. Zsak, K. O. Strother, and J. M. Day, "development of a polymerase chain reaction procedure for detection of chicken and Turkey parvoviruses," *Avian Diseases*, vol. 53, no. 1, pp. 83–88, 2009.

[104] R. Johne, D. Enderlein, H. Nieper, and H. Müller, "Novel polyomavirus detected in the feces of a chimpanzee by nested broad-spectrum PCR," *Journal of Virology*, vol. 79, no. 6, pp. 3883–3887, 2005.

β-Glucosidases from the Fungus *Trichoderma*: An Efficient Cellulase Machinery in Biotechnological Applications

Pragya Tiwari,[1] B. N. Misra,[2] and Neelam S. Sangwan[1]

[1] *Metabolic and Structural Biology Department, CSIR-Central Institute of Medicinal and Aromatic Plants (CSIR-CIMAP), P.O. CIMAP, Lucknow 226015, Uttar Pradesh, India*
[2] *Department of Biotechnology, UP Technical University, Lucknow 226021, Uttar Pradesh, India*

Correspondence should be addressed to Neelam S. Sangwan; sangwan.neelam@gmail.com

Academic Editor: Arzu Coleri Cihan

β-glucosidases catalyze the selective cleavage of glucosidic linkages and are an important class of enzymes having significant prospects in industrial biotechnology. These are classified in family 1 and family 3 of glycosyl hydrolase family. β-glucosidases, particularly from the fungus *Trichoderma*, are widely recognized and used for the saccharification of cellulosic biomass for biofuel production. With the rising trends in energy crisis and depletion of fossil fuels, alternative strategies for renewable energy sources need to be developed. However, the major limitation accounts for low production of β-glucosidases by the hyper secretory strains of *Trichoderma*. In accordance with the increasing significance of β-glucosidases in commercial applications, the present review provides a detailed insight of the enzyme family, their classification, structural parameters, properties, and studies at the genomics and proteomics levels. Furthermore, the paper discusses the enhancement strategies employed for their utilization in biofuel generation. Therefore, β-glucosidases are prospective toolbox in bioethanol production, and in the near future, it might be successful in meeting the requirements of alternative renewable sources of energy.

1. Introduction

β-glucosidases are members of cellulase enzyme complex and are promising candidates in biotechnological applications. Fungal species belonging to genus *Trichoderma* are ubiquitous in nature and classified as imperfect fungi due to absence of sexual reproduction [1]. *Trichoderma* is a saprophyte and produce diverse enzymes, a particular strain being specific for a certain type of enzyme. For example, *T. reesei* is used for cellulase and hemicellulase production, *T. longibratum* is used for xylanase, and *T. harzianum* is used for chitinase [2]. The cellulase system in *T. reesei* constitutes the combined activity of three enzymes: cellobiohydrolase, endo-β-glucanase and β-glucosidases, respectively. Cellobiohydrolases (EC 3.2.1.91) degrade cellobiose residues from the nonreducing end of the glucan, endo-β-glucanase (EC 3.2.1.4) catalyzes the breakdown of internal β-1,4-linkages, while β-glucosidases (EC 3.2.1.21) hydrolyze cellobiose to two molecules of glucose [3]. The conversion of cellulose to glucose is regarded as the rate limiting step in the production of biofuels from lignocellulosic materials, due to high cost of cellulases and their low efficiencies.

β-glucosidases, also named as (β-D-glucoside glucohydrolase, EC 3.2.1.21), catalyze the hydrolysis of the β-glucosidic linkages such as alkyl and aryl β-glucosides, β-linked oligosaccharides as well as several oligosaccharides with release of glucose [4, 5]. β-glucosidases are prominent class of enzymes and catalyze cellulose degradation acting synergistically with cellobiohydrolase and endoglucanase, respectively [6]. The specificity of β-glucosidases is variable towards different substrates depending on the enzyme source. The enzyme is ubiquitously present in nature and found in bacteria [7], fungi [8], yeasts [9], plants [10–13], and animals [14], respectively.

Some *Trichoderma* species amongst cellulolytic fungi have strong cellulose-degrading properties and therefore their cellulase systems have been widely studied. In *T. reesei*, the maximum production of cellulase component is of cellobiohydrolases I (CBHI) which is 60% of the total secreted protein [15], while cellobiohydrolases II (CBHII) and endoglucanases accounts for 20 and 10% of the total secreted protein and this is a major limitation in cellulose saccharification by cellulases [16].

The mechanism of catalysis includes the degradation of cellobiose to glucose resulting in cellulose saccharification and release of the two enzymes from cellobiose inhibition [17, 18]. The enzymes are widely distributed in microbes, plants, and animals and play important roles in biological processes [19]. β-glucosidases, particularly from microorganisms, play a significant role in cellulose saccharification. However, microbes which produce the enzyme in low quantities lead to inefficient degradation of cellulose. While in microorganisms, β-glucosidases are involved in degradation of cellulose as compared to synthesis of beta-glucan during cell wall development, fruit ripening, defense mechanisms, and pigment metabolism [20, 21]. However, β-glucosidase-1 (BGL1) from *T. reesei* hyperproducing strain is produced in very small quantities. Over expression strategies in *T. reesei* or additional incorporation of β-glucosidase from other sources could be a possible option for enhancing and optimizing β-glucosidases mediated cellulose degradation. The products, cellobiose generated by endo- and exoglucanase act as inhibitors of both enzymes and removed by the action of β-glucosidases [22].

Several studies on β-glucosidases, time and again have highlighted their importance in biotechnological applications. Woodward and Wiseman reviewed the research on the fungal enzymes till 1982 [23]. Further, the enzymes from yeast were studied by Leclerc and coworkers [24] and thermostable β-glucosidases from mesophilic and thermophilic fungi [25]. Recently, molecular cloning studies on β-glucosidases were performed by Bhatia and colleagues [26]. Several other studies report on the isolation, cloning, and purification of β-glucosidases [27, 28].

With the present trends in rising the importance of β-glucosidases in industrial applications, this review is an update on fungal β-glucosidases particularly from *Trichoderma* species, an overview of their increasing significance, classification of the enzymes, their structure and properties, and also their prospective role in biotechnological applications. Furthermore, β-glucosidases may serve as a promising tool in meeting the energy crisis by generating an alternative renewable source of biofuels production in future.

2. Phylogenetics and Characteristics of *Trichoderma* Fungus

The genus *Trichoderma* is the best studied among fungi due to its biotechnological prospects and applications. The first report pertaining to the fungus *Trichoderma* dates back to 1794 [29]. Bioinformatics approaches such as oligonucleotide barcode (TrichOKEY) and a similarity search tool (TrichoBLAST) are mostly used in *Trichoderma* studies and can be accessed online at www.isth.info [30, 31]. Phenotype microarrays are the more reliable technique for the identification and characterization of newly isolated *Trichoderma* spp.

The cellulases produced from the *Trichoderma* species are important industrial products for biofuel production from cellulosic waste. *Trichoderma* species is widely present on

cellulosic materials and results in their degradation [32]. At present, 165 records for *Trichoderma* are available in the Index Fungorum database (http://www.indexfungorum.org/Names/Names.asp). The international subcommission on *Trichoderma* includes 104 species characterized at the molecular level (http://www.isth.info/biodiversity/index.php). *Trichoderma* is among the most extensively used fungus species in industrial applications. The whole genome sequencing of the three strains, *T. reesei*, the industrial strain [33] (http://genome.jgi-psf.org/Trire2/Trire2.home.html), *T. atroviride* and *T. virens*, two other important biocontrol species (http://genome.jgi-psf.org/Trive1/Trive1.home.html) is under progress. The results showed that although *T. reesei* is considered as an important industrial strain for cellulose degradation, its genome consists of fewer genes encoding hemicellulolytic and cellulolytic enzymes [34].

Several species of *Trichoderma*, namely, *T. reesei*, *T. atroviride*, *T. virens*, *T. asperellum*, *T. harzianum*, *T. citrinoviride*, and *T. koningii* are considered important and used in various industrial applications. Studies on β-glucosidases from *Trichoderma* species ranging from protein purification and characterization and overexpression in different fungal strains to site-directed mutagenesis and molecular biology studies have been summarized in Table 1.

3. Structure of β-Glucosidases

With the increasing significance of β-glucosidases and their application in industrial biotechnology, efforts have been made to isolate a wide range of β-glucosidases from different sources and also, on the improvement of enzyme activity and thermostability. The structure of *T. reesei* β-glucosidase 2 (TrBgl2) has been elucidated by Lee and coworkers in 2012 [35] with a PDB code-3AHY. The structure of TrBgl2 consists of Glu165 as the catalytic acid/base and Glu367 as the catalytic nucleophile [36] and utilizes a β-retaining mechanism for its activity. The enzyme adopts a $(\alpha/\beta)_8$-TIM barrel fold typical of GH1 enzymes, with the active site including a deep pocket from enzyme's surface to the barrel core of the protein. Two conserved motifs, namely TFNEP and VTENG comprising of catalytic acid/base E165 and catalytic nucleophile E367 are situated opposite to each other at the bottom of active site. The amino acid residues supposed to be involved in substrate binding are as follows: glycone-binding residues: Q16, H119, W120, N164, N296, W417, N422, E424, W425, T431, and F433; aglycone binding residues: C168, N225, F228, Y298, T299, and W339) [36]. Mutational studies were carried out to determine the functional role of amino acids in active site. Two mutants (F250A and P172L/F250A) with increased enzyme catalytic efficiency and two mutants (L167W and P172L) with enhanced thermostability were generated [35]. Structural studies using bioinformatics approaches, are a key platform to decode the structural aspect of β-glucosidases and to understand its catalytic mechanisms.

4. Classification and Properties of Fungal β-Glucosidases

β-glucosidase are classified in glycosyl hydrolase family, and include 132 families according to CAZY web server [47].

TABLE 1: Studies on β-glucosidase from different strains of *Trichoderma* fungus.

S. no.	*Trichoderma* strain	β-glucosidase	Isolation strategies	References
1	*T. citrinoviride*	Extracellular β-Glucosidase	Protein purification, biochemical and proteomic characterization	[28]
2	*T. reesei*	TrBgl2	Mutational studies involving active site residues of the enzyme	[35]
3	*T. reesei* QM9414	bgl1	Overexpression of bgl1 from *Periconia* sp. in *T. reesei* QM9414 under *T. reesei* teflα promoter	[37]
4	Recombinant *T. reesei* strain, X3AB1	bgl1	Construction of *T. reesei* strain expressing *A. aculeatus* bgl under control of xyn3 promoter	[38]
5	*T. reesei*	bgl I	Molecular cloning and expression in *Pichia pastoris*	[39]
6	*T. reesei* CL847	BGL1	Protein purification and kinetic characterization	[3]
7	*T. reesei*	β-Glucosidase (cel3a)	Molecular cloning and expression in *T. reesei*	[40]
8	*T. ressei*	β-Glucosidase BGLII (CellA)	Molecular cloning, expression in *E. coli*, and characterization	[41]
9	*T. harzianum* C-4	—	Protein purification and biochemical characterization	[42]
10	*T. reesei*	BGL2	Molecular cloning and expression in *Aspergillus oryzae*	[43]
11	*T. harzianum* strain P1	1,3-β-Glucosidase	Protein purification and characterization	[44]
12	*T. reesei* QM9414	Aryl-β-D-glucosidase	Protein purification and characterization	[45]
13	*T. viride*	β-Gluc I	Protein purification and biochemical characterization	[46]
14	*T. viride* QM9414 mutants	—	Biochemical studies (pH control)	[16]

β-glucosidases from archeabacteria, plants, and mammals are found in family 1 and usually exhibit β-galactosidase activity while family 3 consists of β-glucosidases from bacteria, fungi and plants [48]. Family 1 and family 3 include retaining enzymes that hydrolyze the substrates with retention of anomeric carbon via a double-displacement method [49, 50].

Cellulose constitute one of the most abundant organic biopolymers on earth, and the cleavage of glycosidic bonds plays a crucial role in a wide range of biological processes in all living organisms. β-glucosidases comprise of a major enzyme group and are classified into 1st and 3rd families and hydrolyze either S-linked β-glycosidic bonds (myrosinase or β-D-thioglucoside glucohydrolase, EC 3.2.3.1) or O-linked-glycosidic bonds (β-D-glucoside glucohydrolase, EC 3.2.1.21) [51].

Based on substrate specificity, β-glucosidases are classified in three classes: class I (aryl β-glucosidases), class II (true cellobiases), and class III (broad substrate specificity enzymes). Mostly, β-glucosidases belong to class III with diverse catalytic mechanisms including cleavage of β 1,4; β 1,6; β 1,2 and α 1,3; α 1,4; α 1,6 glycosidic bonds [26, 52]. The enzymes exhibit functional diversity in terms of substrate specificity and no specific catalytic mechanism has been observed. However, the fungal enzymes are classified on the basis of their relative activities toward cellobiose and P(O)NPG into two groups, namely, (1) cellobiases—enzymes which have higher activity towards cellobiose, and (2) Aryl-β-glucosidases—higher relative activities towards P(O)NPG than cellobiose or negligible activity towards cellobiose. These are further classified according to their affinities towards cellobiose and P(O)NPG into three groups: (1) β-glucosidases with higher affinities for P(O)NPG, (2) β-glucosidases which show higher affinity (lower K_m) for cellobiose and (3) β-glucosidases with affinities (K_m) similar for both substrates

[53]. The values of K_m range from 0.031 (*Neocallimastix frontalis*) [54] to 340 mM (β-glucosidase II from *P. infestans*) [55] for cellobiose and from 0.055 mM (*Stachybotrys atra*) [56] to 34 mM (β-glucosidase II from *P. infestans*) [55] for P(O)NPG substrate.

β-glucosidases are biologically important enzymes and catalyze the transfer of glycosyl group between oxygen nucleophiles. Also, these enzymes exhibit activity for both natural (plant) or synthetic aryl-glucosides and a variety of aglycons [53]. A β-glucosidase purified from *A. niger* showed catalytic activities towards the disaccharides gentiobiose (β 1-6), sophorose (β 1-2), laminaribiose (β 1-3), and salicin (salicyl-glucose) [57]. The glucosidase from *P. herquei*, G1 β-glucosidase demonstrated relative activities of 82.7 and 70.3% toward gentiobiose and salicin (100% for PNPG) while the G2 isoenzymes are 8.7 and 54.5%, respectively [58]. This indicates that variations exist between enzymes from different species as well as between isoenzymes of the same microorganism. These enzymes possess high activity towards oligosaccharides with β $(1 \rightarrow 4)$ linkages; several studies indicated a higher activity towards glucans with β $(1 \rightarrow 2)$ and β $(1 \rightarrow 3)$ linkages. Examples include enzymes from *T. koningii* [59] and *A. fumigates* [60] with activity towards sophorose and laminaribiose than cellobiose. Although, the enzymes exhibit greater variability towards β-1,2/1,3 β-glucans, aryl-glucosides, and cellooligosaccharides, these enzymes are specific for β-anomeric configuration (exception β-glucosidase from *Thermomyces lanuginosus*, shows α-glucosidase activity) [5].

Mainly, β-glucosidases display optimum pH over the range 4.0 to 5.5 but enzyme activity has also been observed in low pH range (pH 2.5) to very high range (pH 8.0). The optimum temperature range for enzyme activity is from 35° to 80°C. The extracellular β-glucosidases from mesophilic fungi are thermostable enzymes (up to 60°C). Example includes a β-glucosidase purified from *T. reesei* QM 9414 strain which shows high stability of 50–55°C [61]. Several reports indicated the role of the carbohydrates in thermostability of the enzymes as cellulases are mostly glycoproteins. Examples are β-glucosidases I, III, and IV from *T. emersonii* [62] and β-glucosidases of *Mucor miehei* [63].

Glucono-δ-lactone is a potent competitive inhibitor of many β-glucosidases, and values of Ki ranging from 0.0083 μM to 12.5 mM have been reported [53]. Steric similarities between the enzyme-bound substrate and Glucono-δ-lactone might explain the competitive inhibition by this compound [64]. Other inhibitors of the enzyme include nojirimycin and deoxy nojirimycin [65] and heavy metals such as Hg^{2+}, Cu^{2+}, Pb^{2+} and Co^{2+}, and p-chloromercuribenzoate [66].

β-glucosidases from *T. reesei* are found bound to the cell wall or cell membrane or in supernatants with pI ranging from 4.4 to 8.7. In *T. reesei*, most of the enzyme is bound to the cell wall [67] during fungal growth and therefore low quantities of β-glucosidase are secreted into the medium [68]. Kubicek [69] reported that the membrane-bound β-glucosidase plays a role in the formation of sophorose which acts as a potent inducer of cellulases. Studies also indicated that the enzyme may act in cell-wall metabolism during conidiogenesis and therefore, not really a true component of cellulolytic enzyme system [67]. Inglin and coworkers [70] isolated an intracellular β-glucosidase and postulated that the enzyme might be involved in transportation across cell membrane as a proenzyme and in metabolic regulation of cellulose induction.

5. Studies on β-Glucosidases from Trichoderma Species

Numerous studies on *Trichoderma* have indicated its importance in biotechnological perspectives. Several molecular biology and biochemical techniques have reported the improved isolation of β-glucosidases from different species of *Trichoderma* namely *T. reesei* [37–41, 43, 71], *T. atroviride* [72], *T. harzianum* [42, 44], *T. viride* [46, 73], *T. koningii* [59], and *T. citrinoviride* [28], respectively (Table 1). Some of the key studies on β-glucosidase from *Trichoderma* fungus are as follows.

5.1. Protein Purification. Biochemical studies resulting in purification and characterization of a β-glucosidase from Type C-4 strain of *T. harzianum* was performed by Yun et al. [42]. A β-glucosidase with high cellulolytic activity was purified to homogeneity through Sephacryl S-300, DEAE-Sephadex A-50, and Mono P column chromatographic steps. SDS-PAGE analysis revealed that the protein was a monomer with a molecular mass of 75 kDa. The enzyme properties were established in terms of optimum activity at pH 5.0 and 45°C. p-Nitrophenyl-β-D-cellobioside and p-Nitrophenyl-β-glucopyranoside served as substrates and glucose and gluconolactone acted as competitive inhibitors, respectively. Similar studies by Chandra and coworkers [28]

reported the homogenous purification, kinetics, and MALDI-TOF assisted proteomic analysis of an extracellularly secreted β-glucosidase of *T. citrinoviride*. The enzyme had a molecular weight of 90 kDa, consisted of a single polypeptide chain, optimal activity at pH 5.5 and 55°C. Further, the enzyme was not inhibited by glucose (5 mM) and possess transglycosylation activity (catalyze conversion of geraniol to its glucoside).

Another study reported the comparative kinetic analysis of two fungal strains, β-glucosidase from *Aspergillus niger* and BGL1 from *T. reesei* through an efficient FPLC technique. 95% purification was obtained for BGL1 from *T. reesei* and cellobiose was used as substrate for kinetic characterization of the enzyme. The study revealed that β-glucosidase, SP188 from *Aspergillus niger* (K_m = 0.57 mM; K_p = 2.70 mM), has a lower specific activity than BGL1 (K_m = 0.38 mM; K_p = 3.25 mM) and more sensitive to glucose inhibition. Furthermore, a Michaelis-Menten model was generated and revealed comparative substrate kinetics of β-glucosidase activity of both enzymes [3]. Chirico and Brown [45] purified a β-glucosidase from the culture filtrate of *T. reesei* QM9414 strain to homogeneity and the purified enzyme exhibited activity towards cellobiose, p-nitrophenyl β-D-glucopyranoside and 4-methylumbelliferyl β-D-glucopyranoside.

A new type of aryl-β-D-glucosidase with no activity towards cellobiose was isolated and purified from a commercial cellulase preparation derived from *T. viride*. The purification techniques included Bio-Gel gel filteration, anion exchange on DEAE-Bio-Gel A, cation exchange on SE-Sephadex, and affinity chromatography on crystalline cellulose. The enzyme had a molecular weight of 76,000 Dalton and showed high activity with on p-nitrophenyl-β-D-glucose and p-nitrophenyl-β-D-xylose and moderate activity towards crystalline cellulose, xylan, and carboxymethyl cellulose [46].

5.2. Genomics Studies

5.2.1. Promoter Analysis. Although *T. reesei* have been explored extensively for cellulase production, the major limitations are the low β-glucosidase activity and inefficient biomass degradation, respectively. The xyn3 and egl3 promoters were used to enhance the expression of β-glucosidase 1 (BGL1) through homologous recombination. The recombinant strains showed 4.0- and 7.5-fold higher β-glucosidase activity under the control of egl3 and xyn3 promoters as compared to native strains. Furthermore, Matrix assisted laser desorption ionization-time of flight (MALDI-TOF) mass spectrometry determination revealed that BGL1 was over expressed. The increased level of BGL1 was adequate for cellobiose and cellotriose degradation [74].

5.2.2. Mutational Studies. The mutants of *T. reesei* capable of cellulase overproduction have been considered significant and economical for saccharification of pretreated cellulosic biomass [75]. Low BGL activity in *T. reesei* results in cellobiose accumulation leading to reduced biomass conversion efficiency and cellobiose-mediated product inhibition of CBH I (Cel7A) [76]. Exogenous supplementation of BGL in *T. reesei* cellulase preparations has been used as an alternative strategy to overcome this problem [77, 78].

Nakazawa and coworkers [38] constructed a recombinant *T. reesei* strain, X3AB1 that was capable of expressing an *Aspergillus aculeatus* β-glucosidase 1 with high specific activity under xyn3 promoter control. The study involved the isolation and harvesting of the culture supernatant from *T. reesei* X3AB1 grown on 1% Avicel (as carbon source). It exhibited 63- and 25-fold higher β-glucosidase activity against cellobiose compared to those of the parent strain PC-3-7 and *T. reesei* recombinant strain expressing an endogenous β-glucosidase I, respectively. The study further demonstrated that xylanase activity was 30% less when compared to due to the absence of xyn3 promoter. X3AB1 strain when grown on 1% Avicel-0.5% xylan medium, produced 2.3- and 3.3-fold more xylanase and β-xylosidase, respectively, than X3AB1 grown on 1% Avicel.

Furthermore, a mutant strain of *T. citrinoviride* was developed by multiple exposures to ethidium bromide and ethyl methyl sulphonate [79]. The mutants secreted FPase, endoglucanase, β-glucosidase and cellobiase 0.63, 3.12, 8.22, and 1.94 IU mL^{-1} which was found to be 2.14-, 2.10-, 4.09-, and 1.73-fold higher compared to the parent strain. Further studies indicated that under submerged fermentation conditions, glucose (upto 20 mM) did not led to inhibition of enzyme production. Comparative fingerprinting revealed the presence of two unique amplicons suggesting genetic uniqueness of the mutants.

5.2.3. Molecular Cloning and Heterologous Expression.
A novel fungal β-glucosidase gene (bgl4) and its homologue (bgl2) have been cloned from *T. reesei* [43]. This enzyme reportedly showed homology with plant β-glucosidases classified in β-glucosidase A (BGA) family. The BGL2 protein from *T. reesei* showed an amino acid composition of 466 on SDS PAGE and exhibited 73.1% identity with β-glucosidase from fungus *Humicola grisea*. Both the genes have been expressed in *Aspergillus oryzae* and purified. Furthermore, β-glucosidases of *Humicola grisea* have been used in combination with *Trichoderma* cellulases to improve the saccharification of cellulose. The study also demonstrated that the recombinant BGL4 from *Humicola grisea* showed strong activity towards cellobiose and the incorporation of the recombinant BGL4 led to improvement in cellulose saccharification by 1.4–2.2 times. Overexpression of recombinant BGL4 gene from *Humicola grisea* in *T. reesei* or *T. viride* has been reported *to* improve the saccharification of cellulose by cellulases complex [43].

A β-glucosidase cloned from *T. reesei* and its expression studies have been reported in *Pichia pastoris* GS115 strain [39]. *T. reesei* produced β-glucosidase in very low amounts [27] which acted as a limiting factor in cellulose degradation. To overcome this, it has been reported that a β-glucosidase from *T. reesei* (bglI) was over expressed in *Pichia pastoris* GS115 under the control of methanol-inducible alcohol oxidase (AOX) promoter and *S. cerevisiae* secretory signal peptide (a-factor). The expression of β-glucosidase in the culture medium has been reported to reach the productivity of 0.3 mg/mL and the maximum activity was reported as 60 U/mL. Furthermore, the protein purification yielded a recombinant β-glucosidase of molecular weight 76 kDa, a 1.8-fold purification with 26% yield, and a specific activity of 197 U/mg was achieved. The optimum activity of the enzyme was at 70°C and pH 5.0.

Several studies aimed at the improvement of the fungus *T. reesei* for β-glucosidase production since the yield is reported to be quite low and it is also required for conversion of cellobiose to glucose which hampers cellulase production. Dashtban and Qin [37] successfully engineered a β-glucosidase gene from the fungus *Periconia* spp. into the genome of *T. reesei* QM9414 strain. As compared to the parent strain (2.2 IU/mg), the *T. reesei* strain showed about 10.5-fold (23.9 IU/mg) higher β-glucosidase activity after 24 h of incubation. The recombinant enzyme was thermotolerant and was completely active when incubated at 60°C for two hours. Also, a very high total cellulase activity (about 39.0 FPU/mg) was found in comparison to the parent strain which did not show any total cellulase activity at 24 h of incubation. Furthermore, enzyme hydrolysis assay using untreated NaOH or Organosolv pretreated barley straw showed that the recombinant *T. reesei* strains released more reducing sugars compared to the parental strain. Such studies would benefit the bioconversion techniques, namely, biomass conversion using cellulases.

5.3. Bioinformatics Studies

5.3.1. Site Directed Mutagenesis.
Another approach of mutational studies was performed by Lee and coworkers [35] and it showed that induced mutations in the active site of β-glucosidase from *T. reesei* lead to improved enzyme activity and thermostability of the enzyme. The study involved mutations in the outer channel of the active site of the enzyme. The mutants, P172L and P172L/F250A showed enhanced enzyme activity in terms of 5.3- and 6.9-fold increase in K_m and k_{cat} values towards 4-nitrophenyl-b-D-glucopyranoside (p-NPG) substrate at 40°C as compared to the wild type. Also, L167W or P172L mutations lead to higher thermostability of the enzyme as demonstrated by their melting temperature, Tm. Furthermore, the mutant, L167W, showed an effective synergistic activity together with cellulases in cellulose degradation. These mutational studies hold prospects in engineering enzymes having industrial applications such as biofuel production.

5.3.2. Biochemical Studies.
Several inhibitors were used to study the enzyme activity of β-glucosidase from *T. reesei* QM 9414 strain. Diethylpyrocarbonate (DEP) at a concentration above 10 mM completely inhibited the enzyme activity while the presence of substrate or analog protected the enzyme from inactivation. The enzyme showed a pseudo-first-order reaction kinetics, having a second-order rate constant of 0.02 mM^{-1} min^{-1}. The presence of 1 M hydroxylamine restored the enzyme activity which indicated the modification of histidine residues. Also, statistical analysis of residual fractional activity compared to the number of modified histidine residues exhibited that presence of one histidine residue is important for catalysis. Other inhibitors of β-glucosidase include *p*-hydroxymercuribenzoate which completely inhibited the enzyme at concentration above

2 mM. The modified enzyme when treated with 5,5´-dithio-bis (2-nitrobenzoic acid) (DTNB) showed that presence of one cysteine residue was essential for enzyme activity. Also, various other inhibitors like 2-ethoxy-l-ethoxycarbonyl-1,2-dihydroquinoline (EEDQ) were used to study the effect of chemical modifications on enzyme kinetics [79].

6. Biotechnological Applications of β-Glucosidases

Studies on β-glucosidases have been carried out from different sources, namely, microbes, plants, and animals [7–14]. Amongst these, fungal sources are immensely explored due to their better prospects in commercial applications. β-glucosidases are promising candidates of glycosyl hydrolase family and catalyze the selective cleavage of glucosidic bonds. The enzyme is found in all living organisms and involved in diverse biological processes, namely, cellular signaling, oncogenesis, host pathogen interactions, degradation of structural and storage polysaccharides, and processes of industrial relevance [26]. Due to the rising significance of β-glucosidases in industrial biotechnology, emerging trends focus on the maximum exploitation of this category of enzymes. In plants, the enzyme catalyzes the beta-glucan synthesis during cell wall development, fruit ripening, pigment metabolism, and defence mechanisms [20, 21] while in microorganisms, these are involved in cellulose induction and hydrolysis [80, 81]. In humans and mammals, the enzyme catalyzes the hydrolysis of glucosyl ceramides [82]. Biosynthesis of glycoconjugates such as aminoglycosides, alkyl glucosides, and fragments of phytoalexin-elicitor oligosaccharides which play a role in microbial and plant defence mechanism is an important application of β-glucosidases [26]. However, the saccharification of cellulosic biomass for biofuel production is the most extensive area of research and application. The fungal β-glucosidases, being an efficient biocatalysts, finds applications in various industrial processes. The major applications of β-glucosidases from *Trichoderma species* are as follows.

Bioethanol Production. The rising energy demands and depletion of fossil fuels initiated research on alternative sources for energy production. Lignocellulosic biomass is the abundant component of plants and renewable in nature therefore utilized for bioethanol production. Cellulase enzyme complex catalyzes cellulose degradation and comprises of three different enzymes: exoglucanase, endoglucanase, and β-glucosidase (BGL) which acts synergistically for complete hydrolysis of cellulose [83, 84]. The initial steps include the cleavage of cellulose fibers by endoglucanase releasing small cellulose fragments which are acted upon by exoglucanase resulting in small oligosaccharides, cellobiose which is hydrolysed into glucose by β-glucosidases. The cellulolytic enzyme complex secreted by fungus, T. reesei is most widely used in industrial bioethanol applications. The conversion of cellobiose to glucose is regarded as the rate limiting step in bioethanol production from lignocellulosic biomass due to low efficiency and high costs of cellulases. Also, hyperproducing strains of T.reesei produce β-glucosidase in very

low amounts [27]. Alternative methods such as cocultivation fungal strains producing cellulose and β-glucosidase, namely, T. reesei and A. phoenics or A. niger was used to enhance the activity of β-glucosidase [85].

Several alternatives strategies have been utilized such as heterologous expression of β-glucosidase in other systems for enhanced production [85], supplementation of exogenous β-glucosidase to the cellulase complex of T. reesei [86], engineering β glucosidase for overexpression and production [37], promoter use for enhanced expression [74], and site directed mutagenesis [35]. Enzyme preparations consisting of extracellular β-glucosidase produced by T. atroviride mutants and cellulase producing T. reesei were found to be better than commercial preparations for saccharification and of pretreated spruce [87]. Furthermore, studies also indicated that enzyme mixtures from different fungal strains exhibited better activity than commercial preparations namely celluclast 1.5 L, novozyme 188, and accellerase 1000 [86]. Delignified bioprocessings from *Artemisia annua* (known as marc of Artemisia) and citronella (*Cymbopogon winterianus*) have been utilized for bioconversion by six species of *Trichoderma* and cellulase production [88]. Among six species, T. citrinoviride was found to be most efficient producer of cellulases and a high amount of β glucosidase. Also, T. virens was not capable of producing complete cellulase enzyme complex on any test waste or pure cellulose, except on marc of *Artemisia*, where it produced all three enzymes of the complex [89]. Table 2 exhibits various enhancement studies and the possible outcome in terms of fold enhancement obtained for production of β-glucosidases from different strains of *Trichoderma*.

7. Conclusion

Biofuel production from lignocellulosic biomass comprising cellulose complex is the most important application and accounts for maximum exploitation of enzyme in industrial processes. However, slow enzymatic degradation rate and feed-back inhibition of the enzyme (particularly β-glucosidase) limit their commercialization. Current β-glucosidase applications involve manipulation strategies such as development of glucose-tolerant β-glucosidase and external administration together with other cellulases. Development of mutants and genetic engineering studies is an emerging area with good prospects in enzyme development with desired properties.

Commercially, companies such as Novozymes and Genencor have developed cellulolytic enzymes cocktails for biomass hydrolysis such as Cellic series of enzymes [90] and Accellerase series of enzymes [91]. Although, the details of enzyme mixture is not disclosed, but it was assumed that the enzymes preparations were from genetically modified T. reesei with high β-glucosidase activity. With the tremendous progress on β-glucosidases with an aim to improve its production and catalytic activity, it is likely that in near future, these would cease to be a limiting factor in biofuel production. Further, expectedly with the ongoing research efforts in this field, the management of energy crisis and fuel

TABLE 2: Studies comprising of the enhancement strategies used for β-glucosidase production.

S. no.	Strain used and enzymes	Enhancement strategies	Conclusion	Reference
1	*Aspergillus aculeatus* β-glucosidase 1	A recombinant *T. reesei* strain, X3AB1 under the control of xyn3 promoter	63- and 25-fold higher β-glucosidase activity against cellobiose	[38]
2	β-glucosidase from *Periconia* spp.	Heterologous expression in *T. reesei*	Around 10.5-fold (23.9 IU/mg) higher β-glucosidase activity A very high total cellulase activity (about 39.0 FPU/mg)	[63]
3	*T. reesei*, Bgl2	Mutational studies and engineering of active site residues	Mutants, P172L, and P172L/F250A showed enhanced k_{cat}/K_m and k_{cat} values by 5.3- and 6.9-fold Also, mutant L167W had the best synergism with *T. reesei* in cellulosic biomass degradation	[32]
4	*T. reesei* (bglI)	Overexpression in *P. pastoris* GS115 under methanol-inducible alcohol oxidase promoter and *S. cerevisiae* secretory signal peptide.	β-glucosidase expression was 0.3 mg/mL and the maximum activity was 60 U/mL	[39]
5	β-glucosidase 1 (BGL1)	Use of xyn3 and egl3 promoters through homologous recombination	4.0- and 7.5-fold higher β-glucosidase activity	[70]
6	*T. citrinoviride* mutants	Mutational studies, use of ethidium bromide and ethyl methyl sulphonate as mutagens	Secretion of endoglucanase, β-glucosidase and cellobiase was found to be 2.14-, 2.10-, 4.09-, and 1.73-fold higher	[28]
7	Thermostable β-glucosidase (cel3a)	cel3a from *Talaromyces emersonii* was expressed in *T. reesei*	High specific activity against p-nitrophenyl-β-D-glucopyranoside (V_{max}, 512 IU/mg) and was competitively inhibited by glucose (ki, 0.254 mM) and displayed transferase activity	[40]
8	BGL4 from *H. grisea*	Overexpression of BGL4 in *T. reesei* or *T. viride*	Improvement in cellulose saccharification by 1.4–2.2 times	[43]
9	*T. reesei* Rut C-30	Temperature and pH profiling studies	0.02% Tween-80 concentration was optimum, pH 5.0 and temperature (31°C) initially (for 18 h) was optimum for maximum production of cellulase and β-glucosidase	[89]

demands to a certain extent would be balanced by the biofuels generation and management.

Abbreviations

CBHI:	Cellobiohydrolase I
BGL1:	β-Glucosidase-1
CBHII:	Cellobiohydrolase II
CAZY:	Carbohydrate active enzyme database
TrBgl2:	*T. reesei* β-glucosidase 2
GH1:	Glycosyl hydrolase family 1
P(O)NPG:	p-Nitrophenyl b-D glucopyranoside
K_m:	Michaelis constant
pI:	Isoelectric point
Ki:	Inhibitor's dissociation constant
KDa:	Kilo dalton
SDS-PAGE:	Sodium dodecyl sulfate polyacrylamide gel electrophoresis
MALDI-TOF:	Matrix-assisted laser desorption/ionization- time of flight
FPLC:	Fast protein liquid chromatography
V_{max}:	maximum rate of reaction

Acknowledgments

The authors are thankful to Director CIMAP for constant encouragement and CSIR, New Delhi for the NWP09 grant. Pragya Tiwari thanks CSIR for the award of senior research fellowship.

References

[1] G. E. Akinola, O. T. Olonila, and B. C. Adebayo-Tayo, "Production of cellulases by *Trichoderma* species," *Academia Arena*, vol. 4, no. 12, pp. 27–37, 2012.

[2] X. Liming and S. Xueliang, "High-yield cellulase production by *Trichoderma reesei* ZU-02 on corn cob residue," *Bioresource Technology*, vol. 91, no. 3, pp. 259–262, 2004.

[3] M. Chauve, H. Mathis, D. Huc, D. Casanave, F. Monot, and N. L. Ferreira, "Comparative kinetic analysis of two fungal β-glucosidases," *Biotechnology for Biofuels*, vol. 3, article 3, 2010.

[4] P. Béguin, "Molecular biology of cellulose degradation," *Annual Review of Microbiology*, vol. 44, pp. 219–248, 1990.

[5] J. Lin, B. Pillay, and S. Singh, "Purification and biochemical characteristics of β-D-glucosidase from a thermophilic fungus,

Thermomyces lanuginosus-SSBP," *Biotechnology and Applied Biochemistry*, vol. 30, no. 1, pp. 81–87, 1999.

[6] B. Henrissat, H. Driguez, C. Viet et al., "Synergism of cellulases from *Trichoderma reesei* in the degradation of cellulose," *Biotechnology*, vol. 3, pp. 722–726, 1985.

[7] Y. W. Han and V. R. Srinivasan, "Purification and characterization of beta-glucosidase of *Alcaligenes faecalis*," *Journal of Bacteriology*, vol. 100, no. 3, pp. 1355–1363, 1969.

[8] V. Deshpande, K. E. Eriksson, and B. Pettersson, "Production, purification and partial characterization of 1,4-β-glucosidase enzymes from *Sporotrichum pulverulentum*," *European Journal of Biochemistry*, vol. 90, no. 1, pp. 191–198, 1978.

[9] L. W. Fleming and J. D. Duerksen, "Purification and characterization of yeast beta-glucosidases," *Journal of Bacteriology*, vol. 93, no. 1, pp. 135–141, 1967.

[10] R. Heyworth and P. G. Walker, "Almond-emulsin beta-D-glucosidase and beta-D-galactosidase," *The Biochemical Journal*, vol. 83, pp. 331–335, 1962.

[11] S. K. Mishra, N. S. Sangwan, and R. S. Sangwan, "Physico-kinetic and functional features of a novel β-glucosidase isolated from milk thistle (*Silybum marianum* Gaertn.) flower petals," *Journal of Plant Biochemistry and Biotechnology*, 2013.

[12] S. K. Mishra, N. S. Sangwan, and R. S. Sangwan, "Comparative physico-kinetic properties of a homogenous purified β-glucosidase from *Withania somnifera* leaf," *Acta Physiologiae Plantarum*, vol. 35, pp. 1439–1451, 2013.

[13] S. K. Mishra, N. S. Sangwan, and R. S. Sangwan, "Purification and characterization of a gluconolactone inhibition-insensitive β-glucosidase from *Andrographis paniculata* nees. leaf," *Preparative Biochemistry and Biotechnology*, vol. 43, no. 5, pp. 481–499, 2013.

[14] L. G. McMahon, H. Nakano, M.-D. Levy, and J. F. Gregory III, "Cytosolic pyridoxine-β-D-glucoside hydrolase from porcine jejunal mucosa. Purification, properties, and comparison with broad specificity β- glucosidase," *Journal of Biological Chemistry*, vol. 272, no. 51, pp. 32025–32033, 1997.

[15] J. M. Uusitalo, K. M. H. Nevalainen, A. M. Harkki, J. K. C. Knowles, and M. E. Penttila, "Enzyme production by recombinant *Trichoderma reesei* strains," *Journal of Biotechnology*, vol. 17, no. 1, pp. 35–49, 1991.

[16] D. Steinberg, P. Vijayakumar, and E. T. Reese, "β Glucosidase: microbial production and effect on enzymatic hydrolysis of cellulose," *Canadian Journal of Microbiology*, vol. 23, no. 2, pp. 139–147, 1977.

[17] T. M. Enari and M. L. Niku-Paavola, "Enzymatic hydrolysis of cellulose: is the current theory of the mechanism of hydrolysis valid?" *Critical Reviews in Biotechnology*, vol. 5, pp. 67–87, 1987.

[18] T. Yazaki, M. Ohnishi, S. Rokushika, and G. Okada, "Subsite structure of the β-glucosidase from *Aspergillus niger*, evaluated by steady-state kinetics with cello-oligosaccharides as substrates," *Carbohydrate Research*, vol. 298, no. 1-2, pp. 51–57, 1997.

[19] I. Khan and M. W. Akhtar, "The biotechnological perspective of beta-glucosidases," *Nature Preceedings*, 2010.

[20] B. Brzobohaty, I. Moore, P. Kristoffersen et al., "Release of active cytokinin by a β-glucosidase localized to the maize root meristem," *Science*, vol. 262, no. 5136, pp. 1051–1054, 1993.

[21] A. Easen, *β-Glucosidases. Biochemistry and Molecular Biology*, American Chemical Society, Washington, DC, USA, 1993.

[22] M. Mandels, "Cellulases," *Annual Reports on Fermentation Processes*, vol. 5, pp. 35–78, 1982.

[23] J. Woodward and A. Wiseman, "Fungal and other β-d-glucosidases—their properties and applications," *Enzyme and Microbial Technology*, vol. 4, no. 2, pp. 73–79, 1982.

[24] M. Leclerc, A. Arnaud, R. Ratomahenina et al., "Yeast β-glucosidases," *Biotechnology and Genetic Engineering Reviews*, vol. 5, pp. 269–295, 1987.

[25] F. Stutzenberger, "Thermostable fungal β-glucosidases," *Letters in Applied Microbiology*, vol. 11, no. 4, pp. 173–178, 1990.

[26] Y. Bhatia, S. Mishra, and V. S. Bisaria, "Microbial β-glucosidases: cloning, properties, and applications," *Critical Reviews in Biotechnology*, vol. 22, no. 4, pp. 375–407, 2002.

[27] I. Herpoël-Gimbert, A. Margeot, A. Dolla et al., "Comparative secretome analyses of two *Trichoderma reesei* RUT-C30 and CL847 hypersecretory strains," *Biotechnology for Biofuels*, vol. 1, article 18, 2008.

[28] M. Chandra, A. Kalra, N. S. Sangwan, and R. S. Sangwan, "Biochemical and proteomic characterization of a novel extracellular β-glucosidase from *Trichoderma citrinoviride*," *Molecular Biotechnology*, vol. 53, pp. 289–299, 2013.

[29] C. H. Persoon, "Disposita methodica fungorum," *Romer's Neues Magazine of Botany*, vol. 1, pp. 81–128, 1794.

[30] I. S. Druzhinina, A. G. Kopchinskiy, M. Komoń, J. Bissett, G. Szakacs, and C. P. Kubicek, "An oligonucleotide barcode for species identification in *Trichoderma* and *Hypocrea*," *Fungal Genetics and Biology*, vol. 42, no. 10, pp. 813–828, 2005.

[31] A. Kopchinskiy, M. Komoń, C. P. Kubicek, and I. S. Druzhinina, "TrichoBLAST: a multilocus database for *Trichoderma* and *Hypocrea* identifications," *Mycological Research*, vol. 109, no. 6, pp. 658–660, 2005.

[32] C. P. Kubicek, M. Komon-Zelazowska, and I. S. Druzhinina, "Fungal genus Hypocrea/*Trichoderma*: from barcodes to biodiversity," *Journal of Zhejiang University*, vol. 9, no. 10, pp. 753–763, 2008.

[33] D. Martinez, R. M. Berka, B. Henrissat et al., "Genome sequencing and analysis of the biomass-degrading fungus *Trichoderma reesei* (syn. *Hypocrea jecorina*)," *Nature Biotechnology*, vol. 26, no. 5, pp. 553–560, 2008.

[34] M. Schmoll and A. Schuster, "Biology and biotechnology of *Trichoderma*," *Applied Microbiology and Biotechnology*, vol. 87, no. 3, pp. 787–799, 2010.

[35] H. L. Lee, C. K. Chang, W. Y. Jeng et al., "Mutations in the substrate entrance region of β-glucosidase from *Trichoderma reesei* improve enzyme activity and thermostability," *Protein Engineering, Design and Selection*, vol. 25, no. 11, pp. 733–740, 2012.

[36] W.-Y. Jeng, N.-C. Wang, M.-H. Lin et al., "Structural and functional analysis of three β-glucosidases from bacterium *Clostridium cellulovorans*, fungus *Trichoderma reesei* and termite *Neotermes koshunensis*," *Journal of Structural Biology*, vol. 173, no. 1, pp. 46–56, 2011.

[37] M. Dashtban and W. Qin, "Overexpression of an exotic thermotolerant β-glucosidase in *Trichoderma reesei* and its significant increase in cellulolytic activity and saccharification of barley straw," *Microbial Cell Factories*, vol. 11, no. 63, pp. 1–15, 2012.

[38] H. Nakazawa, T. Kawai, N. Ida et al., "Construction of a recombinant *Trichoderma reesei* strain expressing *Aspergillus aculeatus* β-glucosidase 1 for efficient biomass conversion," *Biotechnology and Bioengineering*, vol. 109, no. 1, pp. 92–99, 2012.

[39] P. Chen, X. Fu, T. B. Ng, and X.-Y. Ye, "Expression of a secretory β-glucosidase from *Trichoderma reesei* in *Pichia pastoris* and its characterization," *Biotechnology Letters*, vol. 33, no. 12, pp. 2475–2479, 2011.

[40] P. Murray, N. Aro, C. Collins et al., "Expression in *Trichoderma reesei* and characterisation of a thermostable family 3 β-glucosidase from the moderately thermophilic fungus *Talaromyces emersonii*," *Protein Expression and Purification*, vol. 38, no. 2, pp. 248–257, 2004.

[41] M. Saloheimo, J. Kuja-Panula, E. Ylösmäki, M. Ward, and M. Penttilä, "Enzymatic properties and intracellular localization of the novel *Trichoderma reesei* β-glucosidase BGLII (Cel1A)," *Applied and Environmental Microbiology*, vol. 68, no. 9, pp. 4546–4553, 2002.

[42] S.-I. Yun, C.-S. Jeong, D.-K. Chung, and H.-S. Choi, "Purification and some properties of a β-glucosidase from *Trichoderma harzianum* type C-4," *Bioscience, Biotechnology and Biochemistry*, vol. 65, no. 9, pp. 2028–2032, 2001.

[43] S. Takashima, A. Nakamura, M. Hidaka, H. Masaki, and T. Uozumi, "Molecular cloning and expression of the novel fungal β-glucosidase genes from *Humicola grisea* and *Trichoderma reesei*," *Journal of Biochemistry*, vol. 125, no. 4, pp. 728–736, 1999.

[44] M. Lorito, C. K. Hayes, A. Di Pietro, S. L. Woo, and G. E. Harman, "Purification, characterization, and synergistic activity of a glucan 1,3-beta-glucosidase and an N-acetyl-beta-glucosaminidase from *Trichoderma harzianum*," *Phytopathology*, vol. 84, no. 4, pp. 398–405, 1994.

[45] W. J. Chirico and R. D. Brown Jr., "Purification and characterization of a β-glucosidase from *Trichoderma reesei*," *European Journal of Biochemistry*, vol. 165, no. 2, pp. 333–341, 1987.

[46] G. Beldman, M. F. Searle-Van Leeuwen, F. M. Rombouts, and F. G. Voragen, "The cellulase of *Trichoderma viride*. Purification, characterization and comparison of all detectable endoglucanases, exoglucanases and beta-glucosidases," *European Journal of Biochemistry*, vol. 146, no. 2, pp. 301–308, 1985.

[47] http://www.cazy.org/.

[48] J. N. Varghese, M. Hrmova, and G. B. Fincher, "Three-dimensional structure of a barley β-D-glucan exohydrolase, a family 3 glycosyl hydrolase," *Structure*, vol. 7, no. 2, pp. 179–190, 1999.

[49] S. G. Withers and I. P. Street, "β-Glucosidase: mechanism and inhibition," in *Plant Cell Wall Polymers: Biogenesis and Biodegradation*, N. G. Lewis, Ed., pp. 597–607, American Chemical Society, Washington, DC, USA, 1989.

[50] S. G. Withers, "Mechanisms of glycosyl transferases and hydrolases," *Carbohydrate Polymers*, vol. 44, no. 4, pp. 325–337, 2001.

[51] B. Henrissat and G. Davies, "Structural and sequence-based classification of glycoside hydrolases," *Current Opinion in Structural Biology*, vol. 7, no. 5, pp. 637–644, 1997.

[52] C. Riou, J.-M. Salmon, M.-J. Vallier, Z. Günata, and P. Barre, "Purification, characterization, and substrate specificity of a novel highly glucose-tolerant β-glucosidase from *Aspergillus oryzae*," *Applied and Environmental Microbiology*, vol. 64, no. 10, pp. 3607–3614, 1998.

[53] J. Eyzaguirre, M. Hidalgo, and A. Leschot, "β-Glucosidases from filamentous fungi: properties, structure, and applications," in *Handbook of Carbohydrate Engineering*, CRC Taylor and Francis group, 2005.

[54] C. A. Wilson, S. I. McCrae, and T. M. Wood, "Characterisation of a β-D-glucosidase from the anaerobic rumen fungus *Neocallimastix frontalis* with particular reference to attack on cello-oligosaccharides," *Journal of Biotechnology*, vol. 37, no. 3, pp. 217–227, 1994.

[55] J. Bodenmann, U. Heiniger, and H. R. Hohl, "Extracellular enzymes of *Phytophthora infestans*: endo-cellulase, β-glucosidases, and 1,3-β-glucanases," *Canadian Journal of Microbiology*, vol. 31, no. 1, pp. 75–82, 1985.

[56] R. L. De Gussem, G. M. Aerts, M. Claeyssens, and C. K. De Bruyne, "Purification and properties of an induced β-D-glucosidase from *Stachybotrys atra*," *Biochimica et Biophysica Acta*, vol. 525, no. 1, pp. 142–153, 1978.

[57] T. Unno, K. Ide, T. Yazaki et al., "High recovery purification and some properties of a β-glucosidase from *Aspergillus niger*," *Bioscience Biotechnology and Biochemistry*, vol. 57, pp. 2172–2173, 1993.

[58] T. Funaguma and A. Hara, "Purification and properties of two β-glucosidases from *Penicillium herquei* Banier and Sartory," *Agricultural and Biological Chemistry*, vol. 52, pp. 749–755, 1988.

[59] T. M. Wood and S. I. McCrae, "Purification and some properties of the extracellular β-d-glucosidase of the cellulolytic fungus *Trichoderma koningii*," *Journal of General Microbiology*, vol. 128, no. 12, pp. 2973–2982, 1982.

[60] M. J. Rudick and A. D. Elbein, "Glycoprotein enzymes secreted by *Aspergillus fumigatus*. Purification and properties of β glucosidase," *Journal of Biological Chemistry*, vol. 248, no. 18, pp. 6506–6513, 1973.

[61] G. Schmid and C. Wandrey, "Purification and partial characterization of a cellodextrin glucohydrolase (β-glucosidase) from *Trichoderma reesei* strain QM9414," *Biotechnology and Bioengineering*, vol. 30, no. 4, pp. 571–585, 1987.

[62] A. McHale and M. P. Coughlan, "The cellulolytic system of *Talaromyces emersonii*. Purification and characterization of the extracellular and intracellular β-glucosidases," *Biochimica et Biophysica Acta*, vol. 662, no. 1, pp. 152–159, 1981.

[63] H. Yoshioka and S. Hayashida, "Relationship between carbohydrate moiety and thermostability of β-glucosidase from *Mucor miehei* YH-10," *Agricultural and Biological Chemistry*, vol. 45, pp. 571–577, 1981.

[64] K. Iwashita, K. Todoroki, H. Kimura, H. Shimoi, and K. Ito, "Purification and characterization of extracellular and cell wall bound β-glucosidases from *Aspergillus kawachii*," *Bioscience, Biotechnology and Biochemistry*, vol. 62, no. 10, pp. 1938–1946, 1998.

[65] E. T. Reese, F. W. Parrish, and M. Ettlinger, "Nojirimycin and d-glucono-1,5-lactone as inhibitors of carbohydrases," *Carbohydrate Research*, vol. 18, no. 3, pp. 381–388, 1971.

[66] X. Li and R. E. Calza, "Purification and characterization of an extracellular β-glucosidase from the rumen fungus *Neocallimastix frontalis* EB188," *Enzyme and Microbial Technology*, vol. 13, no. 8, pp. 622–628, 1991.

[67] M. A. Jackson and D. E. Talburt, "Mechanism for β-glucosidase release into cellulose-grown *Trichoderma reesei* culture supernatants," *Experimental Mycology*, vol. 12, no. 2, pp. 203–216, 1988.

[68] M. Nanda, V. S. Bisaria, and T. K. Ghose, "Localization and release mechanism of cellulases in Trichoderma reesei QM 9414," *Canadian Journal of Microbiology*, vol. 4, no. 10, pp. 633–638, 1982.

[69] C. P. Kubicek, "Involvement of a conidial endoglucanase and a plasma-membrane-bound β-glucosidase in the induction of endoglucanase synthesis by cellulose in *Trichoderma reesei*," *Journal of General Microbiology*, vol. 133, no. 6, pp. 1481–1487, 1987.

[70] M. Inglin, B. A. Feinberg, and J. R. Loewenberg, "Partial purification and characterization of a new intracellular beta-glucosidase of *Trichoderma reesei*," *Biochemical Journal*, vol. 185, no. 2, pp. 515–519, 1980.

[71] C. W. Bamforth, "The adaptability, purification and properties of exo-beta 1,3-glucanase from the fungus *Trichoderma reesei*," *Biochemical Journal*, vol. 191, no. 3, pp. 863–866, 1980.

[72] K. Kovács, L. Megyeri, G. Szakacs, C. P. Kubicek, M. Galbe, and G. Zacchi, "*Trichoderma atroviride* mutants with enhanced production of cellulase and β-glucosidase on pretreated willow," *Enzyme and Microbial Technology*, vol. 43, no. 1, pp. 48–55, 2008.

[73] G. Okada, "Enzymatic studies on a cellulase system of *Trichoderma viride*—II. Purification and properties of two cellulases," *Journal of Biochemistry*, vol. 77, no. 1, pp. 33–42, 1975.

[74] Z. Rahman, Y. Shida, T. Furukawa et al., "Application of *Trichoderma reesei* cellulase and xylanase promoters through homologous recombination for enhanced production of extracellular β-glucosidase i," *Bioscience, Biotechnology and Biochemistry*, vol. 73, no. 5, pp. 1083–1089, 2009.

[75] T. Nakari-Setälä, M. Paloheimo, J. Kallio, J. Vehmaanperä, M. Penttilä, and M. Saloheimo, "Genetic modification of carbon catabolite repression in *Trichoderma reesei* for improved protein production," *Applied and Environmental Microbiology*, vol. 75, no. 14, pp. 4853–4860, 2009.

[76] F. Du, E. Wolger, L. Wallace, A. Liu, T. Kaper, and B. Kelemen, "Determination of product inhibition of CBH1, CBH2, and EG1 using a novel cellulase activity assay," *Applied Biochemistry and Biotechnology*, vol. 161, pp. 313–317, 2010.

[77] A. Berlin, V. Maximenko, N. Gilkes, and J. Saddler, "Optimization of enzyme complexes for lignocellulose hydrolysis," *Biotechnology and Bioengineering*, vol. 97, no. 2, pp. 287–296, 2007.

[78] M. Chen, J. Zhao, and L. Xia, "Enzymatic hydrolysis of maize straw polysaccharides for the production of reducing sugars," *Carbohydrate Polymers*, vol. 71, no. 3, pp. 411–415, 2008.

[79] M. Chandra, A. Kalra, N. S. Sangwan, S. S. Gaurav, M. P. Darokar, and R. S. Sangwan, "Development of a mutant of *Trichoderma citrinoviride* for enhanced production of cellulases," *Bioresource Technology*, vol. 100, no. 4, pp. 1659–1662, 2009.

[80] I. De la Mata, M. P. Castillon, J. M. Dominguez, R. Macarron, and C. Acebal, "Chemical modification of β-glucosidase from *Trichoderma reesei* QM 9414," *Journal of Biochemistry*, vol. 114, no. 5, pp. 754–759, 1993.

[81] V. S. Bisaria and S. Mishra, "Regulatory aspects of cellulase biosynthesis and secretion," *Critical reviews in biotechnology*, vol. 9, no. 2, pp. 61–103, 1989.

[82] P. Tomme, R. A. J. Warren, and N. R. Gilkes, "Cellulose hydrolysis by bacteria and fungi," *Advances in Microbial Physiology*, vol. 37, pp. 1–81, 1995.

[83] N. W. Barton, F. S. Furbish, G. T. Murray et al., "Therapeutic response to intravenous infusions of glucocerebrosidase in patients with Gauchers disease," *Proceedings of the National Academy of Sciences USA*, vol. 87, pp. 1913–1916, 1990.

[84] R. R. Singhania, A. K. Patel, R. K. Sukumaran et al., "Role and significance of beta glucosidases in the hydrolysis of cellulose for bioethanol production," *Bioresource Technology*, vol. 127, pp. 500–507, 2013.

[85] Z. Wen, W. Liao, and S. Chen, "Production of cellulase/β-glucosidase by the mixed fungi culture *Trichoderma reesei* and *Aspergillus phoenicis* on dairy manure," *Process Biochemistry*, vol. 40, no. 9, pp. 3087–3094, 2005.

[86] K. Kovács, G. Szakács, and G. Zacchi, "Enzymatic hydrolysis and simultaneous saccharification and fermentation of steam-pretreated spruce using crude *Trichoderma reesei* and *Trichoderma atroviride* enzymes," *Process Biochemistry*, vol. 44, no. 12, pp. 1323–1329, 2009.

[87] M. Chandra, A. Kalra, P. K. Sharma, and R. S. Sangwan, "Cellulase production by six *Trichoderma* spp. fermented on medicinal plant processings," *Journal of Industrial Microbiology and Biotechnology*, vol. 36, no. 4, pp. 605–609, 2009.

[88] M. Chandra, A. Kalra, P. K. Sharma, H. Kumar, and R. S. Sangwan, "Optimization of cellulases production by *Trichoderma citrinoviride* on marc of *Artemisia annua* and its application for bioconversion process," *Biomass and Bioenergy*, vol. 34, no. 5, pp. 805–811, 2010.

[89] S. K. Tangnu, H. W. Blanch, and C. R. Wilke, "Enhanced production of cellulase, hemicellulase, and β-Glucosidase by *Trichoderma reesei* (Rut C-30)," *Biotechnology and Bioengineering*, vol. 23, pp. 1837–1849, 1981.

[90] http://www.genencor.com/.

[91] http://www.novozymes.com/.

Permissions

The contributors of this book come from diverse backgrounds, making this book a truly international effort. This book will bring forth new frontiers with its revolutionizing research information and detailed analysis of the nascent developments around the world.

We would like to thank all the contributing authors for lending their expertise to make the book truly unique. They have played a crucial role in the development of this book. Without their invaluable contributions this book wouldn't have been possible. They have made vital efforts to compile up to date information on the varied aspects of this subject to make this book a valuable addition to the collection of many professionals and students.

This book was conceptualized with the vision of imparting up-to-date information and advanced data in this field. To ensure the same, a matchless editorial board was set up. Every individual on the board went through rigorous rounds of assessment to prove their worth. After which they invested a large part of their time researching and compiling the most relevant data for our readers. Conferences and sessions were held from time to time between the editorial board and the contributing authors to present the data in the most comprehensible form. The editorial team has worked tirelessly to provide valuable and valid information to help people across the globe.

Every chapter published in this book has been scrutinized by our experts. Their significance has been extensively debated. The topics covered herein carry significant findings which will fuel the growth of the discipline. They may even be implemented as practical applications or may be referred to as a beginning point for another development. Chapters in this book were first published by Hindawi Publishing Corporation; hereby published with permission under the Creative Commons Attribution License or equivalent.

The editorial board has been involved in producing this book since its inception. They have spent rigorous hours researching and exploring the diverse topics which have resulted in the successful publishing of this book. They have passed on their knowledge of decades through this book. To expedite this challenging task, the publisher supported the team at every step. A small team of assistant editors was also appointed to further simplify the editing procedure and attain best results for the readers.

Our editorial team has been hand-picked from every corner of the world. Their multi-ethnicity adds dynamic inputs to the discussions which result in innovative outcomes. These outcomes are then further discussed with the researchers and contributors who give their valuable feedback and opinion regarding the same. The feedback is then collaborated with the researches and they are edited in a comprehensive manner to aid the understanding of the subject.

Apart from the editorial board, the designing team has also invested a significant amount of their time in understanding the subject and creating the most relevant covers. They scrutinized every image to scout for the most suitable representation of the subject and create an appropriate cover for the book.

The publishing team has been involved in this book since its early stages. They were actively engaged in every process, be it collecting the data, connecting with the contributors or procuring relevant information. The team has been an ardent support to the editorial, designing and production team. Their endless efforts to recruit the best for this project, has resulted in the accomplishment of this book. They are a veteran in the field of academics and their pool of knowledge is as vast as their experience in printing. Their expertise and guidance has proved useful at every step. Their uncompromising quality standards have made this book an exceptional effort. Their encouragement from time to time has been an inspiration for everyone.

The publisher and the editorial board hope that this book will prove to be a valuable piece of knowledge for researchers, students, practitioners and scholars across the globe.

List of Contributors

Laura Held, Thomas Kurt Eigentler, Ulrike Leiter and Claus Garbe
Department of Dermatology, Center for Dermatooncology, University Hospital Tübingen, 72076 Tübingen, Germany

Mark Jürgen Berneburg
Department of Dermatology, Center for Dermatooncology, University Hospital Tübingen, 72076 Tübingen, Germany
Department of Dermatology, Eberhard Karls University, Liebermeisterstraße 25, 72076 Tübingen, Germany

Giuseppe Servillo, Maria Vargas, Antonio Pastore and Michele Iannuzzi
Medical Intensive Care Unit, Department of Surgical and Anesthesiological Sciences, University Federico II of Naples, 80129 Naples, Italy

Alfredo Capuano, Andrea Memoli, Alfredo Procino, Eleonora Riccio and Bruno Memoli
Department of Nephrology, University Federico II of Naples, 80129 Naples, Italy

Kazuya Ogawa
Graduate School of Materials Science, Nara Institute of Science and Technology, 8916-5 Takayama, Ikoma, Nara 630-0101, Japan
Interdisciplinary Graduate School of Medicine and Engineering, University of Yamanashi, 4-3-11 Takeda, Kofu, Yamanashi 400-8511, Japan

Yoshiaki Kobuke
Graduate School of Materials Science, Nara Institute of Science and Technology, 8916-5 Takayama, Ikoma, Nara 630-0101, Japan
Institute of Advanced Energy, Kyoto University, Gokasho, Uji, Kyoto 611-0011, Japan

Peixin Dong and Masanori Kaneuchi
Department of Women's Health Educational System, Hokkaido University School of Medicine, Hokkaido University, N15, W7, Sapporo 060-8638, Japan

Yosuke Konno, Hidemichi Watari, Satoko Sudo and Noriaki Sakuragi
Department of Gynecology, Hokkaido University School of Medicine, Hokkaido University, N15, W7, Sapporo 060-8638, Japan

Pradyut Kundu, Anupam Debsarkar and Somnath Mukherjee
Environmental Engineering Division, Civil Engineering Department, Jadavpur University, Kolkata 32, India

Youjin Hu, Xionghao Liu, Panpan Long, Di Xiao, Jintao Cun, Zhuo Li, Jinfeng Xue, Yong Wu, Sha Luo, Lingqian Wu and Desheng Liang
State Key Laboratory of Medical Genetics, Central South University, 110 Xiangya Road, Changsha, Hunan 410078, China

Neuza Mariko Aymoto Hassimotto and Franco Maria Lajolo
Laboratório de Química e Bioquímica e Biologia Molecular de Alimentos, Departamento de Alimentos e Nutrição Experimental, FCF, Universidade de São Paulo, Avenida Professor Lineu Prestes 580, Bloco 14, 05508-000 São Paulo, SP, Brazil

Vanessa Moreira, Neide Galvão do Nascimento, Pollyana Cristina Maggio de Castro Souto and Catarina Teixeira
Laboratório de Farmacologia, Unidade de Inflamação, Instituto Butantan, Avenida Vital Brasil 1500, 05503-000 São Paulo, SP, Brazil

Aleksandra Taraszkiewicz, Grzegorz Fila, Mariusz Grinholc and Joanna Nakonieczna
Laboratory of Molecular Diagnostics, Department of Biotechnology, Intercollegiate Faculty of Biotechnology, University of Gdansk and Medical University of Gdansk, Kladki 24, 80-822 Gdansk, Poland

Subash C. B. Gopinath
Center for Advanced Studies in Botany, University of Madras, Guindy Campus, Chennai, Tamil Nadu 600025, India
Electronics and Photonics Research Institute, National Institute of Advanced Industrial Science and Technology, Central 5, 1-1-1 Higashi, Tsukuba, Ibaraki 305-8565, Japan

Periasamy Anbu
Department of Biological Engineering, College of Engineering, Inha University, Incheon 402-751, Republic of Korea

Thangavel Lakshmipriya
Department of Mathematics, SBK College, Madurai Kamaraj University, Aruppukottai, Tamil Nadu 626101, India

Azariah Hilda
Center for Advanced Studies in Botany, University of Madras, Guindy Campus, Chennai, Tamil Nadu 600025, India

Daye Cheng
Department of Transfusion, The First Hospital of China Medical University, North Nanjing Street No. 155, Shenyang 110001, China

Bin Liang
High Vocational Technological College, China Medical University, Shenyang 11001, China

Yunhui Li
Department of Clinical Laboratory, No. 202 Hospital, Shenyang 110003, China

Vigya Kesari, Aadi Moolam Ramesh and Latha Rangan
Department of Biotechnology, Indian Institute of Technology Guwahati, Assam 781 039, India

Hua Cong, Min Zhang and Shenyi He
Department of Human Parasitology, School of Medicine, Shandong University, No. 44 Wenhuaxi Road, Jinan, Shandong 250012, China

Qingli Zhang
Laboratory of Morphology, School of Medicine, Shandong University, No. 44 Wenhuaxi Road, Jinan, Shandong 250012, China

Jing Gong
Cancer Research Center, School of Medicine, Shandong University, No. 44 Wenhuaxi Road, Jinan, Shandong 250012, China

Haizi Cong
Department of College English, Shandong University, No. 44 Wenhuaxi Road, Jinan, Shandong 250012, China

Qing Xin
School Hospital of Shandong University, No. 73 Jingshi Road, Jinan, Shandong 250012, China

Baljinder Kaur, Praveen P. Balgir, Bharti Mittu, Balvir Kumar and Neena Garg
Department of Biotechnology, Punjabi University, Patiala, Punjab 147002, India

Norlaily Mohd Ali, Wan Yong Ho and Soon Guan Tan
Department of Cell and Molecular Biology, Faculty of Biotechnology and Biomolecular Sciences, University Putra Malaysia, Serdang, 43300 Selangor, Malaysia

Swee Keong Yeap and Sheau Wei Tan
Institute of Bioscience, University Putra Malaysia, Serdang, 43300 Selangor, Malaysia

Boon Kee Beh
Department of Bioprocess Technology, Faculty of Biotechnology and Biomolecular Sciences, University Putra Malaysia, Serdang, 43300 Selangor, Malaysia

Kathryn M. Silk, Alison J. Leishman and Paul J. Fairchild
Sir William Dunn School of Pathology, University of Oxford, South Parks Road, Oxford, OX1 3RE, UK

Kevin P. Nishimoto and Anita Reddy
Translational Research and Immunology, Geron Corporation, 230 Constitution Drive, Menlo Park, CA 94025, USA

Lin-Tong Yang
Department of Agricultural Resources and Environmental Sciences, College of Resources and Environmental Sciences, Fujian Agriculture and Forestry University, Fuzhou 350002, China
Institute of Horticultural Plant Physiology, Biochemistry, and Molecular Biology, Fujian Agriculture and Forestry University, Fuzhou 350002, China

Huan-Xin Jiang
Institute of Horticultural Plant Physiology, Biochemistry, and Molecular Biology, Fujian Agriculture and Forestry University, Fuzhou 350002, China
Department of Life Sciences, College of Life Sciences, Fujian Agriculture and Forestry University, Fuzhou 350002, China

Yi-Ping Qi
Institute of Materia Medica, Fujian Academy of Medical Sciences, Fuzhou 350001, China

Li-Song Chen
Department of Agricultural Resources and Environmental Sciences, College of Resources and Environmental Sciences, Fujian Agriculture and Forestry University, Fuzhou 350002, China
Institute of Horticultural Plant Physiology, Biochemistry, and Molecular Biology, Fujian Agriculture and Forestry University, Fuzhou 350002, China
Department of Horticulture, College of Horticulture, Fujian Agriculture and Forestry University, Fuzhou 350002, China

Mohamed A. Hassan and Amro A. Amara
Protein Research Department, Genetic Engineering and Biotechnology Research Institute, City of Scientific Research and Technological Applications, New Borg Al-Arab, P.O. Box. 21934, Alexandria, Egypt

Bakry M. Haroun
Botany and Microbiology Department, Faculty of Science (Boys), Al-Azhar University, Cairo, Egypt

Ehab A. Serour
King Abdulaziz City for Science and Technology, Riyadh, Saudi Arabia

Michael J. Heller
Department of Bioengineering, University of California San Diego, La Jolla, CA 92093-0412, USA
Department of Nano Engineering, University of California San Diego, La Jolla, CA 92093-0412, USA

Alexander P. Hsiao
Department of Bioengineering, University of California San Diego, La Jolla, CA 92093-0412, USA

Keisuke Matsui and Yoshikazu Tanaka
Research Institute, Suntory Global Innovation Center Limited, 1-1-1Wakayama-dai, Shimamoto-cho, Mishima-gun, Osaka 618-8503, Japan

Junichi Togami
Safety Science Institute, Quality Assurance Division, Suntory Business Expert Limited, 57 Imaikami-cho, Nakahara-ku, Kanagawa Kawasaki 211-0067, Japan

John G. Mason
Biosciences Research Division, Department of Environment & Primary Industries, AgriBio, Centre for AgriBioscience, 5 Ring Road, La Trobe University, Bundoora, VIC 3083, Australia

Stephen F. Chandler
School of Applied Sciences, RMIT University, P.O. Box 71, Bundoora, VIC 3083, Australia

Sílvia Bofill-Mas, Marta Rusiñol, Xavier Fernandez Cassi, Anna Carratalà, Ayalkibet Hundesa and Rosina Girones
Laboratory of Viruses Contaminants of Water and Food, Department of Microbiology, Faculty of Biology, Avenida Diagonal 643, 08028 Barcelona, Catalonia, Spain

Pragya Tiwari and Neelam S. Sangwan
Metabolic and Structural Biology Department, CSIR-Central Institute of Medicinal and Aromatic Plants (CSIR-CIMAP), P.O. CIMAP, Lucknow 226015, Uttar Pradesh, India

B. N. Misra
Department of Biotechnology, UP Technical University, Lucknow 226021, Uttar Pradesh, India

www.ingramcontent.com/pod-product-compliance
Lightning Source LLC
Chambersburg PA
CBHW080637200326
41458CB00013B/4662